THE CALCULUS COMPANION VOL. 1 TO ACCOMPANY

CALCULUS
HOWARD ANTON
SECOND EDITION

WILLIAM H. BARKER
JAMES E. WARD

BOWDOIN COLLEGE

JOHN WILEY & SONS

NEW YORK/CHICHESTER
BRISBANE/TORONTO
SINGAPORE

ISBN0-471-09230-4
Printed in the United States of America

10 9 8 7 6 5 4 3

PREFACE

No calculus text can contain everything that each individual student, or each individual instructor, would like to have included. First of all, there would not be enough room for all this material. More fundamental, however, is that while some readers need a great deal of detail, that detail would be unnecessary and burdensome for others.

Hence The Calculus Companion. It supplements Anton's Calculus with extras that simply could not go into that book. In particular, Volume I of The Companion contains:
- additional explanations of difficult concepts;
- additional computational examples with very detailed solutions;
- suggested procedures for attacking specific types of complex problems;
- numerous warnings concerning common mistakes and trouble spots;
- answers to the often-raised question "Why is this topic important?";
- numerous helpful figures and diagrams;
- a detailed 100-page Algebra and Trigonometry Review, concentrating on those concepts which are relevant to the calculus;
- special emphasis on the solution of word problems, including a special section in the Algebra and Trigonometry Review;
- optional sections on supplementary topics such as error analysis in numerical integration, and the general use of Riemann sums in applications of the definite integral.

The Companion stresses conceptual understanding and computational skill and, for that reason, it rarely elaborates on proofs or derivations. The writing style is informal and chatty, both to make the reading more appealing and to make the material less intimidating.

For ease of use, The Companion is organized in modular fashion: each section in Anton's Calculus has a corresponding section in The Companion, and different sections of The Companion rarely refer to each other. Thus readers can refer to those sections which are useful to them, and skip those which are not. Moreover, The Companion has a detailed index, so finding a particular topic is easy.

The Companion has been used in classes at Bowdoin College since 1981 and we are indebted to the many students and colleagues who made helpful suggestions about it. We are also grateful to Gary W. Ostedt and Robert W. Pirtle, mathematics editors at John Wiley and Sons, who have been cooperative and encouraging throughout this project, and to Barbara J. Moody of "North Country Technical Typing" who performed the Herculean task of typing the manuscript so beautifully. Finally, we must thank Howard Anton, both for requesting us to write The Calculus Companion, and for his willingness to be spoofed occasionally.

William H. Barker
James E. Ward

Bowdoin College
Brunswick, Maine
December, 1983

TO THE INSTRUCTOR

The Companion is a book students can turn to when in need of help on specific topics; for this reason, students having difficulty with the calculus should find it particularly useful. Stronger (or simply more curious) students will obtain a fuller understanding of the calculus by reading The Companion as a regular supplement to Anton's text. The Companion should also prove to be a useful tool when reviewing for examinations.

In addition,
- the numerous lists of suggested procedures and warnings about common trouble spots should contain new lecture ideas even for the most seasoned calculus instructor;
- the optional supplementary sections can be the basis for interesting additions to the standard calculus course, or as enrichment reading for the better students;
- the Algebra and Trigonometry Review can be a valuable resource for reference in lectures and tutorial sessions;
- The Companion should answer many of the simpler questions that students often bring to an instructor. By using it for this purpose, the instructor can spend more time helping students with fundamental problems, and less on routine matters.

The Companion has been tested, and has proved to be very effective, in a self-paced calculus program with no regular classroom lectures. When used in this way, it takes the place of the missing lectures.

The Companion can be offered as an optional, not required, text for a calculus course. This is probably a very sensible procedure in many situations.

TO THE STUDENT

You should use The Companion as you would your own, private tutor. After reading a section in Anton's Calculus, consult The Companion if you need additional explanation or if you would like to know more about a particular topic. The preface gives a detailed listing of the sorts of things you can expect to find in this volume.

There is a section in The Companion for each section in Anton's Calculus and, for the most part, the sections are self-contained. Thus you can skip around, reading only those sections which are important to you. The index should prove to be useful in this regard.

Our students have found The Companion to be very helpful. We hope you will too.

W.H.B. and J.E.W.

CONTENTS OF VOLUME I

ALGEBRA AND TRIGONOMETRY REVIEW

Contents

Appendix A :   Elementary geometry formulas

    Time and again in calculus you will need formulas and techniques from elementary geometry; for your reference we will summarize the most important of these results. You should memorize these formulas, although most you probably already know.

1.   Plane Geometry

       i. Rectangle with   base = b , height = h

          Area   A = bh

          Perimeter   p = 2b + 2h

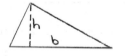

       ii. Parallelogram with   base = b , height = h

          Area   A = bh

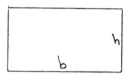

       iii. Trapezoid with   bases = a , b , height = h

          Area   A = $\frac{1}{2}$ (a + b)h

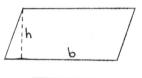

       iv. Triangle with   base = b , height = h

          Area   A = $\frac{1}{2}$ bh

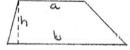

       v. Circle with   radius = r

          Area   A = πr^2

          Circumference   C = $2\pi r$

2.   Solid Geometry

       i. Rectangular box with   length = a , width = b , height = h

          Volume   V = abh

          Surface area   S = 2ab + 2ah + 2bh

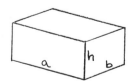

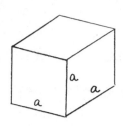

ii. Cube with side length = a (special case of i)

Volume $V = a^3$

Surface area $S = 6a^2$

iii. Right circular cylinder with radius = r , height = h

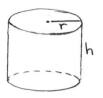

Volume $V = \pi r^2 h$

Surface area $S = 2\pi r^2 + 2\pi r h$

top and bottom   side

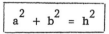

iv. Right circular cone with radius = r , height = h

Volume $V = \frac{1}{3}\pi r^2 h$

v. Sphere with radius = r

Volume $V = \frac{4}{3}\pi r^3$

Surface area $S = 4\pi r^2$

3. Pythagorean Theorem

Suppose Δ is a right triangle with hypotenus = h
and remaining sides = a, b . Then

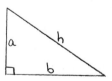

$$a^2 + b^2 = h^2$$

The importance of this result cannot be overstated; it occurs over and over again in calculus.
You should constantly be on the lookout for this relationship, especially in applied problems of
a geometric nature.

4.   Similar Triangles

Suppose Δ_1 and Δ_2 are similar triangles (i.e., the three angles of Δ_1 are the same as the three angles of Δ_2) with corresponding side lengths a_1, b_1, c_1 and a_2, b_2, c_2 respectively

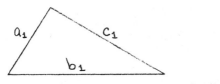 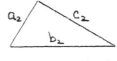

Then the corresponding sides are proportional, i.e.,

$$\frac{a_1}{a_2} = \frac{b_1}{b_2} = \frac{c_1}{c_2} \quad .$$

The importance of this result also cannot be overstated; however, unlike the Pythagorean theorem, for which people tend to develop a sharp eye, in applied problems similar triangle relationships are quite commonly overlooked.

Example.   A 4 foot high fence is 3 feet away from the side of a building. A ladder is propped up on the fence with its foot on the ground and its top against the building side. Express the height h   of the top of the ladder as a function of x , the distance of the foot of the ladder from the fence.

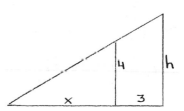

Solution.   There are two similar triangles to be used:

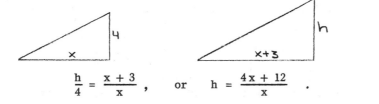

Thus $\dfrac{h}{4} = \dfrac{x + 3}{x}$,   or $h = \dfrac{4x + 12}{x}$.   □

B. 1

Appendix B: Algebra of fractions

Many beginning calculus students are unsure of, or are hesitant in using, the algebraic rules governing fractions.   Proficiency in the use of these rules is crucial for success in calculus.

1.   Multiplication.   Multiplying fractions together is very easy.   Simply multiply together the two numerators and then the two denominators:

$$B1: \qquad \left(\frac{a}{b}\right) \cdot \left(\frac{c}{d}\right) = \frac{ac}{bd}$$

As a special case, consider a fraction $\frac{c}{d}$ multiplied by an arbitrary number x.   Since $x = \frac{x}{1}$ we have

$$x \cdot \left(\frac{c}{d}\right) = \left(\frac{x}{1}\right) \cdot \left(\frac{c}{d}\right) = \frac{xc}{1d} = \frac{xc}{d} ,$$

so x just multiplies the numerator of $\frac{c}{d}$.   Seldom do people have trouble with multiplication of fractions.

Example 1.   Combine $\left(\frac{3+x}{x}\right) \cdot \left(\frac{3-x}{1+x}\right)$ into one fraction.

Solution. $\left(\frac{3+x}{x}\right) \cdot \left(\frac{3-x}{1+x}\right) = \frac{(3+x)(3-x)}{x(1+x)} = \frac{9-x^2}{x+x^2}$. $\square$

2.   Addition.   This is more complicated than multiplication.   Consider first the addition of two fractions with the same denominator.   The rule is: add the numerators together, i.e.,

$$B2: \qquad \frac{a}{h} + \frac{b}{h} = \frac{a+b}{h}$$.

(This makes sense: two-fifths of a pie plus one-fifth of a pie equals three-fifths of a pie.)   For fractions with different denominators, say $\dfrac{a}{b}$ and $\dfrac{c}{d}$, we must first alter the fractions to have a <u>common denominator</u>, say $h = bd$. We do this by multiplication:

$$\frac{a}{b} + \frac{c}{d} = \left(\frac{a}{b}\right) \cdot 1 + 1 \cdot \left(\frac{c}{d}\right)$$

$$= \left(\frac{a}{b}\right) \cdot \left(\frac{d}{d}\right) + \left(\frac{b}{b}\right) \cdot \left(\frac{c}{d}\right)$$

$$= \frac{ad}{bd} + \frac{bc}{bd} \qquad\qquad \text{rule} \quad B1$$

$$= \frac{ad + bc}{bd} \qquad\qquad \text{rule} \quad B2 \quad .$$

We have thus derived the basic addition formula:

$$B3: \quad \boxed{\; \frac{a}{b} + \frac{c}{d} = \frac{ad + bc}{bd} \;} \qquad .$$

The formula is easily remembered as "cross multiplication:"

gives the numerator $\quad ad + bc$,

gives the denominator $\quad bd$.

<u>DANGER</u>: Notice that $\dfrac{a}{b} + \dfrac{c}{d}$ does <u>NOT</u> equal $\dfrac{a+c}{b+d}$!! Memorize rule B3 as though your life depends on it. ... 'cause in calculus it does. As a special case, consider a fraction $\dfrac{c}{d}$ added to an arbitrary number x. Rule B3 still applies by writing $x = \dfrac{x}{1}$, i.e.,

$$x + \frac{c}{d} = \frac{x}{1} + \frac{c}{d} = \frac{xd + c}{d} \quad .$$

<u>Example 2.</u>   Combine $\dfrac{h+3}{1-h} + \dfrac{1}{1+h}$ into one fraction.

Solution.
$$\frac{\bar{h} + 3}{1 - h} + \frac{h}{1 + h} = \frac{(h + 3)(1 + h) + (1 - h)h}{(1 - h)(1 + h)}$$

$$= \frac{h^2 + 4h + 3 + h - h^2}{1 - h^2} = \frac{5h + 3}{1 - h^2} \quad . \qquad \square$$

Example 3. Combine $\dfrac{x + 2h}{x - h} + 2$ into one fraction.

Solution.
$$\frac{x + 2h}{x - h} + 2 = \frac{x + 2h}{x - h} + \frac{2}{1}$$

$$= \frac{(x + 2h) + 2(x - h)}{x - h} = \frac{3x}{x - h} \quad . \qquad \square$$

Note: In B3 we used the denominator $h = bd$ because it was a _common denominator_ for both a/b and c/d; however, in many situations, a smaller common denominator will exist, and computations will be greatly simplified if it is used in conjunction with B1 and B2. The following is a good example.

Example 4. Combine $\dfrac{3 - 2x^2}{4z^3 (x + 3)^3 (x - 1)^2} + \dfrac{x + 1}{2z^3 (x + 3)^3 (x - 1)}$.

Solution. We _could_ use B3 with common denominator $h = 8z^6 (x + 3)^6 (x - 1)^3$ ... but that would be very silly when we notice that the first denominator is merely $2(x - 1)$ times the second. We can, instead, give both fractions a common denominator by multiplying the second fraction by $\dfrac{2(x - 1)}{2(x - 1)}$, as we now show:

$$\frac{3 - 2x^2}{4z^3(x+3)^3(x-1)^2} + \frac{x+1}{2z^3(x+3)^3(x-1)} \cdot \frac{2(x-1)}{2(x-1)}$$

$$\underset{\text{rule B1}}{=} \frac{3-2x^2}{4z^3(x+3)^3(x-1)^2} + \frac{2(x+1)(x-1)}{4z^3(x+3)^3(x-1)^2}$$

$$\underset{\text{rule B2}}{=} \frac{(3-2x^2)+(2x^2-2)}{4z^3(x+3)^3(x-1)^2} = \frac{1}{4z^3(x+3)^3(x-1)^2} \quad . \qquad \square$$

3. <u>Subtraction.</u> If you know how to add fractions then you know how to subtract them; the

"trick" is to convert $-\left(\dfrac{c}{d}\right)$ into $\dfrac{(-c)}{d}$ as follows:

$$\frac{a}{b} - \frac{c}{d} = \frac{a}{b} + \frac{(-c)}{d} \underset{B3}{=} \frac{ad + b(-c)}{bd}$$

$$= \frac{ad - bc}{bd} \quad .$$

Thus our subtraction formula:

$$B4: \qquad \boxed{\frac{a}{b} - \frac{c}{d} = \frac{ad - bc}{bd}} \quad .$$

<u>Example 5.</u> Combine $\dfrac{x-y}{x+y} - \dfrac{x+y}{x-y}$ into one fraction.

<u>Solution.</u> $\dfrac{x-y}{x+y} - \dfrac{x+y}{x-y} \underset{B4}{=} \dfrac{(x-y)(x-y) - (x+y)(x+y)}{(x+y)(x-y)}$

$$= \frac{(x^2 - 2xy + y^2) - (x^2 + 2xy + y^2)}{x^2 - y^2}$$

$$= \frac{-4xy}{x^2 - y^2} = \frac{4xy}{y^2 - x^2} \quad . \qquad \square$$

4. <u>Division.</u> Here is the big pitfall. People are often very careless with division of

fractions, and consequently make horrendous errors in calculus problems. <u>Go over these</u>

<u>rules carefully.</u>

Consider the division of $\frac{a}{b}$ by $\frac{c}{d}$. The "trick" is to remember that $\frac{c}{d} \cdot \frac{d}{c} = \frac{cd}{cd} = 1$.

Thus

$$\frac{\left(\dfrac{a}{b}\right)}{\left(\dfrac{c}{d}\right)} = \frac{\left(\dfrac{a}{b}\right)}{\left(\dfrac{c}{d}\right)} \cdot \frac{\left(\dfrac{d}{c}\right)}{\left(\dfrac{d}{c}\right)} = \frac{\left(\dfrac{a}{b}\right)\left(\dfrac{d}{c}\right)}{\left(\dfrac{c}{d}\right)\left(\dfrac{d}{c}\right)} \qquad \text{rule B1}$$

$$= \frac{\dfrac{ad}{bc}}{1} \qquad \text{rule B1, used in both the numerator and the denominator.}$$

$$= \frac{ad}{bc} \quad .$$

Our basic division formula is thus

$$\boxed{\text{B5}: \quad \frac{\left(\dfrac{a}{b}\right)}{\left(\dfrac{c}{d}\right)} = \frac{ad}{bc}}$$

The formula is easily remembered by "swinging arcs":

$$\frac{a}{b} \atop \frac{c}{d} \qquad \text{gives the numerator} \quad ad \quad ,$$

$$\frac{a}{b} \atop \frac{c}{d} \qquad \text{gives the denominator} \quad bc \quad .$$

The formula can also be remembered by "invert the denominator, then multiply," i.e.,

$$\frac{a}{b} \div \frac{c}{d} = \frac{a}{b} \cdot \frac{d}{c} = \frac{ad}{bc} \quad .$$

All fraction divisions can be handled by B5; however, it is easy to be careless in applying the formula in "less obvious" situations. For example, people tend to confuse the two expressions

$$\frac{\frac{a}{h}}{d} \quad \text{and} \quad \frac{a}{\frac{h}{d}} \quad .$$

These are very different, as we now show:

$$\frac{\frac{a}{h}}{d} = \frac{\left(\frac{a}{h}\right)}{\left(\frac{d}{1}\right)} \underset{D1}{=} \frac{a \cdot 1}{dh} = \frac{a}{dh} \quad , \qquad$$

$$\frac{a}{\frac{h}{d}} = \frac{\left(\frac{a}{1}\right)}{\left(\frac{h}{d}\right)} \underset{D1}{=} \frac{a\,d}{1 \cdot h} = \frac{a\,d}{h} \quad .$$

unequal expressions

These formulas are common enough to be highlighted; keep in mind, however, that they are still just subcases of B5:

B6: $\boxed{\dfrac{\left(\dfrac{a}{h}\right)}{d} = \dfrac{a}{dh}}$, i. e., $\dfrac{a}{h} \div d = \dfrac{a}{h} \cdot \dfrac{1}{d}$

B7: $\boxed{\dfrac{a}{\left(\dfrac{h}{d}\right)} = \dfrac{a\,d}{h}}$, i. e., $a \div \dfrac{h}{d} = a \cdot \dfrac{d}{h}$.

Example 6. Simplify the expression $\dfrac{\dfrac{(n + 1)(x - 1)^{n+1}}{n}}{\dfrac{n(x - 1)^{n}}{n - 1}}$.

Solution. $\dfrac{\dfrac{(n + 1)(x - 1)^{n+1}}{n}}{\dfrac{n(x - 1)^{n}}{n - 1}} \underset{B5}{=} \dfrac{(n - 1)(n + 1)(x - 1)^{n+1}}{n^{2}(x - 1)^{n}}$

$$= \frac{(n^{2} - 1)(x - 1)}{n^{2}} = \left(1 - \frac{1}{n^{2}}\right)(x - 1) \quad . \qquad \square$$

5. **Summary of rules.**

$$B1: \quad \left(\frac{a}{b}\right) \cdot \left(\frac{c}{d}\right) = \frac{ac}{bd}$$

$$B2: \quad \frac{a}{h} + \frac{b}{h} = \frac{a+b}{h}$$

$$B3: \quad \frac{a}{b} + \frac{c}{d} = \frac{ad+bc}{bd}$$

$$B4: \quad \frac{a}{b} - \frac{c}{d} = \frac{ad-bc}{bd}$$

$$B5: \quad \frac{\left(\dfrac{a}{b}\right)}{\left(\dfrac{c}{d}\right)} = \frac{ad}{bc}$$

$$B6: \quad \frac{\left(\dfrac{a}{h}\right)}{d} = \frac{a}{dh}$$

$$B7: \quad \frac{a}{\left(\dfrac{h}{d}\right)} = \frac{ad}{h}$$

Exercises.

In Problems 1 through 8 combine the given expressions into one simple fraction.

1. $\dfrac{2}{3} - \dfrac{4}{x}$

2. $\dfrac{1}{3}(a + 1) + \dfrac{2}{a-1}$

3. $\dfrac{\left(\dfrac{x}{x+1}\right)}{3} + 1$

4. $\dfrac{x}{\left(\dfrac{x+1}{3}\right)} + 1$

5. $\dfrac{5 - 3x^2}{3(a+1)^2(x-1)^2} + \dfrac{x+2}{(a+1)^2(x-1)}$

6. $\dfrac{x+1}{x-1} + \dfrac{x-1}{x+1}$

7. $\dfrac{\left(\dfrac{-2}{h}\right)}{h+1} + \dfrac{2}{h}$.

8. $\dfrac{\dfrac{1}{x+1} - \dfrac{1}{x}}{h}$

9. Solve for x : $\qquad \dfrac{x+1}{x-3} - \dfrac{10}{x+3} = 1$

Answers:

1. $\dfrac{2x-12}{3x}$

2. $\dfrac{a^2+5}{3a-3}$

3. $\dfrac{4x+3}{3x+3}$

4. $\dfrac{4x+1}{x+1}$

5. $\dfrac{3x-1}{3(a+1)^2(x-1)^2}$

6. $\dfrac{x+1}{x-1} + \dfrac{x-1}{x+1} \underset{B3}{=} \dfrac{(x+1)^2 + (x-1)^2}{(x-1)(x+1)}$

$$= \dfrac{x^2+2x+1+x^2-2x+1}{x^2-1} = \dfrac{2x^2+2}{x^2-1}$$

7. $\dfrac{\left(\dfrac{-2}{h}\right)}{h+1} + \dfrac{2}{h} \underset{B5}{=} \dfrac{\left(\dfrac{-2}{h}\right)h + 2(h+1)}{(h+1)h} = \dfrac{-2+2h+2}{(h+1)h}$

$$= \dfrac{2h}{(h+1)h} = \dfrac{2}{h+1}$$

8. $\dfrac{\dfrac{1}{x+h} - \dfrac{1}{x}}{h} \underset{B4}{=} \dfrac{\left(\dfrac{x-(x+h)}{x(x+h)}\right)}{h}$

$$\underset{B6}{=} \dfrac{-h}{x(x+h)h} = -\dfrac{1}{x(x+h)}$$

9. $\dfrac{x + 1}{x - 3} - \dfrac{10}{x + 3} = 1$

$\dfrac{(x + 1)(x + 3) - 10(x - 3)}{(x - 3)(x + 3)} = 1$ by B4 ,

$\dfrac{x^2 + 4x + 3 - 10x + 30}{x^2 - 9} = 1$,

$x^2 - 6x + 33 = x^2 - 9$,

$- 6x + 42 = 0$,

$x = 7$.

Appendix C :   Algebra of exponents

The rules governing exponents are crucial for calculus and its applications, and yet many calculus students are unsure of them. What follows is a brief description of the algebra of exponents. The examples and exercises which we present are similar to the more difficult problems you will see in calculus.

Suppose $\underline{a}$ is any <u>positive real</u> number, and r is any <u>rational</u> number, i. e., $r = m/n$, the quotient of two <u>integers</u> m and n , where $n \neq 0$. We wish to recall the definition of a^r , read "$\underline{a}$ raised to the r-th power." We do this in the following stages:

1.   <u>What you should already know.</u>

Suppose n is a <u>positive integer.</u>

Then a^n is simply n copies of $\underline{a}$ multiplied together, i. e.,

$$a^n = \underbrace{a \cdot a \cdot \ldots \cdot a}_{n \text{ copies of } \underline{a}}$$

Examples: $2^3 = 2 \cdot 2 \cdot 2 = 8$, $\left(\frac{3}{4}\right)^2 = \frac{3}{4} \cdot \frac{3}{4} = \frac{9}{16}$

Also recall that any number raised to the "zero-th" power is defined to be 1 , i. e.,

$$a^0 = 1 .$$

Examples: $3^0 = 1$, $\pi^0 = 1$, $\left(\frac{1}{2}\right)^0 = 1$

You are probably also comfortable with the "n-th root" of a positive number $\underline{a}$: it is that positive real number $a^{1/n}$ (also written $\sqrt[n]{a}$) which, when raised to the n-th power, gives back $\underline{a}$, i. e.,

$$\left(a^{1/n}\right)^n = a$$

In particular: $\left(a^{1/2}\right)^2 = a$, $\left(a^{1/3}\right)^3 = a$.

As an example, $8^{1/3}$ is that positive number which, when raised to the 3-rd power, gives 8. Since $2^3 = 8$, then 2 is the number we are looking for, i.e., $8^{1/3} = 2$.

2. What you might have forgotten ...

We are ready for the major definition: what is $a^{m/n}$, where m and n are any two positive integers? Well, we know how to define the n-th root of a (from the previous paragraph) and we know how to raise any number to the m-th power; thus the expression $\left(a^{1/n}\right)^m$ makes sense. This is our definition for $a^{m/n}$, i.e.,

---

Definition. Suppose a is any positive real number, and m, n are any positive integers. Then

$$a^{m/n} = \left(a^{1/n}\right)^m .$$

Examples of this definition are

$$8^{2/3} = \left(8^{1/3}\right)^2 = 2^2 = 4$$

$$2^{3/2} = \left(2^{1/2}\right)^3 = \sqrt{2} \cdot \sqrt{2} \cdot \sqrt{2} = 2\sqrt{2}$$

$$\left(\frac{1}{16}\right)^{3/4} = \left[\left(\frac{1}{16}\right)^{1/4}\right]^3 = \left[\frac{1}{2}\right]^3 = \frac{1}{8} .$$

This definition allows us to raise any positive real number a (the "base") to any non-negative rational power $r = m/n$ (the "exponent"). To allow negative rational powers we use the familar "invert the base" rule:

$$\text{CO:} \quad \boxed{a^{-r} = \left(\frac{1}{a}\right)^{r} = \frac{1}{a^{r}}}$$

As examples of this rule,

$$8^{-2/3} = \frac{1}{8^{2/3}} = \frac{1}{4}$$

$$2^{-3/2} = \frac{1}{2^{3/2}} = \frac{1}{2\sqrt{2}} = \frac{1}{2\sqrt{2}} \cdot \frac{\sqrt{2}}{\sqrt{2}} = \frac{\sqrt{2}}{4} \quad .$$

$$\left(\frac{1}{16}\right)^{-3/4} = 16^{3/4} = \left(16^{1/4}\right)^{3} = 2^{3} = 8 \quad .$$

At this point you might feel a bit overwhelmed by the definitions; don't be!! The rules for exponents which we are about to develop "contain" all the above definitions. Learn the rules (which you must do...) and you will have also mastered the definitions. (We could, of course, derive all our rules from the definitions, but, to save time, we won't.)

3. Rules of exponent algebra.   MEMORIZE CAREFULLY!

C1. Addition and subtraction of exponents.

$$\boxed{a^{r+s} = a^{r} a^{s} \quad \text{and} \quad a^{r-s} = \frac{a^{r}}{a^{s}} = \frac{1}{a^{s-r}}}$$

Examples:
$$2^{1+r} = 2(2^{r})$$

$$4^{(1/2)-r} = \frac{4^{1/2}}{4^{r}} = \frac{2}{4^{r}} \quad .$$

C2. Multiplication and division of exponents.

$$\boxed{a^{rs} = \left(a^{r}\right)^{s} = \left(a^{s}\right)^{r} \quad \text{and} \quad a^{r/s} = \left(a^{1/s}\right)^{r} = \left(a^{r}\right)^{1/s}} \quad .$$

Examples:
$$2^{2r} = \left(2^{2}\right)^{r} = 4^{r}$$

$$8^{-r/3} = \left(8^{1/3}\right)^{-r} = 2^{-r} = \frac{1}{2^{r}} \quad .$$

C3.　Multiplication and division of bases.

$$(ab)^r = a^r b^r \qquad \text{and} \qquad \left(\frac{a}{b}\right)^r = \frac{a^r}{b^r}$$

Examples: 　　　$(4a)^{1/2} = 4^{1/2} a^{1/2} = 2\sqrt{a}$

$$\left(\frac{a^2}{8}\right)^{2/3} = \frac{\left(a^2\right)^{2/3}}{8^{2/3}} = \frac{a^{4/3}}{4}$$

NON-FORMULA C4.　Addition and subtraction of bases.

$$(a + b)^r = ? \qquad \text{and} \qquad (a - b)^r = ?$$

There are NO simple formulas in these cases. For instance, $(a + b)^r$ does NOT equal $a^r + b^r$. This represents a very common error, so be careful! There are a number of observations to be made here.

1.　Memorizing these rules is not so difficult if you keep in mind the integer exponent case. For example, suppose you need to simplify $\left(3^{r+1}\right)^{r-1}$ but cannot recall if the rule for $\left(a^r\right)^s$ is a^{rs} or a^{r+s}. Then check the formulas in a simple case, say with $a = 3$, $r = 2$ and $s = 3$:

$$\left(a^r\right)^s = \left(3^2\right)^3 = 9^3 = 729 \quad , \quad \Bigg] \text{EQUAL}$$
$$a^{rs} = 3^{2 \cdot 3} = 3^6 = 729$$
$$a^{r+s} = 3^{2+3} = 3^5 = 243 \quad .$$

This should pretty quickly make you remember $\left(a^r\right)^s = a^{rs}$. Thus

$$\left(3^{r+1}\right)^{r-1} = 3^{(r+1)(r-1)} = 3^{r^2-1} \quad .$$

2. Although a^r has only been defined for $\underline{a}$ a $\underline{positive}$ real number, in $\underline{some}$ cases a^r

does make sense when $\underline{a}$ is negative. For example, $(-2)^3 = -8$, and thus $(-8)^{1/3} = -2$.

On the other hand, $(-8)^{1/2}$ is $\underline{not\ defined}$ (unless we allow ourselves to use complex numbers).

In general, $a^{m/n}$ will be defined for negative $\underline{a}$ if n is an $\underline{odd}$ integer, but not if n is

an $\underline{even}$ integer. For these reasons (and others) rules C1 through C3 are not always valid

for negative bases $\underline{and\ must\ be\ handled\ with\ care}$! As an example, when $\underline{a}$ is negative (say

$a = -2$) it is NOT true that $\left(a^2\right)^{1/2} = a^{2(1/2)} = a$; instead $\left(a^2\right)^{1/2} = -a$. See

Examples 5 and 6 below.

3. A major goal of calculus is to define a^r for r any $\underline{real}$ number, not just for r a

$\underline{rational}$ number. This is done in Chapter 7 of the text. $\underline{It\ is\ then\ amazing\ but\ true\ that\ all}$

$\underline{the\ rules\ CO\ through\ C3\ remain\ valid.}$

4. Rule CO is quite useful in allowing us to eliminate negative exponents in fractions.

Observe its use in the following simplification:

$$\frac{(x+2)^{-3}(x-2)^3}{(x^2-4)^{-2}} \underset{CO}{=} \frac{(x-2)^3(x^2-4)^2}{(x+2)^3}$$

where the terms with negative exponents have "switched" between the numerator and denominator

$$= \frac{(x-2)^3\left[(x-2)(x+2)\right]^2}{(x+2)^3}$$

by factoring (x^2-4)

$$\underset{C3}{=} \frac{(x-2)^3(x-2)^2(x+2)^2}{(x+2)^3}$$

$$\underset{C1}{=} \frac{(x-2)^5(x+2)^2}{(x+2)^3}$$

by combining the two $(x-2)$ terms

$$= \frac{(x-2)^5}{(x+2)}$$

by cancelling $(x+2)^2$ from both the numerator and the denominator

n-th Roots. An important special case of exponents occurs when $r = 1/n$ for n a positive integer. In that case we often will use the notation

$$a^{1/n} = \sqrt[n]{a} \; ,$$

and refer to $\sqrt[n]{a}$ as the __n-th root of a__. (When $n = 2$ we simply write $\sqrt{a}$ in place of $\sqrt[2]{a}$.) From its definition the basic formulas

$$C5: \qquad \boxed{\left(\sqrt[n]{a}\right)^n = a \qquad \text{and} \qquad \sqrt[n]{a^n} = a}$$

easily follow for any __positive__ real number a.

All the rules C0 through C3 are of course valid for $\sqrt[n]{a}$; it is, however, instructive to write C3 down in the n-th root notation:

$$C6: \qquad \boxed{\sqrt[n]{ab} = \sqrt[n]{a}\,\sqrt[n]{b} \qquad \text{and} \qquad \sqrt[n]{\frac{a}{b}} = \frac{\sqrt[n]{a}}{\sqrt[n]{b}}}$$

It is also useful to emphasize again the lack of any "C4 rules," i.e., there are NO simple formulas for

$$\sqrt[n]{a+b} \qquad \text{or} \qquad \sqrt[n]{a-b} \; .$$

Don't dream anything up for these expressions.

4. Numerous examples.

__Example 1.__   Simplify $\dfrac{x(x+1)^{1/3} - (x+1)^{4/3}}{x^2 - 1}$.

__Solution.__ Using $4/3 = 1 + (1/3)$, C1 allows us to rewrite our expression as

$$\frac{x(x+1)^{1/3} - (x+1)(x+1)^{1/3}}{x^2 - 1} = \frac{(x+1)^{1/3}(x - (x+1))}{(x+1)(x-1)}$$

by factoring $(x+1)^{1/3}$ from both terms in the numerator

$$\underset{C1}{=} \frac{(x+1)^{1/3}(-1)}{(x+1)^{1/3}(x+1)^{2/3}(x-1)}$$

since $\frac{1}{3} + \frac{2}{3} = 1$

$$= -\frac{1}{(x+1)^{2/3}(x-1)}$$

by cancelling $(x+1)^{1/3}$ from both the numerator and the denominator

$\square$

Example 2. Simplify $\dfrac{(x-2)^{-1/3} + (x-2)^{2/3}}{x-1}$

Solution. This is a fairly common type of expression. As in the previous example we have the same base raised to different fractional powers. In general the way to proceed is to factor out the lowest fractional power -- in this case $(x-2)^{-1/3}$ -- from both terms in the numerator:

$$\frac{(x-2)^{-1/3} + (x-2)^{2/3}}{x-1} \underset{C1}{=} \frac{(x-2)^{-1/3} + (x-2)^{-1/3}(x-2)^1}{x-1}$$

since $\frac{2}{3} = -\frac{1}{3} + 1$

$$= \frac{(x-2)^{-1/3}[1 + (x-2)]}{x-1}$$

by factoring $(x-2)^{-1/3}$ from both terms in the numerator

$$= \frac{1}{(x-2)^{1/3}} \cdot \left(\frac{x-1}{x-1}\right) = \frac{1}{\sqrt[3]{x-2}}$$

$\square$

Example 3. Simplify $\sqrt[3]{8a^6(x-h)^4} + 2a^2h\sqrt[3]{x-h}$.

Solution. The trick in this type of an expression is to "pull" as much out of the cube root as possible. We start by examining the first term:

$$\sqrt[3]{8a^6(x-h)^4} \underset{C6}{=} \sqrt[3]{2^3} \; \sqrt[3]{a^6} \; \sqrt[3]{(x-h)^4}$$

$$\underset{C5}{=} 2a^{6/3}(x-h)^{4/3}$$

$$\underset{C1}{=} 2a^2(x-h)\sqrt[3]{x-h} \qquad \text{since } \frac{4}{3} = 1 + \frac{1}{3}$$

Thus our full expression becomes

$$\sqrt[3]{8a^6(x-h)^4} + 2a^2h\sqrt[3]{x-h}$$

$$= 2a^2(x-h)\sqrt[3]{x-h} + 2a^2h\sqrt[3]{x-h}$$

$$= 2a^2\sqrt[3]{x-h}\,[(x-h)+h] \qquad \text{by factoring } 2a^2\sqrt[3]{x-h}$$
$$\qquad\qquad\qquad\qquad\qquad\qquad \text{from both terms}$$

$$= 2a^2x\sqrt[3]{x-h} \qquad\qquad\qquad\qquad\qquad \square$$

Example 4. Remove all square roots from the denominator of

$$\frac{h}{\sqrt{x+h}-\sqrt{x}}$$

Solution. We must _rationalize the denominator_. This is done by using a variant of the difference of squares law:

$$(\sqrt{c}-\sqrt{d})(\sqrt{c}+\sqrt{d}) = (\sqrt{c})^2 - (\sqrt{d})^2$$
$$= c - d$$

(The terms $\sqrt{c}+\sqrt{d}$ and $\sqrt{c}-\sqrt{d}$ are called _algebraic conjugates_ of each other.) In the case at hand we have

$$\frac{h}{\sqrt{x+h}-\sqrt{x}} = \frac{h}{\sqrt{x+h}-\sqrt{x}} \cdot \frac{\sqrt{x+h}+\sqrt{x}}{\sqrt{x+h}+\sqrt{x}}$$

$$= \frac{h(\sqrt{x+h}+\sqrt{x})}{(x+h)-x} = \frac{h(\sqrt{x+h}+\sqrt{x})}{h}$$

$$= \sqrt{x+h}+\sqrt{x} \qquad\qquad\qquad\qquad\qquad \square$$

Note: This same technique can be used to _rationalize the numerator_ of a fraction. You are asked to do this in Exercise 6.

<u>Example 5.</u>   Is it always true that $\sqrt{a^2} = a$?

<u>Answer.</u>   No. The equation is true only when <u>a</u> is positive. In general the correct formula

is

$$\sqrt{a^2} = |a|$$

(Recall that the <u>absolute value of a</u> , written $|a|$, is defined to be the distance of <u>a</u>

to zero, or intuitively, the "positive" part of <u>a</u> , e.g., $|3.2| = 3.2$, $|-4| = 4$). Many

calculus errors are made by forgetting these absolute value signs!   ☐

<u>Example 6.</u>   Determine all x   values for which

$$\sqrt{(x - 1)^2} = 1$$

<u>Solution.</u>   From the previous example we know our equation is equivalent to

$$|x - 1| = 1$$

Thus $x - 1 = 1$ or $x - 1 = -1$. Hence $x = 0$ or 2 .   ☐

5.   <u>Summary of rules</u>: Suppose $a > 0$ and $b > 0$. Then:

CO: $a^{-r} = \dfrac{1}{a^r}$

C1: $a^{r+s} = a^r a^s$ and $a^{r-s} = \dfrac{a^r}{a^s} = \dfrac{1}{a^{s-r}}$

C2: $a^{rs} = \left(a^r\right)^s = \left(a^s\right)^r$ and $a^{r/s} = \left(a^{1/s}\right)^r = \left(a^r\right)^{1/s}$

C3: $(ab)^r = a^r b^r$ and $\dfrac{a}{b}^r = \dfrac{a^r}{b^r}$

$$C5: \quad \left(\sqrt[n]{a}\right)^n = a \qquad \text{and} \qquad \sqrt[n]{a^n} = a$$

$$C6: \quad \sqrt[n]{ab} = \sqrt[n]{a}\ \sqrt[n]{b} \qquad \text{and} \qquad \sqrt[n]{\frac{a}{b}} = \frac{\sqrt[n]{a}}{\sqrt[n]{b}}$$

Exercises. Simplify each of the following expressions. These are complicated, so don't get discouraged if each one takes some time to solve!

1. $\left[(x+1)^{-1/6}\right]^3 - (x+1)^{1/2}$

2. $\left(\dfrac{x+h}{x-h}\right)^{-2/3} \cdot \dfrac{x+h}{\sqrt[3]{x^2 - h^2}}$

3. $\dfrac{\left(\dfrac{4a^3}{9}\right)^{1/2} - \dfrac{2}{3}\sqrt{a}}{a-1}$

4. $\dfrac{\left(8a^3 b^{-1/2} x\right)^{2/3}}{a^3 \left(b^2\right)^{1/3} \left(4a^{-4} x^{1/3}\right)^{1/2}}$

5. $\dfrac{\sqrt{a+b} + \sqrt{a-b}}{\sqrt{a+b} - \sqrt{a-b}}$ (Rationalize the denominator)

6. $\dfrac{\sqrt{x+h+3} - \sqrt{x+3}}{h}$ (Rationalize the numerator)

Answers (with some intermediate steps provided).

1. $(x + 1)^{-1/2} - (x + 1)^{1/2} = (x + 1)^{-1/2} [1 - (x + 1)]$

$$= - \frac{x}{\sqrt{x + 1}}$$

2. $\left(\dfrac{x - h}{x + h}\right)^{2/3} \cdot \dfrac{x + h}{(x + h)^{1/3} (x - h)^{1/3}} = \dfrac{(x - h)^{\frac{2}{3} - \frac{1}{3}}}{(x + h)^{\frac{2}{3} + \frac{1}{3} - 1}}$

$$= \sqrt[3]{x - h}$$

3. $\dfrac{\frac{2}{3} a^{3/2} - \frac{2}{3} a^{1/2}}{a - 1} = \frac{2}{3} a^{1/2} \left(\dfrac{a - 1}{a - 1}\right) = \frac{2}{3} a^{1/2}$

4. $\dfrac{4 a^2 b^{-1/3} x^{2/3}}{a^3 b^{2/3} \left(2 a^{-2} x^{1/6}\right)} = \dfrac{2 a^{2-3+2} x^{\frac{2}{3} - \frac{1}{6}}}{b^{\frac{2}{3} + \frac{1}{3}}} = \dfrac{2a}{b} \sqrt{x}$

5. $\dfrac{\sqrt{a + b} + \sqrt{a - b}}{\sqrt{a + b} - \sqrt{a - b}} \cdot \dfrac{\sqrt{a + b} + \sqrt{a - b}}{\sqrt{a + b} + \sqrt{a - b}} = \dfrac{(a + b) + 2\sqrt{a^2 - b^2} + (a - b)}{(a + b) - (a - b)}$

$$= \dfrac{a + \sqrt{a^2 - b^2}}{b}$$

6. $\dfrac{\sqrt{x + h + 3} - \sqrt{x + 3}}{h} \cdot \dfrac{\sqrt{x + h + 3} + \sqrt{x + 3}}{\sqrt{x + h + 3} + \sqrt{x + 3}} = \dfrac{(x + h + 3) - (x + 3)}{h\left(\sqrt{x + h + 3} + \sqrt{x + 3}\right)}$

$$= \dfrac{1}{\sqrt{x + h + 3} + \sqrt{x + 3}}$$

Appendix D: Algebra of polynomials

1. **Multiplication.** Most calculus students are comfortable with the multiplication of simple expressions, e.g.,

$$(a + b)(c + d) = ac + ad + bc + bd \ .$$

For multiplication of longer expressions a person can always resort, if confused, to "long multiplication," as we now illustrate.

Example 1. Multiply $3x + 4y - 1$ by $x - 2y + 2$.

Solution. A "long multiplication" would be written out as follows.

$$
\begin{array}{r}
3x + 4y - 1 \\
x - 2y + 2 \\
\hline
6x + 8y - 2 \\
- 6xy - 8y^2 + 2y \\
3x^2 + 4xy - x \\
\hline
3x^2 - 2xy - 8y^2 + 5x + 10y - 2
\end{array}
$$

←— first expression multiplied by 2

←— first expression multiplied by $-2y$

←— first expression multiplied by x

←— adding the previous three lines gives the answer.

The procedure is just that of ordinary mulitplication of numbers; the only difference is in the placement of "like terms" underneath each other to aid in the final addition.

Certain combinations of terms appear so often that you are well-advised to memorize them:

$$
\begin{array}{l}
(a + b)(a - b) = a^2 - b^2 \\
(a + b)^2 = a^2 + 2ab + b^2 \\
(a + b)^3 = a^3 + 3a^2b + 3ab^2 + b^3
\end{array}
\ .
$$

The second and third of these equations are special cases of the <u>binomial formula</u>:

$$(a+b)^n = a^n + \binom{n}{1}a^{n-1}b + \binom{n}{2}a^{n-2}b^2 + \cdots$$

$$\cdots + \binom{n}{n-2}a^2 b^{n-2} + \binom{n}{n-1}ab^{n-1} + b^n \quad .$$

The symbol $\binom{n}{k}$ is read "n choose k" and is defined by

$$\binom{n}{k} = \frac{n!}{k!\,(n-k)!}$$

where $m!$ (read "m factorial") is defined to be

$$0! = 1$$

$$m! = 1 \cdot 2 \cdot 3 \cdot \ldots \cdot (m-1) \cdot m \qquad \text{for} \qquad m = 1, 2, 3, \ldots$$

<u>Example 2.</u>   Expand $(x+2)^4$ by the binomial formula.

<u>Solution</u> $(x+2)^4 = x^4 + \binom{4}{1}x^3 \cdot 2^1 + \binom{4}{2}x^2 \cdot 2^2 + \binom{4}{3}x^1 \cdot 2^3 + 2^4$

$$= x^4 + \frac{4!}{1!\,3!}x^3 \cdot 2 + \frac{4!}{2!\,2!}x^2 \cdot 4 + \frac{4!}{3!\,1!}x \cdot 8 + 16$$

$$= x^4 + 4 \cdot x^3 \cdot 2 + 6 \cdot x^2 \cdot 4 + 4 \cdot x \cdot 8 + 16$$

$$= x^4 + 8x^3 + 24x^2 + 32x + 16 \quad . \qquad\qquad \square$$

2.   <u>Division.</u>   The division of one algebraic expression by another frequently gives people difficulty. However, the technique of "long division" is quite important and is similar to the usual "long division" of decimal numbers.

<u>Example 3.</u>   Divide $x^3 + x^2 + x - 3$ by $x - 1$.

<u>Solution.</u> We'll go through this division very slowly to see precisely what is happening. We start by arranging the divisor $(x - 1)$ and the dividend $(x^3 + x^2 + x - 3)$ in the usual way:

$$x - 1 \overline{\smash{)}x^3 + x^2 + x - 3} \quad .$$

Notice that both terms are written with the powers of x in <u>descending order</u> (i. e., $x^3 + x^2 + x - 3$, not $x + x^3 - 3 + x^2$).

i.   Take the first term (x) in the divisor and divide it into the first term $\left(x^3\right)$ in the dividend; the result is x^2 , which we write above the x^3 as shown:

$$x - 1 \overline{\smash{)}x^3 + x^2 + x - 3} \quad .$$
(with x^2 circled above)

Now multiply the divisor $(x - 1)$ by x^2 , place the result $\left(x^3 - x^2\right)$ under the dividend (with correct positioning of powers of x) , and subtract:

$$
\begin{array}{r}
x^2 \\
x - 1 \overline{\smash{)}x^3 + x^2 + x - 3} \\
\underline{x^3 - x^2 } \\
\boxed{2x^2 + x - 3}
\end{array}
$$

ii.   The bottom term so obtained will be referred to as the "new dividend. " We operate on it just as we did on the original dividend: divide the x in the divisor into the $2x^2$ of the new dividend, and place the resulting $2x$ above the division sign as shown:

$$
\begin{array}{r}
x^2 + 2x \\
x - 1 \overline{\smash{)}x^3 + x^2 + x - 3} \\
\underline{x^3 - x^2 } \\
2x^2 + x - 3
\end{array}
$$
(with $2x$ circled above)

Then multiply the divisor $(x - 1)$ by $2x$, place the result $(2x^2 - 2x)$ under the new dividend, and subtract:

$$
\begin{array}{r}
x^2 + 2x \phantom{{}- 3} \\
x - 1 \overline{\smash{)}\, x^3 + x^2 + x - 3} \\
\underline{x^3 - x^2 \phantom{{}+ x - 3}} \\
2x^2 + x - 3 \\
\underline{2x^2 - 2x \phantom{{}- 3}} \\
\boxed{3x - 3}
\end{array}
$$

iii. The bottom line so obtained is an "even newer dividend"... and we're sure you can guess what to do with it! But if you're still not sure: divide the x in the divisor into the $3x$ of the "even newer dividend," and place the resulting 3 above the division sign. Then multiply the divisor $(x - 1)$ by 3, and subtract the result from the "even newer dividend:"

$$
\begin{array}{r}
x^2 + 2x + 3 \\
x - 1 \overline{\smash{)}\, x^3 + x^2 + x - 3} \\
\underline{x^3 - x^2 \phantom{{}+ x - 3}} \\
2x^2 + x - 3 \\
\underline{2x^2 - 2x \phantom{{}- 3}} \\
3x - 3 \\
\underline{3x - 3} \\
0
\end{array}
$$

In this example we were lucky: we obtained a zero remainder in our subtraction, and thus our division is done. The result is that $x^3 + x^2 + x - 3$, when divided by $x - 1$, yields $x^2 + 2x + 3$, i.e., $x - 1$ divides $x^3 + x^2 + x - 3$ "evenly" and

$$
\frac{x^3 + x^2 + x - 3}{x - 1} = x^2 + 2x + 3 \qquad\qquad \square
$$

Example 4. Divide $x^3 + 2 + 3x + 2x^2$ by $x^2 + 1$.

Solution. Again we set up the divisor and dividend with the terms in decreasing powers of x:

$$x^2 + 1 \overline{\smash{)}x^3 + 2x^2 + 3x + 2}$$

i. Divide x^2 into x^3 to get x; multiply $x^2 + 1$ by x and (<u>with correct positioning of powers of</u> x) subtract the result from the dividend:

$$
\begin{array}{r}
\,\fbox{x} \\
x^2 + 1 \overline{\smash{)}x^3 + 2x^2 + 3x + 2} \\
\underline{x^3 + x} \\
2x^2 + 2x + 2
\end{array}
$$

ii. Divide x^2 into $2x^2$ to obtain 2; multiply $x^2 + 1$ by 2 and subtract the result from the "new dividend":

$$
\begin{array}{r}
\,x + \fbox{2} \\
x^2 + 1 \overline{\smash{)}x^3 + 2x^2 + 3x + 2} \\
\underline{x^3 + x} \\
2x^2 + 2x + 2 \\
\underline{2x^2 + 2} \\
\fbox{$2x$}
\end{array}
$$

iii. Ahh... now we have a major difference from Example 3. We cannot obtain a <u>positive</u> power of x by dividing x^2 into $2x$. Thus our division is finished, but we did not end up with a zero remainder -- we have a remainder of $2x$ (i.e., $x^2 + 1$ does not divide $x^3 + 2x^2 + 3x + 2$ "evenly"). To see what we do with it, consider an ordinary division:

$$
\begin{array}{r}
16 \\
7 \overline{\smash{)}115} \\
\underline{7} \\
45 \\
\underline{42} \\
\fbox{3}
\end{array}
$$

This yields

$$\frac{115}{7} = 16 + \frac{\fbox{3}}{7}$$

We treat our remainder term in the same way as in this numerical example, i. e.,

our division computation yields

$$\frac{x^3 + 2x^2 + 3x + 2}{x^2 + 1} = x + 2 + \frac{2x}{x^2 + 1}$$

To check the accuracy of this answer combine the terms on the right-hand side of the

equation (Rule B3, Appendix B) to obtain

$$x + 2 + \frac{2x}{x^2 + 1} = \frac{(x + 2)(x^2 + 1) + 2x}{x^2 + 1}$$

$$= \frac{(x^3 + 2x^2 + x + 2) + 2x}{x^2 + 1}$$

$$= \frac{x^3 + 2x^2 + 3x + 2}{x^2 + 1}$$

The last fraction is what we started with, and so our answer checks out as correct.   □

Example 5.   Divide $2t^2 - t + t^3$ by $t + 1$.

Solution.   Without the running commentary the solution would look like this:

$$
\require{enclose}
\begin{array}{r}
t^2 + t - 2 \\
t + 1 \enclose{longdiv}{t^3 + 2t^2 - t } \\
\underline{t^3 + t^2 } \\
t^2 - t \\
\underline{t^2 + t } \\
- 2t \\
- 2t - 2 \\
\boxed{+ 2} \\
\end{array}
$$

Thus $\dfrac{t^3 + 2t^2 - t}{t + 1} = t^2 + t - 2 + \dfrac{2}{t + 1}$.   □

3.   **Polynomials.**   A polynomial in x is simply an addition of non-negative integer powers of x multiplied by constants. Six examples of polynomials are

$$x^2 + 5x - 3 \qquad \sqrt{2}\, x + 5 \qquad x^3$$

$$\pi x^8 + 2 \qquad\qquad 3 \qquad \frac{1}{2} x - \frac{3}{2}$$

Three examples of non-polynomials are

$$x^2 + \sqrt{x} \qquad (\sqrt{x} = x^{1/2} \text{ is a } \underline{\text{non-integer}} \text{ power of } x) \ ,$$

$$3 + 1/x \qquad (1/x = x^{-1} \text{ is a } \underline{\text{negative}} \text{ power of } x) \ ,$$

$$x^3 + \sin x \qquad (\sin x \text{ is } \underline{\text{not a power}} \text{ of } x)$$

The general expression for any polynomial in x is

$$p(x) = a_n x^n + a_{n-1} x^{n-1} + \cdots + a_1 x + a_0$$

where $a_n, a_{n-1}, \ldots, a_1$ and a_0 are simply constants, called the coefficients of $p(x)$.
Polynomials are the most elementary functions of a single variable; one major goal of calculus is to "approximate" other functions by polynomials (Taylor series, as done in Chapter 11).

The degree of a polynomial is the highest power of x which it contains. For example, $x^4 - 1$ is of degree 4 , while $\sqrt{3}$ is of degree 0 . Polynomials of low degrees have special names:

degree	form	name
0	a	constant
1	$ax + b$	linear term $(a \neq 0)$
2	$ax^2 + bx + c$	quadratic term $(a \neq 0)$
3	$ax^3 + bx^2 + cx + d$	cubic term $(a \neq 0)$

The multiplication of two polynomials will always yield a polynomial. For example,

$$(ax + b)(cx + d) = acx^2 + (ad + bc)x + bd \ ,$$

$$(x^2 - x)(3x^3 - x^2 + 1) = 3x^5 - 4x^4 + x^3 + x^2 - x$$

Such computations are easily carried out by "long multiplication" as discussed earlier; however, for most calculus applications, the reverse procedure of factoring is much more important.

4.   Factoring.   To factor a polynomial means to break it down into a product of polynomials of smaller degree.   Examples of factored polynomials are as follows:

$$2x^2 + x - 1 \qquad = (2x - 1)(x + 1)$$

$$x^2 - 2 \qquad = (x - \sqrt{2})(x + \sqrt{2})$$

$$2x^3 - \frac{17}{3}x^2 - \frac{5}{3}x + 2 = (x - 3)(2x - 1)(x + 2/3)$$

$$x^3 - 1 \qquad = (x - 1)(x^2 + x + 1)$$

Factoring a polynomial is not always a pleasant or easy operation, but it is important. We will thus study the procedure in some depth, first for quadratic terms, and then for general polynomials.

5.   Factoring quadratic terms.

Sometimes the factorization of a quadratic term can be arrived at "by inspection," as the following example illustrates.

Example 6.   Factor the quadratic term $2x^2 + x - 6$.

Solution.   If we assume that $2x^2 + x - 6$ factors into linear terms with integer coefficients, then we would have

$$2x^2 + x - 6 = (2x + a)(x + b)$$

where $2x$ and x are necessary to obtain the $2x^2$ term, and $\underline{a}$ and $\underline{b}$ are two integers which need to be determined. Multiplying out the right-hand side of this equation shows

$$2x^2 + x - 6 = 2x^2 + (a + 2b)x + ab$$

and thus we must choose $\underline{a}$ and $\underline{b}$ so that $ab = -6$ and $a + 2b = 1$. The first condition gives you a small number of (integer) possibilities to test, i. e., those pairs of integers whose product is the constant term -6 :

a	1	-1	2	-2	-6	6	-3	3
b	-6	6	-3	3	1	-1	2	-2

However, only $a = -3$, $b = 2$ satisfy the second condition, i.e., $a + 2b = 1$. Thus $2x^2 + x - 6 = (2x - 3)(x + 2)$. $\qquad \square$

Factorization of polynomials is related to the $\underline{roots}$ of polynomials: a number r is a root of $p(x)$ if $p(r) = 0$. If a quadratic term can be factored as follows:

$$\boxed{p(x) = ax^2 + bx + c = a(x - r_1)(x - r_2)} \qquad (*)$$

then quite clearly $p(r_1) = p(r_2) = 0$, so that r_1 and r_2 are roots of $p(x)$. Surprisingly, the converse of this statement is also true: if r_1 and r_2 are the roots of $ax^2 + bx + c$, then equation $(*)$ must hold true. Thus, $\underline{\text{factoring a quadratic term is}}$ $\underline{\text{equivalent to finding its roots}}$. This is fortunately made easy by:

The Quadratic Formula.

The roots r_1 and r_2 of the quadratic term

$$a x^2 + b x + c \qquad (a \neq 0)$$

are equal to

$$\frac{- b \pm \sqrt{b^2 - 4 a c}}{2 a} \quad .$$

Thus our two roots r_1 and r_2 are given by

$$r_1 = \frac{- b + \sqrt{b^2 - 4 a c}}{2 a} \qquad \text{and} \qquad r_2 = \frac{- b - \sqrt{b^2 - 4 a c}}{2 a}$$

and, when combined with equation $(*)$, show that any quadratic term can be factored as follows:

$$a x^2 + b x + c = a \left(x - \frac{- b + \sqrt{b^2 - 4 a c}}{2 a} \right) \left(x - \frac{- b - \sqrt{b^2 - 4 a c}}{2 a} \right)$$

If you take a moment and multiply out the right-hand side of this equation, you will indeed obtain $a x^2 + b x + c$ as claimed (this actually _proves_ both equation $(*)$ and the quadratic formula).

Example 7.   Factor the quadratic term $3 x^2 - 5 x + 1$.

Solution.   According to the quadratic formula, the roots are $\dfrac{5 \pm \sqrt{25 - 12}}{6} = \dfrac{5 \pm \sqrt{13}}{6}$.

Thus

$$3 x^2 - 5 x + 1 = 3 \left(x - \frac{5 + \sqrt{13}}{6} \right) \left(x - \frac{5 - \sqrt{13}}{6} \right)$$

You can always _check_ a factorization by multiplying the terms out to see if you obtain what

you started with.   In this example a check would go as follows:

$$3\left(x - \frac{5 + \sqrt{13}}{6}\right)\left(x - \frac{5 - \sqrt{13}}{6}\right)$$

$$= 3\left[x^2 - \left(\frac{5 + \sqrt{13}}{6} + \frac{5 - \sqrt{13}}{6}\right)x + \left(\frac{5 + \sqrt{13}}{6}\right)\left(\frac{5 - \sqrt{13}}{6}\right)\right]$$

$$= 3\left[x^2 - \frac{10}{6}x + \frac{25 - 13}{36}\right]$$

$$= 3\left[x^2 - \frac{5}{3}x + \frac{1}{3}\right]$$

$$= 3x^2 - 5x + 1, \quad \text{as desired.} \qquad \square$$

<u>Example 8.</u>   Factor the quadratic term $x^2 - 2x + 2$.

<u>Solution</u>.   The roots are $\dfrac{2 \pm \sqrt{4 - 8}}{2} = 1 \pm i$, where i is the "imaginary" number $\sqrt{-1}$. Thus $x^2 - 2x + 2 = (x - 1 - i)(x - 1 + i)$. While this is a perfectly correct factorization using <u>complex numbers</u> (i. e., numbers containing i), it is <u>not</u> a factorization using only <u>real numbers</u>. Since in elementary calculus we do not wish to deal with complex numbers, a quadratic such as $x^2 - 2x + 2$ which does not factor into real linear terms will be called an <u>irreducible quadratic term</u>. We will not use the complex factorization of such a term. $\square$

6.   Factoring general polynomials.

It is a theorem of algebra that any polynomial can be factored into real linear and irreducible quadratic terms. <u>It is, however, another matter entirely actually to determine what these factors are!</u> (This is not an uncommon situation in mathematics. Frequently, we can prove that something exists, but we do not have an effective, surefire way to compute what that "something" is.) Generally the factorization (when computable!) is done in stages: a given polynomial is factored into the product of two lower degree polynomials, each of which is then further factored ... , etc., ... until you can go no farther. Many times the start of such a procedure is "by inspection," i. e., there are certain commonly occurring factorizations which a person should remember. The most important of these are as follows:

D1: $\quad a^2 - b^2 = (a - b)(a + b)$

D2: $\quad a^3 - b^3 = (a - b)(a^2 + ab + b^2)$

D3: $\quad a^3 + b^3 = (a + b)(a^2 - ab + b^2)$

Example 9. Factor the cubic term $8x^3 - 27$.

Solution. You should recognize this term as a difference of cubes, i.e.,

$$8x^3 - 27 = (2x)^3 - 3^3$$

Thus D2 applies to give

$$8x^3 - 27 = (2x - 3)(4x^2 + 6x + 9)$$

Since the quadratic formula shows that the quadratic term $4x^2 + 6x + 9$ is irreducible

(i.e., it has complex roots because $\sqrt{b^2 - 4ac} = \sqrt{36 - 144} = \sqrt{-108} = i\sqrt{108}$), we

have obtained as complete a factorization as is possible.   □

Example 10. Factor the 5th degree polynomial $x^5 - 16x$.

Solution. If a polynomial in x has no constant term, then you have one factor for free (!),
namely "x". Thus

$$x^5 - 16x = x\left(x^4 - 16\right)$$
$$= x\left(\left(x^2\right)^2 - 4^2\right) \qquad \text{a difference of squares!}$$
$$= x\left(x^2 - 4\right)\left(x^2 + 4\right) \qquad \text{by} \quad D1 \ ,$$
$$= x\left(x - 2\right)\left(x + 2\right)\left(x^2 + 4\right) \qquad \text{by} \quad D1 \qquad \square$$

Equations D1 - D3 are simply specific cases of the following general rules:

D4:

For any positive integer n ,

$$a^n - b^n = \left(a - b\right)\left(a^{n-1} + a^{n-2}b + \cdots + ab^{n-2} + b^{n-1}\right)$$

D5:

For any <u>odd</u> positive integer m ,

$$a^m + b^m = \left(a + b\right)\left(a^{n-1} - a^{n-2}b + \cdots - ab^{n-2} + b^{n-1}\right)$$

Memorizing these rules is probably not necessary, but at least knowing that $(a - b)$ is always a factor of $a^n - b^n$ and $(a + b)$ is always a factor of $a^m + b^m$ (m odd) can be useful.

When faced with a factoring problem to which D1 - D5 do not apply (a common occurrence...), one generally attempts to use the following result:

The Factor Theorem

the linear term $x - r$ is a factor of a polynomial $p(x)$

if and only if r is a root of $p(x)$, i.e., $p(r) = 0$.

This is very much like the quadratic situation which we discussed earlier. <u>However</u> ... unlike the quadratic case, there is no simple formula which gives the roots of a polynomial of degree greater than 2 . We are thus again reduced to "inspection" methods for finding roots for polynomials.

<u>Example 11.</u> Factor the 3rd degree polynomial $x^3 - 5x^2 + 6x - 2$.

<u>Solution.</u> If you try some small integer values for x you will find that $x = 1$ is a root of $p(x)$ and hence $x - 1$ is a factor (we will give below a good method for "guessing" at the roots of a polynomial). The other factor is now obtained by long division:

$$\begin{array}{r} x^2 - 4x + 2 \\ \hline x - 1 \overline{)x^3 - 5x^2 + 6x - 2} \\ \underline{x^3 - x^2} \\ -4x^2 + 6x - 2 \\ \underline{-4x^2 + 4x} \\ 2x - 2 \end{array}$$

Thus $x^3 - 5x^2 + 6x - 2 = (x - 1)(x^2 - 4x + 2)$. The roots of the quadratic factor are now found to be

$$\frac{4 \pm \sqrt{16 - 8}}{2} = 2 \pm \sqrt{2}$$

Thus

$$x^3 - 5x^2 + 6x - 2 = (x - 1)(x - 2 - \sqrt{2})(x - 2 + \sqrt{2}) \qquad \square$$

Finding the exact roots of an arbitrary polynomial can often times be impossible; numerical approximations via a computer or hand calculator are then called for. However, many situations are helped along by the following result:

The
Rational
Root
Test

> Suppose $p(x) = a_n x^n + a_{n-1} x^{n-1} + \ldots + a_1 x + a_0$ is a polynomial with integer coefficients, and
>
> $r = p/q$ is a rational number where p/q is expressed in lowest terms.
>
> Then $r = p/q$ can be a root of $p(x)$ only if
>
> p   divides the constant term a_0 and
>
> q   divides the "leading coefficient" a_n .

Notice that this only says   " $r = p/q$ can be a root...;" it does not say " $r = p/q$ is a root... ."

Example 12.   Factor the 3rd degree polynomial $3x^3 - 8x^2 + x + 2$.

Solution.   Suppose this polynomial has a rational root $r = p/q$; then p and q are integers such that p divides 2 and q divides 3, i.e., the possibilities are

$$p = \pm 1, \pm 2 \quad , \quad q = \pm 1, \pm 3$$

Hence there are a totoal of eight <u>possible</u> rational roots:

$$\pm 1, \pm 2, \pm 1/3, \pm 2/3$$

By plugging these into the polynomial we find that only $r = 2/3$ actually is a root. Thus $x - 2/3$ is a factor, and long division produces

$$3x^3 - 8x^2 + x + 2 = (x - 2/3)(3x^2 - 6x - 3)$$

The roots of the quadratic factor are now found to be $1 \pm \sqrt{2}$. Thus

$$3x^3 - 8x^2 + x + 2 = 3(x - 2/3)(x - 1 - \sqrt{2})(x - 1 + \sqrt{2}) \qquad \square$$

7.   <u>Completing the square.</u>   With quadratic terms (especially irreducible terms) it is often important to <u>complete the square.</u> Here is how this works on a quadratic term whose x^2 coefficient is 1:

$$x^2 + bx + c = x^2 + bx + \underbrace{\left(\frac{b}{2}\right)^2 - \left(\frac{b}{2}\right)^2}_{} + c$$

add in and subtract out the square
of one half the x-coefficient

$$= \left(x + \frac{b}{2}\right)^2 - \frac{b^2}{4} + c$$

Example 13.   Complete the square in $x^2 + 3x + 4$.

Solution.
$$x^2 + 3x + 4 = x^2 + 3x + \left(\frac{3}{2}\right)^2 - \left(\frac{3}{2}\right)^2 + 4$$

$$= \left(x + \frac{3}{2}\right)^2 + \frac{7}{4} \qquad \square$$

Example 14. Complete the square in $3x^2 - 2x + 1$.

Solution. If the x^2 coefficient is not 1 , then first factor this coefficient out of the whole expression.

$$3x^2 - 2x + 1 = 3\left[x^2 - \frac{2}{3}x + \frac{1}{3}\right]$$

$$= 3\left[x^2 - \frac{2}{3}x + \left(-\frac{1}{3}\right)^2 - \left(-\frac{1}{3}\right)^2 + \frac{1}{3}\right]$$

$$= 3\left[\left(x - \frac{1}{3}\right)^2 + \frac{2}{9}\right]$$

$$= 3\left(x - \frac{1}{3}\right)^2 + \frac{2}{3} \qquad \square$$

Although the usefulness of this operation may not be immediately apparent, its value lies in "eliminating" the x term (i.e., x to the first power). See, for example, §9.6 in Anton. Also see its use in Appendix E (Conic Sections).

Exercises.

1.   Multiply the following expressions.

     a. $\left(\frac{2}{3}x + 4y\right)\left(\frac{1}{2}x - 3y + 1\right)$

     b. $(2x + a)(ax - 1)(2x - a)$

     c. $(2x + 1)^4$

2. Divide the following expressions.

    a. $\dfrac{2x^3 - x^2 - 3x + 14}{x + 2}$ b. $\dfrac{x^3 + 7/8}{2x - 1}$

    c. $\dfrac{2x^4 + x^3 + 1}{x^2 - 2}$

3. Factor the following polynomials.

    a. $3x^2 + 11x - 4$ b. $2x^2 + 2x - 1$

    c. $2x^2 + 2x + 1$ d. $x^4 - 8x$

    e. $2x^3 - 4x^2 + 3x - 1$ f. $10x^4 + 41x^3 + 12x^2 - 7x - 2$

4. Complete the square in 3a, b, c.

Answers.

1. a. $\frac{1}{3}x^2 + \frac{2}{3}x - 12y^2 + 4y$ b. $4ax^3 - 4x^2 - a^3x + a^2$

    c. $16x^4 + 32x^3 + 24x^2 + 8x + 1$

2. a. $2x^2 - 5x + 7$ b. $\frac{1}{2}x^2 + \frac{1}{4}x + \frac{1}{8} + \frac{1}{2x-1}$

    c. $2x^2 + x + 4 + \dfrac{2x+9}{x^2-2}$

3. a. $(3x-1)(x+4)$ b. $2(x + \frac{1}{2} - \frac{1}{2}\sqrt{3})(x + \frac{1}{2} + \frac{1}{2}\sqrt{3})$

    c. $2x^2 + 2x + 1$ (irreducible quadratic term)

    d. $x(x-2)(x^2 + 2x + 4)$ e. $(x-1)(2x^2 - 2x + 1)$

    f. $(2x+1)(5x-2)(x+2-\sqrt{3})(x+2+\sqrt{3})$

    Method: The only possible rational roots for the original polynomial are

    $\pm 2, \pm 1, \pm\frac{1}{2}, \pm\frac{2}{5}, \pm\frac{1}{5}, \pm\frac{1}{10}$. Running down through the list in the order given

    will find $r = -\frac{1}{2}$ as the first of these numbers which is a root. Hence

    $x + \frac{1}{2}$ will factor, but for convenience we use $2x+1$. (If $ax+b$ factors

    a given polynomial $p(x)$, then any constant multiple of $ax+b$ also factors

    $p(x)$.) Long division yields

    $$5x^3 + 18x^2 - 3x - 2 \quad.$$

    Continuing through our list of possible roots will also yield $r = \frac{2}{5}$ as a root.

    Hence $x - \frac{2}{5}$ will factor, but again for convenience we use $5x-2$. Long

    division yields $x^2 + 4x + 1$, to which the quadratic formula applies, giving

    our final answer.

4. a. $3\left(x + \dfrac{11}{6}\right)^2 - \dfrac{169}{12}$

    b. $2\left(x + \dfrac{1}{2}\right)^2 - \dfrac{3}{2}$

    c. $2\left(x + \dfrac{1}{2}\right)^2 + \dfrac{1}{2}$

Appendix E :   Conic sections

In this appendix we review the most basic properties of the four conic section curves:

the circle, ellipse, parabola and hyperbola. They are known as conic sections because they

can all be obtained by slicing a cone with a plane. Pictures of such slices are given in Anton's

Figure 12.1.2.

A much more detailed and sophisticated study of conic sections is undertaken in Chapter

12 of Anton.

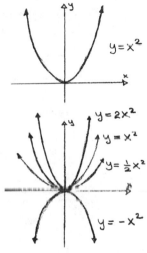

1.   The Parabola.   First consider the simplest equation of a

conic section: $y = x^2$. Its graph is an upward-turning

parabola with vertex at $(0,0)$. Multiplying x^2 by a

non-zero constant α will only change the flatness or

steepness of the curve and, in the case of a negative α ,

will make the curve turn downward.

We can move the vertex of the parabola

E1:
$$y = \alpha x^2$$

from $(0,0)$ to an arbitrary point (h,k) with the equation change

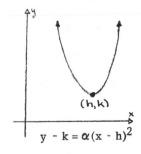

E1':
$$y - k = \alpha(x - h)^2$$

Here we have used

The Translation Principles

i. to move the graph of an equation to the right (resp. , left) by <u>h</u> units, replace x with x - h (resp. , x + h) ;

ii. to move the graph of an equation up (resp. , down) by <u>k</u> units, replace y with y - k (resp. , y + k) .

<u>Optional</u>: To see why these principles are true, consider a curve A whose equation is given by y = f(x) , and the curve B whose equation is given by replacing x and y with x - h and y - k respectively, i. e. ,

$$y - k = f(x - h)$$

Then the following are equivalent statements:

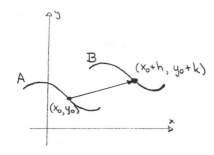

i. (x_0, y_0) lies on the curve A

ii. $y_0 = f(x_0)$

iii. $(y_0 + k) - k = f((x_0 + h) - h)$

iv. $(x_0 + h, y_0 + k)$ lies on the curve B

Thus curve B is obtained by moving curve A to the right by h units and up by k units, as the translation principles claim. □

<u>Example 1.</u> Sketch the graph of $y - 1 = \frac{1}{2}(x - 2)^2$.

<u>Solution.</u> We have only to translate the graph of $y = \frac{1}{2} x^2$ up by k = 1 and to the right by h = 2 units. □

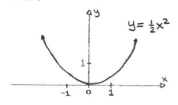

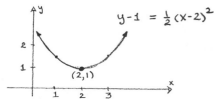

Example 2.   Sketch the graph of $y = -2x^2 - 4x$.

Solution.   We must first put the equation into the standard form El'; this requires completing the square (see Appendix D §7).

$$y = -2(x^2 + 2x)$$

$$= -2(x^2 + 2x + 1 - 1)$$

$$= -2(x + 1)^2 + 2$$

Thus $y - 2 = -2(x + 1)^2$,   so we must translate the graph of $y = -2x^2$ up by $k = 2$ units and to the left by $h = 1$ unit (or, if you prefer, to the right by $h = -1$ unit).   □

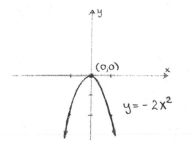
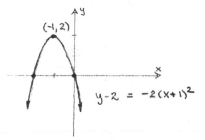

As Example 2 illustrates, any equation of the form

$$\boxed{y = ax^2 + bx + c} \qquad a \neq 0$$

is the equation of an upward or downward turning parabola, and can be put in the form El' by completing the square. To get rightward or leftward turning parabolas we simply reverse the roles of x and y , as the next example shows.

Example 3.   Sketch the graph of $x - y^2 + 2y = 1$.

Solution.   Rearranging terms we obtain

$$x = y^2 - 2y + 1 = (y - 1)^2$$

an equation whose graph is the translation upward by $k = 1$ unit of the graph of $x = y^2$. This is a rightward turning parabola. □

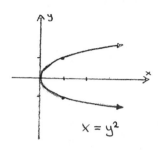

$x = y^2$

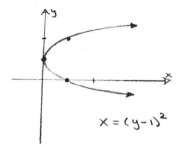

$x = (y-1)^2$

2. The Circle. The simplest equation for a circle is

E2:

$$x^2 + y^2 = r^2 \qquad r > 0$$

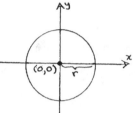

the graph of which is a circle of radius r and center $(0,0)$.

Thus, from the translation principles, we see that a circle with radius $r > 0$ and center (h, k) has as its equation

E2':

$$(x - h)^2 + (y - k)^2 = r^2$$

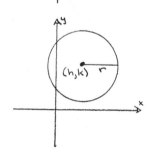

Example 4. Sketch the graph of $x^2 + y^2 - 2x + y + 1 = 0$.

Solution. We must first complete the squares in both the x and the y terms (Appendix D §7). This proceeds as follows. Group the x terms together and the y terms together:

$$\left(x^2 - 2x\right) + \left(y^2 + y\right) + 1 = 0$$

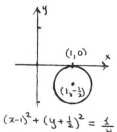

$(x-1)^2 + (y+\frac{1}{2})^2 = \frac{1}{4}$

Then complete the square in each:

$$\left(x^2 - 2x + 1 - 1\right) + \left(y^2 + y + \frac{1}{4} - \frac{1}{4}\right) + 1 = 0$$

$$\left(x - 1\right)^2 - 1 + \left(y + \frac{1}{2}\right)^2 - \frac{1}{4} + 1 = 0$$

$$\left(x - 1\right)^2 + \left(y + \frac{1}{2}\right)^2 = \frac{1}{4}$$

We thus have a circle of ratius $r = \frac{1}{2}$ centered on the point $\left(1, -\frac{1}{2}\right)$.   □

3.   The Ellipse. *   An ellipse centered on $(0,0)$ has an equation of the form

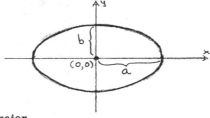

E3:
$$\frac{x^2}{a^2} + \frac{y^2}{b^2} = 1$$

where a and b are positive numbers.

The constants a and b are the length of the <u>semi-major</u>

and <u>semi-minor</u> axes (the larger number representing the semi-major axis). The four

"extreme points," $(\pm a, 0)$ and $(0, \pm b)$, are the <u>vertices</u> of the ellipse. The translation

principles then tell us

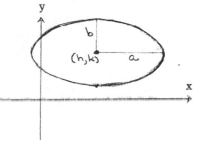

E3' :
$$\frac{(x - h)^2}{a^2} + \frac{(y - k)^2}{b^2} = 1$$

is the equation of an ellipse centered on (h, k) and with

$a, b > 0$ the lengths of the semi-major and semi-minor axes.

If $a = b$, then our ellipse is a circle with radius $r = a$.

Example 5.   Sketch the graph of $x^2 + 4x + 2y^2 - 4y - 2 = 0$.

Solution.   Completing the squares (Appendix D §7) in both x and y yields

$$(x + 2)^2 + 2(y - 1)^2 = 8 , \qquad \text{or}$$

$$\frac{(x + 2)^2}{8} + \frac{(y - 1)^2}{4} = 1$$

_____

*
  For simplicity of development, the use of a and b for the ellipse differ slightly from
  that of Anton's §12.3. In our usage, a denotes the term dividing the x term; in
  Anton, a always denotes the larger of the constants a and b .

E. 6

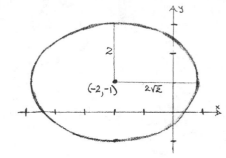

Thus we have an ellipse centered on $(-2, 1)$, with the semi-major and semi-minor axes of lengths $2\sqrt{2}$ and 2 (since we have obtained equation E 3' with $a = 2\sqrt{2}$ and $b = 2$).

4. The Hyperbola. Hyperbolas are the most interesting (and the most complicated) of the conic sections. For starters let's recall the hyperbolas centered on $(0, 0)$ which are given by

E4 :
$$\boxed{\frac{x^2}{a^2} - \frac{y^2}{b^2} = 1}$$

Notice the minus sign!

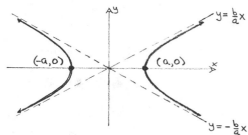

where $\underline{a}$ and $\underline{b}$ are positive numbers.

It is easy to see that the x-intercept points are $(\pm a, 0)$; these are called the vertices of the hyperbola. There are no y-intercepts. Moreover, solving for y in terms of x we obtain

$$\frac{y^2}{b^2} = \frac{x^2}{a^2} - 1$$

$$\frac{y}{b} = \pm\sqrt{\frac{x^2}{a^2} - 1}$$

$$y = \pm b\sqrt{\frac{x^2}{a^2} - 1}$$

Thus, when x becomes large, the "1" in the radical sign becomes insignificant when compared with x^2/a^2. Therefore, for large values of x we obtain

$$y \cong \pm b\sqrt{\frac{x^2}{a^2}} = \pm\frac{b}{a}x$$

The lines $y = \pm \dfrac{b}{a} x$ are the <u>asymptotes</u> of the hyperbola; as x gets large the hyperbola

is approximated very well by these lines (i. e., the hyperbola approaches these lines but never

quite touches them). A handy way to remember the asymptote formulas is to take the hyper-

bola equation, replace the "1" with a "0", and solve for y:

$$\frac{x^2}{a^2} - \frac{y^2}{b^2} = 1 \quad \text{is changed to} \quad \frac{x^2}{a^2} - \frac{y^2}{b^2} = 0$$

Solving for y yields

$$y = \pm \frac{b}{a} x \qquad *$$

the correct asymptote equations.

<u>Example 6.</u> Sketch the graph of $\dfrac{x^2}{4} - \dfrac{y^2}{9} = 1$.

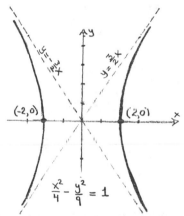

<u>Solution.</u> In this case $a = 2$ and $b = 3$. Thus the

vertices are $(\pm 2, 0)$, and the asymptotes are the

lines $\dfrac{x^2}{4} - \dfrac{y^2}{9} = 0$, i.e., $y = \pm \dfrac{3}{2} x$. $\square$

The hyperbolas given by E4 all turn outward to the <u>left</u> and <u>right</u>; to obtain hyperbolas

turning <u>upward</u> and <u>downward</u> we need to switch the roles of x and y by considering

equations of the form

E5: $$\boxed{\dfrac{y^2}{a^2} - \dfrac{x^2}{b^2} = 1}$$ $\boxed{\begin{array}{l}\text{Notice the}\\ \text{minus sign!}\end{array}}$

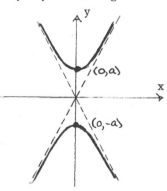

Here there are no x-intercepts, while the y-intercepts

(the <u>vertices</u>) are $(0, \pm a)$. The asymptotes are given by

---

* Note that the asymptotes lie along the diagonals of the rectangle
   with vertices $x = \pm a$ and $y = \pm b$.

changing the "1" to a "0", so $\dfrac{y^2}{a^2} - \dfrac{x^2}{b^2} = 0$, which yields

$$y = \pm \frac{a}{b} x$$

Example 7. Sketch the graph of $y^2 - 4x^2 = 2$.

Solution. We first must place the equation into the standard form E5. To do so divide by 2 to obtain

$$\frac{y^2}{2} - 2x^2 = 1$$

which is equivalent to

$$\frac{y^2}{2} - \frac{x^2}{\left(\frac{1}{2}\right)} = 1$$

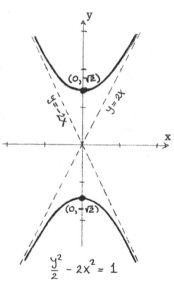

by fraction rule B7. Thus $a = \sqrt{2}$ and

$b = \sqrt{1/2} = \sqrt{2}/2$, giving vertices

$(0, \pm\sqrt{2})$ and asymptotic lines

$\dfrac{y^2}{2} - \dfrac{x^2}{\left(\frac{1}{2}\right)} = 0$, i.e., $y = \pm 2x$. $\square$

As done previously for the parabola, circle and ellipse, we use the translation principles to find the equations of hyperbolas with centers other than $(0,0)$:

E4':

$$\boxed{\dfrac{(x - h)^2}{a^2} - \dfrac{(y - k)^2}{b^2} = 1}$$

is the equation of a "left-right" hyperbola centered on (h, k) with vertices $(h \pm a, k)$ and asymptotes

$$\frac{(x - h)^2}{a^2} - \frac{(y - k)^2}{b^2} = 0 , \quad \text{i. e.,}$$

$$y - k = \pm \frac{b}{a} (x - h) ;$$

E5' :

$$\boxed{\frac{(y - k)^2}{a^2} - \frac{(x - h)^2}{b^2} = 1}$$

is the equation of an "up-down" hyperbola centered on (h, k) with vertices $(h, k \pm a)$ and asymptotes

$$\frac{(y - k)^2}{a^2} - \frac{(x - h)^2}{b^2} = 0 , \quad \text{i. e.,}$$

$$y - k = \pm \frac{a}{b} (x - h)$$

An important note concerning memorization: You really need not memorize equations E1' through E5' for conics with center (h, k) if you remember the simpler equations E1 through E5 for conics with center $(0, 0)$ along with the translation principles!

Example 8. Sketch the graph of $x^2 - 4y^2 + x + 8y + \frac{1}{4} = 0$.

Solution. As in Examples 4 and 5 we must first complete the squares in both the x and y terms.

$$\left(x^2 + x\right) - 4\left(y^2 - 2y\right) + \frac{1}{4} = 0$$

$$\left(x^2 + x + \frac{1}{4}\right) - \frac{1}{4} - 4\left(y^2 - 2y + 1\right) + 4 + \frac{1}{4} = 0$$

$$\left(x + \frac{1}{2}\right)^2 - 4\left(y - 1\right)^2 + 4 = 0$$

$$4\left(y - 1\right)^2 - \left(x + \frac{1}{2}\right)^2 = 4$$

$$\left(y - 1\right)^2 - \frac{\left(x + \frac{1}{2}\right)^2}{4} = 1$$

Thus we are in standard form E5' (or standard form E5 translated by $\left(-\frac{1}{2}, 1\right)$) with

$a = 1$ and $b = 2$; this is an up-down hyperbola

with center $\left(-\frac{1}{2}, 1\right)$, vertices $\left(-\frac{1}{2}, 1 \pm 1\right) =$

$= \left(-\frac{1}{2}, 0\right)$, $\left(-\frac{1}{2}, 2\right)$ and asymptotes

$y - 1 = \pm \frac{1}{2}\left(x + \frac{1}{2}\right)$, i.e.,

$$y = \frac{1}{2}x + \frac{5}{4} \quad \text{and} \quad y = -\frac{1}{2}x + \frac{3}{4} \qquad \square$$

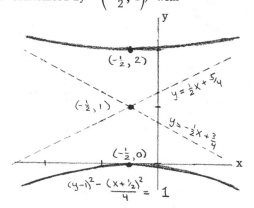

5. "Mixed term" Hyperbolas. There is another very standard way in which hyperbolas

commonly arise:

E6:
$$y = \frac{\alpha}{x} \quad \text{or} \quad xy = \alpha$$

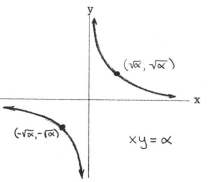

where α is a non-zero constant. The center is at

$(0,0)$, the asymptotes are $x = 0$ and $y = 0$

(the coordinate axes), and the hyperbolas fall in the

1st and 3rd quadrants if $\alpha > 0$, or the 2nd

and 4th quadrants if $\alpha < 0$. By the translation principles the equation

E6':
$$y - k = \frac{\alpha}{x - h} \quad \text{or} \quad (x - h)(y - k) = \alpha$$

gives hyperbolas with center (h, k) and asymptotes $x = h$ and $y = k$.

Example 9. Sketch the graph of $xy + 2x - 4y - 7 = 0$.

Solution. We must transform our equation into standard form E6'. Notice that E6' can be

rewritten as

$$xy - kx - hy + hk - \alpha = 0$$

Thus in our case we must have $k = -2$, $h = 4$ and $hk - \alpha = -7$, i.e.,

$\alpha = 7 + (-2)(4) = -1$. Our equation then becomes

$$(x - 4)(y + 2) = -1$$

which is a hyperbola with center $(4, -2)$ and

asymptotes $x = 4$ and $y = -2$. The <u>vertices</u>

are seen to be

$$(4 - 1, -2 + 1) = (3, -1) \quad \text{and}$$

$$(4 + 1, -2 - 1) = (5, -3)$$

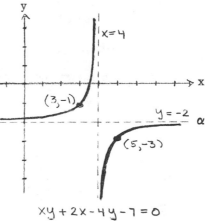

$xy + 2x - 4y - 7 = 0$

6. <u>"Degenerate" cases.</u>   Minor changes in the equations for conic sections can often produce

surprisingly radical changes in the graphs themselves. In some cases perfectly respectable

conic sections can degenerate into pairs of lines, single points, or the "empty set" (i.e., there

are no values of x and y which satisfy the equation under consideration). For example,

$$\frac{(x + 2)^2}{8} + \frac{(y - 1)^2}{4} = 1$$

is the equation for an ellipse; change the 1 to a 0 however, and the graph becomes the

one point $(-2, 1)$. If instead you change the 1 to -1, then the graph is the empty set

(a sum of positive numbers can never be negative). As a final example,

$$(y - 1)^2 - \frac{\left(x + \frac{1}{2}\right)^2}{4} = 1$$

is the equation for a hyperbola; change the 1 to a 0 however, and the graph becomes the

two lines $y = \frac{1}{2} x + \frac{5}{4}$ and $y = -\frac{1}{2} x + \frac{3}{4}$. [When these two lines are graphed, they form

an "X".]

Exercises. Identify each of the following graphs by putting each equation into one of the standard forms discussed in this appendix. Specify the center, and if appropriate, the vertices, the radius, the lengths of the semi-major and semi-minor axes, or the asymptotes. Sketch the curve. [Note: A few "degenerate" cases are also included to keep you alert.]

1. $x^2 + y^2 - 2x + 4y + 2 = 0$

2. $4x^2 - 36y^2 - 8x + 36y + 31 = 0$

3. $x\left(\frac{1}{2}x - \sqrt{2}\right) + y(y + 2) + 1 = 0$

4. $x^2 - y^2 + 4x + 2y + 3 = 0$

5. $3x^2 = y + 3x - 2$

6. $y + 2x = xy$

7. $2x^2 + y^2 - 4x + 6y + 15 = 0$

8. $8x^2 = 4y(y - 1) + 17$

Answers.

1. $(x - 1)^2 + (y + 2)^2 = 3$; a circle

with center $(1, -2)$ and radius $\sqrt{3}$.

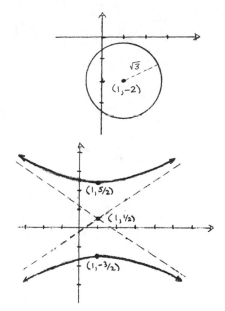

2. $\dfrac{\left(y - \dfrac{1}{2}\right)^2}{4} - \dfrac{(x - 1)^2}{9} = 1$; and "up-down"

hyperbola with $a = 2$, $b = 3$, center

$\left(1, \dfrac{1}{2}\right)$, vertices $\left(1, -\dfrac{3}{2}\right)$, $\left(1, \dfrac{5}{2}\right)$

and asymptotes $y = \dfrac{2}{3} x - \dfrac{1}{6}$ and

$y = -\dfrac{2}{3} x + \dfrac{7}{6}$.

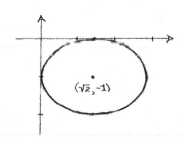

3. $\dfrac{\left(x - \sqrt{2}\right)^2}{2} + (y + 1)^2 = 1$; an ellipse

with semi-major axis length $a = \sqrt{2}$

and semi-minor axis length $b = 1$;

center $\left(\sqrt{2}, -1\right)$, and vertices

$(0, -1)$, $\left(2\sqrt{2}, -1\right)$, $\left(\sqrt{2}, -2\right)$,

and $\left(\sqrt{2}, 0\right)$.

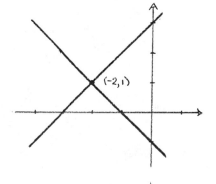

4. $(x + 2)^2 - (y - 1)^2 = 0$; a "degenerate"

hyperbola which reduces to two lines,

      $y = x + 3$ and $y = -x - 1$.

5. $y - \frac{5}{4} = 3\left(x - \frac{1}{2}\right)^2$; a steep upward

turning parabola with vertex $\left(\frac{1}{2}, \frac{5}{4}\right)$.

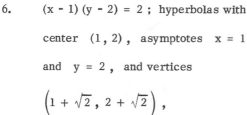

6. $(x - 1)(y - 2) = 2$; hyperbolas with

center $(1, 2)$, asymptotes $x = 1$

and $y = 2$, and vertices

$\left(1 + \sqrt{2}, 2 + \sqrt{2}\right)$,

$\left(1 - \sqrt{2}, 2 - \sqrt{2}\right)$

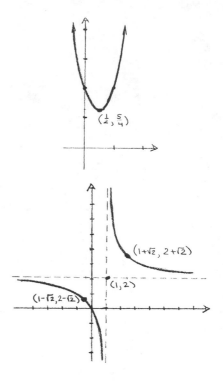

7. $\frac{(x - 1)^2}{2} + \frac{(y + 3)^2}{4} = -1$; there are <u>no</u>

solutions to this equation since the sum

of positive numbers can never be negative.

8. $\frac{x^2}{2} - \frac{\left(y - \frac{1}{2}\right)^2}{4} = 1$; a "left-right"

hyperbola with $a = \sqrt{2}$, $b = 2$,

center $\left(0, \frac{1}{2}\right)$, vertices

$\left(-\sqrt{2}, \frac{1}{2}\right)$, $\left(\sqrt{2}, \frac{1}{2}\right)$ and

asymptotes

$$y = \sqrt{2}\, x + \frac{1}{2}, \quad y = -\sqrt{2}\, x + \frac{1}{2}$$

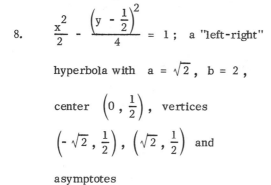

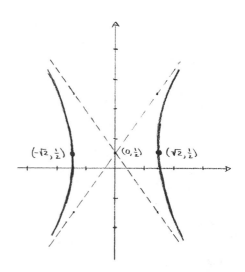

Appendix F : Systems of equations.

1. Two simultaneous equations.   Given two equations in variables x and y , it is often
times necessary to find those values of x and y which satisfy both of the equations
simultaneously; this is naturally referred to as solving a system of simultaneous equations.

Example 1.   Solve the system of equations

$$\begin{cases} x + y = 3 \\ 2x^2 + y^2 = 6 \end{cases}$$

Solution.   A common procedure in solving simultaneous equations is to use one equation to
solve for one variable in terms of the other. This is called the method of substitution.   In
the case at hand we can use the first equation to solve for y in terms of x ,

$$y = 3 - x$$

and then plug this into the second equation (i.e., substitute 3 - x for y) to obtain

$$2x^2 + (3 - x)^2 = 6$$

$$2x^2 + 9 - 6x + x^2 = 6$$

$$3x^2 - 6x + 3 = 0$$

$$x - 2x + 1 = 0$$

$$(x - 1)^2 = 0$$

$$x = 1$$

Thus x = 1 and y = 3 - 1 = 2 ; testing this back in the original two equations shows that
(x , y) = (1 , 2) is indeed a solution.   □

Solving simultaneous equations, while an algebraic operation, has an important geometric meaning: the (x,y) solution values which one finds are the <u>intersection points</u> of the <u>graphs</u> of the two given equations. Thus, in Example 1,

the line $x + y = 3$ intersects the ellipse

$2x^2 + y^2 = 6$ in the one point $(1,2)$, as shown

in the figure to the right.

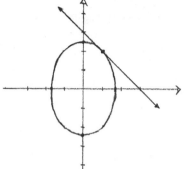

The following list generalizes the procedures used

to solve Example 1.

<u>Method of substitution for solving two simultaneous equations in x and y</u>

   i. Use one of the equations to solve for y in terms of x,

      obtaining $y = f(x)$.

   ii. Use $y = f(x)$ to eliminate y from the remaining equation,

      i.e., <u>substitute</u> $f(x)$ for y in the remaining equation.

      The result is an equation containing only x.

  iii. Solve the new equation for specific values of x.

  iv. Find the corresponding values of y by using $y = f(x)$.

   v. Check all your (x,y) values back in the original two equations.

      (It is <u>not uncommon</u> to find that incorrect solutions have slipped

      in during the solution process.) If feasible, sketch the graphs of

      the two equations and compare their intersection points with the

      (x,y) solutions you discovered algebraically.

<u>Note:</u> The roles of x and y can be (and, in some instances, <u>must</u> be) reversed in this process, i.e., first solve for x in terms of y to get $x = g(y)$, etc.

Example 2. Solve the system of equations

$$\begin{cases} \dfrac{2 - xy}{y + 1} = 1 \\ \\ 2x^2 + 5x - (3x + 5)y + 3y^2 = 0 \end{cases}$$

Solution.

i. Solving the first equation for y in terms of x yields

$$y = 1/(x + 1)$$

ii. Plugging this into the second equation will yield

$$2x^2 + 5x - (3x + 5)/(x + 1) + 3/(x + 1)^2 = 0$$

which simplifies to

$$2x^4 + 9x^3 + 9x^2 - 3x - 2 = 0$$

iii. By the Rational Root Test (Appendix D §6) the only possible rational roots

$x = p/q$ are those for which p divides -2 and q divides 2 , i.e.,

$$x = \pm 2, \quad \pm 1 \quad \text{or} \quad \pm 1/2$$

Only two of these check out to be roots: $x = -2$ and $x = 1/2$. Long division

(Appendix D §2) of our polynomial by the product

$$2(x + 2)(x - 1/2) = 2x^2 + 3x - 2$$

yields $x^2 + 3x + 1$. By the Quadratic Formula (Appendix D §5) the roots of

this term are $-3/2 \pm \sqrt{5}/2$. Thus all the possible x values are

$$-2, \quad 1/2, \quad -3/2 \pm \sqrt{5}/2$$

iv. From $y = 1/(x + 1)$ we see that our possible (x,y) values are $(-2, -1)$,

$(1/2, 2/3)$, and

$$\left(-\frac{3}{2}+\frac{\sqrt{5}}{2}\,,\,\frac{1}{2}+\frac{\sqrt{5}}{2}\right)\quad,\quad\left(-\frac{3}{2}-\frac{\sqrt{5}}{2}\,,\,\frac{1}{2}-\frac{\sqrt{5}}{2}\right)$$

v. The first of these points does <u>not</u> check out in the original equations since it would

   require a division by zero.   The remaining three points check out properly.

   You would ordinarily <u>not</u> graph these two equations to check their solutions

   because of the complexity of the second function.   We do, however, show the two

   graphs below so that you can see our solution values as points of intersection

   of the two graphs.   The graph of our first equation can be shown to be a "mixed

   term" hyperbola with the point $(-2,-1)$ deleted (Appendix E §5).   The

   second equation requires advanced techniques from Anton's §12.5 for its

   analysis: it is a tilted ellipse with center $(-1, 1/3)$.   ☐

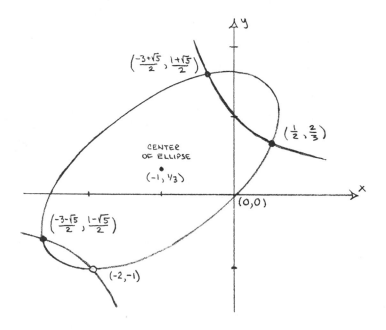

The method just given for solving two simultaneous equations can run into difficulties in two places: in Step i it may be very difficult (or impossible) to use one equation to solve for y in terms of x, or in Step iii, the one variable equation may be difficult (or impossible) to solve for x. In such cases approximate solutions might be sought by numerical methods (which we do not discuss here) or slightly more round-about solution methods might be employed. This latter situation is quite common when Step i proves unpleasant, as illustrated in the next example.

Example 3.   Solve the system of equations

$$\begin{cases} xy + y^2 - x^2 = 5 \\ 2xy + y^2 - x^2 = 2 \end{cases}$$

Solution.   Although either equation can be used to solve for y in terms of x, the result is somewhat unpleasant. It is much easier to notice that subtracting the first equation from the second will yield $xy = -3$, i.e., $y = -3/x$. This is a common type of modification of Step i. The rest of the steps now proceed in the standard way:

$$xy + y^2 - x^2 - 5 \quad \text{becomes} \quad -3 + 9/x^2 - x^2 = 5$$

which is $x^4 + 8x^2 - 9 = 0$. This factors into

$$0 = \left(x^2 - 1\right)\left(x^2 + 9\right) = (x - 1)(x + 1)\left(x^2 + 9\right)$$

so that the only possible x values are 1 and -1; the corresponding y-values are -3 and 3 respectively. Both solutions (1, -3) and (-1, 3) test out correctly in the original equations. $\square$

2. <u>Three simultaneous equations.</u>   There are times when you will run into <u>three</u> simultaneous equations in <u>three</u> unknowns, say x, y and z.   The method of substitution generalizes to this situation in a straightforward manner.   Choose one of the equations to solve for z in terms of x and y,   and use this result to eliminate z from the remaining two equations. You will then be reduced to a system of two equations in two unknowns, and the procedures of the previous section apply.

Example 4.   Solve the system of equations

$$\left\{ \begin{array}{l} x + y + y^2 z = 2 \\ 2x = -yz \\ xy^2z = 2 \end{array} \right\}$$

<u>Solution.</u>   There are numerous ways to begin; we will use the second equation to solve for z in terms of x and y :

$$\boxed{z = -2x/y}$$

The remaining two equations then become

$$\left\{ \begin{array}{l} x + y - 2xy = 2 \\ -2x^2 y = 2 \end{array} \right\}$$

Solve the second of these equations for y in terms of x :

$$\boxed{y = -1/x^2}$$

The remaining equation becomes

$$x - 1/x^2 + 2/x = 2$$

which simplies to

$$(*) \qquad\qquad x^3 - 2x^2 + 2x - 1 = 0$$

By the Rational Root Test (Appendix D §6) the only possible rational roots for this cubic are

$x = \pm 1$; in fact, $x = 1$ does work, while $x = -1$ does not. Dividing our cubic by

$x - 1$ yields

$$x^2 - x + 1$$

an irreducible quadratic term (no real roots). Thus $x = 1$ is the only solution to equation

$(*)$. Tracing back we find

$$y = -1/1^2 = -1$$
$$z = -2(1)/(-1) = 2$$

Thus $x = 1$, $y = -1$ and $z = 2$ give the only possible solution for our system of equations.

Checking these values in our equations show that indeed they do give a solution.   □

Solving simultaneous equations can be very difficult and (in the general case) no one method can

be given which always works. Many times clever tricks need to be employed... , but we've

done enough on the general case for now. We move on, instead, to the special (but important)

case of linear equations.

3.   Systems of linear equations.   An equation of the form

$$ax + by = c$$

where a, b and c are constants, is called a linear equation in the variables x and y .

As long as either a or b is non-zero, the graph of this equation is a line. Systems of

linear equations occur quite often in calculus and, in contrast to the general situation discussed in Section one, there are specific techniques for dealing with linear systems which apply in <u>all</u> situations.

<u>Example 5.</u>   Solve the system of linear equations

$$\begin{cases} 2x + 3y = 4 \\ x - 2y = -5 \end{cases}$$

<u>Solution.</u>   As these equations represent two non-parallel lines in the plane (unequal slopes), there must be one and only one intersection point (i. e., the system has <u>exactly one</u> solution). We find this solution in two different ways:

1st method: The <u>method of substitution</u> described in the previous sections will always work with linear systems.   In the case at hand the first equation yields

$$y = 4/3 - (2/3)x$$

which when placed into the second equation will give

$$x - 2\left(\frac{4}{3} - \frac{2}{3}x\right) = -5$$

This solves to $x = -1$, and hence $y = 2$. The point $(-1, 2)$ checks out correctly in both of the original equations.

2nd method: This is called the <u>elimination method</u>: we <u>eliminate</u> one variable by adding together carefully chosen multiples of the two given equations.   In the case at hand multiplying the second equation by -2 and adding it to the first eliminates x quite neatly:

$$\begin{array}{r} 2x + 3y = 4 \\ -2x + 4y = +10 \\ \hline 0 + 7y = 14 \end{array}$$

so that $y = 2$. The first equation then gives $x = -1$. $\square$

You might, with good reason, question why we bothered to give the <u>elimination method</u> in Example 5 when <u>substitution</u> worked so well itself. The answer lies in generalization to more complicated linear systems, say 3 linear equations in 3 unknowns, or 4 linear equations in 4 unknowns. In these (very common) situations, both of the given methods will work, but the elimination method is much faster and less prone to careless errors (and it is programmable on a computer or hand calculator).

<u>Example 6.</u>   Solve the system of linear equations

$$\begin{cases} x + 2y - z = -3 \\ 2x - y + 3z = 9 \\ -3x + y - z = -6 \end{cases}$$

<u>Solution.</u>   We use the 1st equation to eliminate x from all the equations below it (i.e., from Equations 2 and 3):

-2 times the 1st equation added to the 2nd yields

$$\begin{aligned} -2x - 4y + 2z &= 6 \\ \underline{2x - y + 3z} &= \underline{9} \\ -5y + 5z &= 15, \quad \text{or} \quad -y + z = 3 \end{aligned}$$

3 times the 1st equation added to the 3rd yields

$$\begin{aligned} 3x + 6y - 3z &= -9 \\ \underline{-3x + y - z} &= \underline{-6} \\ 7y - 4z &= -15 \end{aligned}$$

We thus have a new system of linear equations:

$$\begin{cases} x + 2y - z = -3 \\ \quad -y + z = 3 \\ \quad 7y - 4z = -15 \end{cases}$$

in which the x variable has been eliminated from all but the 1st equation. We now use the 2nd equation to eliminate y from all the equations below it (i.e., from Equation 3):

7 times the 2nd equation added to the 3rd yields

$$-7y + 7z = 21$$
$$\underline{7y - 4z = -15}$$
$$3z = 6 \quad , \quad \text{or} \quad z = 2$$

This yields a third system of linear equations,

$$\begin{cases} x + 2y - z = -3 \\ \quad -y + z = 3 \\ \quad\quad z = 2 \end{cases}$$

where each successive equation has one less variable, until only one variable appears. But thus the last variable has been solved for, and <u>back substitution</u> up through the system gives all the solutions:

3rd equation: $z = 2$ gives $z = 2$

2nd equation: $-y + z = 3$

$\quad\quad\quad\quad\quad\quad -y + 2 = 3$ gives $y = -1$

1st equation: $x + 2y - z = -3$

$\quad\quad\quad\quad\quad\quad x + 2(-1) - (2) = -3$ gives $x = 1$

The solution $(2, -1, 1)$ checks out in the three original equations.   □

The following list generalizes the procedures used to solve Example 6.

Elimination method for solving three simultaneous linear equations in x, y and z:

i. Use the 1st equation to eliminate x from the 2nd and 3rd equations. Do this by adding suitable multiples of the 1st equation to suitable multiples of the 2nd and 3rd equations.

ii. Use the 2nd equation (new version) to eliminate y from the 3rd equation (new version). Do this by adding a suitable multiple of the 2nd equation to a suitable multiple of the 3rd equation.

iii. Solve the 3rd equation for z, solve the 2nd equation for y, and solve the 1st equation for x.

iv. Check your answer in the original equations.

Some comments are in order concerning this method:

a. The order of the equations and the order of the elimination process (x, then y, then z) are not sacred and can be changed around for convenience in solving a specific system. In fact, sometimes they *must* be switched around.

b. The generalization of this method to systems of n linear equations in n unknowns should be fairly clear.

c. There are systems of linear equations with no solutions; a simple example is

$$\begin{cases} x + y = 1 \\ x + y = 0 \end{cases}$$

These equations represent two parallel lines, and hence there cannot be a simultaneous solution.

In such cases the elimination method reveals the problem by producing, at some stage, a nonsense equation such as $0 = 1$.

d. There are systems of linear equations with more than one solution (in fact, with an <u>infinite number</u> of solutions); a simple example is

$$\begin{cases} x + y = 1 \\ 2x + 2y = 2 \end{cases}$$

In such cases the method indicates this occurrence by a "simultaneous elimination," e.g., when eliminating y you find that x has also been eliminated. Why this leads to an infinite number of solutions is illustrated in Example 7.

Comments (c) and (d) illustrate an important principle:

There are only three possibilities for a system of two linear equations in two unknowns:

1) It has <u>exactly one</u> solution.

2) It has an <u>infinite number</u> of solutions.

3) It has <u>no solutions</u>.

It can be very helpful to have these three options in mind.

<u>Example 7.</u> Solve the system of linear equations

$$\begin{cases} x + 2y \quad\quad = -1 \\ 2x - y + 5z = 3 \\ 2x \quad\quad + 4z = 2 \end{cases}$$

Solution.   Use the 1st equation to eliminate x in the 2nd and 3rd equations. This gives

$$\begin{cases} x + 2y \qquad\quad = -1 \\ \quad\; -5y + 5z = \quad 5 \\ \quad\; -4y + 4z = \quad 4 \end{cases} \quad \text{or} \quad \begin{cases} x + 2y \qquad\quad = -1 \\ \qquad y \;-z = -1 \\ \qquad y \;-z = -1 \end{cases}$$

Then using the 2nd equation to eliminate y in the 3rd equation will yield

$$\begin{cases} x + 2y \qquad\quad = -1 \\ \qquad y \;-z = -1 \\ \qquad\qquad 0 = \quad 0 \end{cases}$$

This is an example of comment (d) above; we cannot solve for z since z can take on any value. However, once z is specified, then

$$y = z - 1 \quad \text{and} \quad x = -2y - 1 = -2z + 1$$

Any solution of the form $(-2z + 1, z - 1, z)$, where z is _any_ number, checks out correctly in the original set of equations. This is termed the _general solution_ for our system of equations; specific solutions are obtained by fixing a value for z. For example, taking $z = 1$ gives the specific solution

$$(-2(1) + 1, 1 - 1, 1) = (-1, 0, 1) \qquad\qquad \square$$

Exercises.   Solve each of the following system of equations

1. $$\begin{cases} 2x + 4y = -1 \\ 3x + \;\; y = \quad 1 \end{cases}$$

2. $$\begin{cases} xy = 2 \\ xy^2 + x^2y = 6 \end{cases}$$

3. $\begin{cases} xy = x - y \\ 4y + 2xy + x^2 = 4 \end{cases}$

4. $\begin{cases} x^3 + z^2 + xz = 3 \\ x^3 - z + xz = 1 \end{cases}$

5. $\begin{cases} A - B - 5C = 3 \\ A + B + C = 3 \\ 2A + B + 3C = -2 \end{cases}$

6. $\begin{cases} x - y + z = -1 \\ x + y - z = 4 \\ 3x - y + z = 1 \end{cases}$

7. $\begin{cases} x - y - z = 0 \\ 2x + 3y + 6z = 3 \\ x - 2y + z = 0 \end{cases}$

8. $\begin{cases} x + y - (3/2)z = 0 \\ -2x - z = -2 \\ x + y - z = 1 \end{cases}$

9. $\begin{cases} x + 3y + 2xz = -1 \\ 3xyz = -4 \\ x^2 + 3yz = -3 \end{cases}$

Answers.

1. $(x, y) = (1/2, -1/2)$

2. $(x, y) = (1, 2)$ or $(2, 1)$

3. $(x, y) = (-2, 2)$ or $(1, 1/2)$

4. $(x, z) = (1, 1), (1, -2) \left(-\frac{1}{2} + \frac{\sqrt{5}}{2}, -2 \right)$ or $\left(-\frac{1}{2} - \frac{\sqrt{5}}{2}, -2 \right)$

5. $(A, B, C) = (-1, 6, -2)$

6. No solutions

7. $(x, y, z) = (1/2, 1/3, 1/6)$

8. $(x, y, z) = \left(1 - \frac{1}{2}z, -1 + 2z, z \right)$; z is arbitrary

9. $(x, y, z) = (1, 2/3, -2)$ or $(1, -4/3, 1)$.

Appendix G :   Trigonometry refresher.

Appendix 1 in Anton's text is an extensive trigonometry review; this appendix concen-

trates more on the most important trigonometry results and/or those results which tend to

give students the most trouble. Students with serious deficiencies in this topic should go through

Anton's material (and perhaps top it off with this refresher).

1.   Radian measure.   Consider the unit circle

in the xy-plane with center $(0,0)$:

$$x^2 + y^2 = 1$$

Take any angle θ measured <u>counterclockwise</u> from

the x-axis. The length of that portion of the circle

determined by θ (i. e., in the picture the circular arc from A to B) is called the

<u>radian measure</u> of θ .

Since the circumference of the unit circle is 2π , the radian measure of an angle of

360° is 2π . The conversion from degrees to radians is always given by this proportion, i. e.,

G1:

$$\boxed{\dfrac{\theta \;\; \text{in radians}}{\theta \;\; \text{in degrees}} = \dfrac{2\pi}{360^o} = \dfrac{\pi}{180^o}}$$

Listing some common angles in both degrees and radians gives

G2:

degrees ... radians	degrees ... radians
30° ... $\pi/6$	45° ... $\pi/4$
60° ... $\pi/3$	90° ... $\pi/2$

Angles measured in the <u>clockwise</u> direction from the x-axis are given <u>negative</u> radian measure. Thus, for example,

$$- 30^{o} \ldots - \pi/6 \qquad - 180^{o} \ldots - \pi$$

People are frequently mystified as to why this bizarre radian measure is introduced

for measuring angles when measurement by degrees seems so simple. Isn't trigonometry

complicated enough without radian measure to further cloud the issue? The fact is, however,

that radian measure is really a <u>necessity</u> for the <u>calculus</u> of trigonometric functions. With-

out it many of our important calculus formulas would need unpleasant changes in them ...

but we can't prove that to you until "differentiation" is developed in Chapter 3 of the text.

This question is discussed in more detail in §3.3.4 of <u>The Companion.</u>

2. <u>The Trigonometric Functions.</u>   As in the previous section consider the unit circle in the

xy-plane   with center $(0,0)$,

$$x^2 + y^2 = 1$$

THE BASIC PICTURE

$P(\cos\theta, \sin\theta)$

and take any angle θ measured counterclockwise

from the   x-axis. The (x,y) coordinates of the

point   P   on the unit circle corresponding to θ are

defined to be the <u>cosine of</u> θ and the <u>sine of</u> θ

respectively, i.e.,

$$P = (x, y) = (\cos\theta , \sin\theta) \qquad *$$

Since the point   P   is the <u>same</u> for both the angle θ and the angle $\theta + 2\pi$ we immediately

see that the sine and cosine are <u>periodic with period</u> 2π ,   i.e.,

---

* That these definitions for $\sin\theta$ and $\cos\theta$ are the same as the usual ones involving
"opposite, adjacent and hypotenuse" is established in   Theorem 2 of Anton's   Appendix 1.

G3 :

$$\cos (\theta + 2\pi) = \cos \theta$$

$$\sin (\theta + 2\pi) = \sin \theta$$

for all angles θ (when measured in radians of course!) Also, as seen from the diagram

to the right, flipping from θ to $-\theta$

does not change the x-coordinate of P

(the cosine) but negates the y-coordinate

of P (the sine); thus

G4 :

$$\cos (-\theta) = \cos \theta$$

$$\sin (-\theta) = -\sin \theta$$

Finally, since $P = (\cos \theta, \sin \theta)$ lies on the unit circle $x^2 + y^2 = 1$, we immediately

obtain our most well-known identity

G5:

$$\sin^2 \theta + \cos^2 \theta = 1$$

The values of sine and cosine for the angles in G2 occur over and over again in

calculus; they should be memorized:

G6 :

θ	0	$\pi/6$	$\pi/4$	$\pi/3$	$\pi/2$
$\sin \theta$	0	$1/2$	$\sqrt{2}/2$	$\sqrt{3}/2$	1
$\cos \theta$	1	$\sqrt{3}/2$	$\sqrt{2}/2$	$1/2$	0

Here is a very useful method for remembering this table:

i.   Write down the integers   0   1   2   3   4

ii.   Take the square roots: $\sqrt{0}$ $\sqrt{1}$ $\sqrt{2}$ $\sqrt{3}$ $\sqrt{4}$

iii.   Divide by 2: $\dfrac{\sqrt{0}}{2}$ $\dfrac{\sqrt{1}}{2}$ $\dfrac{\sqrt{2}}{2}$ $\dfrac{\sqrt{3}}{2}$ $\dfrac{\sqrt{4}}{2}$

These numbers simplify to $\boxed{\;0 \quad \dfrac{1}{2} \quad \dfrac{\sqrt{2}}{2} \quad \dfrac{\sqrt{3}}{2} \quad 1\;}$ which is the row of sine values

in the above table! The row of cosine values is obtained simply by writing down the sine values

in <u>reverse order</u>, i.e., $\boxed{\;1 \quad \dfrac{\sqrt{3}}{2} \quad \dfrac{\sqrt{2}}{2} \quad \dfrac{1}{2} \quad 0\;}$

Another way to remember these values is to memorize the following two right triangles:

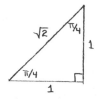 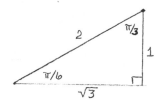

The sine and cosine of the angles $\pi/6$, $\pi/4$ and $\pi/3$ can be read off from these triangles
by

$$\sin\theta = \frac{\text{opposite}}{\text{hypothenuse}} \quad , \quad \cos\theta = \frac{\text{adjacent}}{\text{hypothenuse}}$$

The other four trigonometric functions are all defined in terms of sine and cosine;

these definitions need to be memorized:

G7: $\boxed{\begin{array}{ll} \tan\theta = \dfrac{\sin\theta}{\cos\theta} & \sec\theta = \dfrac{1}{\cos\theta} \\[2ex] \cot\theta = \dfrac{\cos\theta}{\sin\theta} & \csc\theta = \dfrac{1}{\sin\theta} \end{array}}$ *

---

* When the denominators are non-zero, of course.

The secant and cosecant are periodic with period 2π, while the tangent and cotangent are periodic with period π. Notice that, unlike the sine and cosine, these new functions are not defined for all values of θ since the denominator terms can be zero for some values of θ.

3.   Trigonometric Identities.   There are a number of trigonometry formulas which you need to memorize because they are regularly used in calculus and its applications.   But, there is a hierarchy of importance in the formulas so that you can learn the most important ones now and fill in the others later.   Moreover, most of the identities are quickly derived from a small "core" of formulas; memorize the core very well and remember the simple derivation tricks, and you'll be in good shape!

You already have most of the "core": identities G3 - G5 (quickly derived from " The Basic Picture"), the table of values G6 , and definitions G7.   We need only one more pair of formulas:

G8:

$$\cos (\alpha + \beta) = \cos \alpha \cos \beta - \sin \alpha \sin \beta$$
$$\sin (\alpha + \beta) = \sin \alpha \cos \beta + \sin \beta \cos \alpha$$

These are not trivial identities to verify (see Anton's Appendix 1) nor are they "intuitive" in any reasonable sense.   However, they need to be carefully memorized.

It is useful to memorize the rest of the formulas of this section, but all are derivable from our core collection, most in very easy ways.   Learning the derivations will free you from the worries of small memory errors: if unsure of a certain formula, you simply rederive it. We suggest that you try to derive the following formulas from the core formulas yourself before referring to our calculations:

G9:

$$\cos 2\alpha = \cos^2 \alpha - \sin^2 \alpha$$
$$\sin 2\alpha = 2 \sin \alpha \cos \alpha$$

G. 6

These two formulas are obtained from G8 by setting $\beta = \alpha$. An important feature of the second of these equations is that it allows us to express the product of $\sin\alpha$ and $\cos\alpha$ as one single trigonometric function:

G10:

$$\sin\alpha\cos\alpha = \frac{1}{2}\sin 2\alpha$$

This can come in quite handy in calculus (especially in "integration" of trig functions); we would like to have similar formulas for $\sin^2\alpha$ and $\cos^2\alpha$. This is fortunately provided by the first identity in G9 along with G5 :

$$\cos^2\alpha + \sin^2\alpha = 1 \qquad (G5)$$
$$\cos^2\alpha - \sin^2\alpha = \cos 2\alpha \qquad (G9)$$

Adding these two equations and dividing by 2 gives a formula for $\cos^2\alpha$; subtracting the two equations and dividing by 2 gives a formula for $\sin^2\alpha$. The results are

G11:

$$\cos^2\alpha = \frac{1 + \cos 2\alpha}{2}$$
$$\sin^2\alpha = \frac{1 - \cos 2\alpha}{2}$$

These are very important formulas in "integration theory."

More generally, we can express any product of the form $\sin\alpha\cos\beta$, $\sin\alpha\sin\beta$ or $\cos\alpha\cos\beta$ as a sum of single sine and cosine functions. This is done by first establishing the identities

G12:

$$\cos(\alpha - \beta) = \cos\alpha\cos\beta + \sin\alpha\sin\beta$$
$$\sin(\alpha - \beta) = \sin\alpha\cos\beta - \sin\beta\cos\alpha$$

from G8 and G4. Then judicious adding or subtracting between G8 and G12 yield

G13:

$$\sin \alpha \cos \beta = \frac{1}{2}\,[\sin(\alpha - \beta) + \sin(\alpha + \beta)]$$
$$\sin \alpha \sin \beta = \frac{1}{2}\,[\cos(\alpha - \beta) - \cos(\alpha + \beta)]$$
$$\cos \alpha \cos \beta = \frac{1}{2}\,[\cos(\alpha - \beta) + \cos(\alpha + \beta)]$$

It is, in our opinion, better to just know that formulas G12 and G13 exist rather than to memorize them (... but perhaps it's best to check with your calculus instructor on this advice!)

There is just one last pair of formulas which we feel is essential to know for calculus (others which are helpful or only occasionally needed can be found in Anton):

G14:

$$\tan^2\theta + 1 = \sec^2\theta$$
$$1 + \cot^2\theta = \csc^2\theta$$

These are very useful in "integration theory" where it is essential at times to switch between $\tan^2\theta$ and $\sec^2\theta$, or between $\cot^2\theta$ and $\csc^2\theta$. Their derivations are very easy: simply divide G5 by $\cos^2\theta$ for the first identity, and by $\sin^2\theta$ for the second. Here is the first derivation:

$$\frac{\sin^2\theta + \cos^2\theta}{\cos^2\theta} = \frac{1}{\cos^2\theta}$$

i. e., $$\tan^2\theta + 1 = \sec^2\theta$$

4.   Law of Cosines.   Using the cosine function we can obtain an important generalization of the Pythagorean Theorem known as the Law of Cosines:

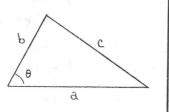

Law of
Cosines

Suppose a triangle has side lengths

a , b   and   c ,   and θ is the angle

between the sides of lengths   a   and   b .

Then

$$c^2 = a^2 + b^2 - 2\,a\,b\,\cos\theta$$

Notice that if θ is a right angle, i.e., $\theta = \pi/2$,   then $\cos\theta = 0$ and the Law of Cosines

becomes the familiar Pythagorean Theorem

$$c^2 = a^2 + b^2$$

A proof of the Law of Cosines can be found in Anton's Appendix 1.

Example 1.   Given a triangle with side lengths

2 , 3   and   4 ,   determine the angle between the

sides of lengths   4   and   2 .

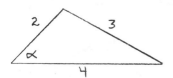

Solution.   Let α be the angle between the sides of lengths   4   and   2 .   Then the Law of

Cosines gives

$$3^2 = 4^2 + 2^2 - 2\,(4)(2)\cos\alpha$$

$$9 = 16 + 4 - 16\cos\alpha$$

so that $\cos\alpha = (16 + 4 - 9)/16 = 11/16 = .6875$

Using either Anton's Table 1 in Appendix 3 , or a hand calculator, we see that

$$\alpha \approx .813 \text{ radians} \approx 46.5^{\circ} \quad .$$

5.   Graphs of Trigonometric Functions.   From the Basic Picture in §2 it can be seen that

the sine function y = sin x   starts at   y = 0   when   x = 0 , increases to y = 1   when

x = π/2 , decreases to y = 0   when   x = π   and further to   y = -1   when   x = 3π/2 ,

and then increases back to   y = 0   when   x = 2π .   Since from G3 the sine function is

periodic with period   2π ,   the graph just described from   x = 0   to   x = 2π   is simply

repeated over every interval   [2πn , 2π(n + 1)]   for   n   any integer.   We thus obtain

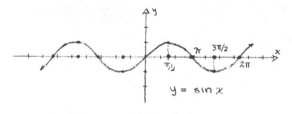

the graph as shown to the left.   The

same analysis for the cosine produces

the same type of graph, except that

$y = 1$ when $x = 0$.   The graph is

shown to the right.   The remaining

four graphs are shown below.

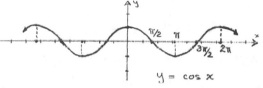

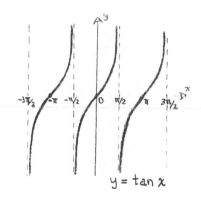

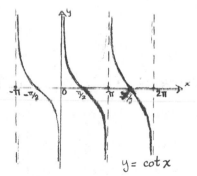

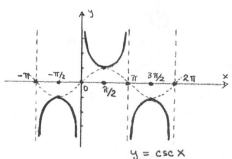

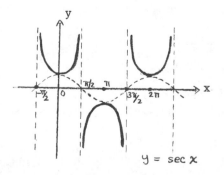

Notice how the graph of y = csc x "hangs off" of the graph of y = sin x, and similarly

for y = sec x relative to y = cos x. Tis provides a convenient method for remembering

the graphs of the secant and the cosecant.

The translation principles, discussed in Appendix E §1, tell us how to obtain quickly

the graphs of such functions as y = 2 + cos(x − 3). The graph of this particular example

would be the graph of y = cos x translated up by 2 units and over to the right by 3 units.

However, with trigonometric functions it is also common to expand or contract their graphs,

as we now show.

Example 2.   Graph the function y = 2 sin 3x.

Solution.   Let's first examine the function y = sin 3x. For each x, the corresponding y

has the value given by the sine of 3x. Thus, in the

x-interval $[0, \frac{2\pi}{3}]$ we will have assumed all the

sine values that are normally taken on the   x-interval

$[0, 2\pi]$. Hence the period* of y = sin 3x is 2π/3,

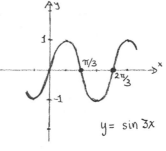

as opposed to period 2π for y = sin x, and the

graph of sin x is contracted by a factor of 1/3

along the x-axis as shown in the graph to the right.

Return to the function y = 2 sin 3x. The 2

merely expands the graph of sin 3x by a factor of

2 along the y-axis as shown in the graph to the right.

Thus the amplitude** of y = 2 sin 3x is 2, as

opposed to amplitude 1 for y = sin x.   □

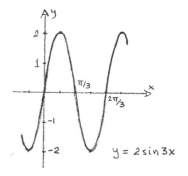

---

* The period is the length of the longest   x-interval possible before a graph begins to repeat itself.
** The amplitude is one-half the vertical distance between the highest and lowest points on a graph.

The general rules illustrated in Example 2 are valid for any equation:

The Expansion / Contraction Principles

> i. to expand/contract the graph of an equation along the x-axis by a factor of $b > 0$, replace x with x/b (the graph expands if $b > 1$ and contracts if $1 > b > 0$);
>
> ii. to expand/contract the graph of an equation along the y-axis by a factor of $a > 0$, replace y with y/a (the graph expands if $a > 1$ and contracts if $1 > a > 0$).

You should compare these with the translation principles as given in Appendix E §1. In Example 2 we took $y = \sin x$ and replaced x with $3x = x/\left(\dfrac{1}{3}\right)$ and y with $y/2$; thus we had a contraction along the x-direction by a factor of $1/3$ and an expansion along the y-direction by a factor of 2, as shown in the final graph of Example 2.

Example 3. Graph the function $y = \dfrac{2}{3} \cos \dfrac{2}{5} x$.

Solution. Here we have taken $y = \cos x$ and replaced x with $\dfrac{2}{5} x = x/\left(\dfrac{5}{2}\right)$ and replaced y with $y/\left(\dfrac{2}{3}\right)$. Thus we have expanded the graph of $y = \cos x$ by a factor of $5/2$ along the x-axis and contracted the graph by a factor of $2/3$ along the y-axis. □

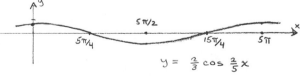

$y = \dfrac{2}{3} \cos \dfrac{2}{5} x$

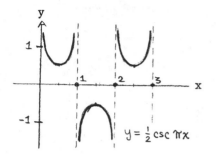

Example 4. Graph the function $y = \frac{1}{2} \csc \pi x$.

Solution. We have taken $y = \csc x$ and replaced x with $x / \left(\frac{1}{\pi}\right)$ and y with $y / \left(\frac{1}{2}\right)$. Thus we need only contract the graph of $\csc x$ by a factor of $1/\pi$ along the x-axis and contract the graph by a factor of $1/2$ along the y-axis. $\square$

Numerous exercises for the material of this appendix can be found at the ends of Units 1 and 2 in Anton's Appendix 1.

Appendix H :   Word problems

Word problems: the very term strikes terror into the hearts of most people,
and their appearance on assignments or exams is enough to make a football hero weep.   Is
all this anguish and gnashing of teeth justified?

No!

Word problems are difficult only because most people do not approach them in a
careful, disciplined and organized fashion.   In fact, word problems are frequently easier than
many calculus problems because the real-world, physical nature of the problem gives you
some extra common-sense tools to use in finding its solution.   However, if you are careless,
lazy, or sloppy, you are going to be in serious trouble.   This is especially true with the first
step in a word problem solution: translation from words into appropriate mathematical
equations.   This must be done slowly, carefully, one-step-at-a-time.   In this appendix we
give a step-by-step procedure which will cover most word problems you'll encounter in calculus.
This will give you the organization ... you must provide the care and discipline.

We should also emphasize how vitally important mastering words problems is for real
life applications of calculus.   Ninety-nine percent of applications start with a "word problem,"
a situation described in English sentences which demands translation into, and solution by the
methods of, mathematics.

The best way (perhaps the only way) to learn how to solve word problems is to do some.
The examples we give are as similar to those in calculus as they can be without actually using
calculus.   Moreover, to illustrate the thinking that goes into a solution, our explanations are
very long-winded.   So get ready ....

<u>Example 1.</u>   Find the dimensions of a rectangle if the diagonal is   2   more than the longer side, which in turn is   2   more than the shorter side.

<u>Step 1.</u>   We must first <u>translate</u> our problem into mathematics; this generally takes a number of readings. A quick overview shows that we are looking for the dimensions x and y of a rectangle (let x be the longer of the two).   The length of the diagonal also plays an important role in the problem, and so we assign it the label <u>d</u> . We now reread the word problem very slowly and carefully, translating every detail into an equation:

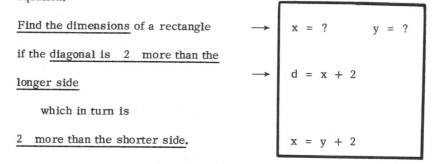

<u>Find the dimensions</u> of a rectangle   → $x = ?$ $y = ?$

if the <u>diagonal is   2   more than the</u>

<u>longer side</u>   → $d = x + 2$

   which in turn is

<u>2   more than the shorter side.</u> $x = y + 2$

Read the problem another time: did we miss anything? It appears that we caught every-thing in either our picture or our equations, and thus the translation phase is complete. You should, from this point on, have little need to refer to the actual word problem again; you have only to use your picture and equations.

<u>Step 2.</u>   Think about your method of solution, or <u>game plan.</u>

Question :   What are you looking for?

Answer :   x   and   y .

Question :   What equations do you have to work with?

Answer : $d = x + 2$
               $x = y + 2$.

Oops! Two equations in three unknowns!

This should immediately cause a bell to ring: in general you need three equations to solve for three unknowns -- two equations are not enough! So we go back and look at our word problem again (we shouldn't have to do this -- but we're worried!): did we miss an equation relating x, y and d? Nope ... well, wait a minute. We certainly did not miss any explicit equations, i.e., relationships which are explicitly stated in the problem. However, are there some implicit equations that we could find, i.e., relationships which, although not directly stated in the problem, are implied by the relationships which are given? Ah ha! Yes! [Before reading on ... can YOU find this relationship? Try it!] Look at our picture: there is a right triangle crying out to be "Pythagorized":

$$\boxed{d^2 = x^2 + y^2}$$

So ... now we do have three equations in three unknowns. Our game plan is thus to solve the following system of simultaneous equations:

$$\begin{cases} d = x + 2 \\ x = y + 2 \\ d^2 = x^2 + y^2 \end{cases}$$

Step 3. It's time to execute our game plan. (This is the easy part, we hope...) Our first two equations give d and y in terms of x:

$$\begin{cases} d = x + 2 \\ y = x - 2 \end{cases}$$

Plug these into the third equation and we get

$$(x + 2)^2 = x^2 + (x - 2)^2$$

which simplifies to

$$0 = x^2 - 8x = x(x - 8)$$

Thus either $x = 0$ (impossible since a rectangle has non-zero side lengths) or $x = 8$. But if $x = 8$, then $y = 8 - 2 = 6$ or $d = 8 + 2 = 10$. Thus $\boxed{(x, y, z) = (8, 6, 10)}$ is the only possible solution, and sure enough, it does check out in all of our equations. □

We summarize our method as follows:

How to tackle a word problem (... and live to tell about it).

Step 1. <u>Translate into mathematics.</u> This major step takes several readings of the problem. The procedures to use (more-or-less in the following order) are

i. obtain a general overview of the problem, sketching a picture or constructing a table if appropriate,

ii. determine the quantities which you desire to compute, and assign labels to them (e. g., $x, y, t,$ etc.). These are the "unknowns"

iii. label other quantities which you believe will be important,

iv. write down equations for relationships between your labelled quantities: use as few variables as possible, preferably only your "unknowns" from ii,

v. check that every piece of information has been translated into an equation and/or has been placed in your picture.

Step 2.   Devise game plan.

    i.   determine if you have enough equations to solve for the unknown quantities,

    ii. if necessary, look for relationships between variables which are <u>implicit</u> in the

        problem; use these to eliminate variables if possible,

    iii. decide how you will solve for your unknowns.

Step 3.   Execute game plan. ...

    ... and then check that your answers are reasonable.

Step 1 , the translation from words to mathematics, is surely the step to concentrate on.

Most people do reasonably well once a word problem has been accurately translated; however,

the translation is often done incompletely or inaccurately, and dooms the solver to failure.

Use our method faithfully and your success rate with word problems is guaranteed to improve.

We give a few more examples to illustrate our method.

Example 2.   Two cars start from the same point and travel in opposite directions with speeds of

45 and 60 miles per hour respectively. In how many hours will they be 490 miles apart?

Step 1.   Translation.   This is a problem involving (constant) <u>speed</u>, hence (from the familiar

formula $\underline{d = rt}$, i.e., distance equals

speed times time) we need to consider

distance and time quantities. We are

looking for the (unknown) time $\underline{t}$ which

it takes for two autos to be a certain distance

apart; we thus label the distances which each car travels in time t:

$$d_1 = \text{distance traveled by } 45 \text{ mph car}$$

$$d_2 = \text{distance traveled by } 60 \text{ mph car}$$

Each of these quantities can be related to our unknown time t (as we always desire to do, when possible) by using the rate equation $d = rt$:

$$\boxed{d_1 = 45t} \quad \text{and} \quad \boxed{d_2 = 60t}$$

Now we reread our problem very carefully for every detail.

Two cars start from the same point and travel in opposite directions with <u>speeds of 45 and 60 miles per hour</u> respectively. In <u>how many hours</u> will they be <u>490 miles</u> part?

$$\left.\begin{array}{l} d_1/t = 45 \\ \\ d_2/t = 60 \end{array}\right\} \begin{array}{l} \text{as} \\ \text{noted} \\ \text{above} \end{array}$$

$$t = ? \quad \text{when}$$

$$d_1 + d_2 = 490$$

Our total collection of equations is therefore:

$$d_1 = 45t, \qquad d_2 = 60t$$

$$t = ? \qquad \text{when} \quad d_1 + d_2 = 490$$

<u>Step 2.</u> <u>Devise game plan.</u> We wish to solve for t, and we have three equations in the three unknowns d_1, d_2 and t. So we solve the equations....

<u>Step 3.</u> <u>Execute game plan.</u>

Since $d_1 = 45t$ and $d_2 = 60t$, then

$$490 = d_1 + d_2 = 45t + 60t = 105t$$

Thus

$$t = \frac{490}{105} = \boxed{4\frac{2}{3} \text{ hours}}$$

This answer is easily checked: $d_1 = 45 \left(4\frac{2}{3}\right) = 210$

$$d_2 = 60 \left(4\frac{2}{3}\right) = 280$$

Thus $d_1 + d_2 = 490$, as desired. $\square$

Example 3.   A 6 foot man, walking at a rate of 5 feet/sec, passes under an 18 foot lamp post.

How long is his shadow 10 seconds after passing the lamp post?

Step 1.   Translate.   We have another speed ("rate") problem, and thus time and distance

variables must be considered. We are looking

for the length of our walker's shadow at a

certain time; label this desired unknown as h .

The picture drawn to the right then screams

out for us to label the distance from

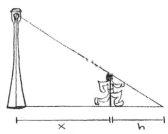

the walker to the lamp post; label this as x . Since his speed is 5 feet/sec and he

covers the distance x in 10 seconds, the rate equation d = r t yields

$$\boxed{x = 5(10) = 50 \text{ feet}}$$

We have one labelled unknown, h , and a basic picture; we reread the problem

for other relationships.

A <u>6 foot man</u>, walking at a

<u>rate of 5 feet/sec</u>, passes

under an <u>18 foot lamp post</u>.

<u>How long is his shadow 10</u>

<u>seconds</u> after passing the

lamp post?

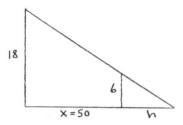

place   6   in the picture,

$5 = x/t = x/10$, as noted above

place 18   in the picture

$h = ?$ when

$t = 10$ seconds have elapsed

Our picture is thus as shown to the

right and our equations are

$$h = ? \quad \text{when} \quad t = 10, \quad x = 50$$

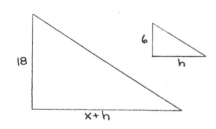

Step 2.   <u>Devise game plan</u>.   We want h , but we have not written down any equation relating

h with x or t .   Clearly there must be some <u>implicit</u> relationship between these

variables.   [Before reading on --- can YOU find this relationship? Try it!]   Let's

turn to the picture .... Ah ha, similar

triangles (Appendix A §4) are staring us

in the face:

$$\boxed{\frac{x + h}{h} = \frac{18}{6}}$$

This is how we relate x and h .   The game plan is simply to solve for h in this

equation.

Step 3. Execute game plan.

$$6(x + h) = 18h$$

$$6x + 6h = 18h$$

$$6x = 12h$$

$$h = \frac{1}{2}x = \frac{1}{2}(50) = \boxed{25} \qquad \square$$

Example 4. A builder has constructed a large window which is

a rectangle of clear glass with a semicircle of red-tinted glass

above it. From his records you see that he used 24 feet of

molding around the outer perimeter of the window, that the

cost of clear glass is $\$1/\text{ft}^2$, the cost of red-tinted glass

is $\$2/\text{ft}^2$, and that the builder's total glass cost was $54. What are the dimensions of the

window?

Step 1. Translate. It's pretty clear that we will need to label the dimensions of the window:

    x = length of rectangle

    h = height of rectangle

    r = radius of semicircle

  24 = perimeter of the window

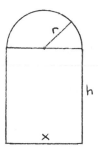

Two relationships which are immediately seen are

$$x = 2r \qquad\qquad\qquad (1)$$

$$24 = x + 2h + \pi r = 2h + (2 + \pi)r \qquad (2)$$

Thus x has been effectively eliminated and we are left with finding h and r.

We now fill in details with a rereading of the problem:

...   24 feet of molding around the

        outer perimeter ...

...   cost of clear glass is $1/ft^2 ...

...   cost of red-tinted glass is $2/ft^2 ...

...   total glass cost was $54 .

     What are the dimensions of the

       window?

> perimeter of 24 , shown above to give the equation
> $$24 = 2h + (2 + \pi) r$$
> $C_1 = \$1/ft^2$
>
> $C_2 = \$2/ft^2$
>
> $C = \$54$
>
> $x = ?, \quad h = ?, \quad r = ?$
> ($x = 2r$, as shown above)

Step 2.   Devise game plan.

We have 2 unknowns (h and r) and desire to compute both of them. But we have only one equation, (2), which relates h and r, and our general knowledge of solving for unknowns should tell us that we'll need at least 2 equations. [Before going on ... can YOU find the 2nd relationship? Try it!] We look to the cost for our second equation: we know that the total cost C of the glass must depend on the cost per square foot of clear and red-tinted glass, along with the dimensions of the window panes. Thus

$$C = A_1 C_1 + A_2 C_2$$

     where A_1 = area of rectangular pane

                  = $xh = 2rh$

     and A_2 = area of semi-circular pane

                  = $\frac{1}{2} \pi r^2$

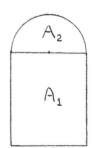

Thus $C = 2rh\,C_1 + \frac{1}{2}\pi r^2 C_2.$ Using the values of C, C_1 and C_2 yields

$$54 = 2rh + \pi r^2 \qquad\qquad (3)$$

The game plan is thus to solve equations (2) and (3) for h and r, and then

compute x from equation (1).

Step 3. Execute game plan. Listing our equations together gives

$$24 = 2h + (\pi + 2)r$$
$$54 = 2rh + \pi r^2$$

Solving the first equation for h in terms of r yields

$$h = 12 - \frac{1}{2}(\pi + 2)r \qquad\qquad (4)$$

Plugging this into the second equation yields

$$54 = r(24 - (\pi + 2)r) + \pi r^2$$

which simplifies to

$$0 = r^2 - 12r + 27 = (r - 3)(r - 9)$$

Thus r = 3 or 9. However, using equation (4) for h we find that

r = 9 gives a negative value for h:

$$h = 12 - \frac{1}{2}(\pi + 2)9 \cong 12 - \frac{9}{2}(5.14) = -11.13$$

Since h, as the height of the rectangle, cannot be negative, this rules out r = 9

as a possible value. Thus turn to r = 3. With this value, x = 6 from (1) and

$h = 9 - \frac{3}{2}\pi \approx 4.29$ from (4). These are physically allowable dimensions, and the

numbers do check out in equations (1), (2) and (3). Thus our solution is

$$r = 3 , \ x = 6 , \ h \approx 4.29 \qquad \square$$

Example 5. A doctor has a 2 liter solution of 3% boric acid. How much of a 10% solution of the acid must she add to have a 4% solution?

Step 1. Translate. Percentages are proportions, and as such, are divisions of one quantity by another. In this case we have percentages of boric acid in a solution, i.e.,

$$\left[\% \text{ of boric acid} \right] = \left[\frac{\text{volume of boric acid}}{\text{total volume of solution}} \right] \tag{5}$$

The quantity we desire to compute is the total volume of added 10% solution; label this as **h**. From (5) we then can compute the volume of boric acid which we will be adding. There are, of course, two other solutions (the initial 3% solution and the final 4% solution), and each has a total volume and a volume of boric acid. We could give labels to each of these, but introducing so many new variable designations seems unwise if it can be avoided. Instead we set up a table for all these quantities which "incorporates" equation (5) in a convenient way, and which contains all the specific information given in the word problem itself:

% of boric acid in solution	X	total volume of solution (liters)	=	volume of boric acid (liters)
3%		2		?
10%		h		?
4%		2 + h		?

The "2 liter" entry is a given piece of information, and the 2 + h reflects the fact that the 4% solution is the sum of the 3% and 10% solutions. We now can use equation (5) to fill in the last column:

% of acid	total volume	volume of acid
3%	2	.06
10%	h	(.10)h
4%	2 + h	(.04)(2 + h)

A rereading of the problem will show that we have not missed any information.

Step 2. <u>Devise a game plan.</u> We have one unknown quantity, <u>h</u>, one table, and no equations; clearly there must be at least one implicit equation for h which can be drawn out of our table. The game plan is therefore to analyze our table to find this relationship. [Before going on ... can YOU find the relationship? Try it!]

Step 3. <u>Execute game plan.</u> The trick lies in checking the consistency of the table. <u>The volume of acid in the final mixture must equal the sum of volumes of acid in the solutions being mixed</u>; after all, every drop of acid in the final mixture came from one of the two! That is, the sum of the first two entries in the 3rd column must equal the third entry in the 3rd column:

$$.06 + (.10)h = (.04)(2 + h)$$

Ah ha! This equation will be true only for one value of h :

$$.06 + (.10)h = .08 + (.04)h$$

$$.06\,h = .02$$

$$\boxed{h = 1/3 \text{ liter}} \qquad \qquad \square$$

<u>Example 6.</u> One pipe takes 30 minutes to fill a tank. After it has been running for 10 minutes, it is shut off. A second pipe is then opened and it finishes filling the tank in 15 minutes. How long would it take the second pipe alone to fill the tank?

Step 1.   Translate.   This is a <u>rate</u> problem, since clearly the difference in the two pipes is that they allow fluid to pass through at differing speeds.   The basic relationship which governs our situation is

$$\left[\begin{array}{c}\text{volume of fluid}\\ \text{passing through pipe}\end{array}\right] = \left[\begin{array}{c}\text{rate of}\\ \text{flow}\end{array}\right] \times \left[\text{time}\right] \qquad \text{i. e.,}$$

$$\left[\text{rate of flow}\right] = \left[\text{volume/time}\right] \qquad\qquad (6)$$

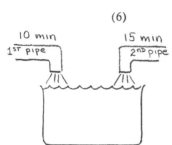

The unknown which we desire to find is

t = the amount of time which the 2nd

pipe would need to fill the tank

Other quantities which should play a role in this problem are:

V = volume of tank

r_1 = rate of flow for 1st pipe

r_2 = rate of flow for 2nd pipe

As in the previous problem, we have one relationship, (6), which applies a number of times.   Thus suggests the use of a table again:

rate of flow $\times$ time = volume

rate of flow	time	volume	
r_1	30	V	1st pipe filling tank alone
r_2	t	V	2nd pipe filling tank alone
r_1	10	V_1 ⎞ sum	⎞
r_2	15	V_2 ⎠ is V	⎭ Both pipes filling tank

Here, of course, V_1 is the part of the volume of the tank filled by the 1st pipe in 10 minutes, and V_2 is the part left for the 2nd pipe to fill.   Notice how all the given information is nicely recorded and organized in such a table.

Step 2. <u>Devise game plan.</u>   Somehow we need to obtain equations from our table to solve for t.
Well, we have a lot of variables in our table; let's cut down the number until perhaps
we'll end up with an equation just involving t. That's our game plan.

Step 3. <u>Execute game plan.</u>   The first two lines of our table will eliminate r_1 and r_2:

$$r_1 = V/30 \qquad \text{and} \qquad r_2 = V/t$$

The third and fourth lines together will then yield

$$V = V_1 + V_2 = 10\,r_1 + 15\,r_2$$

Substituting from above for r_1 and r_2 yields

$$V = 10(V/30) + 15(V/t)$$

$$1 = 1/3 + 15/t \qquad\qquad (V \text{ just cancels out of the problem!})$$

$$\boxed{t = 22.5 \text{ minutes}}$$

<u>Disclaimer.</u>   We hope that our method and examples have shown that word problems can be
effectively tackled in a coherent, step-by-step fashion. The main ingredients are complete and
accurate <u>translation</u> of the problem into mathematics, and a systematic <u>analysis</u> of what is
necessary to solve it. The method we have described is one way to organize these ingredients;
however, it is not the only way, nor will it in all cases provide the most efficient means of
solution. To begin with, no one method can be expected to cover so vast and varied a collection
as "word problems" -- that's almost equivalent to devising a method of solving <u>any type of</u>
<u>mathematical problem</u>! Any such list of procedures is going to have inherent shortfalls; our

H. 16

list is probably a bit too rigid and detailed, and in some instances it might lead you to introduce

more variables than are actually needed (and, in so doing, will make the problem solution more

complicated than it need be). As you gain in experience and confidence, you will be better able

to tailor the method to individual problems, and thus achieve more efficient solutions.

Nonetheless, the solution of any word problem requires the basic procedures of our

method, and you are thus encouraged to follow our outline pretty closely.

Exercises.

1.   Howard Anton takes $7\frac{1}{2}$ hours to make a trip overseas in a prop plane. Later he dis-

covers that he could have taken a jet and saved $2\frac{1}{2}$ hours of flying time. Find the speed of the

jet if the jet travels 225 mph faster than the prop plane.

2.   On another trip, this one an auto trip of 126 miles, Howard Anton calculated that had he

decreased his average speed by 8 m.p.h., his trip would have taken one hour longer. What was

his original rate?

3.   One solution is 20% sulfuric acid while another is 12% sulfuric acid. How much of

each solution must be mixed together to produce 60 milliliters of solution containing 9

milliliters of sulfuric acid?

4.   Flying east between two cities, a plane's speed is 380 mph. On the return trip, it flies

at 420 mph. Find the average speed over the whole round trip. (No, the answer is not 400 mph.)

5.   A rectangle is said to be in the "Divine Proportion" if the ratio of its width to its length

is equal to the ratio of its length to the sum of its length and width. What are the dimensions of

a rectangle in the Divine Proportion if its perimeter is 10 meters?

6.   An offshore oil well is located in the ocean at a point W , which is 3 miles from the closest shorepoint A on a straight shoreline. The oil is to be piped to a shorepoint B that is 9 miles from A by piping it on a straight line underwater from W to some shorepoint P between A and B and then on to B via a pipe along the shoreline. If the cost of laying pipe is $500,000 per mile underwater, $300,000 per mile over land, and the total cost of the pipe installation is $4,000,000, then how far is point P from point A if this distance is known to be at least one mile? [For convenience, calculate with money in units of $100,000.]

7.   In a long-distance race around a 400-meter track, the winner finished the race one lap ahead of the loser. If the average speed of the winner was 6 meters/sec and the average speed of the loser was 5.75 meters/sec, how soon after the start did the winner complete the race?

8.   At 8 a.m., a bus traveling 100 kilometers/hr leaves Philadephia for Boston, a distance of 500 kilometers. At 10 a.m., a bus traveling 80 kilometers/hr leaves Boston for Philadelphia. At what time do the two buses pass each other?

9.   A confectioner has 15 pounds of chocolate worth $3.20 per pound and 12 pounds of caramels worth $2.20 per pound. How many pounds of nougats worth $2.40 per pound should be added to these candies to obtain a mixture that is to sell for $2.60 per pound?

10.   A radiator contains 10 liters of a water and antifreeze solution of which 60% is anti-freeze. How much of this solution should be drained and replaced with water in order for the new solution to be 40% antifreeze?

11.   One painter can paint a house in 40 hours and another painter can paint the same house in 35 hours. How long will it take to paint the house if they work together?

12.   A pipe can fill a swimming pool in 10 hours. If a second pipe is opened, the two pipes together can fill the pool in 4 hours. How long would it take the second pipe alone to fill the pool?

H. 18

Answers.

1.   d = distance of trip

    s_1 = speed of prop plane t_1 = time of prop plane trip

    s_2 = speed of jet t_2 = time of jet trip

$$t_1 = 7\frac{1}{2} \qquad t_2 = 5$$

$$s_1 = d/t_1 \qquad s_2 = d/t_2 \qquad s_2 = s_1 + 225$$

Thus $s_2 = 675$

2.   36 mph

3.   22.5 and 37.5 milliliters

4.   399 mph

5.   length = $5\sqrt{5} - 5$, width = $15 - 5\sqrt{5}$

6.   x = distance from

        A to P

      = 4 miles

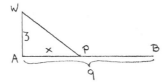

7.   26 minutes and 40 seconds

8.   11:40 a.m.

9.   21 pounds

10. $3\frac{1}{3}$ liters

11. $18\frac{2}{3}$ hours

12. $6\frac{2}{3}$ hours

Appendix I: Mathematical Induction

  Mathematical induction is the sophisticated name given to a simple logical principle that can be used to prove certain types of mathematical statements.

  Suppose we have a series of statements, one about each positive integer. For example, the statement

  "the sum of the first n positive integers is $\frac{n(n+1)}{2}$,"

is, in fact, a series of statements, one for each positive integer n ("the sum of the first 6 positive integers is $\frac{6(6+1)}{2} = 21$," "the sum of the first 7 positive integers is $\frac{7(7+1)}{7} = 28$," etc.)

  In proving statements of this type, we have a fundamental problem: there are an infinite number of the statements and so it is impossible to check them all, one by one! However, there is a pattern to the statements and, by taking advantage of it, we can derive a simple, do-able procedure for proving them.

  To prove statements of this type by mathematical induction, we reason as follows:

The Principle of Mathematical Induction

To verify a statement for every positive integer n such that $n \geq n_0$, where n_0 is some fixed starting integer (usually $n_0 = 1$) :

I. Prove that the statement is true for $n = n_0$;

II. a) (The Induction Hypothesis) Assume that the statement is true for $n = k$ (where k is an arbitrary value of n), and then

   b) Prove that the statement is true for $n = k + 1$.

Then the statement is true for all values of $n \geq n_0$.

  It is really very easy to see why this principle works (and, after all, a principle in mathematics is something that is not proved but is accepted as being obvious, so it had <u>better</u> be easy to see why it works!). In Part I, it has been verified that the statement is true for the first value of n. Say that first value is n = 1. Then, because it is true for n = 1, Part II says that it must be true for n = 2 (by using k = 1 and k + 1 = 2). And then, because it is true for n = 2, Part II says that it must be true for n = 3 (by using k = 2 and k + 1 = 3). And then it must be true for n = 4, and so on. The inescapable conclusion is that the statement must be true for <u>all</u> the positive integers.

The principle of mathematical induction has been described as the "ladder climbing" principle. It says that if you can get on the ladder (at Step 1, usually) and if you can go from any step (Step k) to the <u>next</u> step (Step k + 1), then you can climb the ladder to any step.

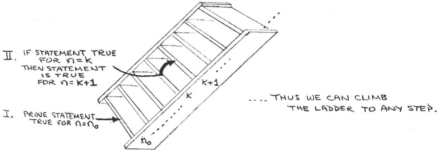

The principle of mathematical induction can also be thought of in terms of falling dominoes. If you can knock over one domino (usually the first) and if the dominoes are arranged so that each domino (the kth) knocks over the <u>next</u> one in line (the (k + 1)st), then all the dominoes will fall.

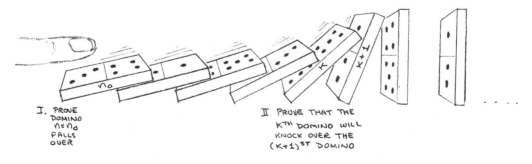

The use of the principle of mathematical induction is illustrated in the proof of the following theorem.

THEOREM (Anton's Theorem 5. 5. 2 (a) , (b)).

(a) The sum of the first n positive integers is $\dfrac{n(n + 1)}{2}$.

(b) The sum of the squares of the first n positive integers is $\dfrac{n(n + 1)(2n + 1)}{6}$.

Proof. (a) Let us define $S_n = 1 + 2 + 3 + \cdots + n$ where n is any positive integer. Then the statement we want to prove is $"S_n = \dfrac{n(n + 1)}{2} . "$

I. First observe that for $n = 1$, the statement is true since $S_1 = 1$ and $\dfrac{1(1 + 1)}{2} = 1 .$

II. (a) Assume that the statement is true for $n = k$. That is, assume that $S_k = \dfrac{k(k + 1)}{2}$. (This is the induction hypothesis.)

II. (b) We want to use the induction hypothesis to show that the statement is true for $n = k + 1$, i.e., that $S_{k+1} = \dfrac{(k + 1)((k + 1) + 1)}{2} = \dfrac{(k + 1)(k + 2)}{2}$. This is accomplished through the following sequence of equations:

$$S_{k+1} = S_k + (k + 1) = \dfrac{k(k + 1)}{2} + (k + 1) = \dfrac{k(k + 1) + 2(k + 1)}{2}$$

$$= \dfrac{(k + 1)(k + 2)}{2}$$

the induction hypothesis

Therefore S_{k+1} has the required form. This completes the induction argument and shows, by the principle of mathematical induction, that $S_n = \dfrac{n(n + 1)}{2}$ for all positive integers n .

(b)   Define $S_n^{(2)} = 1^2 + 2^2 + 3^2 + \cdots + n^2$ where n is any positive integer. Then

the statement we want to prove is " $S_n^{(2)} = \dfrac{n(n + 1)(2n + 1)}{6}$. "

    I.   When n = 1 , $S_1^{(2)} = 1^2 = 1$ and $\dfrac{1(1 + 1)(2(1) + 1)}{6} = 1$ so the statement is

true.

    II.   (a) Assume that the statement is true for n = k , i.e., that $S_k^{(2)} = \dfrac{k(k + 1)(2k + 1)}{6}$.

(This is the induction hypothesis.)

    II.   (b) Under this assumption, we want to show that the statement is true for

n = k + 1 , i.e., that $S_{k+1}^{(2)} = \dfrac{(k + 1)((k + 1) + 1)(2(k + 1) + 1)}{6} = \dfrac{(k + 1)(k + 2)(2k + 3)}{6}$.

This is done as follows:

$$S_{k+1}^{(2)} = S_k^{(2)} + (k + 1)^2 = \frac{k(k + 1)(2k + 1)}{6} + (k + 1)^2 \qquad \text{(the induction}$$
$$\text{hypothesis)}$$

$$= (k + 1)\left[\frac{k(2k + 1)}{6} + (k + 1)\right]$$

$$= (k + 1)\left[\frac{k(2k + 1) + 6(k + 1)}{6}\right]$$

$$= (k + 1)\left[\frac{2k^2 + 7k + 6}{6}\right]$$

$$= \frac{(k + 1)(k + 2)(2k + 3)}{6}$$

Therefore $S_{k+1}^{(2)}$ has the required form, proving by the principle of mathematical induction

that $S_n^{(2)} = \dfrac{n(n + 1)(2n + 1)}{6}$ for all positive integers n .   □

Note: Very often, a proof by mathematical induction is not the only way to prove that statements

are true for all positive integers. For example, in Section 5.5 Anton proves the two results

above by different methods.   However, in general you will find that a proof by mathematical
induction is more straightforward and involves fewer tricks.

Exercises

1.   Prove that $1^3 + 2^3 + 3^3 + \cdots + n^3 = [n^2(n+1)^2]/4$.

       [This is Anton's Theorem 5.5.2(c).]

2.   Prove that if $x \neq 1$,   then $1 + x + x^2 + \cdots + x^n = (1 - x^{n+1})/(1 - x)$.

3.   If a set S contains n elements, show that S has 2^n subsets (counting S
       itself and the empty set ϕ as subsets).

Chapter 1:   Coordinates, Graphs, Lines

Section 1. 1.   Real Numbers, Sets and Inequalities (A Review)

Although this section covers material which you have probably already studied in earlier mathematics courses, there is one topic which deserves very close and careful attention: solving inequalities. As you will see, many calculus operations involve solving inequalities. It is not surprising, therefore, that errors with inequalities are a major source of mistakes in calculus!

1.   Simple inequalities and common errors.   The five basic rules governing inequalities are given in Theorem 1. 1. 2. While all five are important, you should concentrate on rule (d) because it causes by far the most trouble:

> If $a < b$ then $ac < bc$ when c is positive
>
> and $ac > bc$ when c is negative   ←——— Note this!

Forgetting to reverse inequality signs when multiplying by a negative number is one of the most common and costly mistakes in calculus. It occurs most often when a person forgets that an expression containing a variable might take on negative values, i. e., the negative values are "hidden." Here are some typical examples:

Example A.   Solve $\dfrac{x + 2}{x + 1} > 0$.

Incorrect "solution."

$$\frac{x + 2}{x + 1} > 0$$

$$(x + 1)\left(\frac{x + 2}{x + 1}\right) > (x + 1) \cdot 0 \longleftarrow \boxed{\begin{array}{l} \text{OOPS! } \underline{\text{ERROR!}} \\ \text{NOT TRUE WHEN } x + 1 < 0 \ ! \end{array}}$$

$$\left.\begin{array}{l} x + 2 > 0 \\ \\ x > -2 \end{array}\right\} \quad \begin{array}{l} \text{Since the previous step was} \\ \text{incorrect, the inequalities} \\ \text{which follow are incorrect.} \end{array}$$

The mistake in this solution lies in not realizing that $x + 1$ can be negative. We'll now give two correct solutions for this problem.

<u>First solution.</u> In order to multiply the inequality $\dfrac{x + 2}{x + 1} > 0$ by $x + 1$ we need to consider two cases: the case when $x + 1$ is positive and the case when $x + 1$ is negative:

<u>Case 1.</u> $x + 1 > 0$, i.e., $x > -1$.

In this case multiplying our inequality by $x + 1$ will not reverse the inequality sign:

$$\frac{x + 2}{x + 1} > 0$$

$$(x + 1) \cdot \left(\frac{x + 2}{x + 1}\right) > (x + 1) \cdot 0 \qquad \begin{array}{l} \text{inequality in same direction} \\ \text{since } x + 1 > 0 \end{array}$$

$$x + 2 > 0$$

$$x > -2$$

There is <u>one more step</u> needed to obtain the set of solutions: we must discard those x-values which are less than or equal to -1 since we are in Case 1 where we only allow $x > -1$. Thus $x > -2$ is "cut back" to

$$\boxed{x > -1}$$

This is the set of solutions for our inequality in Case 1.

Case 2. $x + 1 < 0$, i. e., $x < -1$.

In this case multiplying our inequality by $x + 1$ will reverse the
inequality sign:

$$\frac{x + 2}{x + 1} > 0$$

$$(x + 1) \cdot \left(\frac{x + 2}{x + 1}\right) < (x + 1) \cdot 0 \qquad \begin{array}{l} \text{reverse inequality} \\ \text{since} \;\; x + 1 < 0 \end{array}$$

$$x + 2 < 0$$

$$x < -2$$

All these solutions are allowed in Case 2 since our only restriction is
$x < -1$. (If x is less than -2, then it is certainly less than -1.)
Hence

$$\boxed{x < -2}$$

is the set of solutions for our inequality in Case 2 .

Putting the two cases together gives the solution set

$$x > -1 \qquad \text{or} \qquad x < -2$$

When written in interval notation this is

$$\boxed{(-\infty , -2) \cup (-1 , +\infty)}$$

Second solution.   As you can see, things can be a bit complicated when you are forced to deal with separate cases. Fortunately there is an alternate method of solution which avoids the use of separate cases.

The idea of the method is simple: the inequality

$$\frac{x + 2}{x + 1} > 0$$

will be true only when the sign of $x + 2$ multiplied by the sign of $x + 1$ is positive.   To determine when this happens, draw two number

lines, one for $x + 2$ and one for

$x + 1$, and indicate by $+$ and $-$

signs where each term is positive

or negative (see the lines to the right).

Then draw a third line for $(x + 2)/(x + 1)$;

the signs along this line are determined by

multiplying the signs of the previous two lines.

For example, above -3 we find $-$ and $-$,

the product of which is $+$. Hence we assign

$+$ to -3. On the other hand, above $-3/2$ we find $+$ and $-$, the product of which is $-$. Hence we assign $-$ to $-3/2$. Determining the signs in this way can be done quickly and the final result shows that $(x + 2)/(x + 1)$ is positive on the set

$$\boxed{(-\infty, -2) \cup (-1, +\infty)}$$

This agrees with the answer in our previous solution.   □

Example B.   Solve $(5 - x)/(x + 2) < 1$.

Solution.   We first move all the terms to one side of the inequality: *

$$0 < 1 - \frac{5 - x}{x + 2} = \frac{(x + 2) - (5 - x)}{x + 2} = \frac{2x - 3}{x + 2}$$

Thus the inequality can be written

$$0 < \frac{2x - 3}{x + 2}$$

Since we now have a zero on one side of the inequality, we can use the solution methods of

Example A.   Lines for each of the terms

$2x - 3$ and $x + 2$ are shown to the right,

and they determine the signs for the quotient

$(2x - 3)/(x + 2)$ as shown on the third line.

From this we can see that $(2x - 3)/(x + 2)$ is

positive whenever $x < -2$ or $x > 3/2$.

Hence the solution set for our original

inequality is

$$(-\omega, -2) \cup (3/2, +\omega)$$

Example C.   Solve $1/(x + 1) < 2/(x - 1)$.

Solution.   As in Example B we first move all the terms to one side of the inequality:

$$0 < \frac{2}{x - 1} - \frac{1}{x + 1} = \frac{2(x + 1) - (x - 1)}{(x - 1)(x + 1)}$$

$$0 < \frac{x + 3}{(x - 1)(x + 1)}$$

---

*
  See Appendix B, Sections 2 and 3, if you find the fraction algebra confusing.

We now have to consider the signs of three terms: $x + 3$, $x - 1$ and $x + 1$. So we draw lines for each of these terms, and then determine the signs for the quotient (by multiplying three $+$ or $-$ signs) as shown on the fourth line. From this

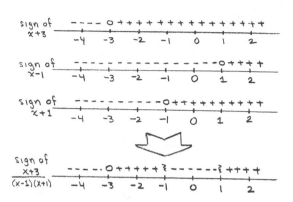

$\frac{x + 3}{(x - 1)(x + 1)}$ is positive whenever $-3 < x < -1$ or $1 < x$. Hence the solution set for our original inequality is

$$(-3, -1) \cup (1, +\infty)$$

□

2. <u>Solving complicated inequalities: the interval method.</u>   As you can see in Examples A, B and C, solving inequalities can be a tricky business. This is especially true when dealing with the inequalities that so often arise in calculus. For this reason we describe another easy-to-use technique that solves nearly all inequalities encountered in applications: the <u>interval method.</u>

Suppose we have an inequality of the form $f(x) > 0$, where $f(x)$ is an expression (i.e., a function*) in the variable x. We first observe that

---

*You may already be familiar with the "function" symbol $f(x)$. If not, do not let it confuse you. Here it is merely shorthand for "an expression in x." For example, in Example C we considered the inequality

$$\frac{x + 3}{(x - 1)(x + 1)} > 0$$

This inequality could also be expressed as $f(x) > 0$ where

$$f(x) = \frac{x + 3}{(x - 1)(x + 1)}$$

Functions will be studied in Chapter 2.

> a normal*   f(x) cannot change sign on an interval
>
> unless it first becomes zero or undefined.

Said another way, any normal f(x) cannot go from below the x-axis to above the x-axis

without <u>crossing</u> the x-axis [f(x) = 0] or jumping over it through a <u>missing point</u>

[f(x) undefined].

    To see this principle in picture form,

consider the f(x) whose graph is shown to

the right.

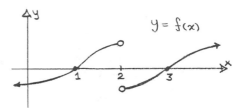

Notice that f(x) changes sign at the points 1, 2 and 3. At predicted above, these are

the points at which f(x) <u>crosses</u> the x-axis [f(1) = f(3) = 0], or at which f(x) is

<u>missing</u> [f(2) is undefined].

    The crossing points (x = 1 and x = 3) and missing points (x = 2) can be used to

divide the real line into four open intervals:

$$(-\infty, 1), \quad (1, 2), \quad (2, 3), \quad (3, +\infty)$$

Here's the critical observation: <u>on each of these open intervals the sign of f(x) must be</u>

<u>always positive or always negative.</u> To determine the solution set for the inequality f(x) > 0

we have only to determine on which of our four intervals f(x) is positive. Since we can

check that f(x) is positive only on (1, 2) and (3, +\infty), then our solution set is

(1, 2) ∪ (3, +∞).

    This example motivates the Interval Method:

---

* f(x) is "normal" if it is "continuous on its domain of definition. " Continuity is defined
and studied in §3. 7.

The Interval Method for solving $f(x) > 0$

Step 1.   Crossing points. Determine those x-values

for which $f(x) = 0$.

Step 2.   Missing points.   Determine those x-values

for which $f(x)$ is undefined.

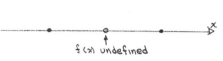

Step 3.   Pick a point x* from each of the open intervals formed by the crossing points

and the missing points.

Plug x* into $f(x)$ to get $f(x^*)$.

If $f(x^*) > 0$, then the corresponding interval is in the solution set for $f(x) > 0$.

Don't let the length of this discussion mislead you: the interval method is an easy-to-use technique. Here are two examples, the first of which is a reworking of Example C:

Example D.   Solve $1/(x+1) < 2/(x-1)$ by the Interval Method.

Solution.   As in Example C we move all the terms to one side of the inequality and obtain

$$0 < \frac{x+3}{(x-1)(x+1)}$$

Now we apply the three steps of the Interval Method to

$$f(x) = \frac{x+3}{(x-1)(x+1)}$$

Step 1.   Crossing points.   We must determine those x-values   for which

$$\frac{x+3}{(x-1)(x+1)} = 0$$

That's easy to do since a fraction can equal zero only when the numerator equals zero. Thus x + 3 = 0 , which gives

$$\boxed{x \; = \; -3 \quad \text{is the only crossing point.}}$$

Step 2. <u>Missing points.</u> We must determine these x-values for which

$$\frac{x+3}{(x-1)(x+1)} \qquad \text{is undefined.}$$

The only thing that could "go wrong" in this fraction is that the denominator would equal zero

if $x = 1$ or -1. Thus

$$\boxed{x \; = \; 1 \; \text{ and } \; -1 \quad \text{are the missing points}}$$

Step 3. Our points $x = -3$, -1 and 1 divide the real line into four open intervals.

From each we choose a point. Then we plug that point into $(x+3)/(x-1)(x+1)$ and

determine if it makes this fraction positive or negative. In this way we determine if the

interval is in our solution for $f(x) > 0$:

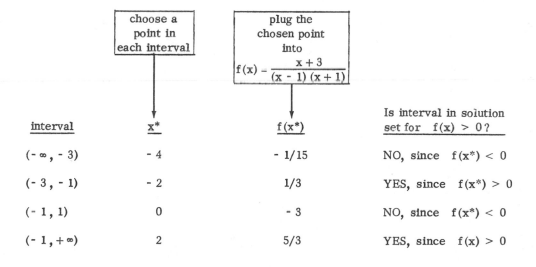

interval	x*	f(x*)	Is interval in solution set for f(x) > 0?
$(-\infty, -3)$	-4	$-1/15$	NO, since $f(x^*) < 0$
$(-3, -1)$	-2	$1/3$	YES, since $f(x^*) > 0$
$(-1, 1)$	0	-3	NO, since $f(x^*) < 0$
$(-1, +\infty)$	2	$5/3$	YES, since $f(x) > 0$

(Notice how we chose the x^*-values so that each $f(x^*)$ is easy to compute.) Thus the solution set is $\boxed{(-3, -1) \cup (1, +\infty)}$, which agrees with the answer arrived at in Example C. $\square$

<u>Example E.</u>   Solve $0 \leq (\sqrt{x} - 2)/(4 - \sqrt{x + 8})$.

<u>Solution.</u>   Since the expression

$$f(x) = \frac{\sqrt{x} - 2}{4 - \sqrt{x + 8}}$$

contains the terms $\sqrt{x}$ and $\sqrt{x + 8}$, we must have $x \geq 0$ and $x + 8 \geq 0$ (since we cannot take the square root of negative numbers).   Hence we are restricted to considering only those x-values for which $\boxed{x \geq 0}$.   (These are the x-values for which <u>both</u> $x \geqq 0$ and $x \geqq -8$ are true.)

<u>Step 1.</u>   <u>Crossing points.</u>   We can have

$$\frac{\sqrt{x} - 2}{4 - \sqrt{x + 8}} = 0$$

only when the numerator equals zero, i. e. , when

$$\sqrt{x} - 2 = 0$$

$$\sqrt{x} = 2$$

$$\boxed{x = 4 , \text{ the only crossing point}}$$

<u>Step 2.</u>   <u>Missing points.</u>   Having already restricted our attention to $x \geq 0$, we will not have any negative numbers occurring under square root signs (which would make the expression undefined).   Thus our expression is undefined only when the denominator is zero, i. e. , when

$$4 - \sqrt{x+8} = 0$$

$$\sqrt{x+8} = 4$$

$$x + 8 = 16$$

$$\boxed{x = 8 \text{, the only missing point}}$$

Step 3.   The points $x = 4$ and 8 divide the positive x-axis into three open intervals:

interval	x^*	$f(x^*)$	Is interval in solution set for $f(x) \geq 0$?
$(0, 4)$	1	-1	NO, since $f(x^*) < 0$
$(4, 8)$	4.25	$2(\sqrt{4.25} - 2)$	YES, since $f(x^*) > 0$
$(8, +\infty)$	17	$2 - \sqrt{17}$	NO, since $f(x^*) < 0$

(The x^*-values were chosen so that $\sqrt{x+8}$ is easy to compute: $\sqrt{1+8} = \sqrt{9} = 3$,
$\sqrt{4.25+8} = \sqrt{12.25} = 3.5$, $\sqrt{17+8} = \sqrt{25} = 5$.) Thus the solution set to Example F
contains the open interval $(4, 8)$. However, notice that our inequality is of the form
$f(x) \geq 0$, i.e., $f(x) = 0$ is allowed. Thus we must also include the crossing point
$x = 4$, so our solution set is

$$\boxed{[4, 8)}$$ □

3.   <u>Reversible Steps (Optional).</u>   Anton's remark (preceding Example 16) concerning reversible
steps is important, although it might not appear so at first glance. Here is an example of a
<u>non</u>-reversible series of steps: (a) $x > 1$

(b) $x^2 > 1$

Notice that if (a) holds true, then (b) must be true. However, if (b) is true, then it does not follow that (a) is true. For example, $x = -2$ is in the solution set of (b), but it is not in the solution set of (a). Thus "(a) implies (b)" is not a reversible step because it is not true that (b) implies (a).

Example F. Solve $\sqrt{3x - 2} < x$

Solution.
$$\sqrt{3x - 2} < x$$
$$3x - 2 < x^2 \quad \text{(if } 0 \le a < b, \text{ then } a^2 < b^2\text{)}$$
$$0 \quad < x^2 - 3x + 2$$
$$0 \quad < (x - 2)(x - 1)$$

In order for this last inequality to be true, either $x - 2$ and $x - 1$ must both be positive (i.e., $x > 2$ and $x > 1$) or must both be negative (i.e., $x < 2$ and $x < 1$). Thus

$$x < 1 \quad \text{(both negative) or} \quad x > 2 \quad \text{(both positive)}$$

One might be tempted to conclude that the solution set for our inequality is $(-\infty, 1) \cup (2, +\infty)$. This is false. The very first step is not reversible (e. g., $x = 0$ satisfies the second equation, but not the first). The step is reversible only when $3x - 2 \ge 0$, i.e., when $x \ge 2/3$. For values of x less than $2/3$ the term $\sqrt{3x - 2}$ is not even defined. Thus only part of our initial "solution set" is correct. The actual solution set is

$$[2/3, 1) \cup (2, +\infty)$$

As this example shows, you must check your answers, especially when your computations involve possibly non-reversible steps. □

The lesson to be learned is this: when solving inequalities (or equalities) it is not at all uncommon "accidentally" to introduce some incorrect solutions along the way. This is especially true when squares of quantities are involved. Thus, when solving an equation, do not claim that your answer is the

solution set until

> i. you have checked that all your solution steps
>
>   are reversible, or
>
> ii. if even one step is non-reversible, you have
>
>   checked all your solutions back in the original
>
>   equation.

If the steps used to show that a statement A implies a statement B are all reversible,

then we can say that A is true if and only if B is true. The term "if and only if" is

commonly abbreviated to "iff."

Section 1. 2.   Absolute Value

1. The definition of absolute value. When asked to find the absolute value of a specific number --

for example,   5 , - 2/3 , - π -- the chance that you'll make a mistake is very small:

$$|5| = 5 , \qquad |-2/3| = 2/3 , \qquad |-\pi| = \pi$$

Perhaps it is because the intuitive idea of absolute value is so simple that many people never

bother to learn the precise definition:

$$|a| = \begin{cases} a & \text{if} & a \geq 0 \\ -a & \text{if} & a < 0 \end{cases}$$

Notice how the definition captures the intuitive idea that   "$|a|$ is a made positive" :

  i. if a is already non-negative, then $|a|$ is just a ,

 ii. if a is negative, then $|a|$ is defined to be   -a , and this is a positive

      quantity since "the negative of a negative is positive."

The precise definition is important for uses in calculus; you must learn it carefully. The following two examples are typical of what you will encounter in many calculus applications:

Example A. Simplify the expression $\dfrac{2x - |x|}{x}$.

Solution. In order to deal with $|x|$, we must consider two cases: $x \geq 0$ and $x < 0$.

Case 1: $x \geq 0$.

If $x = 0$, then the expression is undefined since we can't divide by zero.

If $x > 0$, then $|x| = x$ by its definition. Thus

$$\frac{2x - |x|}{x} = \frac{2x - x}{x} = \frac{x}{x} = 1$$

Case 2: $x < 0$.

If $x < 0$, then $|x| = -x$ by its definition. Thus

$$\frac{2x - |x|}{x} = \frac{2x - (-x)}{x} = \frac{3x}{x} = 3$$

In summary, our answer becomes

$$\frac{2x - |x|}{x} = \begin{cases} 1 & \text{when} \quad x > 0 \\ \text{undefined} & \text{when} \quad x = 0 \\ 3 & \text{when} \quad x < 0 \end{cases} \qquad \square$$

Example B. Simplify the expression $x - |x + 1|$.

Solution. The two cases here are $x + 1 \geq 0$ and $x + 1 < 0$.

Case 1: $x + 1 \geq 0$, i.e., $x \geq -1$. Thus $|x + 1| = x + 1$, which gives

$$x - |x + 1| = x - (x + 1) = -1$$

Case 2: $x + 1 < 0$, i.e., $x < -1$. Thus $|x + 1| = -(x + 1)$, which gives

$$x - |x + 1| = x - (-(x + 1)) = 2x + 1$$

In summary, our answer becomes

$$x - |x + 1| = \begin{cases} -1 & \text{if} \quad x \geq -1 \\ 2x + 1 & \text{if} \quad x < -1 \end{cases} \qquad \square$$

2. <u>Algebraic properties of absolute value.</u>   The behavior of absolute values with <u>products,</u> <u>quotients</u> and <u>powers</u> are all quite predicable and easily memorized (Theorem 1.2.3). We can add to that list the simple but useful equality

$$|-a| = |a|$$

Unfortunately the rules governing <u>addition</u> and <u>subtraction</u> are not so nice; in fact, their misuse occurs very often in calculus:

> <u>Common Errors:</u> $|a + b|$ is NOT equal to $|a| + |b|$
>
> $|a - b|$ is NOT equal to $|a| - |b|$

There are, in fact, no simple rules which equate $|a + b|$ or $|a - b|$ with expressions involving only $|a|$ and $|b|$. The best we can do with either $|a + b|$ or $|a - b|$ is to bound them by inequalities as follows:

$$\left| \, |a| - |b| \, \right| \leq |a + b| \leq |a| + |b|$$
$$\left| \, |a| - |b| \, \right| \leq |a - b| \leq |a| + |b|$$

Of the four inequalities shown here, by far the <u>most important</u> is the upper bound for $|a + b|$:

$$\left|\, a + b \,\right| \; \leq \; \left|\, a \,\right| \; + \; \left|\, b \,\right|$$

Known as the <u>triangle inequality,</u> it is proved in Anton's Theorem 1.2.5. The other three inequalities can be derived from clever applications of the triangle inequality. This is the subject of Anton's Exercises 33 - 35.

Here is another common (and oftentimes costly) mistake involving absolute values:

<u>Common Error:</u> $\sqrt{a^2}$ is NOT always equal to   <u>a</u>

The correct result uses absolute values (Theorem 1.2.2):

$$\sqrt{a^2} \; = \; \left|\, a \,\right| \tag{A}$$

This occurs because the symbol $\sqrt{x}$ denotes the <u>non-negative</u> square root of x, e.g.,

$$\sqrt{(-3)^2} \; = \; \sqrt{9} \; = \; 3 \; = \; \left|\, -3 \,\right|$$

Carefully read over Anton's discussion of this preceding Theorem 1.2.2.

<u>Example C.</u>   Simplify the expression $1 - \sqrt{x^2 - 2x + 1}$.

<u>Solution.</u> $1 - \sqrt{x^2 - 2x + 1} \; = \; 1 - \sqrt{(x-1)^2}$

$$= \; 1 - \left|\, x - 1 \,\right| \quad \text{by (A)}$$

$$= \; \begin{cases} 1 - (x-1) = 2 - x & \text{if} \quad x \geq 1 \\ 1 + (x-1) = x & \text{if} \quad x < 1 \end{cases} \qquad \square$$

using the method of Example B

<u>Example D.</u>   Solve for x: $\left|\, x - 2 \,\right| + 1 = 2x$.

<u>Solution.</u>   First we isolate the absolute value on one side of the equation: $\left|\, x - 2 \,\right| = 2x - 1$.

Now, depending on whether x - 2 is positive or negative, this equation can be written as

$$x - 2 = 2x - 1 \qquad \text{or} \qquad -(x - 2) = 2x - 1$$

Solving each of these equations we obtain

$$x = -1 \qquad \text{or} \qquad x = 1$$

Don't jump to conclusions, however! It is not uncommon "accidentally" to introduce some incorrect solutions while solving an absolute value problem. Hence we must <u>check our two answers in the original equation.</u> In this case x = 1 solves the original equation but x = - 1 does NOT. Thus $\boxed{x = 1}$ is the desired answer.   □

3. <u>Absolute value and distance.</u>   The fundamental fact which relates the <u>algebraic</u> concept of absolute value with the <u>geometric</u> notion of distance is the following:

<table>
<tr><td>

the quantity $|a - b|$ equals the distance

between <u>a</u> and <u>b</u> when <u>a</u> and <u>b</u>

are considered as points on the x-axis.

</td></tr>
</table>
                                                                                        (B)

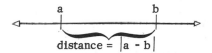

$$\text{distance} = |a - b|$$

In particular, $|a|$ equals the distance from <u>a</u> to the origin since $|a| = |a - 0|$.

You must become entirely comfortable with statement (B). By choosing some specific values of <u>a</u> and <u>b</u> , convince yourself that it is true regardless of whether <u>a</u> and <u>b</u> are positive or negative, and regardless of whether a < b or b < a .

Statement (B) allows us to interpret inequalities such as $|x - 2| < 3$ in a geometric fashion:

the solution set for $|x - 2| < 3$ consists of

all those x values which are less than 3

units away from the point 2 , i. e. ,

$$-1 < x < 5$$

Thus the expression $|x - 2| < 3$ can be read in two ways:

algebraic: the absolute value of x minus 2 is less than 3

geometric: the distance between x and 2 is less than 3

Inequalities such as $|x - 2| < 3$ are important in calculus because of their interpretation in terms of distance.

4. Common absolute value inequalities. One general type of inequality which occurs quite often in calculus is

$$|f(x)| < k$$

where f(x) is an expression involving x (i. e. , a function of x). Using the geometric interpretation of absolute value this inequality can be read as "the distance between f(x) and the origin is less than k units. "

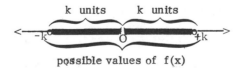

possible values of f(x)

Thus we see that $-k < f(x) < k$, which is equivalent to the two (simultaneous) inequalities $-k < f(x)$ and $f(x) < k$. Summarizing:

> The following expressions are equivalent:
>
> (1) $\left| f(x) \right| < k$
>
> (2) $- k < f(x) < k$
>
> (3) $- k < f(x)$ <u>and</u> $f(x) < k$

Anton's Theorem 1. 2. 4 merely applies this principle to the two inequalities $\left| x \right| < k$ and $\left| x - a \right| < k$. Given an inequality of the form $\left| f(x) \right| < k$ your first <u>automatic reaction</u> should be to convert it to $- k < f(x) < k$, and, if necessary, to convert it further to the pair of simultaneous inequalities $- k < f(x)$ and $f(x) < k$. Here are some examples:

<u>Example E</u>.   Solve for x: $2 \left| x - 1 \right| - 3 < x$.

<u>Solution</u>.   First we isolate the absolute value: $\left| 2x - 2 \right| < x + 3$.

Ah ha! We now have an inequality of the form $\left| f(x) \right| < k$, and our next two steps are automatic reactions:

$$- x - 3 < 2x - 2 < x + 3 \qquad \qquad \text{Automatic}$$

$$- x - 3 < 2x - 2 \quad \text{and} \quad 2x - 2 < x + 3 \qquad \qquad \text{Automatic}$$

$$- 1 < 3x \qquad \text{and} \qquad x < 5 \qquad\qquad \left. \begin{array}{c} \text{Solve} \\ \text{both} \end{array} \right.$$

$$- \frac{1}{3} < x \qquad \text{and} \qquad x < 5 \qquad\qquad \text{inequalities} \ldots$$

$$- \frac{1}{3} < x < 5 \qquad\qquad \begin{array}{l} \ldots \text{ and combine} \\ \text{the two answers} \end{array}$$

All the steps we used are reversible, and hence all the x-values just found are indeed solutions of our original inequality. Our solution set is thus the interval $(- 1/3, 5)$.   □

Example E can also be solved by using the Interval Method of §1.1.2 of

<u>The Companion.</u> Try it!

<u>Example F.</u> Solve for x: $|2x - 3| < |x + 4|$.

<u>First solution.</u> Division by $|x + 4|$ yields the inequality

$$\left| \frac{2x - 3}{x + 4} \right| < 1$$

which is of the form $|f(x)| < k$. Thus our next two steps are automatic reactions:

$$- 1 < \frac{2x - 3}{x + 4} < 1$$

$$- 1 < \frac{2x - 3}{x + 4} \quad \underline{\text{and}} \quad \frac{2x - 3}{x + 4} < 1$$

Now we solve each inequality as we did in §1.1:

1st inequality: $0 < \frac{2x - 3}{x + 4} + 1 = \frac{3x + 1}{x + 4}$

Thus $\boxed{(- \infty, \ -4) \cup (-\frac{1}{3}, \ +\infty)}$ is the solution

set for the 1st inequality.

2nd inequality:
$$0 < 1 - \frac{2x - 3}{x + 4} = \frac{-x + 7}{x + 4}$$

sign of
$-x+7$

sign of
$x+4$

sign of
$(-x+7)/(x+4)$

Thus $\boxed{(-4, 7)}$ is the solution set for the 2nd inequality.

The solution set for the original inequality is the set of points which are in <u>both</u> sets, i.e.,
it is the <u>intersection</u> of the solution sets just obtained:

$$(-\infty, -4) \cup (-1/3, +\infty) \text{ intersected with } (-4, 7)$$

This intersection is the interval $\boxed{(-1/3, 7)}$

<u>Second solution.</u>   Using the Interval Method described in §1.1.2 of <u>The Companion</u> will
give a simpler solution for the inequality $|2x - 2| < |x + 4|$. We move all the terms to
one side of the inequality to obtain

$$0 < |x + 4| - |2x - 3|$$

Now we can apply the three steps of the Interval Method to

$$f(x) = |x + 4| - |2x - 3|$$

Step 1. Crossing points. $|x + 4| - |2x - 3| = 0$

$$|x + 4| = |2x - 3|$$

$$x + 4 = 2x - 3 \quad \text{or} \quad x + 4 = -(2x - 3)$$

$$\boxed{x = 7} \quad \text{or} \quad \boxed{x = -1/3}$$

(Here we have used the same solution techniques employed in Anton's Example 5.)

Step 2. Missing points. There are none since $|x + 4| - |2x - 3|$ is defined for all values of x.

Step 3. The points $x = -1/3$ and 7 give us three intervals:

interval	x^*	$f(x^*)$	Is interval in solution set for $f(x) > 0$?
$(-\infty, -1/3)$	-1	-2	NO
$(-1/3, 7)$	0	1	YES
$(7, +\infty)$	10	-9	NO

Thus the solution set is $\boxed{(-1/3, 7)}$, agreeing with the first solution. ☐

5. Another absolute value inequality. Inequalities of the form

$$k < |f(x)| ,$$

although not quite as common as those studied in the previous subsection, do occur in applications. In general they are best handled by the Interval Method as in the following example:

Example G.   Solve for x: $\quad x \leq 2\,|x - 2|$.

Solution.   We rewrite the inequality as

$$0 \leq 2\,|x - 2| - x$$

so that we can apply the Interval Method to

$$f(x) = 2\,|x - 2| - x$$

Step 1. Crossing points. $2\,|x - 2| - x = 0$

$$|2x - 4| = x$$

$$2x - 4 = x \quad\text{or}\quad -2x + 4 = x$$

$$\boxed{x = 4} \quad\text{or}\quad \boxed{x = 4/3}$$

Step 2. Missing points.   There are none.

Step 3. The points $x = 4/3$ and 4 give us three intervals:

interval	x^*	$f(x^*)$	Is interval in solution set for $f(x) \geq 0$?
$(-\infty, 4/3)$	0	4	YES
$(4/3, 4)$	2	-2	NO
$(4, +\infty)$	5	1	YES

Since our inequality allows $f(x) = 0$, we must include the two crossing points $x = 4/3$ and 4 in the solution set for our inequality. Our final answer is therefore

$$\boxed{(-\infty, 4/3] \cup [4, +\infty)}$$

Section 1.3.   Coordinate Planes; Distance; Graphs.

1. <u>Analytic geometry: geometry into algebra.</u>   Most beginning calculus students are familiar and

comfortable with the   xy-coordinate plane, so much so that they overlook the incredible power

and importance of the technique. Identifying points on the plane with ordered pairs of real

numbers converts very hard <u>geometry</u> problems into generally much easier <u>algebra</u> problems:

<u>this is the power of "analytic geometry. "</u> For example, for the early Greeks the study of

parabolas (Appendix E) involved intricate and tricky geometric techniques; however, through

analytic geometry we can simply work with the algebraic properties of equations of the form

$y = \alpha x^2 + \beta$. This conversion of geometry into algebra is the cornerstone of calculus.

2. <u>The distance formula.</u>   Perhaps the most basic result of analytic geometry is the <u>distance</u>

formula:

the distance between two points $P_1(x_1, y_1)$

and $P_2(x_2, y_2)$ is given by

$$d = \sqrt{(x_2 - x_1)^2 + (y_2 - y_1)^2} \qquad\qquad (A)$$

(Anton's Theorem 1.3.1). Unfortunately it also seems to be one of the first formulas that

people forget! This is unfortunate for two reasons:

  i. The distance formula is <u>vital for calculus</u> for, as you will soon learn, calculus

   is built on the concept of distance.

  ii. The distance formula is <u>easy to remember:</u>

   it is really nothing but the Pythagorean

   Theorem (Appendix A §3) applied to

   the picture to the right.

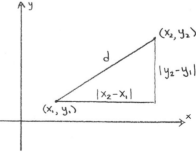

<u>One small remark</u>: since $\left(x_1 - x_2\right)^2 = \left(x_2 - x_1\right)^2$, the order in which x_1 and x_2

appear in (A) is unimportant. The same is true of y_1 and y_2. However, be sure to

group x variables with x variables and y variables with y variables.

<u>Example A.</u>   Is the triangle with vertices $(-2, 1)$, $(-1, 3)$, $(2, -1)$ an equilateral triangle?
An isosceles triangle? A right triangle?

<u>Solution.</u>   From the sketch of the triangle

we strongly suspect that it is a right

triangle. However, a rough sketch proves

nothing (are you <u>sure</u> that the angle at

$(-2, 1)$ is a full 90°? Might it not be 88°?)

To prove that we have a right triangle let us

compute the length of each side, i.e.,

compute the distances between the vertices:

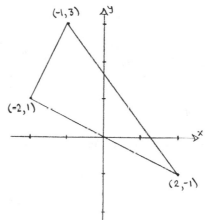

The distance between $(-2, 1)$ and $(-1, 3)$ is

$$\sqrt{(-1 - (-2))^2 + (3 - 1)^2} = \sqrt{1^2 + 2^2} = \sqrt{5}$$

The distance between $(-1, 3)$ and $(2, -1)$ is

$$\sqrt{(2 - (-1))^2 + (-1 - 3)^2} = \sqrt{3^2 + (-4)^2} = \sqrt{25} = 5$$

The distance between $(2, -1)$ and $(-2, 1)$ is

$$\sqrt{(-2 - 2)^2 + (1 - (-1))^2} = \sqrt{(-4)^2 + 2^2} = \sqrt{20}$$

Since the side lengths $\sqrt{5}$, 5 and $\sqrt{20}$ are all different, then we do not have an

equilateral or an isosceles triangle. However,

$$(\sqrt{5})^2 + (\sqrt{20})^2 = 5 + 20 = 25 = (5)^2$$

Thus the Pythagorean Theorem is valid for this triangle, proving it to be a right triangle

with right angle at the vertex $(-2, 1)$ (the vertex between the two sides of lengths $\sqrt{5}$

and $\sqrt{20}$). ◻

3. Solution sets. The solution set of an equation in two variables x and y is merely the

collection of all ordered pairs of real numbers (a, b) such that the equation is satisfied when

we substitute $x = a$ and $y = b$. The equation itself is an algebraic object; however, we

can turn its solution set into a geometric object by identifying each solution with the corre-

sponding point in the xy-plane, thus obtaining the graph of the equation.

The interplay between equations (algebraic objects) and their graphs (geometric objects)

is important in real-world applications. For example, suppose a chemist determines that two

quantities, x and y, are related according to some particular equation. To be specific,

suppose the equation is

$$y = x^{4/3} - x^{1/3} + 1$$

How then does y vary when x is changed

from 0 to 1? What value of x gives the

lowest value of y? Is there a maximum value

of y? All of these questions can be answered

by determining the graph of the equation; it is

drawn very accurately to the right. How did we

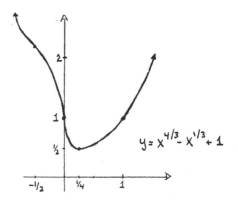

get this graph? Black magic? No... CALCULUS (which in some cases is more powerful than

black magic!) The accurate sketching of graphs is an important application of calculus, and will

be developed in §§4.4 and 4.5.

Anton's Exercises 20-36 ask you to do some simple "curve sketching" -- without calculus of course! The methods you'll be using in these problems compare with the calculus methods to be given in Chapter 4 in much the same way that a horse-and-buggy compares with the Space Shuttle.

§4.   The uses of symmetry.   In the discussion surrounding the symmetry tests of Theorem 1.3.2, Anton illustrates their uses in curve sketching:

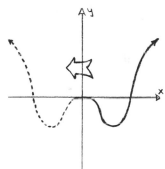

if a plane curve is symmetric about the y-axis

(i.e., if replacing x by -x leaves the

equation unchanged) then we have only to

graph that portion of the curve for which

x $\geq$ 0 , and then reflect it about the y-axis

if a plane curve is symmetric about the x-axis

(i.e., if replacing y by -y leaves the

equation unchanged) then we have only to

graph that portion of the curve for which

y $\geq$ 0 , and then reflect it about the x-axis

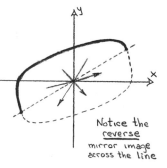

if a plane curve is symmetric about the origin

(i.e., if replacing (x,y) with (-x,-y)

leaves the equation unchanged) then given any

line through (0,0), we have only to graph that

portion of the curve which lies on one side of

the line, and then reflect it through the origin.

Notice the reverse mirror image across the line

Simply put, symmetry can cut our graphing work almost in half, as the following example shows:

<u>Example B.</u>   Sketch the graph of $y = \dfrac{4x}{1 + x^2}$.

<u>Solution.</u>   We first look for symmetries: replacing x by - x will change the equation, as will replacing y with - y. However, replacing x <u>and</u> y with - x <u>and</u> - y will not change the equation:

$$(-y) = \frac{4(-x)}{1 + (-x)^2} = -\frac{4x}{1 + x^2}$$

so $y = \dfrac{4x}{1 + x^2}$,   the original equation.

Hence, by our symmetry tests, the graph will be symmetric about the origin (but not about the x-axis or the y-axis). We therefore have only to graph that portion of the curve which lies on one side of a line through (0,0), and then reflect it through the origin. We will graph the curve for $x \geq 0$ (i. e., on the right side of the y-axis) and then reflect it through (0,0).

Using a hand calculator to compute y-values from given x-values we obtain the following table:

x	0	0.5	1.0	1.5	2.0	3.0	4.0
$y = \dfrac{4x}{1 + x^2}$	0	1.6	2.0	1.85	1.6	1.2	0.94

Plotting these points leads us to

fill in the curve shown to the right;

the dotted portion of the curve

(for $x < 0$) is obtained by reflection

through the origin.   □

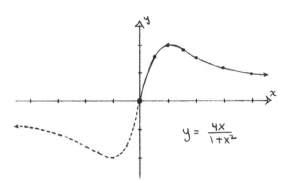

$$y = \frac{4x}{1+x^2}$$

Warning.   Anton's Remark following Example 5 is important: sketching a graph by plotting

points will only yield an approximation to the graph. In §§4.4 and 4.5 we will use the

techniques of differential calculus to obtain exact graphs of curves.

Section 1.4.   Slope of a Line.

⎢   Since trigonometry is used in this section, Anton recommends that people who need
⎢
⎢   a trigonometry review consult his Appendix 1. A somewhat shorter trigonometry
⎢
⎢   refresher can also be found in Appendix G of The Companion.

1.   The definition of slope of a line.   As mentioned in §1.3 , the power of analytic geometry

lies in turning geometry problems into algebra problems. The slope of a line is a good

example. To any non-vertical line (a geometric object) in the xy-coordinate plane we can

associate a number called the slope (an algebraic object). This number contains a vast amount

of information about the line.

You cannot learn calculus if you do not understand slope. Indeed, the first major

concept of calculus, the derivative, is defined in terms of slopes.

Anton provides two basic ways to define the slope m of a line:

I. The slope is the difference in the
    y-values over the difference in
    the x-values , i. e. ,

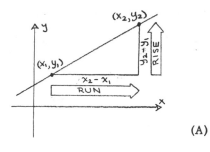

$$m = \frac{\text{rise}}{\text{run}} = \frac{y_2 - y_1}{x_2 - x_1}$$

(A)

where (x_1, y_1) and (x_2, y_2) are <u>any</u> two distinct points on the line.

II. The slope is the tangent of the
     angle of inclination ϕ made
     between the line and the x-axis , i. e. ,

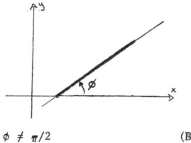

$$m = \tan \phi$$, $0 \leq \phi < \pi$, $\phi \neq \pi/2$

(B)

Anton proves that these two formulations of slope are
equivalent in Theorem 1.4.4. Essentially the proof
is nothing but applying formula (A) to the picture
to the right:

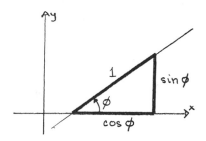

$$m \underset{(A)}{=} \frac{\text{rise}}{\text{run}} = \frac{\sin \phi}{\cos \phi} = \tan \phi$$

Two remarks are in order concerning formula (A):

i. <u>Any</u> two points on a line can be used to compute its slope (that's established by
Anton's Theorem 1.4.2). Thus, if you have a choice of points ... choose the two which
are most convenient for the calculation at hand!

ii. In formula (A) it is important that each y-value be placed over its corresponding

x-value, i.e.,

$$\frac{y_2 - y_1}{x_2 - x_1} \quad \text{and} \quad \frac{y_1 - y_2}{x_1 - x_2} \quad \text{are both equal to the slope, but}$$

$$\frac{y_2 - y_1}{x_1 - x_2} \quad \text{and} \quad \frac{y_1 - y_2}{x_2 - x_1} \quad \text{are \underline{INCORRECT}.}$$

$$\boxed{\text{NO!}} \qquad\qquad \boxed{\text{NO!}}$$

This is a common mistake, so be careful!

You must know equations (A) and (B) very well! Moreover you must "see" and "feel"

the slope of a line as a measure of the <u>steepness</u> of the line. In this regard be sure you

completely understand Anton's Figure 1.4.7. In particular:

1.   A horizontal line has slope 0.

2.   A vertical line has an undefined slope (speaking informally, we sometimes say

an "infinite" slope).

3.   A line which "points upward" has a positive slope.

4.   A line which "points downward" has a negative slope.

2.   <u>Slopes of pairs of lines.</u> Suppose L_1 and L_2 are two non-vertical lines with slopes

m_1 and m_2. Anton proves the following two results:

THEOREM 1.4.5. L_1 and L_2 are <u>parallel</u> if and only if $m_1 = m_2$

THEOREM 1.4.6. L_1 and L_2 are <u>perpendicular</u> if and only if $m_2 = -\dfrac{1}{m_1}$

These results are perfect illustrations of our claim that the slopes (algebraic quantities) of lines contain geometric information. These are important results and you must know them well. When someone says "Lines L_1 and L_2 are parallel," you must automatically convert this information into "Lines L_1 and L_2 have the same slope." When someone says "Lines L_1 and L_2 are perpendicular," you must automatically convert this information into "Lines L_1 and L_2 have slopes which are the negative reciprocals of each other."

<u>Example</u> A. Use slopes to determine if $(-2, 1)$, $(-1, 3)$ and $(2, -1)$ are the vertices of a right triangle.

<u>Solution.</u> We have a right triangle only if two of the three lines forming the sides of our triangle are perpendicular to each other. By Theorem 1.4.6 this is the case only if two of the three slopes are the negative reciprocals of each other. Thus we have translated a <u>geometric</u> problem into a simpler <u>algebraic</u> problem. Now we compute our three slopes:

The slope of the line connecting $(-2, 1)$ to $(-1, 3)$ is

$$\frac{3 - 1}{-1 - (-2)} = \frac{2}{1} = \boxed{2}$$

The slope of the line connecting $(-1, 3)$ to $(2, -1)$ is

$$\frac{-1 - 3}{2 - (-1)} = \frac{-4}{3} = \boxed{-4/3}$$

The slope of the line connecting $(2, -1)$ to $(-2, 1)$ is

$$\frac{1 - (-1)}{-2 - 2} = \frac{2}{-4} = \boxed{-1/2}$$

Ah ha! Our first and third lines have slopes 2 and $-1/2$ respectively.

Since these numbers are the negative reciprocals of each other, the corresponding lines are perpendicular. Hence we do have a right triangle, with the right angle at the vertex $(-2, 1)$ (the vertex which lies on both of the perpendicular sides).   □

Note: The previous example was solved by different methods in Example A of §1.3.2.

3. <u>Why is Theorem 1.4.6 true?</u>   Theorem 1.4.5 on parallel lines is easily remembered and intuitively clear; its proof is quite straightforward. However, none of this is true for Theorem 1.4.6 on perpendicular lines. In particular, the proof given by Anton, depending on numerous trigonometric identities, does not shed light on <u>why</u> the result is true.

The reason is not that mysterious. Suppose L_1 and L_2 are two perpendicular lines. As shown in the diagram to the right, things can be arranged so that

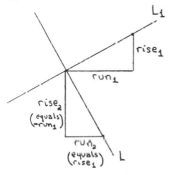

> the rise and run of L_1 become
> _____
> the run and negative rise of L_2 , respectively.

When written in terms of slope this yields:

$$m_1 = \left\{ \begin{array}{c} \text{slope} \\ \text{of } L_1 \end{array} \right\} = \left\{ \frac{\text{rise of } L_1}{\text{run of } L_1} \right\}$$

$$= \left\{ \frac{\text{run of } L_2}{-\text{ rise of } L_2} \right\} = -\left\{ \frac{1}{\begin{array}{c} \text{slope} \\ \text{of } L_2 \end{array}} \right\} = -\frac{1}{m_2}$$

This is the formula of Theorem 1.4.6!

(This informal justification of Theorem 1.4.6 can be turned into a rigorous proof by using elementary geometry to prove the statement between the double vertical lines.)

<u>Section 1.5.</u>   <u>Equations of Straight Lines</u>

      Straight lines and their equations play a central role in the development and application

of calculus, and for that reason you must learn the material in §1.5 very well.

1. <u>Standard forms.</u> When trying to find the equation of a line, you need only determine certain

<u>geometric information.</u> In fact, the basic concept which you should learn from this section is:

To determine the equation of a line you need ───────────────

     i.   a <u>point</u> on the line and its <u>slope</u>, <u>OR</u>

    ii.   the <u>y-intercept</u> of the line and its <u>slope</u>, <u>OR</u>

   iii.   <u>two distinct points</u> on the line.

The information in any one of these three groups is enough to <u>determine a line</u>, i.e., given

values for the quantities in i, ii or iii, there is <u>one and only one</u> line in the coordinate plane

which has those given specifications. For example, we learn in geometry that two distinct

points determine a line (i.e., that iii determines a line). The same principle holds true for

i and ii . Anton further shows how the <u>equation</u> of the line can be derived in each case.

These methods (for non-vertical lines) are summarized as follows:

<u>Name</u>	Necessary <u>quantities</u>	Equation of <u>the line</u>
<u>Point-slope form</u>	The <u>slope</u> m The coordinates (x_1, y_1) of <u>one point</u> on the line	$y - y_1 = m(x - x_1)$
<u>y-Intercept form</u>	The <u>slope</u> m The <u>y-intercept</u> <u>b</u>	$y = mx + b$

Two point form	The coordinates (x_1, y_1) and (x_2, y_2) of two points on the line	$y - y_1 = m(x - x_1)$ where $$m = \frac{y_2 - y_1}{x_2 - x_1}$$

Memorization of these forms is easy:

    i. The point-slope form is just a rewrite of the slope definition formula

$$m = \frac{y - y_1}{x - x_1}$$

    ii. The y-intercept form is a well-known analytic geometry formula.

    iii. The two point form is just the point-slope form combined with the slope definition formula.

---

*Remember: When determining the equation of a line, always focus on finding the quantities needed for one of the three standard forms.

Example A. Determine the equation of the line which passes through the point $(2, -1)$ and has slope $-2/3$.

Solution. We are given the exact specifications required for the point-slope form of a line:

$$m = \frac{y - y_1}{x - x_1} \qquad \text{(point-slope form of a line)}$$

$$-2/3 = \frac{y + 1}{x - 2} \qquad \text{(plugging in given information)}$$

$$y + 1 = (-2/3)(x - 2)$$

$$y = -(2/3)x + 1/3 \qquad\qquad\qquad\qquad \square$$

Example B.   Determine the equation of the line   L   which contains all the points (x, y) equidistant from $(4, -2)$ and $(3, 1)$.

Solution.   There is more work to do here than in the first example because we are not given
the exact specifications for any one of our three standard forms.   Nonetheless we can compute
the quantities needed for the point-slope form from our given information.

A point on   L   is easily found: the midpoint (x_1, y_1) of the line segment from
$(4, -2)$ to $(3, 1)$:

$$x_1 = (4 + 3)/2 = 7/2$$
$$y_1 = (-2 + 1)/2 = -1/2$$

Thus $(x_1, y_1) = (7/2, -1/2)$

To determine the slope of   L ,   notice that

L   is perpendicular to the line segment

from $(3, 1)$ to $(4, -2)$;   since the slope

of this line segment is

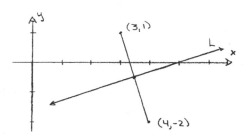

$$m_0 = \frac{-2 - 1}{4 - 3} = -3/1 = -3$$

then (from §1.4) the slope of   L   is given by

$$m = -1/m_0 = -1/(-3) = 1/3$$

Using the point-slope form of a line we thus have

$$m = \frac{y - y_1}{x - x_1}$$ (point-slope form of a line)

$$1/3 = \frac{y + 1/2}{x - 7/2}$$ (plugging in information
calculated above)

$$y + 1/2 = (1/3)x - 7/6$$

$$\boxed{y = (1/3)x - 5/3}$$ □

Vertical lines. Since <u>vertical lines</u> have an undefined slope, the point-slope and y-intercept

forms are irrelevant to such lines. When given two points, however, it is always easy to see

if they determine a vertical line: <u>they will</u>

<u>have the same x-coordinate</u> , say x_0 .

The equation of the line is then

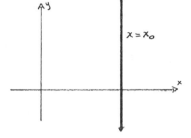

2.   <u>The general form of a line.</u> Every equation for a line, including that of a vertical line, can be

put in the form of a <u>first degree equation</u> in x and y ,

$$\boxed{Ax + By + C = 0}$$ (A)

where at least one of the constants A or B is not zero. This result is part of Anton's

Theorem 1. 5. 6 and is easy to prove. However, the converse of this result is even more

interesting: <u>any first degree equation in x and y is an equation for a line,</u> i. e. , has a line

as its graph. (This is also part of Theorem 1. 5. 6.) Thus equations such as

$$3x - 4y + 2 = 0, \quad \pi x + \sqrt{2}\, y - 2 = 0, \quad \text{or} \quad 4x - \sqrt{3} = 0$$

are all equations for lines. To determine the important quantities associated with the corre-

sponding line (e. g. , slope,   y-intercept, etc.) we simply <u>convert our first degree equation</u>

<u>into an appropriate standard form</u> and then "read off" the desired quantities. The y-intercept

form is particularly useful in this regard.

1.5.5

<u>Example C.</u>   Determine the slope, the y-intercept, and at least two points of the line given by

the first degree equation

$$\pi x + \sqrt{2}\, y - 2 = 0 \quad .$$

<u>Solution.</u>   We convert our equation to y-intercept form   simply by solving for y in terms

of x :

$$\pi x + \sqrt{2}\, y - 2 = 0$$

$$\sqrt{2}\, y = -\pi x + 2$$

$$y = (-\pi/\sqrt{2})x + (2/\sqrt{2})$$

$$y = (-\pi\sqrt{2}/2)x + \sqrt{2}$$

Compare this result with $y = \qquad m x \quad + \quad b$ (the point-slope formula).

From the comparison we see that the slope is $m = -\pi\sqrt{2}/2$ and the y-intercept is

$b = \sqrt{2}$. To find specific points on the line we have only to choose convenient x-values   and

compute the corresponding y-values:

$$x = 0 \quad \text{gives} \quad y = \sqrt{2}$$

$$x = 1 \quad \text{gives} \quad y = -\pi\sqrt{2}/2 + \sqrt{2}$$

Thus $(0, \sqrt{2})$ and $(1, -\pi\sqrt{2}/2 + \sqrt{2})$ are two points on the line.   □

3. <u>Problems involving lines.</u>   We'll end our section with two examples of analytic geometry

problems which require the material developed in this section about lines. Concentrate, in

particular, on the problem solving techniques used in Example E.

<u>Example D.</u>   Determine whether the following two lines are parallel, perpendicular, or neither:

$2x + 3y - 6 = 0$ and $3x - 2y - 4 = 0$.

<u>Solution.</u>   Determining whether two lines are parallel or perpendicular requires comparing their slopes ($\S 1.4$). We'll determine these slopes as in Example C by converting each equation to the   y-intercept form:

$$2x + 3y - 6 = 0 \quad \text{becomes} \quad y = -(2/3)x + 2$$

$$3x - 2y - 4 = 0 \quad \text{becomes} \quad y = (3/2)x - 2$$

As we now see, the slope of one line is the negative reciprocal of the other (i.e.,

$3/2 = -\dfrac{1}{(-2/3)}$). Thus the lines are perpendicular by Theorem 1.4.6.   □

<u>Example E.</u>   Find the (shortest) distance from the point $P(-2, 9)$ to the line   L   given by

$2x - 3y + 5 = 0$.

<u>Solution.</u>   To determine the plan of attack in such a problem it is often useful to

> think backwards from where you want to end.

In this problem our ultimate goal (where we want to end) is to

1.   Compute the shortest distance from   P   to   L ,
     which is the length of the shortest possible
     line segment joining   P   to   L .

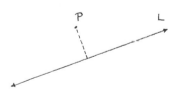

A little reflection should convince you that, to compute this length we must

2.   Determine the point   Q   on   L   such
     that the line segment $\overline{PQ}$ is
     perpendicular to   L .

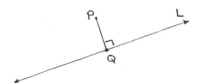

We see, however, that Q is the intersection of L with the line L' that passes through
P and is perpendicular to L . If we can find an equation for L' , then Q will be found
by solving the equations for L and L' simultaneously. Thus, to find Q it is sufficient
to

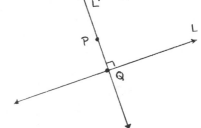

    3.   Find an equation for the line L'

         which contains P and is

         perpendicular to L .

If we knew the slope of L' we could use the point-slope formula to compute the equation for
L' . Since L and L' are perpendicular, Theorem 1.4.6 says that to compute the slope
of L' it is sufficient to

    4.   Compute the slope of L.

Ahh... Our "backwards reasoning" has finally bumped into a computation which <u>we know how</u>
<u>to do.</u> Our game plan for the solution of Example E is thus to perform all the computations
we laid out above... but in the order 4 - 3 - 2 - 1 of course! Here's the execution of our
plan:

4.   L is given by $2x - 3y + 5 = 0$. So we solve for y in terms of x to obtain the

         slope of L:

$$3y = 2x + 5$$

$$y = (2/3)x + 5/3$$

    Thus the slope of L is $m = 2/3$.

3.   The slope of L' is $m' = -1/m$ (Theorem 1.4.6)

$$= -3/2$$

    A point on L' is $P(-2, 9)$.

Thus the point-slope equation for L' is

$$-3/2 = \frac{y - 9}{x + 2}, \quad \text{or} \quad y = -(3/2)x + 6$$

2.   The intersection point Q of L and L' is found by solving the simultaneous equations

$$\begin{cases} y = (2/3)x + 5/3 \\ y = -(3/2)x + 6 \end{cases}$$

Equating these two different expressions for y yields

$$(2/3)x + 5/3 = (-3/2)x + 6$$

$$(13/6)x = 13/3$$

$$x = 2, \quad \text{which gives} \quad y = 3$$

Thus $Q = (2, 3)$.

1.   The distance between P and Q is

$$d = \sqrt{(-2 - 2)^2 + (9 - 3)^2} = \sqrt{52} = 2\sqrt{13} \qquad \square$$

Section 1.6.   Circles; Equations of the Form $y = ax^2 + bx + c$

1.   **The general pattern for lines.**   In its logical framework, Anton's developement of equations for circles and parabolas is very similar to his development of equations for lines in §1.5. From that section, given certain geometric data about a line (e. g. , the slope m and the coordinates (x_1, y_1) of one point on the line) we can write down a standard form for the line (e. g. , the point-slope form $y - y_1 = m(x - x_1)$) and then a general equation for the

line $(Ax + By + C = 0)$. The reverse procedure was also possible: given a <u>general equation</u> for a line we can convert to a <u>standard form</u> and then read off <u>geometric data</u> about the line (see Example C in §1.5.2 of <u>The Companion</u>). This can be summarized in a diagram:

$$\left\{ \begin{array}{c} \text{geometric} \\ \text{information} \\ \text{for a line} \end{array} \right\} \longleftrightarrow \left\{ \begin{array}{c} \text{standard} \\ \text{forms} \\ \text{for a line} \end{array} \right\} \longleftrightarrow \left\{ \begin{array}{c} \text{general} \\ \text{equation} \\ \text{for a line} \end{array} \right\}$$

Remember that there were three standard forms for a line studied in §1.5 (see the table in §1.5.1 of <u>The Companion</u>).

2. <u>Circles.</u> The above pattern continues to hold for circles: we can go back and forth between <u>geometric data</u> for a circle, a <u>standard equation</u> for a circle, and a <u>general equation</u> for a circle.

Geometric data:

> To determine the equation of a circle you need the coordinates (x_0, y_0) of the <u>center</u> and the value r of the <u>radius</u>.

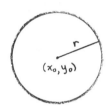

Given values for the center and the radius, there is one and only one circle which has these specifications:

<u>Standard equation</u> for a circle with center (x_0, y_0) and radius r:

$$(x - x_0)^2 + (y - y_0)^2 = r^2$$

<u>General equation</u> for a circle:

$$Ax^2 + Ay^2 + Dx + Ey + F = 0, \quad A \neq 0$$

> Note that the coefficients of x^2 and y^2 are equal

Here are examples of how to go back and forth between these objects:

Example A.   Find a general equation for the circle with center $(2, -1)$ and radius 3.

Solution.   We are given geometric data and wish to find the general equation; the procedure is to find the standard equation as an intermediate step. Since the standard equation is

$$(x - x_0)^2 + (y - y_0)^2 = r^2$$

we merely plug in the values $(x_0, y_0) = (2, -1)$ and $r = 3$ to obtain

$$(x - 2)^2 + (y + 1)^2 = 9$$

To go from this standard equation to a general equation only requires expanding the squared terms and combining the constants:

$$(x^2 - 4x + 4) + (y^2 + 2y + 1) - 9$$

$$\boxed{x^2 + y^2 \quad 4x + 2y - 4 = 0}$$

                                                                                 □

Example B.   Determine the center and radius of the circle given by the general equation $2x^2 + 2y^2 + 4x - 2y + 2 = 0$.

Solution.   As in Example A, finding the standard equation is our intermediate step. To go from a general to a standard equation requires completing the square in both x and y (Appendix D §7):

$$(2x^2 + 4x) + (2y^2 - 2y) + 2 = 0$$

$$(x^2 + 2x) + (y^2 - y) + 1 = 0$$

$$(x^2 + 2x + 1) + (y^2 - y + \tfrac{1}{4}) + 1 - 1 - \tfrac{1}{4} = 0$$

1.6.4

$$(x + 1)^2 + (y - \tfrac{1}{2})^2 - \tfrac{1}{4} = 0$$

$$(x + 1)^2 + (y - \tfrac{1}{2})^2 = \tfrac{1}{4}$$

This is the standard equation for a circle with center $(-1, 1/2)$ and radius $1/2$. $\square$

Warning: As Anton points out in Theorem 1.6.1, an equation of the general form $Ax^2 + Ay^2 + Dx + Ey + F = 0$ does NOT necessarily have a circle for a graph; in some instances a single point or no graph at all will result. For example, in Example B, if the constant term 2 is replaced by $5/2$, then the graph will be the single point $(-1, 1/2)$; if 2 is replaced by any number higher than $5/2$, then there will be no graph at all.

Example C. Find the standard equation for the circle with center $(-3, -4)$ which passes through the origin.

Solution. The geometric data needed in order to determine a circle are its center and radius; when given any other type of data, immediately try to determine the center and radius. We know that the radius must be

$$r = \text{distance from} \quad (-3, -4) \quad \text{to} \quad (0, 0)$$

$$= \sqrt{(-3 - 0)^2 + (-4 - 0)^2} = \sqrt{25} = 5$$

Thus, using $(x_0, y_0) = (-3, -4)$ and $r = 5$, we obtain the standard equation

$$(x + 3)^2 + (y + 4)^2 = 25$$
$\square$

3. **Parabolas.**   Anton's discussion of parabolas can be organized in the way we did for circles.

A quadratic equation in   x,

$$y = ax^2 + bx + c \qquad (a \neq 0) \qquad\qquad (A)$$

is the analogue for parabolas of the general equation for a circle discussed in the previous

section.   To get a standard equation for the parabola we complete the square (Appendix D §7):

$$y = a(x^2 + \frac{b}{a} x) + c$$

$$y = a\left(x^2 + \frac{b}{a} x + \frac{b^2}{4a^2}\right) - \frac{b^2}{4a^2} + c$$

$$y = a\left(x + \frac{b}{2a}\right)^2 + \left(c - \frac{b^2}{4a^2}\right)$$

Let $$x_0 = - \frac{b}{2a}$$ and $$y_0 = c - \frac{b^2}{4a^2}$$

Then

$$y = a(x - x_0)^2 + y_0 \qquad (a \neq 0) \qquad\qquad (B)$$

An equation of type (B) is called a **standard equation for a parabola.**   It is not hard

to show that the vertex of such a parabola is (x_0, y_0) (for a justification of this statement,

see Appendix E §1); since $x_0 = - \frac{b}{2a}$ from above, this corresponds to Anton's equation (5).

The **geometric data** which determine a parabola are the **vertex** (x_0, y_0) and the

number   **a**,   which we will call the **parabolic steepness.**   As shown in Anton's Figure 1.6.3 ,

if   **a**   is positive, then the parabola turns upward; if   **a**   is negative, then the parabola turns downward. Moreover, as $|a|$ gets large, the steepness of the parabola increases (this is shown in Appendix E §1).

Here's a summary of our approach to parabolas:

Geometric data:

> To determine the equation of a parabola you need the coordinates (x_0, y_0) of the <u>vertex</u> and the value $\underline{a}$ of the <u>parabolic steepness.</u>

Standard equation for a parabola with vertex (x_0, y_0) and parabolic steepness $\underline{a}$:

$$y = a(x - x_0)^2 + y_0$$

General equation for a parabola (quadratic equation):

$$y = ax^2 + bx + c \qquad (a \neq 0)$$

Here are examples showing how these are used:

Example D.   Find an equation for the parabola passing through $(4, 1)$ with vertex at $(2, -1)$. Sketch the graph.

<u>Solution.</u>   We are given the vertex $(x_0, y_0) = (2, -1)$, and thus the standard equation becomes

$$y = a(x - 2)^2 - 1$$

We have only to determine the value of the parabolic steepness a ; to do this we note that

$(x, y) = (4, 1)$ must satisfy our standard equation:

$$1 = a(4 - 2)^2 - 1$$

$$1 = 4a - 1$$

$$a = 1/2$$

Thus our equation is

$$y = \frac{1}{2}(x - 2)^2 - 1$$

Expanding the squared term leads to the quadratic equation

$$y = \frac{1}{2}x^2 - 2x + 1$$

To graph the equation we first plot the vertex $(2, -1)$ and then find other points on the graph using either of our above equations:

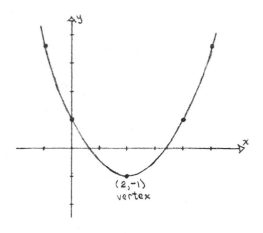

$(2,-1)$
vertex

x	- 1	0	2	4	5
y	7/2	1	- 1	1	7/2

These points enable us to make the sketch of our parabola shown to the right.   □

__Example E.__ Find the vertex and the intersections with the x-axis of the parabola given by

$$y = -2x^2 + 2x + 1$$

__Solution.__ Anton's method of solution would be to use the equation $x_0 = -b/2a$ to find the x-coordinate of the vertex (i.e., $x_0 = -b/2a = -(2)/2(-2) = 1/2$), and then to use the quadratic formula (Appendix D §5) on $0 = -2x^2 + 2x + 1$ to find the intersections with the x-axis (i.e., $x = -b \pm \sqrt{b^2 - 4ac})/2a = (-2 \pm \sqrt{4+8})/(-4) = (1 \pm \sqrt{3})/2)$. We'll give an alternate method: first determine the standard equation for the parabola. This is done by completing the square (Appendix D §7):

$$\begin{aligned}
y &= -2x^2 + 2x + 1 \\
&= -2(x^2 - x) + 1 \\
&= -2(x^2 - x + 1/4 - 1/4) + 1 \\
&= -2(x - 1/2)^2 + 1/2 + 1 \\
&= \underset{\underset{x_0}{\uparrow}}{-2(x - 1/2)^2} + \underset{\underset{y_0}{\uparrow}}{3/2}
\end{aligned}$$

Thus the vertex is $(x_0, y_0) = (1/2, 3/2)$. To obtain the intersections with the x-axis we merely set $y = 0$:

$$0 = -2(x - 1/2)^2 + 3/2$$

$$(x - 1/2)^2 = 3/4$$

$$x - 1/2 = \pm \sqrt{3}/2$$

$$x = (1 \pm \sqrt{3})/2 \approx -.366 \text{ and } 1.366 \qquad \square$$

Chapter 2 :   Functions and Limits

Section 2.1:   Functions

   The main character in calculus is the <u>function</u>: it is what we "differentiate," "integrate,"
"graph," "maximize," etc. . In this section Anton introduces the function concept and develops
the basic notions connected with functions. In this section of <u>The Companion</u> we merely
expand upon some of the trickier areas and discuss some common pitfalls.

1. <u>Functions as machines.</u> One useful and intuitive way to view a function f is as a machine.
You put an x-value into the f machine, the machine chews and hammers on it, remolding
it into a new value which is sent out with the label f(x) .

Only one f(x) is returned for each x which is

entered (a Coke machine gives <u>one</u> can for your money,

not two or three). This is the <u>single-valued</u> property of

functions.

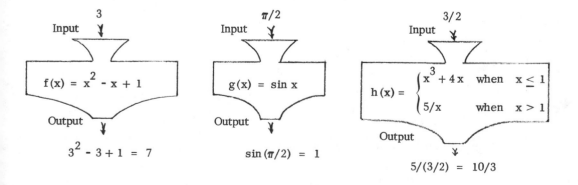

   Most of our calculus "function machines" are specified by an equation (or equations)
which say what to do with a given x-value in order to obtain the corresponding f(x) value.
The following are examples of such functions, set up as "machines":

Input 3

$$f(x) = x^2 - x + 1$$

Output

$$3^2 - 3 + 1 = 7$$

Input $\pi/2$

$$g(x) = \sin x$$

Output

$$\sin(\pi/2) = 1$$

Input 3/2

$$h(x) = \begin{cases} x^3 + 4x & \text{when } x \le 1 \\ 5/x & \text{when } x > 1 \end{cases}$$

Output

$$5/(3/2) = 10/3$$

There is nothing sacred about using the letter   x   to specify the rule defining   f   --
any letter will do. For example, our functions   f, g   and   h   could have been defined by

$$f(u) = u^2 - u + 1, \quad g(t) = \sin t, \quad h(\beta) = \begin{cases} \beta^3 + 4\beta & \text{when} \quad \beta \le 1 \\ \\ 5/\beta & \text{when} \quad \beta > 1 \end{cases}$$

More generally, we have the following important, but often overlooked fact:

<table>
<tr><td>The Function
Principle</td><td>A function   f   gives equal treatment

to anything put inside the function brackets.</td></tr>
</table>

Let's illustrate this with our specific example $f(x) = x^2 - x + 1$. In words, given   x ,   f(x)
is obtained by

"subtracting   x   from the square of   x   and adding   1 ."

Thus, by the Function Principle, given   u ,   f(u)   is obtained by

"subtracting   u   from the square of   u   and adding   1 ."

In a more substantial example, given   2t + 1 ,   f(2t + 1)   is obtained by

"subtracting   (2t + 1)   from the square of   (2t + 1)   and adding   1 ."

In symbols this becomes

$$f(2t + 1) = (2t + 1)^2 - (2t + 1) + 1$$
$$= 4t^2 + 4t + 1 - 2t - 1 + 1$$
$$= 4t^2 + 2t + 1$$

Complicated expressions like this appear very often in Calculus, and many errors are made

with them.   These errors are all easily avoided!   For instance, let's examine one frightening-
looking expression that will arise soon (when we define "the derivative of a function

f   at   x"):

$$\frac{f(x+h) \ - \ f(x)}{h} \qquad (A)$$

Example A.   Evaluate (A) when $f(x) = x^2 - x + 1$.

Solution.

this is   f(x + h)   this is   f(x)

$$\frac{f(x+h) \ - \ f(x)}{h} = \frac{[(x+h)^2 - (x+h) + 1] - [x^2 - x + 1]}{h}$$

The Function
Principle
applied to
f(x + h)

$$= \frac{x^2 + 2hx + h^2 - x - h + x - x^2 + x - 1}{h}$$

$$= \frac{2hx \ - \ h \ + \ h^2}{h} = \frac{h(2x \ - \ 1 \ + \ h)}{h}$$

$$= 2x \ - \ 1 \ + \ h \qquad \square$$

You'll see many more of these types of calculations in §3. 1.

2.   The domain of a function.   If a coin machine is designed to take quarters, it will probably
jam if fed a nickel.   The same is true of functions: they won't necessarily accept every
real number as an input.   For example,

$$f(t) \ = \ \sqrt{t+1}$$

produces a real number only when $t + 1 \geq 0$, i.e., when $t \geq -1$.   Hence   f   rejects
$t = -2$ because $f(-2) = \sqrt{-2+1} = \sqrt{-1}$ is not a real number.   The domain of   f   consists
of all real numbers which   f   will accept, in this case all real numbers which are greater than

2. 1. 4

or equal to minus one. In symbols:

$$\text{domain } f = \{t \mid t \geq -1\}$$

Here we have used the rule which Anton mentions in a low-key fashion just prior to his Example 3; it is important enough to mention again in highlighted form:

The Implicit Domain Rule

> If the rule for evaluating a function is given by a formula, and <u>if there is no mention of the domain,</u> it is understood that the domain consists of all real numbers for which the formula makes sense and yields a real value.

Anton's Examples 3 and 4 illustrate the use of this rule.

<u>A common mistake</u> made with the Implicit Domain Rule is to ignore the significant changes that can result in a function when its defining equation is "rearranged" or "simplified." Here's an example:

<u>Example B.</u> Simplify the defining equation for

$$g(x) = \frac{x-2}{x^2 - 5x + 6}$$

<u>Solution.</u> Factoring the denominator and cancelling yields

$$\frac{x-2}{x^2 - 5x + 6} = \frac{x-2}{(x-3)(x-2)} = \frac{1}{x-3}$$

But BE CAREFUL! By the Implicit Domain Rule we find

$$g(x) = \frac{x-2}{(x-3)(x-2)} \quad \text{has domain} \quad \{x \mid x \neq 2, 3\}$$

$$\text{while} \quad \frac{1}{x-3} \quad \text{has domain} \quad \{x \mid x \neq 3\}$$

Thus the two versions of our expression have <u>different</u> domains (and so do not represent the same function). What happened? Well, when we cancelled x - 2 we technically divided both the numerator and the denominator by x - 2. <u>But when</u> <u>x = 2</u>, then x - 2 equals 0 and dividing by zero is highly illegal. Thus we must indicate that x = 2 is <u>not</u> an allowable value for our simplified version of g(x):

$$g(x) = \frac{1}{x - 3} \qquad \text{when} \qquad x \neq 2, 3 \qquad\qquad \square$$

Example B may seem overly picky and pedantic. It is not. Correctly noting the domains of functions is vitally important in calculus. The two most common reasons for having to eliminate an x-value from the domain of a function are:

(1) the x-value results in a negative value in a square root, or

(2) the x-value results in a division by zero.

3. <u>The graph of a function.</u>   Here we again see that interplay between algebra and geometry which is the cornerstone of analytic geometry and calculus (§1.3). An <u>equation</u> defining a function is an <u>algebraic</u> object, while the <u>graph</u> of the function is a <u>geometric</u> object (recall the discussion in §1.3.3 of The Companion). Presently you have the ability to graph only the simplest of functions. In §4.4 and 4.5 we will apply very powerful calculus techniques to obtain complicated function graphs.

Nonetheless, once you do have the graph of a function, then graphs for all <u>translations</u> of this function are easily obtained by the "Translation Principles." Anton gives these principles in Table 2.1.2; here's a rewording so that they apply to graphs of <u>any equation,</u> not simply to graphs of function equations:

The Translation Principles

---

i. To move the graph of an equation to the right (respectively, left) by $\underline{h}$ unit, replace x with $x - h$ (respectively, $x + h$);

ii. To move the graph of an equation up (respectively, down) by $\underline{k}$ units, replace y with $y - k$ (respectively, $y + k$).

Thus, to move the graph of $y = f(x)$ $\underline{up}$ by $k = c$ units, we replace y with $y - c$ to obtain $y - c = f(x)$, or

$$y = f(x) + c$$

This is the way Anton describes our principle (ii) in his table. Our description has the slight advantage of treating both x and y in the same fashion, and, moreover, our principles apply to non-function equations such as

$$x^2 + y^2 = 1$$

Our Translation Principles are discussed at length in Appendix E. In particular, see Examples 1, 2, 3 and 4 in Appendix E.

Example C. Sketch the graph of $y + 2 = \sqrt[3]{x - 3}$.

Solution. The graph of the simpler equation $y = \sqrt[3]{x}$ is shown to the right; it can be found in Anton's Catalog of Basic Curves (§1.3) or sketched fairly easily by plotting a few points. To get our equation we take $y = \sqrt[3]{x}$ and

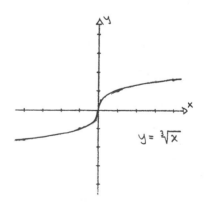

$y = \sqrt[3]{x}$

replace   x   with   x - 3 ,

replace   y   with   y + 2

Hence, by the Translation Principles,   the graph of $y + 2 = \sqrt[3]{x - 3}$ is obtained by moving

the graph of $y = \sqrt[3]{x}$

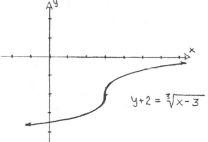

to the right by   h = 3   units

down by   k = 2   units

The resulting graph is shown to the right.   □

4. <u>When is a curve the graph of a function?</u>   This is the question Anton asks in   Problem 2. 1. 3 ,

and which is answered by the <u>Vertical Line Test</u>:

| The Vertical Line Test | a curve is the graph for some function $y = f(x)$

if and only if <u>no</u> vertical line intersects the curve

<u>more than once.</u> |

This is a very easy test to apply, as Anton's Examples 18 and 19 show. Remember that the

Vertical Line Test is merely the <u>geometric</u> version of the following <u>algebraic</u> property of a

function:

| The Single Value Property | an equation defines   <u>y   as a function of   x</u>

if to each   x-value   there is <u>at most one</u>

corresponding   y-value. |

Anton employs the Single Value Property (although he does not name it) in Examples 18 and 19.

Example D.   Does the equation $y^2 x = 4$ define y as a function of x? [Equivalent question: is the curve defined by $y^2 x = 4$ the graph for some function $y = f(x)$?]

First solution.   The curve defined by the equation $y^2 x = 4$ is shown to the right. Since any vertical line $x = a$ $(a > 0)$ intersects the curve twice, then the Vertical Line Test proves that the curve is not the graph of a function $y = f(x)$.

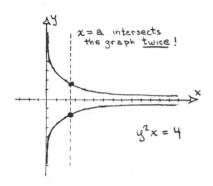

x = a intersects the graph twice!

$y^2 x = 4$

Second solution.   Solving our equation for y in terms of x yields $y = \pm 2/\sqrt{x}$, so that for each positive value of x we have two values of y; by the Single Value Property, this proves that $y^2 x = 4$ does not define y as a function of x .   □

As Anton points out prior to Example 20, we sometimes reverse the roles of x and y by studying x as a function of y: $x = g(y)$. In such cases the Vertical Line Test is replaced by the Horizontal Line Test (just replace $y = f(x)$ by $x = g(y)$, and "vertical" by "horizontal"), and the roles of x and y are reversed in the Single Value Property. These altered tests are illustrated in Examples 21 and 22. You might try the tests on the equation $y^2 x = 4$ of Example D: although it does not define y as a function of x, it does define x as a function of y.

Section 2. 2 :   Operations on Functions; Classifying Functions

1.   <u>Algebraic operations on functions.</u>   Given two functions f and g, their <u>sum</u> f + g is

defined by

$$(f + g)(x) = f(x) + g(x)$$ (A)

(Anton's Definition 2. 2. 1.)   Let's be sure we know what this means -- it is easy not to

analyze Equation (A) deeply enough and thus not to understand what is being defined.

We start with <u>two functions</u> f and g and wish to define a <u>third function,</u> called the

<u>sum</u> of f and g , and denoted by f + g . To define the function f + g we have to tell you

what value f + g assigns to a real number x , i. e., what (f + g)(x) is defined to be.

<u>That's exactly what Equation (A) does!</u> (f + g)(x) is defined by Equation (A) to be the

(ordinary arithmetic) sum of the two numbers f(x) and g(x) .

The machine analogy that we gave for functions in the previous section is useful in

explaining sums. Given the f and g

machines, we build the f + g machine

as shown to the right. An x-value fed into

the f + g machine is immediately run

through the f and g machines separately.

The two separate outputs f(x) and g(x) are

then <u>added together,</u> the result being the output

of the full f + g machine, i. e.,

(f + g)(x) = f(x) + g(x) .

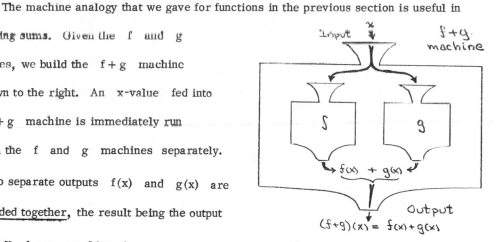

If you think this is a crazy way to explain the sum of two functions ... well, it's not.

This is essentially the logic you would use to program a computer to add functions together.

Example A. Determine a formula for the sum f + g of the two functions $f(x) = 2x^2 - 1/x$

and $g(x) = 1/x - x^2$.

Solution. By (A) our new function f + g is given by

$$(f + g)(x) \quad = \quad f(x) + g(x)$$
$$\text{(A)}$$
$$= (2x^2 - 1/x) + (1/x - x^2) = x^2$$

Thus $(f + g)(x) = x^2$ --- well, almost. We do have to be careful about the domain of f + g .

Since f + g is defined only at x-values where both f and g are defined, and neither

f or g is defined at x = 0 , we see that x = 0 must be excluded from the domain of

f + g , i.e.,

$$(f + g)(x) = x^2 \qquad \underline{\text{for} \quad x \neq 0} \qquad\qquad \square$$

Our discussion concerning f + g might seem like overkill. After all, Equation (A)

is hardly complicated. Nonetheless, we've seen a surprisingly large number of people who

thought they understood f + g ... but in fact did not.

The discussion given for the sum f + g of two functions could very well be repeated

for the difference f - g , the product f · g , and the quotient f/g (see Anton's

Definition 2.2.1). We'll let you fill in the details. We especially recommend that you

construct "machine diagrams" for these new functions just as we did for the sum f + g .

Remember, the definitions all look deceptively easy; be sure you understand them.

2. Composition of functions. Given two functions f and g, the composition of f with g,

denoted by f ∘ g, is defined by

$$(f \circ g)(x) = f(g(x)) \qquad \text{(B)}$$

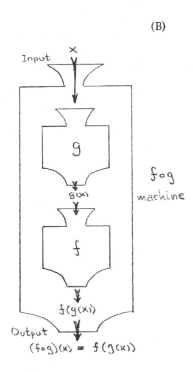

$f \circ g$
machine

Thus, as with the sum $f + g$, the composition

operation starts with two functions,   f and g ,

and produces a third function, $f \circ g$, according

to formula (B). The machine analogy (shown to

the right) is particularly useful: the $f \circ g$

machine is simply the g machine followed by

the f machine, i.e., the g and f machines

in series. Note that it is the function g which

is first applied to x , with the function f then

applied to g(x) .

   In computational examples people sometimes have difficulty in dealing with $f(g(x))$ --

no such problems should occur if you keep in mind the Function Principle, introduced in the

last section:

> A function f gives equal treatment to
>
> anything put inside its function brackets.

In this case the "anything" inside the brackets is   g(x).

Example B. Determine $f \circ g$ for

$$f(x) = x^2/(1 + x^2) \qquad \text{and} \qquad g(x) = \sqrt{x^2 - 1}$$

__Solution.__ By the Function Principle, to obtain $f(g(x))$ from $f(x)$ we have only to take the equation for $f(x)$ and replace every x with $g(x)$:

$$f(x) = x^2 \Big/ \left(1 + x^2\right)$$

$$(f \circ g)(x) \underset{(B)}{=} f(g(x)) = [g(x)]^2 \Big/ \left(1 + [g(x)]^2\right) \qquad \text{by the Function Principle}$$

$$= \left[\sqrt{x^2 - 1}\right]^2 \Big/ \left(1 + \left[\sqrt{x^2 - 1}\right]^2\right) \qquad \text{since} \qquad g(x) = \sqrt{x^2 - 1}$$

$$= (x^2 - 1) / (1 + x^2 - 1)$$

$$= (x^2 - 1) / x^2$$

$$= 1 - 1/x^2$$

Thus $(f \circ g)(x) = 1 - 1/x^2$ ... well, almost. We must check the __domain of $f \circ g$__. Let's leave our example for a moment and talk in general terms.

The restrictions on the domain of $f \circ g$ are easily read off the $f \circ g$ machine diagram: the only allowable x-values are those for which

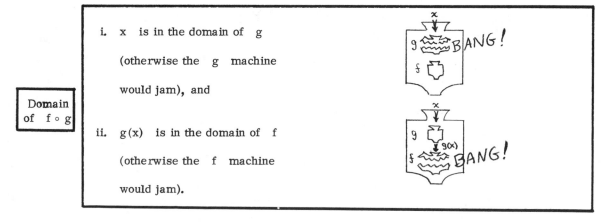

__Domain of $f \circ g$__

i. x is in the domain of g (otherwise the g machine would jam), and

ii. $g(x)$ is in the domain of f (otherwise the f machine would jam).

Returning to our example, the domain of g   consists of all x-values   such that

$x^2 - 1 \geq 0$, i. e.,

$$x \leq -1 \qquad \text{or} \qquad 1 \leq x$$

Since   f(x)   is defined for all real   x   (and hence for all values   g(x)   that   g   could possibly

assume) there are no further restrictions on the domain of   f ∘ g .   Thus

$$(f \circ g)(x) = 1 - 1/x^2 \quad \text{for} \quad x \leq -1 \quad \text{or} \quad 1 \leq x$$ □

Example C.   Determine   f ∘ g   for $f(x) = \sqrt{1/x}$ and   g(x) = csc x .

Solution. $(f \circ g)(x) \underset{(B)}{=} f(g(x))$

$= \sqrt{1/g(x)}$ by the Function Principle

$= \sqrt{1/\csc x} = \sqrt{\sin x}$

Now we must check the domain conditions.   (i)   The domain of   g(x) = csc x = 1/sin x

consists of all those   x-values   for which   sin x ≠ 0 ,   i. e., $x \neq 0, \pm\pi, \pm 2\pi, \ldots$.

(ii)   The domain of $f(x) = \sqrt{1/x}$ consists of all   x > 0.   Thus   g(x)   is in the domain of

f(x)   if and only if   g(x) > 0 ,   i. e.,

$$\csc x > 0 \qquad \text{since} \qquad g(x) = \csc x ,$$

$$\sin x > 0 \qquad \text{since} \qquad \csc x = 1/\sin x$$

By examining the graph of   y = sin x   (Figure A. 24 in Anton's Appendix 1, or The

Companion's Appendix G. 5) we see that   sin x   is positive on (all of) the intervals

$$(0, \pi), \quad (2\pi, 3\pi), \quad (4\pi, 5\pi), \ldots$$

$$\text{or} \quad (-2\pi, -\pi), \quad (-4\pi, -3\pi), \quad (-6\pi, -5\pi), \ldots$$

A convenient shorthand for listing these intervals is

$$\boxed{\quad 2n\pi \; < \; x \; < \; 2(n+1)\pi \quad \text{for any} \quad n = 0, \; \pm 1, \; \pm 2, \ldots \quad}$$

These are the x-values for which $g(x)$ is in the domain of $f(x)$.

As the domain restriction in (ii) contains the restrictions in (i), we have

$$\boxed{\quad (f \circ g)(x) \; = \; \sqrt{\sin x} \quad \text{for} \quad 2n\pi \; < \; x \; < \; (2n+1)\pi, \; \text{with} \; n \; \text{any integer} \quad} \qquad \square$$

3. <u>Construction and decomposition of functions.</u> You might, with good reason, be puzzled as to the uses of the function operations defined thus far. There are two important uses which we will describe.

Our operations of forming sums, compositions, etc. allow us to <u>construct</u> complicated functions out of simpler ones. This is what Anton does in §2.2 to construct <u>polynomials</u> and <u>rational</u> functions:

Start with $\underline{h(x) \; = \; x}$

<u>Products</u> of h with itself yield <u>powers</u> $(x^2, x^7, \ldots)$

<u>Products</u> of powers with constant functions yield <u>monomials</u> $(3x^2, \; -\pi x^7, \ldots)$

<u>Summations</u> of monomials and constant functions yield <u>polynomials</u> $(3x^2 - \pi x^7, \ldots)$

<u>Quotients</u> of polynomials yield <u>rational functions</u> $\left(\dfrac{3x^2 - \pi x^7}{2x + x^3}, \; \ldots \right)$

Going one step further, suppose f is a rational function and $g(x) = \sin x$. Then the composition of f with g , i. e.,

$$(f \circ g)(x) = f(g(x)) = f(\sin x)$$

is known as a rational function of sin x .

Example D.   Is $h(x) = (1 + \sin^2 x)/(2 - \sin x)$ a rational function of sin x ?

Solution.   Yes. It is the composition of

$$f(x) = (1 + x^2)/(2 - x) \qquad \text{with} \qquad g(x) = \sin x \qquad \qquad \square$$

This example illustrates the other use of operations on functions: the decomposition of complicated functions into simpler ones.   It turns out that

> many calculus operations can be carried out
>
> on a complicated function if it can be
>
> decomposed into simpler functions !!

This is the primary reason for our interest in the function operations of this section! For example, it is a trivial matter to compute "the derivative" (whatever that is!) of the function $h(x) = x$. But because any rational function can be decomposed into combinations of h , we are thus able to compute "the derivative" of any rational function!

The decomposition of a function into sums, differences, products or quotients of other functions is generally easy (when such a decomposition exists). Anton's Example 5 illustrates such a decomposition; here is another example:

<u>Example E.</u>   Decompose the following function into simpler functions:

$$f(x) = (3x^2 + \sin x)/\tan x$$

<u>Solution.</u> $f(x) = (3g(x) + h(x))/k(x)$

where $g(x) = x^2$, $h(x) = \sin x$ and $k(x) = \tan x$.   □

Decomposing complicated functions through <u>compositions</u> tends to give people more trouble, primarily because compositions in a given function equation are not always as readily apparent as are sums and products. Seeing them takes practice -- and some concentration. <u>But it is important</u>!! Keep in mind the   "f ∘ g   machine" picture given in Subsection 2 : <u>if your function performs an operation on an   x-value,   and then performs a second operation on the output of the first operation</u> ... well, you have two "function machines" in series, which is a <u>composition.</u>   Here are two examples which are similar to Anton's Exercises 10 through 15 :

<u>Example F.</u>   Decompose the following function as a composition:

$$f(x) = (x - 3)^3$$

<u>Solution.</u>   Given an   x-value,   our first operation is to subtract 3   from x. We then take this result and cube it: this is a second operation on the output of the first operation.   Thus let

$$h(x) = x - 3 \qquad \text{(the first operation)}$$
$$g(x) = x^3 \qquad \text{(the second operation)}$$

so that $f(x) = g(h(x))$,   or $f = g \circ h$.   □

<u>Example G.</u>   Decompose the following function as a composition: $f(x) = \sin(x^4 - x)$.

Solution.   The first operation is

$$h(x) = x^4 - x$$

which is followed by

$$g(x) = \sin x$$

Thus $f = g \circ h$.   □

Warning: The "1st operation" is the _inner_ function while the "2nd operation" is the _outer_ function.

This means that in the composition notation $f = g \circ h$ the two functions g and h are listed

in an _apparently_ backwards order, i. e. ,

$\quad\quad\quad$ h is the 1st operation , but it is the $\underline{2^{nd}}$ listed function ,

$\quad\quad\quad$ g is the $\underline{2^{nd}}$ operation, but it is the $\underline{1^{st}}$ listed function

The full notation $f(x) = (g \circ h)(x) = g(h(x))$ should help minimize confusion on this point.   The

h is the $\underline{2^{nd}}$ written function, but it is the 1st to get a crack at x .

Composition of three functions (or more) is also common in calculus.

$\quad$ Here's an example.

Example H.   Decompose the following function as a triple composition:

$$f(x) = \sin \sqrt{x^2 - 1}$$

Solution.   The first operation performed on x is

$$k(x) = x^2 - 1$$

This is followed by

$$h(x) = \sqrt{x}$$

which in turn is followed by

$$g(x) = \sin x$$

Thus $f(x) = g(h(k(x)))$, which we denote by

$$f = g \circ h \circ k$$ □

Section 2.3. Introduction to Calculus: Tangents and Velocity

1. The Tangent and Area Problems.   Your initial reaction to the Tangent and Area Problems could very well have been "If this is the central concern of calculus, then why is calculus so important?!" It is true that, although they are interesting geometry problems in and of them- selves, the Tangent and Area Problems would hardly merit the time, effort and general whoop-de-doo that we dedicate to calculus. So how do we explain Anton's first sentence: "Calculus centers around the following two fundamental problems..."?

Well... calculus really centers around two fundamental abstract mathematical operations: differentiation (defined in §3.1) and (definite) integration (defined in §5.7). These abstract operations have, via analytic geometry, the Tangent and the Area Problems as their geometric models, i.e.,

abstract operation	geometric model
differentiation	tangent lines to the graph of a function.
integration	area under the graph of a function.

The geometric formulations will enable us to "see" differentiation and integration in a very natural and concrete fashion, while the abstract operations are harder to get a feel for (abstract notions are always like that). Seen in this light, it is easy to understand why Anton directs your attention to the Tangent and Area Problems.

However, here is the crucial observation: the abstract formulations of differentiation and integration are what occur over and over again in applications: engineering and all the sciences, economics, statistics, etc. ... the list is almost endless. In his discussion Anton gives one such important application: velocity.

In this Companion section we will try to make clear the abstract operation (differentiation) which lies at the core of both the Tangent Problem and the development of velocity. We will do this as follows:

Section 2 summarizes Anton's basic solution to the Tangent Problem. The discussion is intended in part to clarify the solution, but of more importance to us, it is arranged for easy comparison against the subsequent discussion of velocity.

Section 3 reformulates Anton's discussion of velocity in such a way as to parallel completely the calculation of slopes of tangent lines given in Section 2. The similarity between the two procedures is made as clear and striking as possible.

Section 4 extracts from the previous two sections that abstract operation which lies at the heart of each: differentiation. The parallel developments given in Sections 2 and 3 ease the difficulty of this task.

2. Tangent lines. We will summarize Anton's solution to

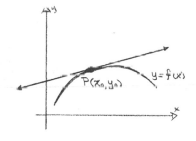

The Tangent Problem

Given a function f and a point $P(x_0, y_0)$ on its graph, find the equation of the tangent to the graph at P .

An intuitive definition for this tangent line is that line L through $P(x_0, y_0)$ which best approximates the graph of $y = f(x)$ near the point $P(x_0, y_0)$. Anton describes how this

line should be the "limit" of secant lines: as x_1 approaches x_0 , the secant line $\overleftrightarrow{PQ}$ will approach the tangent line L .

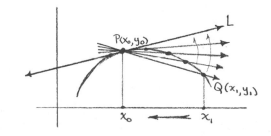

We already know one point on the tangent line: $P(x_0, y_0)$. Thus, to get an equation for the tangent line (via the point-slope formula) we need only determine its slope, m_{tan}. But this slope should be the limit of the <u>secant slopes</u> m_{sec}:

$$m_{tan} = \text{limiting value as } x_1 \text{ approaches } x_0 \text{ of } m_{sec}$$

or more concisely,

THE KEY
EQUATION

$$m_{tan} = \lim_{x_1 \to x_0} m_{sec} \qquad\qquad (A)$$

However, we are able to compute m_{sec} because we have two points on the secant line: $P(x_0, y_0)$ and $Q(x_1, y_1)$. Thus, since $y_0 = f(x_0)$ and $y_1 = f(x_1)$, we have

$$m_{sec} = \frac{y_1 - y_0}{x_1 - x_0} = \frac{f(x_1) - f(x_0)}{x_1 - x_0}$$

so that (A) becomes

$$m_{tan} = \lim_{x_1 \to x_0} \frac{f(x_1) - f(x_0)}{x_1 - x_0} \qquad\qquad (B)$$

<u>Equations (A) and (B) are variations on the same principle.</u> Equation (A) <u>is</u> the principle (slope of tangent = limit of slopes of secants), while Equation (B) is the useful computational formulation. These equations solve the Tangent Problem <u>in simple</u> situations, as the next example shows.

<u>Example A.</u> Find an equation for the tangent line to the graph of $f(x) = 1/x$ at the point $P(1/2, 2)$.

Solution. By (B) the slope of the tangent line is

$$m_{tan} = \lim_{x_1 \to x_0} \frac{f(x_1) - f(x_0)}{x_1 - x_0}$$

$$= \lim_{x_1 \to 1/2} \frac{f(x_1) - f(1/2)}{x_1 - (1/2)} \qquad \text{since } x_0 = 1/2$$

$$= \lim_{x_1 \to 1/2} \frac{\left(\dfrac{1}{x_1}\right) - (2)}{x_1 - (1/2)} \qquad \text{since } f(x) = \frac{1}{x}$$

$$\text{and } \frac{1}{1/2} = 2$$

$$= \lim_{x_1 \to 1/2} \frac{\dfrac{1 - 2x_1}{x_1}}{\dfrac{2x_1 - 1}{2}} \qquad \text{finding common denominators}$$

$$= \lim_{x_1 \to 1/2} \frac{2(1 - 2x_1)}{x_1(2x_1 - 1)} \qquad \text{inverting denominator}$$

$$\text{and multiplying}$$

$$= \lim_{x_1 \to 1/2} \left(\frac{2}{x_1}\right)(-1)$$

$$= -\frac{2}{1/2} = \boxed{-4}$$

Here we use the intuitive fact that the limiting value of $-\dfrac{2}{x_1}$ as x_1 approaches $1/2$ must be $-\dfrac{2}{1/2} = -4$. That's pretty easy to accept!

Thus $m_{tan} = -4$, and a point on the tangent line is $P(1/2, 2)$. Thus an equation for our tangent line (using the point-slope form of a line from Theorem 1.5.2) is

$$y - 2 = (-4)(x - 1/2), \quad \text{or} \quad \boxed{y = -4x + 4} \qquad \square$$

Warning. Do not be deceived into believing that we have mastered the Tangent Problem! Solving Equations (A) or (B) for anything but the simplest of functions requires techniques to be developed in Chapter 3! If you don't believe us, just try $f(x) = \sin x$ at $x_0 = \pi/3$, or $f(x) = (x^{1/2} - 1/x)/(x^4 + 1)$ at $x_0 = 2$! (Really do try it --- you'll see the problems soon enough! And then you'll appreciate the techniques of Chapter 3 when you finally get there.)

3. Velocity in straight line motion. We can now give a development of (instantaneous) velocity which completely parallels the §2 method for finding slopes for tangent lines.

We are considering the motion of an object, say a car, along a straight line. For convenience we label this straight line as

the s-axis. Then at any time t the car is located at some point s on the s-axis -- thus s is a function of t, which we will write as $\boxed{s = f(t).}$

What is the instantaneous velocity v_{inst} of our car at time $t = t_0$? Anton defines this to be the limit of average velocities v_{av} over time intervals $[t_0, t_1]$:

$$v_{inst} = \text{limiting value as } t_1 \text{ approaches } t_0 \text{ of } v_{av},$$

or more concisely

THE KEY EQUATION

$$\boxed{v_{inst} = \lim_{t_1 \to t_0} v_{av}} \qquad (C)$$

(Compare this with Equation (A)!) Intuition suggests that this is a sensible definition: instantaneous velocity at t_0 is the limit of average velocities over a shortening time interval $[t_0, t_1]$. However, we are able to compute the average velocity v_{av} over the time interval $[t_0, t_1]$ by using the familiar formula $r = d/t$ (rate equals distance divided by time):

$$v_{av} = \frac{f(t_1) - f(t_0)}{t_1 - t_0}$$

Here $t_1 - t_0$ is the time elapsed, while $f(t_1) - f(t_0)$ is the change in position during the given time interval. Equation (C) now becomes

$$v_{inst} = \lim_{t_1 \to t_0} \frac{f(t_1) - f(t_0)}{t_1 - t_0} \qquad \text{(D)}$$

(Compare this with Equation (B)!)

Equations (C) and (D) are all variations on the same definition.   Equation (C) is the definition (instantaneous velocity = limt of average velocities), while Equation (D) is the useful computational formulation.

   The parallel development of compuation of tangent line slopes and velocities should be readily apparent . . . .

Example B.   A rocket is propelled straight up so that in   t sec   it reaches a height of $2t^2 + t$ ft. Determine the instantaneous velocity of the rocket at time $t = t_0$.

Solution.

$$v_{inst} = \lim_{t_1 \to t_0} \frac{f(t_1) - f(t_0)}{t_1 - t_0} \qquad \text{by (D)}$$

$$= \lim_{t_1 \to t_0} \frac{[2t_1^2 + t_1] - [2t_0^2 + t_0]}{t_1 - t_0} \qquad \text{since } f(t) = 2t^2 + t$$

$$= \lim_{t_1 \to t_0} \frac{2\left(t_1^2 - t_0^2\right) + (t_1 - t_0)}{t_1 - t_0} \qquad \text{regrouping terms}$$

$$= \lim_{t_1 \to t_0} \frac{2(t_1 + t_0)(t_1 - t_0) + (t_1 - t_0)}{t_1 - t_0} \qquad \text{since } t_1^2 - t_0^2 \text{ is a difference}$$

of squares (see Equation D1 in Appendix D §6)

$$= \lim_{t_1 \to t_0} 2(t_1 + t_0) + 1 \qquad \text{by cancelling the term } t_1 - t_0$$

from the numerator and denominator

$$= 2(t_0 + t_0) + 1 = \boxed{4t_0 + 1 \ \text{ft/sec}} \qquad \qquad \square$$

Example C. Determine, for the rocket in Example B :

(a)   the average velocity of the rocket during the first 78 feet of its flight,

(b)   the instantaneous velocity of the rocket after rising 210 feet.

Solution. (a) To find the average velocity during the first 78 feet of the flight we must determine the time interval $[t_0, t_1]$. Using $s = f(t) = 2t^2 + t$ we find

$$0 = f(t_0) = 2t_0^2 + t_0 \qquad \text{and} \qquad 78 = f(t_1) = 2t_1^2 + t_1$$

The first equation factors to $t_0(2t_0 + 1) = 0$, which gives $\boxed{t_0 = 0}$ since $t_0 = -1/2$ makes no sense in our problem. The second equation factors to

$$(2t_1 + 13)(t_1 - 6) = 0$$

(see Appendix D if this factoring troubles you) which yields $\boxed{t_1 = 6}$ since $t_1 = -13/2$ makes no sense in our problem.   Thus

$$v_{av} = \frac{f(t_1) - f(t_0)}{t_1 - t_0} = \frac{78 - 0}{6 - 0} = \boxed{13 \text{ ft/sec}}$$

    (b)   The instantaneous velocity at time t_0 was computed in Example B to be $4t_0 + 1$ ft/sec. We must determine t_0 when the rocket has risen 210 feet, i. e. ,

$$210 = f(t_0) = 2t_0^2 + t_0$$

$$(2t_0 + 21)(t_0 - 10) = 0$$

or $t_0 = 10$ sec . Thus $v_{inst} = \lfloor 4(10) + 1\rfloor$ ft/sec $= \boxed{41 \text{ ft/sec.}}$ □

4.   The abstract operation: differentiation.   Suppose we are considering a straight line motion problem governed by the equation $s = f(t)$. If we set up a position verses time graph as shown to the right, then comparing

Equation (B) with Equation (D)

shows that

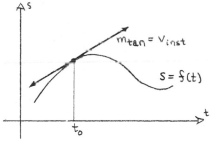

the instantaneous velocity v_{inst} at $t = t_0$
equals the slope m_{tan} of the tangent line
to the graph of $s = f(t)$ at $t = t_0$.

This is merely a rewording of Anton's "Geometric Interpretation of Instantaneous Velocity. " Thus velocity problems have been converted to geometry problems.

However, the most basic observation is the following: calculating slopes of tangent lines and instantaneous velocities are both applications of the same abstract operation: <u>differentiation</u>. Given a function $y = f(x)$, the <u>derivative of f at $x = x_0$</u>, denoted by $f'(x_0)$, is defined to be

$$f'(x_0) = \lim_{x_1 \to x_0} \frac{f(x_1) - f(x_0)}{x_1 - x_0}$$

Ah ha: from Equation (B) we have $m_{tan} = f'(x_0)$ and from Equation (D) we have $v_{inst} = f'(t_0)$! Thus: <u>our central concern is really the abstract operation of differentiation!</u>

In order to work with derivatives we must first obtain a better understanding of the <u>limit</u> operation. So far we have dealt with limits only in an intuitive way; this is OK for simple examples, but not for the more complicated situations which lie ahead. The rest of Chapter 2 will therefore be concerned with limits. Differentiation will be taken up in Chapter 3... .

<u>Section 2.4.</u>   <u>Limits (An Intuitive Introduction).</u>

You cannot possibly understand differentiation (or, later on, integration) without a solid intuitive understanding of limits. Once you catch on to the general principles, limits are very easy; nonetheless, many students seem to miss the general principles, so read what follows very carefully.

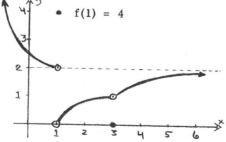

<u>Example A.</u>   Consider the function f graphed to the right, and compute

$$\lim_{x \to 1^-} f(x)$$

i.e., the limit of $f(x)$ as x approaches 1 from the left.

Solution.   We'll use an insect friend to help visualize our limits: Howard Ant. Consider an

x   value slightly to the left of   x = 1 ,

say,   x = 0 , and (in your mind) place

Howard Ant on the graph over   x = 0 , as

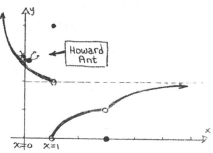

shown. Now slowly move your   x-value   from

x = 0   to   x = 1   and see where Howard goes:

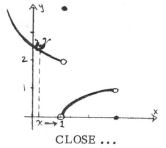

CLOSE ...

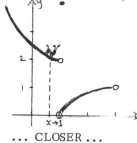

... CLOSER ...

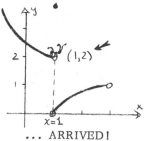

... ARRIVED !

As you can see, as   x   moves from   x = 0   to   x = 1,   Howard crawls to the point   (1 , 2),

i. e. , to a   y-value   of   2. Notice that   (1, 2)   is not itself a point on the graph; that's OK,

for nonetheless it is the point to which the graph of   y = f(x)   for   x < 1   leads. Thus, as

x   moves from   x = 0   to   x = 1,   our function values   f(x)   converge toward the value

y = 2 , so that

$$\lim_{x \to 1^-} f(x) = 2$$ □

There is a major principle which is illustrated by Example A :

Limits from the left	$\lim_{x \to x_0^-} f(x)$ does not depend on the value of $f(x)$ at $x = x_0$	(A)

For, as you can see, Howard Ant did not use the point $(1, f(1)) = (1, 4)$ to arrive at his final destination of $(1, 2)$.

<u>Example B.</u> Consider the same function f as

in Example A , and compute

$$\lim_{x \to 1^{+}} f(x)$$

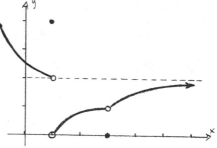

<u>Solution.</u>   This time we wish to compute a limit from the right at $x = 1$. Thus (in your mind)

place <u>Howard Ant on the graph</u> over $x = 2$,

and slowly move the x-values from $x = 2$

to $x = 1$. Clearly Howard crawls to the point

$(1, 0)$ (which is not on the graph ... but who

cares?!), and so our function values $f(x)$ converge

toward $y = 0$, i.e.,

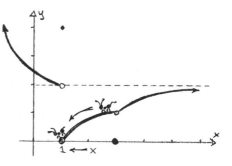

$$\lim_{x \to 1^{+}} f(x) = 0 \qquad\qquad \square$$

This illustrates the limit-from-the-right analogue of (A):

| Limits
from the
right | $\lim_{x \to x_0^{+}} f(x)$ does <u>not</u> depend on the value

of $f(x)$ at $x = x_0$ | (B) |

Examples A and B prove that, for the function f considered, the (two-sided) limit

of $f(x)$ as x approaches 1 <u>does not exist</u> since the two one-sided limits have different

values, i.e.,

$$\lim_{x \to 1^{-}} f(x) = 2 \neq 0 = \lim_{x \to 1^{+}} f(x)$$

The general principle illustrated here is

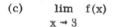

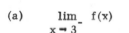

$$\lim_{x \to x_0} f(x) \text{ exists and equals } L \text{ if and only if}$$

i. $\lim_{x \to x_0^-} f(x)$ exists and equals L, <u>and</u>

ii. $\lim_{x \to x_0^+} f(x)$ exists and equals L

(C)

Said in words, <u>the   (2-sided)   limit exists if and only if both one-sided limits exist and are</u>

<u>equal.</u>

<u>Example C.</u>   Consider the same function f as

in Example A and compute

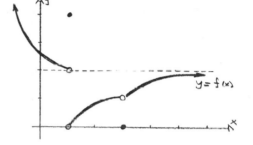

(a) $\lim_{x \to 3^-} f(x)$ (b) $\lim_{x \to 3^+} f(x)$

(c) $\lim_{x \to 3} f(x)$ (d) $f(3)$.

<u>Solution.</u>   Placing <u>Howard Ant on the graph</u> both to the left and to the right of $x = 3$ and

moving him toward 3 we see

(a) $\lim_{x \to 3^-} f(x) = 1$,   and   (b) $\lim_{x \to 3^+} f(x) = 1$

Since both one-sided limits equal 1 , principle (C) gives

(c) $\lim_{x \to 3} f(x) = 1$

Notice, however, that (d) $f(3) = 0 \neq 1 = \lim_{x \to 3} f(x)$.   □

Example C illustrates one last important principle, the analogue for (2 - sided)

limits of (A) and (B):

$$\boxed{\begin{array}{l} \lim_{x \to x_0} f(x) \quad \text{does } \underline{\text{not}} \text{ depend on the} \\ \qquad\qquad \text{value of } \ f(x) \ \text{ at } \ x = x_0 \end{array}} \qquad \text{(D)}$$

Hammer this fact into your mind: $f(x_0)$ is irrelevant to the computation of $\lim_{x \to x_0} f(x)$. In fact, many times $f(x)$ will not be defined at all for $x = x_0$!

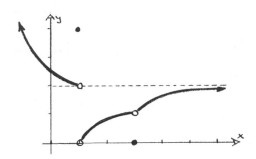

Example D.   Consider the same function f

as in Example A and compute

(a) $\lim_{x \to -\infty} f(x)$, and (b) $\lim_{x \to +\infty} f(x)$

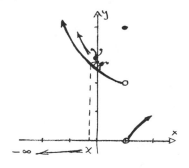

Solution.   Place Howard Ant on the graph

near $x = 0$ and see where he goes as

x runs off toward $-\infty$. Clearly his

y-value gets arbitrarily large, and thus

(a) $\lim_{x \to -\infty} f(x) = +\infty$

On the other hand, when headed out in the direction of $+\infty$, Howard Ant's y coordinate seems to approach the value $y = 2$. Then

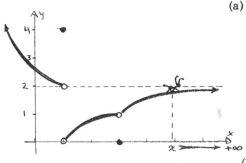

(b) $\lim_{x \to \infty} f(x) = 2$ $\qquad \square$

Section 2.5.   Limits (Computational Techniques)

The basic theme of this section is easy to state: limits for many complicated functions can be computed by decomposing the function into simpler ones, computing the limits of the simpler functions, and then recombining the limits to obtain the desired answer. Anton states the limit rules which make this decomposition/reassembly procedure possible, and then uses the rules on numerous complicated functions. In particular, Anton shows how to compute limits of any polynomial or rational function.

1.   Two basic limits.   The two most elementary (but important) limits that we need are computed in Anton's Examples 1 and 2:

Two
Basic
Limits

$$\lim_{x \to a} k = k$$

$$\lim_{x \to a} x = a$$

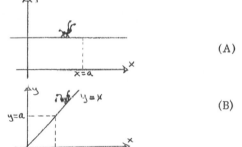

(A)

(B)

The "Howard Ant on the graph" technique of the previous section should convince you of the truth of these limits. Amazingly enough, many mistakes are made with the first limit!

Example A.   Compute $\lim_{x \to 3} 6$.

Solution. $\lim_{x \to 3} 6 = 6$. What more can be said? However, the number of times that answers such as

"$\lim_{x \to 3} 6 = 3$"   or   "$\lim_{x \to 3} 6 = 0$"

          FALSE   FALSE

are given is astounding. Don't make such blunders!   □

Using the limit rules which we're about to discuss, the two Basic Limits will allow us to compute limits for all polynomial and rational functions.

2. **The basic limit rules.** Recall the algebraic operations on two functions   f   and   g   which were defined in §2.2:

> sums   f + g
>
> products   f · g
>
> multiplication by a constant   k f
>
> composition   f ∘ g

> differences   f - g
>
> quotients   f/g

Theorem 2.5.1 lists the limit rules for the first four of these operations.   They are all simple, unsurprising, and easily memorized (especially in the "word versions" which Anton gives following the Theorem itself).   It would be useful if the theorem contained the limit rule for multiplication by a constant.   We'll label this as   Theorem 2.5.1 (f), and we advise you to add it to your text:

THEOREM 2.5.1 (f). $\lim k f(x) = k \lim f(x)$

> The limit of a multiplication by a constant is the
>
> multiplication of the limit by the constant.

This is Anton's Equation (4).

Limit rules for compositions are more difficult to state.   The best rule of this sort (Theorem 3.7.6) requires the concept of continuity, which is not introduced until §3.7.

Warning: All six of the rules just discussed assume the existence of <u>finite limits</u> for the functions   f   and   g   in order for the stated conclusions to follow, e.g.,

$$\lim \left[f(x) \, g(x) \right] = \lim f(x) \cdot \lim g(x)$$

when the two limits $\lim f(x)$ and $\lim g(x)$

are finite numbers.

Here is a very typical incorrect solution given for a problem, the error being ignoring our

finite limit warning:

Example B.   Compute $\displaystyle \lim_{x \to 1} \frac{x^2 - 1}{x - 1}$

Incorrect Solution.

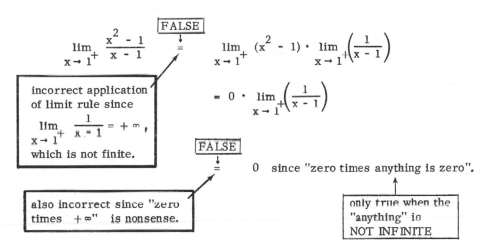

Correct Solution.   (Using the "factor and cancel" technique. See Example D below.)

$$\lim_{x \to 1^+} \frac{x^2 - 1}{x - 1} = \lim_{x \to 1^+} \frac{(x + 1)(x - 1)}{(x - 1)} = \lim_{x \to 1^+} (x + 1) = 2 \qquad \square$$

For another example like this, see the Warning following Example D.

2.5.4

3. Limits of polynomials. From §2.2.3 of The Companion we know that all polynomials can be constructed out of $f(x) = k$ and $g(x) = x$ by function addition, subtraction and multiplication. Since our Basic Limits (A) and (B) give us the limits for $f(x) = k$ and $g(x) = x$, our basic limit rules (Theorem 2.5.1) allow us to compute limits for any polynomial! Anton does so in Example 6, and thus proves the following important result:

$$\boxed{\begin{array}{c} \text{If} \quad p(x) \quad \text{is any polynomial, then} \\[2mm] \lim_{x \to a} p(x) = p(a) \qquad (C) \\[2mm] \text{when} \quad \underline{a} \quad \text{is any finite real number.} \end{array}}$$

Example C. Compute $\lim_{x \to \sqrt{2}} (x^2 - 4x + 3)$.

Solution. $\lim_{x \to \sqrt{2}} (x^2 - 4x + 3) \underset{(C)}{=} (\sqrt{2})^2 - 4\sqrt{2} + 3 = 5 - 4\sqrt{2}$ $\qquad \square$

We will learn later (Theorem 3.7.2) that result (C) proves any polynomial to be a "continuous function."

4. Limits of rational functions. All rational functions are quotients of polynomials $p(x)/q(x)$. The techniques for handling limits at $x = a$ for such functions depend on whether $p(x)$ or $q(x)$ are zero at $x = a$.

Case 1: If the denominator is non-zero at $x = a$, i.e., $q(a) \neq 0$, then

$$\lim_{x \to a} \frac{p(x)}{q(x)} = \frac{\lim_{x \to a} p(x)}{\lim_{x \to a} q(x)} \underset{(C)}{=} \frac{p(a)}{q(a)}$$

$$\boxed{\text{Theorem 2.5.1d}}$$

Thus:

If   p(x)/q(x)   is any rational function,

and $q(a) \neq 0$, then   (D)

$$\lim_{x \to a} \frac{p(x)}{q(x)} = \frac{p(a)}{q(a)}$$

Anton's Example 7 is an illustration of (D). The difficulty with limits of rational functions

occurs when the denominator is zero at the limit point, i.e.,   q(a) = 0 . This situation occurs

quite often in calculus and needs to be carefully studied. The next two cases discuss this

situation.

Case 2:   If the denominator and numerator are both zero at   x = a , we then have a rational function

limit of indeterminant form   0/0 . The correct elementary method of handling such limits is to

Factor
and
Cancel

factor both numerator and denominator

by   x - a , where   a   is the limit point, and

cancel these factors out of the function.

Anton's Example 8 illustrates the factor and cancel technique. Here is another illustration.

Example D.   Find $\displaystyle\lim_{x \to -1} \frac{x^2 + 4x + 3}{x^2 - 1}$

Solution.   Both the numerator and denominator are zero at   a = - 1 , and thus we have a

limit of indeterminant form   0/0 . We thus factor and cancel:

$$\lim_{x \to -1} \frac{x^2 + 4x + 3}{x^2 - 1} = \lim_{x \to -1} \frac{(x + 3)(x + 1)}{(x - 1)(x + 1)} \qquad \boxed{\text{FACTOR}}$$

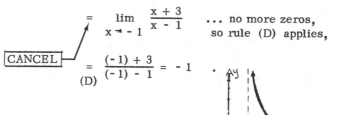

$$= \lim_{x \to -1} \frac{x + 3}{x - 1} \qquad \text{... no more zeros,}$$

so rule (D) applies,

CANCEL

$$\underset{(D)}{=} \frac{(-1) + 3}{(-1) - 1} = -1 \cdot$$

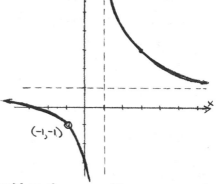

It is instructive to see the graph of

$$f(x) = \frac{x^2 + 4x + 3}{x^2 - 1}$$

As you can see, the point $(-1, -1)$ fits

naturally into the graph (although it is <u>not</u>

actually part of it). "Howard Ant on the graph" would surely agree with our answer of

$$\lim_{x \to -1} f(x) = -1$$

(Note: the techniques used to obtain this graph - calculus, of course! - will be given in §4.5.)

<u>Warning</u>: Here is an all-too-frequent incorrect solution to Example D.

$$\lim_{x \to -1} \frac{x^2 + 4x + 3}{x^2 - 1} = \frac{\lim_{x \to -1} (x^2 + 4x + 3)}{\lim_{x \to -1} (x^2 - 1)} \overset{?}{=} \frac{0}{0} \overset{?!?}{=} 0$$

FALSE application of
limit rule since
denominator is zero

GARBAGE, since
you cannot
divide by zero

NO! NO! NO!
COMPLETE
FANTASY!!

<u>Case 3</u>:   If the denominator is zero at $x = a$, but the numerator is not zero, then we have a

rational function limit of <u>indeterminant form   c/0</u>. Such limits will always be undefined, but

<u>might</u> tend to $+\infty$ or $-\infty$. To determine the exact behavior of the limit we must

$$\boxed{\text{compute both one-sided limits at } x = a}$$

Anton's Example 9 illustrates this procedure; here is another.

Example E.   Find $\displaystyle \lim_{x \to 3} \frac{x + 5}{(x - 3)^2}$.

Solution.   Although the denominator is zero at $a = 3$, the numerator is not, so we have a

limit of indeterminant form $c/0$. Thus we compute the one-sided limits at $x = 3$. When

x approaches $a = 3$ from the right, the numerator is a positive quantity approaching 8 ,

while the denominator is a positive quantity approaching 0 . Thus the ratio is a positive

quantity that increases without bound.   Exactly the same reasoning applies to the case where

x approaches $a = 3$ from the left.   Thus

$$\lim_{x \to 3^+} \frac{x + 5}{(x - 3)^2} = \lim_{x \to 3^-} \frac{x + 5}{(x - 3)^2} = +\infty$$

Hence the (two-sided) limit also tends toward positive infinity, i.e.,

$$\lim_{x \to 3} \frac{x + 5}{(x - 3)^2} = +\infty \qquad\qquad \square$$

5.   Limits at $\pm\infty$ for rational functions.   The basic limits needed for these computations are

$$\lim_{x \to -\infty} \frac{1}{x} = \lim_{x \to +\infty} \frac{1}{x} = 0 \qquad \text{(Anton's Example 4)} \qquad\qquad \text{(E)}$$

These limits should be obvious from the graph

of $y = 1/x$. From the product rule we can

deduce from (E) that, for any positive integer n ,

$$\boxed{\lim_{x \to -\infty} \frac{1}{x^n} = \lim_{x \to +\infty} \frac{1}{x^n} = 0} \qquad\qquad \text{(F)}$$

These, along with our limit rules, are enough to compute the limits as x approaches $+\infty$

or $-\infty$ for any rational function. The method which Anton gives is the following:

Limits at $+\infty \atop -$ of rational functions	divide both numerator and denominator by the highest power of x that occurs in the function. Then use the limit rules along with (F).

Example F.   Find $\lim\limits_{x \to \infty} \dfrac{3x^3 - 2x - 1}{-2x^3 + 1}$.

Solution.   Divide both numerator and denominator by x^3 :

$$\lim_{x \to \infty} \frac{3x^3 - 2x - 1}{-2x^3 + 1} = \lim_{x \to \infty} \frac{3 - 2/x^2 - 1/x^3}{-2 + 1/x^3}$$

$$= \frac{\lim\limits_{x \to \infty} (3 - 2/x^2 - 1/x^3)}{\lim\limits_{x \to \infty} (-2 + 1/x^3)}$$

$$\underset{(F)}{=} \frac{3}{-2} = -3/2 \qquad \square$$

A little bit of thought, along with studying our Example F and Anton's Examples 10, 11

and 13 , should convince you of the following:

cases for p(x)/q(x)	values for $\lim\limits_{x \to +\infty}$ or $\lim\limits_{x \to -\infty}$
degree p(x) < degree q(x)	0
degree p(x) = degree q(x)	a finite, non-zero number
degree p(x) > degree q(x)	$+\infty$ or $-\infty$

6.   <u>Limits of more general functions.</u>   Limits of functions which contain   n-th   roots are generally

handled easily by Theorem 2.5.1 (e) : the limit of an   n-th   root is the   n-th   root of the limit.

See, for example, Anton's Example 12. Difficulties arise, however, when "indeterminant forms"

are encountered:

<u>Example G.</u>   Find $\lim\limits_{x \to 2} \dfrac{\sqrt{x} - \sqrt{2}}{x - 2}$.

<u>Solution.</u>   Both the numerator and denominator assume the value zero at $x = 2$. The trick for

solving this is to multiply both numerator and denominator by $\sqrt{x} + \sqrt{2}$:

$$\lim_{x \to 2} \frac{\sqrt{x} - \sqrt{2}}{x - 2} = \lim_{x \to 2} \frac{(\sqrt{x} - \sqrt{2})(\sqrt{x} + \sqrt{2})}{(x - 2)(\sqrt{x} + \sqrt{2})}$$

$$= \lim_{x \to 2} \frac{x - 2}{(x - 2)(\sqrt{x} + \sqrt{2})} - \lim_{x \to 2} \frac{1}{\sqrt{x} + \sqrt{2}}$$

$$\boxed{\begin{array}{l}\text{legal since} \\ \text{denominator} \\ \text{is NOT zero}\end{array}} \nearrow \quad \frac{\lim\limits_{x \to 2} 1}{\lim\limits_{x \to 2} (\sqrt{x} + \sqrt{2})} = \frac{1}{2\sqrt{2}} - \sqrt{2}/4 \qquad \square$$

Note the purpose of our trick: to get a factor of $x - 2$ in the numerator to cancel the trouble-

some $x - 2$ in the denominator.

       Indeterminant forms such as this arise all the time in establishing "derivative" formulas,

and generally some sort of algebraic "trick" is necessary each time. We wish we could give

you one slick method which always would work --- but that can't be done until "L' Hôpital's Rule"

in Chapter 10. And besides, "L' Hôpital's Rule <u>requires</u> differentiation -- which puts the

cart-before-the-horse at this stage!

2.5.10

One last type of function limit should be mentioned: limits for functions defined by different formulas over separate intervals (so-called "bracket functions"). Here we use Principle (C) from §2.4 of the Companion.

Example H. Find $\lim_{x \to 1} f(x)$ for

$$f(x) = \begin{cases} \dfrac{x^2 - 3x + 2}{x - 1} & x < 1 \\[4mm] \sqrt{x^2 + 3} & x \geq 1 \end{cases}$$

Solution. You must calculate both of the one-sided limits when the definition of $f(x)$ "splits" at the desired limit point, in this case at $x = 1$.

i. $\lim_{x \to 1^+} f(x) = \lim_{x \to 1^+} \sqrt{x^2 + 3}$ since $f(x) = \sqrt{x^2 + 3}$ when $x > 1$

$$= \sqrt{\lim_{x \to 1^+} (x^2 + 3)} = \sqrt{4} = 2$$

ii. $\lim_{x \to 1^-} f(x) = \lim_{x \to 1^-} \dfrac{x^2 - 3x + 2}{x - 1}$ since $f(x) = \dfrac{x^2 - 3x + 2}{x - 1}$ when $x < 1$

$$= \lim_{x \to 1^-} \dfrac{(x - 2)(x - 1)}{x - 1} = \lim_{x \to 1^-} (x - 2) = -1$$

Since $2 \neq -1$, then the two-sided limit $\lim_{x \to 1} f(x)$ does not exist. □

There are many other types of function limits which we have yet to consider (for example, limits of trigonometric functions). These will be considered at appropriate places in the chapters to follow.

<u>Section 2.6</u>:   <u>Limits: A Rigorous Approach (Optional)</u>

1.   <u>Why read this section</u>? Calculus is built upon the rigorous formulation of <u>limit</u>. It is the

single most important concept in all of calculus, and is the cornerstone for most of applied

mathematics.

Thus, a reasonable question to ask is, "Why is this only an optional section?!"

For a number of reasons. First, the rigorous formulation of limits is a very difficult

and sophisticated concept that often takes a long time to master. Moreover, in elementary

calculus many of the more difficult proofs are omitted in order to make the subject less

forbidding; instead, we appeal to intuition at certain key places in the development. It's at

precisely these points that the rigorous limit formulation would be needed. Thus you can get

through elementary calculus without reading this section and you will probably not notice its

absence (well, ... almost. It depends on how deeply you think about it).

Then what of value is lost in skipping "rigorous" limits? Well, if you go on past the

elementary calculus sequence you will run across this concept with increasing frequency –

there is no free lunch when it comes to dealing with limits. You have to confront the rigorous

concept sooner or later. But even if you are not intending to go beyond elementary calculus,

isn't something lost if you study the subject for months and months and yet avoid confronting

its most basic concept in a complete and honest fashion? Yes, it's hard, but the rigorous

formulation for limits is a truly powerful and beautiful invention, one of the great ideas in the

development of mathematics. For this reason alone it is worth your attention.

2.   <u>Understand the definition</u>.

Suppose   f   is a function which is defined for all   x-values   in an open interval

containing $x = a$, with the possible exception that   f   might not be defined

at $x = a$. Then

Definition of limit

$$\lim_{x \to a} f(x) = L \quad \text{if and only if}$$

given <u>any</u> $\epsilon > 0$ there exists a $\delta > 0$ such that

$$|f(x) - L| < \epsilon \quad \text{whenever} \quad 0 < |x - a| < \delta$$

That is quite a mouthful to swallow! Anton spends quite a lot of time motivating the definition; we won't repeat that material, but we will highlight a number of important aspects of the definition.

i.   Given <u>any</u> $\epsilon > 0$ you must be able to find a number $\delta > 0$ for which the stated condition is valid. The crucial word here is "<u>any</u>". The ϵ measures how close $f(x)$ is to its supposed limit L , and the idea is that $f(x)$ should be getting <u>arbitrarily close</u> to L . Thus, given <u>any</u> $\epsilon > 0$, we should be able to make the distance $|f(x) - L|$ smaller than ϵ .

To drive this last point home, suppose you wish to prove to a skeptical friend that you can walk over to a classroom door (some people won't believe anything now-a-days). Let your initial distance to the door be L , and let $f(x)$ equal the distance walked to the door after x

seconds. If you can convince your friend that you can come within $\epsilon = 1$ foot of the door (i. e., $|f(x) - L| < 1$ foot), then must he accept that you will reach the door? Of course not; you could still be almost a foot away! What if you can come within $\epsilon = 1/16$ inch (i. e, $|f(x) - L| < 1/16$ inch)? Nope, still not good enough; you could still be almost 1/16 inches away! You must show that you can come within ϵ of the door <u>for any given</u> $\epsilon > 0$. Your doubting friend would then have to believe you ... unless he is devoid of usual common sense!

ii.   Given an $\epsilon > 0$ ("I wish $f(x)$ to be within $\epsilon = .03$ of L "), the $\delta > 0$ is chosen <u>in response to</u> ϵ ("OK. If x is within $\delta = .0013$ of <u>a</u> , then $f(x)$ is within

ϵ = .03 of L"). Every different $\epsilon > 0$ has its own $\delta > 0$. Or, more precisely, every

different $\epsilon > 0$ has its own <u>collection</u> of associated δ's :

> if a particular δ works for a given ϵ , then
>
> every δ' which is less than δ also works for ϵ

This is really a simple observation. For example, if

$f(x)$ is within $\epsilon = 1$ of L whenever x is within $\delta = 1/2$ of a ,

then it is surely true that

$f(x)$ is within $\epsilon = 1$ of L whenever x is within $\delta = 1/4$ of a ,

since if x is within 1/4 of <u>a</u> , it is certainly within 1/2 of <u>a</u>! Thus, given an $\epsilon > 0$,

there will be an infinite number of associated δ's ; however, <u>we only need to find one of them!</u>

     iii. Notice that the statement $\lim\limits_{x \to a} f(x) = L$ makes no claim about the value $f(a)$

(a number which might not even exist!). The restriction of $0 < |x - a|$ in our definition assures

us that we are looking at the behavior of $f(x)$ for x <u>near</u> a , but <u>not equal</u> to a .

(Remember: $0 < |x - a|$ is merely a convenient way to write $x \ne a$.) This is a vitally

important restriction, for in many of the important applications of limits (e. g. , in the definition

of the derivative) we need to consider limits of quantities which are not defined at the point at

which the limit is being considered.

3. <u>Computations with specific ϵ's</u>. To prove that a function f has a certain limit at x = a we

must deal with all $\epsilon > 0$ at once (after all, we can't deal with each possible ϵ value

individually; there are an infinite number of them!). However, in trying to comprehend the limit definition and master its use, it is instructive to start with <u>specific</u> values for ε. Thus, if $\lim\limits_{x \to a} f(x) = L$, we will consider the problem of

given a specific number $\varepsilon_0 > 0$, find a $\delta > 0$ such that

$$|f(x) - L| < \varepsilon_0 \quad \text{whenever} \quad |x - a| < \delta$$

There are three important general remarks to make before giving a specific example.

i. We can use any sneaky, underhanded, disreputable and/or vaguely immoral method of obtaining our δ ... just so long as it works! Being able to find δ is the art of limit problems ... but the real <u>proof</u> comes next, when you verify that indeed your δ does work.

ii. (The art: finding δ.) We want our first inequality (the ε inequality) to be <u>implied by</u> the second inequality (the δ inequality). This is a <u>reverse implication.</u> Hence, when starting with the ε inequality, you must use nothing but reverse implications in reaching the δ inequality. In other words, between each of the steps in your derivation you should be able to write "this inequality will be true whenever the following inequality is true," or "this inequality is implied by the following one." (If certain steps are <u>equivalences,</u> i.e., forward as well as reverse implications, that's fine. But it's not a necessity.)

iii. (The proof: verifying δ.) Once you have found δ, your formal proof write-up begins: show that the δ-inequality does in fact imply the ε-inequality. These steps are done in the usual way, by (ordinary) forward implications.

Example A. Consider the limit $\lim\limits_{x \to 4} \sqrt{x} = 2$.

Given $\epsilon = 1/4$, find an associated $\delta > 0$.

Solution. We must find $\delta > 0$ such that

$$\left| \sqrt{x} - 2 \right| < 1/4 \quad \text{whenever} \quad 0 < \left| x - 4 \right| < \delta \tag{A}$$

We will proceed from the first inequality to the second by the use of reverse implications

(Remark ii above), and will employ any time-saving tricks that we can find (Remark i).

We wish to go from $\left| \sqrt{x} - 2 \right|$ to $\left| x - 4 \right|$. Thus multiplication by $\left| \sqrt{x} + 2 \right| = \sqrt{x} + 2$

(remember: $\sqrt{x} \geq 0$ for all x)   seems a logical starting place:

$$\left.\begin{array}{c} \left| \sqrt{x} - 2 \right| < 1/4 \\[2mm] \left| \sqrt{x} - 2 \right| \cdot \left| \sqrt{x} + 2 \right| < (1/4)(\sqrt{x} + 2) \\[2mm] \left| x - 4 \right| < (\sqrt{x} + 2)/4 \end{array}\right\} \quad \begin{array}{l}\text{equivalent}\\\text{statements}\end{array} \tag{B}$$

In comparing (B) with (A) you might be tempted to say "Ahh ... let $\delta = (\sqrt{x} + 2)/4$!"

WRONG! We cannot let δ depend on x , because the allowable x-values are supposed to

depend on δ ! We have to be more clever with inequality (B). Remember, all we need is for

inequality (B),

$$\left| x - 4 \right| < (\sqrt{x} + 2)/4$$

to be implied by an inequality of the form

$$0 < \left| x - 4 \right| < \delta$$

This implication will certainly be true if $\delta > 0$ is a number chosen so that

$$\delta \leq (\sqrt{x} + 2)/4 \quad \text{for all} \quad x \geq 0$$

But wait a minute ... such a δ is easily found (do you see it?)! The quantity $(\sqrt{x} + 2)/4$ is smallest when $x = 0$, at which point it equals $1/2$, i.e., $(\sqrt{x} + 2)/4 \geq 1/2$. (Convince yourself of this fact!) So pick $\delta = 1/2$. Then

$$|x - 4| < \delta = 1/2$$

should imply

$$|x - 4| < (\sqrt{x} + 2)/4$$

which in turn is equivalent to

$$|\sqrt{x} - 2| < 1/4$$

as desired.

Our formal proof is simply to verify the claims in our last sentence. Thus start with $0 < |x - 4| < \delta = 1/2$:

$$0 < |x - 4| < 1/2$$
$$= (0 + 2)/4$$
$$\leq (\sqrt{x} + 2)/4$$

Thus

$$|x - 4| < (\sqrt{x} + 2)/4$$
$$|\sqrt{x} - 2| \, |\sqrt{x} + 2| < (\sqrt{x} + 2)/4$$
$$|\sqrt{x} - 2| < 1/4 = \epsilon$$

Thus $|\sqrt{x} - 2| < 1/4$ whenever $0 < |x - 4| < 1/2$, just as desired. $\square$

Notice how all the work goes into finding δ by reverse implications. The final proof (which is really all that needs to be written down!) is then simply a stringing together of your reverse implications, but now in forward order.

4.   <u>Computations with general ε's.</u>   To actually prove that $\lim_{x \to 4} \sqrt{x} = 2$ we cannot just deal

with a particular $\epsilon = 1/4$;   we must handle <u>all</u> $\epsilon > 0$ simultaneously.   Except for a some-

what more cumbersome notation, the procedure is the same as that used for a fixed ϵ .   The

proof is given in Example B; you should compare the procedures used against those in Example A.

<u>Example B.</u>   Prove $\lim_{x \to 4} \sqrt{x} = 2$.

<u>Solution.</u>   Given any $\epsilon > 0$ we must find a $\delta > 0$ such that

$$|\sqrt{x} - 2| < \epsilon \quad \text{whenever} \quad 0 < |x - 4| < \delta$$

We will proceed from the ϵ-inequality   to the δ-inequality   by the use of reverse implications.

$$\left. \begin{array}{c} |\sqrt{x} - 2| < \epsilon \\[2mm] |\sqrt{x} - 2| \cdot |\sqrt{x} + 2| < (\sqrt{x} + 2)\epsilon \\[2mm] |x - 4| < (\sqrt{x} + 2)\epsilon \end{array} \right\} \begin{array}{l} \text{equivalent} \\ \text{statements} \end{array} \qquad (C)$$

However, we want (C) to be implied by $0 < |x - 4| < \delta$.   Since $2\epsilon \leq (\sqrt{x} + 2)\epsilon$ we have

only to choose $\delta = 2\epsilon$.   We have thus found a $\delta > 0$ for any given $\epsilon > 0$.

We now write out the formal proof.   Given any $\epsilon > 0$,   choose $\delta > 0$ to be $\delta = 2\epsilon$.

Then, the inequality

$$0 < |x - 4| < \delta = 2\epsilon$$
$$= (0 + 2)\epsilon$$
$$\leq (\sqrt{x} + 2)\epsilon$$

implies
$$|x - 4| < (\sqrt{x} + 2)\epsilon$$

$$|\sqrt{x} - 2| \, |\sqrt{x} + 2| < (\sqrt{x} + 2)\epsilon$$

$$|\sqrt{x} - 2| < \epsilon$$

Thus $|\sqrt{x} - 2| < \epsilon$ whenever $0 < |x - 4| < \delta$, as desired. $\qquad\square$

Note that we have given a <u>formula</u> ($\delta = 2\epsilon$) for determining δ when given ϵ. Here's a somewhat harder example to sharpen your teeth on.

<u>Example C.</u> Prove $\lim\limits_{x \to 2} x^3 = 8$.

<u>Solution.</u> Given any $\epsilon > 0$ we must find a $\delta > 0$ such that

$$|x^3 - 8| < \epsilon \quad \text{whenever} \quad 0 < |x - 2| < \delta$$

We will proceed from the ϵ-inequality to the δ-inequality by the use of reverse implications. The major observation to make is that $x^3 - 8$ can be factored by $x - 2$ to obtain

$$x^3 - 8 = (x - 2)(x^2 + 2x + 4)$$

(See Appendix D if you are unclear as to how this factoring is obtained.) In this way $|x^3 - 8|$ can be related to $|x - 2|$.

$$\boxed{\begin{array}{c} |x^3 - 8| < \epsilon \\[2mm] |x - 2| \, |x^2 + 2x + 4| < \epsilon \end{array}} \left.\begin{array}{c} \\ \\ \end{array}\right\} \begin{array}{l} \text{equivalent} \\ \text{statements} \end{array} \qquad \text{(D)}$$

Following Anton's methods, we would like to find a constant k which is an upper bound for $|x^2 + 2x + 4|$, i.e.,

$$|x^2 + 2x + 4| \leq k \qquad \text{(E)}$$

In such a case, inequality (D) would be implied by

$$k |x - 2| < \epsilon$$

or $\qquad\qquad\qquad |x - 2| < \epsilon/k$

letting us choose $\delta = \epsilon/k$.

There's a snag however (we warned you that this was a hard example!). There is no constant k which satisfies (E) <u>for all x</u>, i.e., the quadratic term $|x^2 + 2x + 4|$ can assume arbitrarily large values as x ranges over all real numbers. <u>But</u> ... we don't care about x-values except for those near 2 (since x is approaching 2). So let's restrict ourselves right away to being within 1 of $x_0 = 2$:

consider only x-values for which $|x - 2| < 1$

i.e., $1 < x < 3$

(We could just as well have restricted ourselves to $|x - 2| < 24.2$. Any restriction will work here.) It now seems reasonable that the term $|x^2 + 2x + 4|$ will have an upper bound k on the interval $1 < x < 3$. We find it as follows:

$$1 < x < 3$$

implies

$$|x^2 + 2x + 4| \leq |x|^2 + 2|x| + 4 \qquad \text{(by the triangle inequality)}$$
$$< 9 + 2(3) + 4$$
$$= 19$$

Thus we have proven

$$\boxed{\text{if} \quad |x - 2| < 1, \quad \text{then} \quad |x^2 + 2x + 4| < 19}$$ (F)

which is our (corrected) refinement of inequality (E).

After that extended digression (whew!) we can return to inequality (D):

$$|x - 2| \, |x^2 + 2x + 4| < \epsilon$$

In view of (F), inequality (D) will be true whenever

$$19 \, |x - 2| < \epsilon \quad (\text{if} \quad |x - 2| < 1)$$

or $$|x - 2| < \epsilon/19 \quad (\text{if} \quad |x - 2| < 1)$$

We thus would like to make $\delta = \epsilon/19$ ... er, but what about our restriction $|x - 2| < 1$? No problem! This restriction merely states that our desired δ had better be less than or equal to 1. So we take

$$\delta = \min \{1, \epsilon/19\}$$

Our formal proof now reads as follows: given any $\epsilon > 0$, choose $\delta > 0$ by $\delta = \min \{1, \epsilon/19\}$. Then the inequality $0 < |x - 2| < \delta \leq \epsilon/19$

implies $$19 \, |x - 2| < \epsilon$$ (G)

However, $|x - 2| < \delta \leq 1$

implies $|x - 2| < 1$

which by (F) gives $|x^2 + 2x + 4| \leq 19$

Thus (G) implies $$|x^2 + 2x + 4| \cdot |x - 2| < \epsilon$$

which is equivalent to

$$|x^3 - 8| < \epsilon$$

Thus $\left|x^3 - 8\right| < \epsilon$ whenever $0 < \left|x - 2\right| < \delta$, as desired.   □

If you are still alive and breathing, go off and tackle some of Anton's exercises. The only way to learn the $\epsilon - \delta$ definition for limits is to jump in head first and start swimming.

5. Verifying non-limits. How do we show that $\lim\limits_{x \to a} f(x)$ is not equal to some number L? Well, we could show perhaps that $\lim\limits_{x \to a} f(x)$ is equal to a number $L' \neq L$. But what if we don't know what L' is? Or worse, what if $\lim\limits_{x \to a} f(x)$ does not exist? Anton illustrates the procedure to use in his Example 4; we give one more illustration:

Example D. Prove or disprove: $\lim\limits_{x \to 0} \dfrac{1}{x} = 1$.

Solution. If by now you don't realize (intuitively) that $\lim\limits_{x \to 0} \dfrac{1}{x}$ is undefined ... well, you better head back fast to §2.4! We want to disprove the stated limit.

The statement " $\lim\limits_{x \to 0} \dfrac{1}{x} = 1$ " says that for every $\epsilon > 0$, we can find a $\delta > 0$, etc. ... Thus, to prove " $\lim\limits_{x \to 0} \dfrac{1}{x} \neq 1$ " we must only show that there is at least one $\epsilon_0 > 0$ for which it is not true, i.e.,

there is at least one $\epsilon_0 > 0$ for which no $\delta > 0$ exists to

make the following statement true: (H)

$$0 < \left|x - 0\right| < \delta \quad \text{implies} \quad \left|\frac{1}{x} - 1\right| < \epsilon_0$$

Let's try $\epsilon_0 = 1$. That seems like an ϵ choice which will not work since we know (intuitively) that $f(x) = 1/x$ hightails itself off to $+\infty$, at least when x approaches 0 from the right.

2.6.12

Let's see what would happen if the inequality $\left|\dfrac{1}{x} - 1\right| < \epsilon_0 = 1$ were true. (To keep things simple we will just deal with the positive values of x):

$$\left|\frac{1}{x} - 1\right| < 1$$

$$\left|\frac{1 - x}{x}\right| < 1$$

$$\left|1 - x\right| < x$$

$$-x < \underbrace{1 - x} < x$$

$$1 < 2x$$

$$\boxed{1/2 < x}$$

Thus, in order for the inequality $\left|\dfrac{1}{x} - 1\right| < 1$ to be valid for positive x-values, we must have $1/2 < x$. Hence, there is no value of δ for which the inequality $\left|x - 0\right| < \delta$ can imply $\left|\dfrac{1}{x} - 1\right| < 1$, since $\left|x - 0\right| < \delta$ includes plenty of positive x-values for which $x \leq 1/2$. This verifies statement (H), and hence proves that $\displaystyle\lim_{x \to 0} \frac{1}{x} \neq 1$. $\qquad\square$

Chapter 3:   Differentiation

Section 3. 1:   The Derivative

1. <u>The formal definition.</u>   The major concept of the next two chapters is given immediately in

Definition 3. 1. 2:

<table>
<tr><td>The Derivative
of   f(x)</td></tr>
</table>

> The <u>derivative</u> of a function   f   is the function   f'   defined by
>
> $$f'(x) = \lim_{h \to 0} \frac{f(x+h) - f(x)}{h}$$
>
> The domain of   f'   consists of all   x   where this limit exists.

The process of taking a given function   f(x)   and calculating its derivative   f'(x)   is called

<u>differentiation.</u>

It is hard to overstate the importance of differentiation.   It occurs in so many applications

that a list would consume several pages.   In §2. 3 two important applications have already been

discussed:

1. The <u>slope of the tangent line</u> to the graph

of   f   at $x = x_0$ is given by

$$m_{tan} = f'(x_0)$$

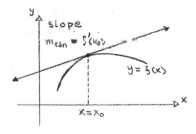

2. The <u>(instantaneous) velocity</u> for a straight

line motion $s = f(t)$ at time $t = t_0$

is given by

$$v_{inst} = f'(t_0)$$

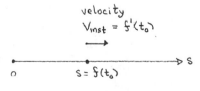

3. 1. 2

Although slope and velocity would at first appear to have little in common, as you can see they are both examples of differentiation!

$$* \quad * \quad * \quad * \quad *$$

As you can see from its definition, differentiation depends on the computation of limits. And, folks, ... that's why we spent so much time on limits in §§2.4 - 2.6! Given a simple function f , <u>you must learn how to compute the derivative f' from the definition in terms of limits</u>! Examples 1 and 3 illustrate the general technique. Here is another illustration:

<u>Example A.</u> Find f'(x) if f(x) = $\sqrt{x + 1}$.

<u>Solution.</u> $f'(x) = \lim_{h \to 0} \dfrac{f(x + h) - f(x)}{h}$ from the definition of f'

> WARNING: You MUST use a limit here. For some reason, unknown to us, many people leave it off, and then pay dearly later on.

$$= \lim_{h \to 0} \frac{\sqrt{(x+h) + 1} - \sqrt{x + 1}}{h}$$

Hmmm... Where to now? We have to find some way to eliminate the h in the denominator .... Well, let's multiply the numerator and denominator by
$\sqrt{(x+h) + 1} + \sqrt{x + 1}$:

$$= \lim_{h \to 0} \frac{\sqrt{(x+h) + 1} - \sqrt{x + 1}}{h} \cdot \frac{\sqrt{(x+h) + 1} + \sqrt{x + 1}}{\sqrt{(x+h) + 1} + \sqrt{x + 1}}$$

$$= \lim_{h \to 0} \frac{[(x+h) + 1] - [x + 1]}{h[\sqrt{x + h + 1} + \sqrt{x + 1}]} \qquad (\text{since } (a - b)(a + b) = a^2 - b^2)$$

$$= \lim_{h \to 0} \frac{h}{h[\sqrt{x+h+1} + \sqrt{x+1}]}$$

Ahhh, our multiplication trick worked and we are left with an h in both the numerator and denominator. CANCEL THEM OUT!

$$= \lim_{h \to 0} \frac{1}{\sqrt{x+h+1} + \sqrt{x+1}}$$

Now we have no trouble taking the limit by substituting h = 0 ...

$$= \frac{1}{\sqrt{x+1} + \sqrt{x+1}}$$

$$= \frac{1}{2\sqrt{x+1}} \qquad\qquad \square$$

As this example shows, the problem you must focus on when computing derivatives from the definition in terms of limits is:

"How do I get rid of that *!$?!#! little

h in the denominator of $\lim\limits_{h \to 0} \dfrac{f(x+h) - f(x)}{h}$? "

In most cases the technique is to

use algebraic operations to manipulate the numerator until

you can factor out an h , i. e., until the numerator looks like

$$f(x+h) - f(x) = h \cdot [\text{stuff in } x \text{ and } h] .$$

This factored h will then cancel the troublesome h in

the denominator.

However, when using this technique, be careful not to change the value of the fraction. For example, if you multiply the numerator by a quantity, you must also multiply the denominator by that same quantity.

This "factoring out" process can at times be very easy (as in Example 1), moderately easy (Example 3) or rather tricky (Example A). You will master the process only by practice, practice, practice....

Computing derivatives by using the definition in terms of limits is time consuming, but TAKE HEART! After a few basic derivatives are computed using limits, we will be able to compute further derivatives <u>without using limits</u> (which is a major time saver!). Such techniques will be developed in later sections.

2. <u>Some Comments on notation.</u>   For both historical and practical reasons there are a number of different notations used for the one process of differentiation. You will gradually see why it is convenient to know them all; <u>they must become second nature to you.</u> Anton discusses notation following Example 4; we illustrate his discussion with a specific case:

In Example A we took $y = f(x) = \sqrt{x+1}$ and showed

$$\frac{dy}{dx} = f'(x) = \frac{d}{dx} [\sqrt{x+1}] = \frac{1}{2\sqrt{x+1}}$$

(All these symbols denote the derivative of f)

To evaluate at a specific point $x = x_0$ we have

$$\frac{dy}{dx}\bigg|_{x=x_0} = f'(x_0) = \frac{d}{dx} [\sqrt{x+1}]\bigg|_{x=x_0} = \frac{1}{2\sqrt{x_0+1}}$$

(All these symbols denote the derivative of f evaluated at $x = x_0$)

Thus, taking $x_0 = 3$, we obtain

$$\left. \frac{dy}{dx} \right|_{x=3} = f'(3) = \left. \frac{d}{dx} [\sqrt{x+1}] \right|_{x=3} = \frac{1}{2\sqrt{4}} = \frac{1}{4}$$

3. Tangent lines and derivatives.   The use of the derivative to compute tangent lines to graphs

was illustrated in Example 1 of Anton's §2. 3 and Example A of §2.3. 2 in The Companion.

We advise you to look over both of these examples -- they are similar to Anton's Exercises 11 - 16.

Here is an illustration of a different sort, similar to Anton's Exercises 27 - 34 :

Example B.   Sketch the graph of the derivative

  f' of the function f whose graph is

shown to the right .

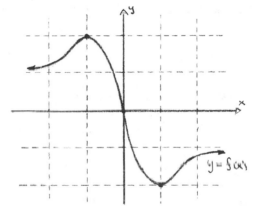

Solution.   At first this may seem like a very

bizarre problem: we're not given any algebraic

expression for f(x) , so how can we compute

f'(x) ? Well, we can't compute f'(x) precisely, but we can get a pretty good sketch for f'(x)

by remembering that

$\Vert$ f'(x) is the slope of the tangent

 line to the graph of y = f(x)

 at the value x .

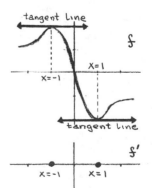

Now we can read off some of these slopes very easily. For

example, at x = ± 1 the tangent lines are seen to be

horizontal. Thus they have slope 0 , proving f'(± 1) = 0 .

3. 1. 6

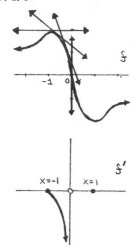

Now what happens, for example, when x moves from -1 to 0? Well, the tangent lines gradually bend down from the horizontal to the vertical. Thus the slope (i. e. , <u>derivative</u>!) <u>decreases</u> from 0 at x = -1 to -∞ (undefined) at x = 0. Hence we have filled in more of the derivative graph as shown to the left.

What about when x moves from 1 to ∞? The tangent line starts out as horizontal (f'(x) = 0), bends up (f'(x) > 0) and then bends back down, with the horizontal as its limiting case $\left(\lim\limits_{x \to +\infty} f'(x) = 0 \right)$

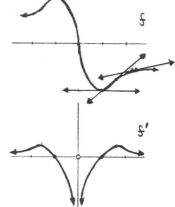

Continuing in this way will yield the sketch for

f' as shown. Make sure that you can justify the rest of this sketch! □

4. <u>Functions which are not differentiable.</u> It is important to realize that a function may be defined at a point x = x_0 <u>but might not be differentiable there!</u> In Example 4 Anton carefully shows that f(x) = |x| is not differentiable at x = 0 .

Although a function may be <u>non-differentiable</u> at a value x = x_0 for all sorts of esoteric reasons, the most common garden-variety problems are:

1. $y = f(x)$ has a <u>break</u> at $x = x_0$. If
    there is a break in the graph, then
    there cannot be a tangent line, and hence
    there is no derivative. *

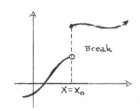

2. $y = f(x)$ has a <u>corner</u> at $x = x_0$.
    Geometrically this means that there
    appear to be two tangent lines at $x = x_0$,
    one for each side of the value $x = x_0$.

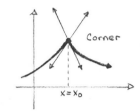

3. $y = f(x)$ has a <u>vertical tangent</u> at $x = x_0$.
    In such a case the tangent line has an
    infinite slope, and hence $f'(x_0)$ is not
    defined.

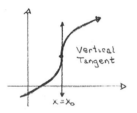

Here are illustrations of each type of problem.

<u>Example C.</u> Is $f(x) = \begin{cases} x^2 & x \leq 0 \\ x + 1 & x > 0 \end{cases}$ differentiable at $x = 0$?

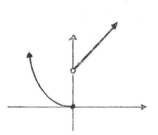

<u>Solution.</u> As the sketch of f shows, f has a
"break" at $x = 0$ and therefore is not differentiable
there. However, we can also establish this fact analytically,
i. e. , without relying on the graph. Look at the definition for $f'(0)$:

$$f'(0) = \lim_{h \to 0} \frac{f(h) - f(0)}{h} \qquad (\text{using } x = 0)$$

_____

* In the terminology to be learned in §3.7, $y = f(x)$ has a "break" at $x = x_0$ means f
  is <u>discontinuous</u> at $x = x_0$.

In this case f(h) is defined differently on either side of h = 0 , so we must look at the <u>one-</u>

<u>sided limits.</u> Let's consider the right-sided limit:

$$\lim_{h \to 0^+} \frac{f(h) - f(0)}{h} = \lim_{h \to 0^+} \frac{[h+1] - [0^2]}{h}$$

$$= \lim_{h \to 0^+} 1 + \frac{1}{h} = \infty \ ? \quad \text{OOPS!}$$

We need go no further! Since the right-sided limit does not exist, then the full limit (and hence

f'(0)) cannot exist. Thus f is not differentiable at x = 0 .   □

<u>Example D.</u>   Is $f(x) = \begin{cases} x^2 & x \leq 0 \\ x & x > 0 \end{cases}$ differentiable at x = 0 ?

<u>Solution.</u> As the sketch of f indicates, f appears

to have a "corner" at x = 0 and therefore should

not be differentiable there. For an analytic argument,

(i. e. , one which does not rely on the graph), we must

again look at the two one-sided limits:

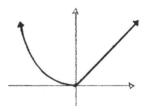

$$\lim_{h \to 0^+} \frac{f(h) - f(0)}{h} = \lim_{h \to 0^+} \frac{h - 0^2}{h} = \lim_{h \to 0^+} 1 = 1 \longleftarrow$$

$$\text{OOPS!}$$

$$\lim_{h \to 0^-} \frac{f(h) - f(0)}{h} = \lim_{h \to 0^-} \frac{h^2 - 0^2}{h} = \lim_{h \to 0^-} h = 0 \longleftarrow$$

Since the two one-sided limits do not agree, then the full (two-sided) limit does not exist, i. e. ,

f'(0) does not exist. *   □

_____

*
  <u>Note</u>: If both one-sided limits <u>did</u> agree, then f'(0) would exist and would equal the value of
        the two limits. See Anton's Exercise 36.

Example E.   Is $f(x) = x^{1/3}$ differentiable at $x = 0$?

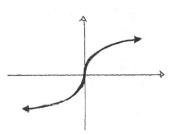

Solution.   As the sketch of f indicates, f appears
to have a vertical tangent line at $x = 0$ and therefore
should not be differentiable there.   For an analytic
argument we need to consider the limit

$$f'(0) = \lim_{h \to 0} \frac{f(h) - f(0)}{h} = \lim_{h \to 0} \frac{h^{1/3} - 0^{1/3}}{h}$$

$$= \lim_{h \to 0} \frac{1}{h^{2/3}} = \infty \quad (!?) \quad \text{OOPS!}$$

Since the limit is undefined, then $f'(0)$ is undefined, i. e., f is not differentiable at

$x = 0$.   □

5.   **Rates of change.**   We now come to the single most important interpretation of the derivative:
a measurement of how fast one quantity is changing with respect to another quantity.   In fact,
every application of the derivative is some form of a rate of change!

    According to Anton's Definition 3. 1. 3 , the quantity

$$\frac{y_1 - y_0}{x_1 - x_0} = \frac{f(x_1) - f(x_0)}{x_1 - x_0}$$

is the average rate at which $y = f(x)$ changes with x over the interval $[x_0, x_1]$.   Then
the limit of this quantity as $x_1 \to x_0$,

$$\lim_{x_1 \to x_0} \frac{f(x_1) - f(x_0)}{x_1 - x_0} = \frac{dy}{dx}\bigg|_{x = x_0}$$

is naturally defined to be the <u>instantaneous rate at which</u> y <u>changes with</u> x at x_0 (Definition 3. 1. 4).

<u>Example E.</u>   According to the Ideal Gas Law, the pressure P exerted by a confined gas is related to its volume V and its temperature T (measured in degrees Kelvin) by

$$PV = kT$$

where k is a constant depending on the amount of gas present.

      a.   Find the average rate at which the volume V of a confined gas changes with the pressure P as the pressure increases from P = 5 to P = 8, and the temperature is held constant at $T = 200^\circ$.

      b.   Find the instantaneous rate at which the volume changes with the pressure when P = 5 and $T = 200^\circ$.

<u>Solution.</u>   We first write volume V as a function of pressure P :

$$V = \frac{kT}{P} = \frac{200\,k}{P} \qquad \text{when} \quad T = 200^\circ K$$

a.   The average rate of change we are looking for is

$$\left\{ \begin{array}{c} \text{average rate at} \\ \text{which } V \text{ changes with} \\ P \text{ over } [P_0, P_1] \end{array} \right\} = \frac{V_1 - V_0}{P_1 - P_0}$$

where
$$P_0 = 5, \qquad V_0 = \frac{200\,k}{P_0} = 40\,k$$

$$P_1 = 8, \qquad V_1 = \frac{200\,k}{8} = 25\,k$$

Thus
$$\frac{V_1 - V_0}{P_1 - P_0} = \frac{25\,k - 40\,k}{8 - 5} = \frac{-15\,k}{3} = \boxed{-5\,k}$$

(The minus sign indicates that volume decreases as the pressure increases.)

b.   "Find the instantaneous rate at which   V   changes with   P"   immediately translates

into "Find $\dfrac{dV}{dP}$. "   Thus

$$\frac{dV}{dP} = \frac{d}{dP} [200\,k\,P^{-1}] = -200\,k\,P^{-2} = -\frac{200\,k}{P^2}$$

Hence, at   P = 5   we obtain

$$\left\{ \begin{array}{l} \text{instantaneous rate of} \\ \text{change of   V   at   } P = 5 \end{array} \right\} = \left. \frac{dV}{dP} \right|_{P = 5} = -\frac{200\,k}{25} = \boxed{-8k}$$ ☐

(Once again, since   k   is positive, then the volume decreases as the pressure increases.)

6.   <u>Applications of the derivative: a summary thus far.</u>   There are certain phrases that are

encountered in applications (i. e. , "word problems") that must immediately ring the bell

"DERIVATIVE"!!   These are the ones we have already seen:

"... the <u>slope of the tangent line</u> to   y = f(x)   at $x = x_0$ ... "

-- this is simply $\left. \dfrac{dy}{dx} \right|_{x = x_0}$

"... the <u>velocity</u> of an object moving along the   s-axis   at time $t = t_0$ ... "

-- this is simply $\left. \dfrac{ds}{dt} \right|_{t = t_0}$

"... the <u>instantaneous rate</u> at which   y   changes with   x   at $x = x_0$ ... "

-- this is simply $\left. \dfrac{dy}{dx} \right|_{x = x_0}$.

<u>Remember these phrases carefully</u>, and always associate them with derivatives!   You will see

them frequently in our subsequent work.

Section 3.2 :   Techniques of Differentiation

Using the definition of the derivative to compute derivatives is very laborious and time consuming; in fact, for complicated functions, such computations can be nearly impossible. So in this section we develop a series of tricks -- "techniques of differentiation" -- which allow us to compute derivatives without using the definition.

The overall procedure has two steps:

Step 1.   Memorize the derivatives of simple functions.

Step 2.   Given functions whose derivatives we know, learn rules for taking derivatives of algebraic combinations of these functions, e. g. , sums, differences, products, quotients.

Thus we first learn the derivatives of some simple functions, calculated using the definition in terms of limits. We will then "break down" not-so-simple functions into combinations of the simple ones; our rules will tell us how to obtain the derivatives of the complicated functions from combinations of the simple derivatives.

The process of building complicated functions out of simpler ones (and the reverse process of breaking down the complicated functions) was discussed at length in §2.2.3 of The Companion.

1. Specific derivatives.   In this section Anton computes the derivatives of two families of simple functions:

| Constant Rule | 1. | $\dfrac{d}{dx} [c] = 0$ | for any constant c | Theorem 3.2.1 |
| Power Rule | 2. | $\dfrac{d}{dx} [x^n] = n x^{n-1}$ | for any integer n | Theorems 3.2.2 , 3.2.7 |

In English these equations state that:

---

1.   the derivative of a constant function is zero,

2.   to take the derivative of a power, drop the power by 1 and multiply

        the result by the original power.

These two derivatives are basic to everything that follows, and must become second nature to you!

In particular, when n = 1 the Power Rule yields

$$\frac{d}{dx} [x] = 1 \cdot x^{1-1} = 1 \cdot x^{0} = 1$$

Example A.   Compute the following derivatives:

a. $\frac{d}{dx} [17]$

b. $f'(x)$,   where $f(x) = -3$

c. $\frac{d}{dx} [x^{3}]$

d. $\frac{d}{dx} [x^{-3}]$

e. $\frac{d}{dt} [t^{-5}]$

f. $f'(s)$,   where $f(s) = 1/s^{6}$.

Solution.   a. $\frac{d}{dx} [17] = ?$ Since "17" is a constant, then the Constant Rule applies:

$$\frac{d}{dx} [17] = 0$$

Simple enough ... , but there is one possible pitfall! Sometimes people mix up the rule for the derivative of a constant (which is zero) with the rule for the limit of a constant (which is the constant), i.e.,

$$\boxed{\frac{d}{dx} [\, c \,] \ = \ 0} \qquad \boxed{\lim_{x \to a} \ c \ = \ c}$$

Be sure you keep these straight!

b. $f'(x) = ?$ when $f(x) = -3$. Ah ha! This is just the Constant Rule in function notation:

$$f'(x) \ = \ \frac{d}{dx} [\, f(x) \,] \qquad \text{since these are equivalent notations,}$$

$$= \ \frac{d}{dx} [\, -3 \,] \qquad \text{since} \ \ f(x) = -3 \, ,$$

$$= \ 0 \qquad\qquad\quad \text{by the Constant Rule}$$

c. $\frac{d}{dx} [\, x^3 \,] = ?$ This is a direct application of the Power Rule:

$$\frac{d}{dx} [\, x^3 \,] \ = \ 3 x^{3-1} \ = \ 3 x^2$$

d. $\frac{d}{dx} [\, x^{-3} \,] = ?$ This too is a direct application of the Power Rule:

$$\frac{d}{dx} [\, x^{-3} \,] \ = \ (-3) x^{-3-1} \ = \ -3 x^{-4}$$

However, beware of the following common but deadly mistake:

$$\frac{d}{dx} [\, x^{-3} \,] \ = \ ? \ = \ -3 x^{-2} \ ? \quad \text{NO!}$$

When we "drop the power -3 by 1" we get -4, NOT -2

e. $\frac{d}{dt} [\, t^{-5} \,] = ?$ We're not going to let a minor change of notation confuse us: the Power

Rule remains valid when   x   is replaced, in every instance, by another variable. * Thus

$$\frac{d}{dt} [t^{-5}] = (-5) t^{-5-1} \qquad \text{by the Power Rule with a}$$
$$\text{t   replacing every   x .}$$

$$= -5 t^{-6}$$

f.   f'(s) = ? when $f(s) = 1/s^6$. Here we must deal with both function notation and a

variable different from   x . But it's not difficult:

$$f'(s) = \frac{d}{ds} [f(s)] \qquad \text{since these are equivalent notations,}$$

$$= \frac{d}{ds} [s^{-6}] \qquad \text{since   } f(s) = 1/s^6 = s^{-6} ,$$

$$= (-6) s^{-6-1} \qquad \text{by the Product Rule with an}$$
$$\text{s   replacing every   x}$$

$$= -6 s^{-7} \qquad\qquad\qquad\qquad\qquad \square$$

2.   General rules of differentiation.   We'll start with a summary of all the basic rules of this

section. Suppose f and g are differentiable at x , and c is a constant. Then the

following "combinations" of f and g are also differentiable at x :

$$cf, \quad f + g, \quad f - g, \quad f \cdot g \quad \text{and} \quad f/g \quad (\text{if} \quad g(x) \neq 0)$$

Moreover, the derivatives are:

---

*

   This is true for any rule of differentiation! Replacing x "in every instance" by another

   variable also includes replacing the x in $\frac{d}{dx}$ with the new variable.

3. 2. 5

Constant Factor Rule	1. $\frac{d}{dx}[cf(x)] = c\frac{d}{dx}[f(x)]$

Constant Factor Rule   1. $\dfrac{d}{dx}[\,cf(x)\,] = c\,\dfrac{d}{dx}[\,f(x)\,]$ Theorem 3.2.3

Sum Rule   2. $\dfrac{d}{dx}[\,f(x)+g(x)\,] = \dfrac{d}{dx}[\,f(x)\,] + \dfrac{d}{dx}[\,g(x)\,]$ Theorem 3.2.4

Difference Rule   3. $\dfrac{d}{dx}[\,f(x)-g(x)\,] = \dfrac{d}{dx}[\,f(x)\,] - \dfrac{d}{dx}[\,g(x)\,]$

Product Rule   4. $\dfrac{d}{dx}[\,f(x)g(x)\,] = f(x)\dfrac{d}{dx}[\,g(x)\,] + g(x)\dfrac{d}{dx}[\,f(x)\,]$ Theorem 3.2.5

Quotient Rule   5. $\dfrac{d}{dx}\left[\dfrac{f(x)}{g(x)}\right] = \dfrac{g(x)\dfrac{d}{dx}[\,f(x)\,] - f(x)\dfrac{d}{dx}[\,g(x)\,]}{[\,g(x)\,]^2}$ Theorem 3.2.6

Now that's a lot to remember! But just as we stated for the Constant Rule and the Power Rule, these rules of differentiation are basic to everything that follows, and must become second nature to you! *

The first three of these rules are pretty easy to believe and to remember. It is helpful to remember their statements in English:

1. A constant (multiplicative) factor can be moved through a derivative sign.

2. The derivative of a sum is the sum of the derivatives.

3. The derivative of a difference is the difference of the derivatives.

---
* Believe us! We cannot exaggerate their importance!

Example B.   Compute the following derivatives.

   a. $\dfrac{d}{dx}\,[\,3x^{-4} + 4x\,]$,

   b. $\dfrac{d}{ds}\,[\,s^4 - 6s^2 + 3\,]\Big|_{s\,=\,2}$

   c. $h'(u)$, where $h(u) = \pi u^2 - u^3$

Solution.

   a. $\dfrac{d}{dx}\,[\,3x^{-4} + 4x\,] = \dfrac{d}{dx}\,[\,3x^{-4}\,] + \dfrac{d}{dx}\,[\,4x\,]$ by the Sum Rule

$$= 3\,\dfrac{d}{dx}\,[\,x^{-4}\,] + 4\,\dfrac{d}{dx}\,[\,x\,]$$ by the Constant Factor Rule

$$= 3\,(-4)\,x^{-5} + 4\,(1)$$ by the Power Rule

$$= -12x^{-5} + 4$$

This is clearly a much faster way to compute the derivative of $3x^{-4} + 4x$ than using the definition of the derivative!

   b.   To compute $\dfrac{d}{ds}\,[\,s^4 - 6s^2 - 3\,]\Big|_{s\,=\,2}$ we must first compute the derivative at a general value of s :

$$\frac{d}{ds}[s^4 - 6s^2 + 3] = \frac{d}{ds}[s^4] - \frac{d}{ds}[6s^2] + \frac{d}{ds}[3]$$

since the Sum and Difference Rules apply to <u>any number</u> of terms!

$$= \frac{d}{ds}[s^4] - 6\frac{d}{ds}[s^2] + \frac{d}{ds}[3]$$

by the Constant Factor Rule

$$= 4s^3 - 6(2)s + 0$$

by the Power Rule and the Constant Rule

$$= 4s^3 - 12s$$

Then $\frac{d}{ds}[s^4 - 6s^2 + 3]\Big|_{s=2} = 4(2)^3 - 12(2) = 32 - 24 = 8$.

c. $h'(u) = ?$ where $h(u) = \pi u^2 - u^3$. We're not going to let function notation mess us up:

$$h'(u) = \frac{d}{du}[h(u)]$$ since these are equivalent notations

$$= \frac{d}{du}[\pi u^2 - u^3]$$ since $h(u) = \pi u^2 - u^3$

$$= \frac{d}{du}[\pi u^2] - \frac{d}{du}[u^3]$$ by the Difference Rule

$$= \pi \frac{d}{du}[u^2] - \frac{d}{du}[u^3]$$ by the Constant Factor Rule
(Remember: π is a constant!)

$$= 2\pi u - 3u^2$$ by the Power Rule   □

* * * * *

Now let's consider rules 4 and 5, the less-than-intuitive Product and Quotient Rules:

4. $\dfrac{d}{dx} [f(x) g(x)] = f(x) \dfrac{d}{dx} [g(x)] + g(x) \dfrac{d}{dx} [f(x)]$

5. $\dfrac{d}{dx} \left[\dfrac{f(x)}{g(x)} \right] = \dfrac{g(x) \dfrac{d}{dx} [f(x)] - f(x) \dfrac{d}{dx} [g(x)]}{[g(x)]^2}$

They are certainly <u>NOT</u> what you would expect, i. e.,

<u>FALSE !!</u> $\dfrac{d}{dx} [f(x) g(x)] = \overset{NO!}{?} = \dfrac{d}{dx} [f(x)] \cdot \dfrac{d}{dx} [g(x)]$ <u>FALSE !!</u>

<u>FALSE !!</u> $\dfrac{d}{dx} \left[\dfrac{f(x)}{g(x)} \right] = \underset{NO!}{?} = \dfrac{\dfrac{d}{dx} [f(x)]}{\dfrac{d}{dx} [g(x)]}$ <u>FALSE !!</u>

It is extremely helpful to remember the (correct!) statements for the Product and Quotient Rules in English:

> 4. The derivative of a product is the first function times the derivative of the second plus the second function times the derivative of the first.
>
> 5. The derivative of a quotient is the denominator times the derivative of the numerator <u>MINUS</u> the numerator times the derivative of the denominator, all divided by the denominator squared.

<u>Example C.</u>   Compute the following derivatives.

a. $\dfrac{d}{dx} \left[\dfrac{x^3 - 6x}{x - 1} \right] \Big|_{x = 2}$

b. $\dfrac{d\lambda}{d\alpha}$, where $\lambda = (3 \alpha^{-1} + \alpha^3)(2 \alpha^2 - \alpha^4)$

3.2.9

Solution.

NOTE THE MINUS SIGN !! IT IS A COMMON MISTAKE TO USE A "PLUS" INSTEAD OF A "MINUS"

a.
$$\frac{d}{dx}\left[\frac{x^3 - 6x}{x - 1}\right] = \frac{(x-1)\frac{d}{dx}[x^3 - 6x] - (x^3 - 6x)\frac{d}{dx}[x-1]}{(x-1)^2}$$

by the Quotient Rule. Using the previous rules to compute

$$\frac{d}{dx}[x^3 - 6x] = 3x^2 - 6 \quad \text{and} \quad \frac{d}{dx}[x-1] = 1,$$

we obtain

$$= \frac{(x-1)(3x^2 - 6) - (x^3 - 6x)}{(x-1)^2}$$

This expression could be further simplified, but that is hardly necessary since we only want the derivative when $x = 2$:

$$\frac{d}{dx}\left[\frac{x^3 - 6x}{x - 1}\right]\Bigg|_{x=2} = \frac{(1)(6) - (-4)}{(1)^2} = 10$$

b.
$$\frac{d\lambda}{d\alpha} = \frac{d}{d\alpha}\left[(3\alpha^{-1} + \alpha^3)(2\alpha^2 - \alpha^4)\right]$$

$$= (3\alpha^{-1} + \alpha^3)\frac{d}{d\alpha}(2\alpha^2 - \alpha^4) + (2\alpha^2 - \alpha^4)\frac{d}{d\alpha}(3\alpha^{-1} + \alpha^3)$$

by the Product Rule. Using the previous rules we obtain

$$\frac{d}{d\alpha}(2\alpha^2 - \alpha^4) = 4\alpha - 4\alpha^3 \quad \text{and} \quad \frac{d}{d\alpha}(3\alpha^{-1} + \alpha^3) = -3\alpha^{-2} + 3\alpha^2.$$

Hence

$$\frac{d\lambda}{d\alpha} = (3\alpha^{-1} + \alpha^3)(4\alpha - 4\alpha^3) + (2\alpha^2 - \alpha^4)(-3\alpha^{-2} + 3\alpha^2)$$

This answer can be further simplified if so desired. □

* * * * *

Here's an illustration of a slightly more interesting problem than the simple computation of a derivative. It is similar to Anton's Exercises 35 through 38.

<u>Example D.</u> Find the values of a and b if the tangent line to $y = \dfrac{a + x}{b - x}$ at x = 1 has slope

m = 6 and the curve intersects the y-axis at y = 2.

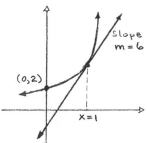

Slope
m = 6

(0,2)

x = 1

<u>Solution.</u> A sketch (of necessity, very rough) of our situation is helpful and is shown to the right. We have two pieces of information:

$$(1) \quad 6 = m = \left.\dfrac{dy}{dx}\right|_{x = 1} \; ,$$ using the derivative as the slope, and

(2)   when x = 0 , then y = 2

This will give us two simultaneous equations in the two unknowns a and b . Using (1) requires our differentiation rules:

$$\boxed{\text{NOTE THE MINUS SIGN}}$$

$$(1) \qquad \dfrac{dy}{dx} = \dfrac{d}{dx}\left[\dfrac{a + x}{b - x}\right] = \dfrac{(b - x)\dfrac{d}{dx}[a + x] - (a + x)\dfrac{d}{dx}[b - x]}{(b - x)^2}$$

by the Quotient Rule. Using the previous rules to compute

$$\dfrac{d}{dx}[a + x] = 1 \quad \text{and} \quad \dfrac{d}{dx}[b - x] = -1$$

(remember: a and b are constants!) we obtain:

$$= \dfrac{(b - x) + (a + x)}{(b - x)^2}$$

$$= \dfrac{a + b}{(b - x)^2}$$

Thus $\qquad 6 = \left.\dfrac{dy}{dx}\right|_{x = 1} = \dfrac{a + b}{(b - 1)^2}$

3.2.11

(2)   When x = 0 , then

$$2 = y = \frac{a + 0}{b - 0} = \frac{a}{b}$$

Hence our two simultaneous equations are

$$(1) \quad \frac{a + b}{(b - 1)^2} = 6$$

$$(2) \quad \frac{a}{b} = 2$$

Solving (2) for a we have a = 2b ; plugging into (1) yields

$$3b = 6(b - 1)^2 = 6b^2 - 12b + 6$$

$$0 = 6b^2 - 15b + 6$$

$$0 = 2b^2 - 5b + 2 \qquad \text{(dividing both sides by 3)}$$

Factoring this equation* we obtain

$$0 = (2b - 1)(b - 2)$$

so that $b = \frac{1}{2}$ or 2 . Then a = 1 or 4 , respectively.

We thus have two possible solutions to our problem:

$$y = \frac{1 + x}{(1/2) - x} \qquad \text{or} \qquad y = \frac{4 + x}{2 - x} \qquad \square$$

3. <u>Polynomials and rational functions.</u> In §2.2 Anton gave the general expression for a <u>polynomial,</u>

$$p(x) = a_n x^n + a_{n-1} x^{n-1} + \cdots + a_1 x + a_0$$

---

* See Appendix D.5 if this gives you trouble.

and for a <u>rational function</u>,

$$r(x) \; = \; \frac{p(x)}{q(x)}$$

where p and q are both polynomials. Thus, since polynomials and rational functions are just sums, differences, products and quotients of powers of x , then

<table>
<tr>
<td>

Derivatives
of polynomials
and rational
functions

</td>
<td>

any polynomial or rational function is

differentiable (where it is defined),

and its derivative can be computed

by the rules of this section.

</td>
</tr>
</table>

In fact, every example in this section of Anton and <u>The Companion</u> is a polynomial or a rational function, although in some cases the functions are not written in a standard way. *

Thus, you now are able to compute the derivative of any polynomial or rational function. In the next section we'll add the trigonometric functions to our list ... , and then even more functions in the subsequent sections ....

4. <u>Higher derivatives</u>.   Not much needs to be added to Anton's discussion of higher derivatives. To compute an n-th derivative, you simply repeat the differentiation process n times as shown in Anton's Example 9 .

---

*   For instance, in Example B(a), $f(x) = 3x^{-4} + 4x$ is <u>not</u> a polynomial since negative powers are not allowed in a polynomial. However, multiplication by (x^4/x^4) yields

$$f(x) \; - \; \frac{3 + 4x^5}{x^4}$$

which shows f to be a rational function.

You do need to become familiar with the various notations for the n-th derivative:
if $y = f(x)$, then

$$f^{(n)}(x) = \frac{d^n}{dx^n}[f(x)] = \frac{d^n y}{dx^n} = y^{(n)}$$

all stand for the n-th derivative.

Applications of higher derivatives will be discussed in later chapters, for example
with convexity (§4. 2), maximization problems (§4. 3), numerical integration (Exercise 10
in §9. 9) and infinite series (§11. 9).

Example E.   Show that $y = x^2 - x^{-2}$ satisfies the equation

$$x^2 y'' + xy' - 4y = 0$$

Solution.   First we compute the first and second derivatives of y :

$$y = x^2 - x^{-2}$$
$$y' = 2x + 2x^{-3} \qquad \text{using} \quad y' = \frac{d}{dx}[y]$$
$$y'' = 2 - 6x^{-4} \qquad \text{using} \quad y'' = \frac{d}{dx}[y']$$

Thus we obtain

$$-4y = -4x^2 + 4x^{-2}$$
$$xy' = 2x^2 + 2x^{-2}$$
$$x^2 y'' = 2x^2 - 6x^{-2}$$

Adding these three equations gives

$$x^2 y'' + xy' - 4y = 0$$

as we wanted to show.   □

Section 3.3:   Derivatives of Trigonometric Functions

$$\begin{bmatrix} \text{In addition to Anton's trigonometry review in Appendix I,} \\ \underline{\text{The Companion}} \text{ has a somewhat shorter review section} \\ \text{in Appendix G.} \end{bmatrix}$$

Perhaps you'd like to know why we put so much time and effort into trigonometry. Too often people come away from trigonometry courses with the idea that trigonometry is important primarily because of its uses in surveying and navigation. Well, trigonometry plays the fundamental role in these subjects, but the real importance of trigonometry in mathematics, science and engineering lies in its description of "periodic phenomena," i.e., quantities which have a regular, repetitive nature. Vibrating strings, the orbits of planets, alternating electrical current, sound and light waves -- all these physical entities need trigonometric functions to describe them.

But we need more than just the trigonometric functions themselves; we also need their calculus properties. In this section we start this study by obtaining the derivatives of the trigonometric functions.

1. <u>Trigonometric derivatives.</u> The major item of this section is the list of six trigonometric derivatives:

$$\frac{d}{dx}[\sin x] = \cos x \qquad\qquad \frac{d}{dx}[\cos x] = -\sin x$$

$$\frac{d}{dx}[\tan x] = \sec^2 x \qquad\qquad \frac{d}{dx}[\cot x] = -\csc^2 x$$

$$\frac{d}{dx}[\sec x] = \sec x \tan x \qquad\qquad \frac{d}{dx}[\csc x] = -\csc x \cot x$$

As with the formulas of the previous section, <u>you must learn these well.</u> This will take a bit of concentration, although in Exercise 51 Anton gives a method that helps considerably in remembering the formulas. Also, if you ever get stuck, you can derive all the formulas from just the $\sin x$ and $\cos x$ derivatives. Anton does this for $\tan x$; we'll do it for $\sec x$:

<u>Example A.</u>   Derive the formula for $\dfrac{d}{dx} [\sec x]$.

<u>Solution.</u> $\dfrac{d}{dx} [\sec x] = \dfrac{d}{dx} \left[\dfrac{1}{\cos x} \right]$ by definition of $\sec x$

$$= \frac{(\cos x) \dfrac{d}{dx} [1] - 1 \cdot \dfrac{d}{dx} [\cos x]}{\cos^2 x} \qquad \text{by the Quotient Rule (Theorem 3.2.6)}$$

$$= \frac{(\cos) \cdot 0 - (-\sin x)}{\cos^2 x} \qquad \text{since } \dfrac{d}{dx} [1] = 0$$
$$\text{and } \dfrac{d}{dx} [\cos x] = -\sin x ,$$

$$= \left(\frac{1}{\cos x} \right) \left(\frac{\sin x}{\cos x} \right) = \sec x \tan x \qquad \square$$

*   *   *   *   *

All the differentiation rules of §3.2 can be applied to trigonometric functions (in fact, the Quotient Rule was just used in Example A). Hence we are able to compute the derivatives of sums, differences, products and quotients of any mixture of trig functions, polynomials, and rational functions.

<u>Example B.</u>   Compute the following derivatives:

a. $f'(x)$, where $f(x) = x^3 \sec x$

b. $\dfrac{d}{d\theta} \left[\dfrac{\theta^2 \sin \theta}{1 + \cos \theta} \right] \Bigg|_{\theta = \pi/2}$

Solution.

a. $f'(x) = \dfrac{d}{dx}[f(x)]$ since these are equivalent notations

     $= \dfrac{d}{dx}[x^3 \sec x]$

       $= x^3 \dfrac{d}{dx}[\sec x] + (\sec x)\dfrac{d}{dx}[x^3]$ by the Product Rule

       $= x^3 \sec x \tan x + 3x^2 \sec x$

b. $\dfrac{d}{d\theta}\left[\dfrac{\theta^2 \sin\theta}{1 + \cos\theta}\right] = \dfrac{(1 + \cos\theta)\dfrac{d}{d\theta}[\theta^2 \sin\theta] - (\theta^2 \sin\theta)\dfrac{d}{d\theta}[1 + \cos\theta]}{[1 + \cos\theta]^2}$

     Quotient Rule

However,

       Product Rule

       $\dfrac{d}{d\theta}[\theta^2 \sin\theta] = \theta^2 \dfrac{d}{d\theta}[\sin\theta] + (\sin\theta)\dfrac{d}{d\theta}[\theta^2]$

          $= \theta^2 \cos\theta + 2\theta \sin\theta$

       $\dfrac{d}{d\theta}[1 + \cos\theta] = 0 - \sin\theta = -\sin\theta$

       Thus we have

       $= \dfrac{(1 + \cos\theta)[\theta^2 \cos\theta + 2\theta \sin\theta] - (\theta^2 \sin\theta)[-\sin\theta]}{[1 + \cos\theta]^2}$

We wish to evaluate this derivative at $\theta = \pi/2$. Since $\cos(\pi/2) = 0$ and

$\sin(\pi/2) = 1$ we obtain

       $\dfrac{d}{d\theta}\left[\dfrac{\theta^2 \sin\theta}{1 + \cos\theta}\right]\Bigg|_{\theta = \pi/2} = \dfrac{(1)[\pi] - (\pi^2/4)[-1]}{1^2} = \pi + \dfrac{\pi^4}{4}$

$\square$

2. <u>Trigonometric limits.</u>   The calculation of the derivatives for   sin x   and   cos x   were based

on two important limits:

$$
\begin{array}{lll}
(1) & \displaystyle\lim_{h \to 0} \frac{\sin h}{h} = 1 & \text{Theorem 3.3.4} \\[4mm]
(2) & \displaystyle\lim_{h \to 0} \frac{\cos h - 1}{h} = 0 & \text{Theorem 3.3.2}
\end{array}
$$

These can in turn be used to show:

$$
\begin{array}{lll}
(3) & \displaystyle\lim_{h \to 0} \sin h = 0 & \text{Exercise 42*} \\[4mm]
(4) & \displaystyle\lim_{h \to 0} \cos h = 1 & \text{Corollary 3.3.3}
\end{array}
$$

These latter two limits should not be surprising given our knowledge of the graphs of   sin x   and

cos x .   However, the first two limits are certainly <u>not</u> obvious, since in each case both the

numerator and denominator have limit zero as   h → 0 .   Memorize these limits well; they will

be used again and again!

Given the above results, many other trigonometric limits of this type can be computed by

"jiggling" them into combinations of limits (1) through (4).

<u>Example C.</u>   Compute the limits which exist.

---

* The result is easily established as follows:

$$
\lim_{h \to 0} \sin h = \lim_{h \to 0} \left(\frac{\sin h}{h} \cdot h \right) = \lim_{h \to 0} \frac{\sin h}{h} \cdot \lim_{h \to 0} h = 1 \cdot 0 = 0
$$

a. $\lim\limits_{\theta \to 0} \dfrac{\theta}{\tan \theta}$

b. $\lim\limits_{\theta \to 0} \dfrac{\sin 4\theta}{\sin 3\theta}$

c. $\lim\limits_{x \to 0} \dfrac{\sin x}{x^2}$

d. $\lim\limits_{h \to 0} \dfrac{1 - \cos h}{\tan h}$

Solution.

a. $\lim\limits_{\theta \to 0} \dfrac{\theta}{\tan \theta} = \lim\limits_{\theta \to 0} \dfrac{\theta \cos \theta}{\sin \theta}$, using the definition of $\tan \theta$. Certainly $\cos \theta$ is

no problem since, from (4) above, it goes to a <u>non-zero</u> limit as $\theta \to 0$. If $\cos \theta$ is

thus removed, then we are left with $\dfrac{\theta}{\sin \theta}$ ... oh, that's just the reciprocal of $\dfrac{\sin \theta}{\theta}$,

whose limit as $\theta \to 0$ from (1) is known to be 1 . We thus have

$$\lim\limits_{\theta \to 0} \frac{\theta}{\tan \theta} = \lim\limits_{\theta \to 0} \cos \theta \left[\frac{1}{\frac{\sin \theta}{\theta}}\right] \xleftarrow{\boxed{\text{THEOREM 2.5.1}}} = \lim\limits_{\theta \to 0} \cos \theta \cdot \frac{1}{\lim\limits_{\theta \to 0} \frac{\sin \theta}{\theta}}$$

$$= 1 \cdot \frac{1}{1} = 1$$

b. $\lim\limits_{\theta \to 0} \dfrac{\sin 4\theta}{\sin 3\theta}$. The trick here is to observe that, by replacing h in limit (1) by 4θ

and 3θ respectively, we obtain

$$\lim\limits_{4\theta \to 0} \frac{\sin 4\theta}{4\theta} = 1 \qquad \text{and} \qquad \lim\limits_{3\theta \to 0} \frac{\sin 3\theta}{3\theta} = 1$$

However, since $4\theta \to 0$ and $\theta \to 0$ are equivalent (as are $3\theta \to 0$ and $\theta \to 0$),

this gives

$$\lim_{\theta \to 0} \frac{\sin 4\theta}{4\theta} = 1 \qquad \text{and} \qquad \lim_{\theta \to 0} \frac{\sin 3\theta}{3\theta} = 1$$

We can now "jiggle" the given limit into these forms as follows:

$$\lim_{\theta \to 0} \frac{\sin 4\theta}{\sin 3\theta} = \lim_{\theta \to 0} \frac{\left(\frac{4\theta}{4\theta}\right)\sin 4\theta}{\left(\frac{3\theta}{3\theta}\right)\sin 3\theta} = \lim_{\theta \to 0} \left(\frac{4}{3}\right)\left(\frac{\sin 4\theta}{4\theta}\right)\left(\frac{3\theta}{\sin 3\theta}\right)$$

$$= \frac{4}{3}\left(\lim_{\theta \to 0} \frac{\sin 4\theta}{4\theta}\right)\left(\frac{1}{\lim\limits_{\theta \to 0} \frac{\sin 3\theta}{3\theta}}\right) = \frac{4}{3}\,(1)\,(1) = \frac{4}{3}$$

c. $\lim\limits_{x \to 0} \dfrac{\sin x}{x^2} = \lim\limits_{x \to 0} \left(\dfrac{\sin x}{x}\right)\left(\dfrac{1}{x}\right)$ ... hmm!? We would like to continue with

$$\ldots = \lim_{x \to 0} \left(\frac{\sin x}{x}\right) \cdot \lim_{x \to 0} \left(\frac{1}{x}\right) = 1 \cdot \lim_{x \to 0} \left(\frac{1}{x}\right) = \ldots \ ?$$

but this is not a correct use of Theorem 2.5.1(c) (the limit of a product) because

the second limit, $\lim\limits_{x \to 0} \dfrac{1}{x}$, is <u>not defined.</u> Since the first <u>non-zero</u> limit cannot

counterbalance the second non-existent limit, then the full limit is <u>undefined.</u>

d. $\lim\limits_{h \to 0} \dfrac{1 - \cos h}{\tan h} = \lim\limits_{h \to 0} \left(\dfrac{1 - \cos h}{\sin h}\right) \cos h$. The multiplicative $\cos h$ term gives

no worries since it goes to 1 as $h \to 0$. For the remaining terms, we observe that

division by h would be very helpful since it yields limits we know:

$$\lim_{h \to 0} \frac{1 - \cos h}{\tan h} = \lim_{h \to 0} \left(\frac{1 - \cos h}{h}\right)\left(\frac{h}{\sin h}\right) \cos h$$

$$= (0)\,(1)\,(1) = 0 \qquad\qquad \square$$

In general, given limits like those in Example C , there are very few choices to be made in their

solutions: you <u>must</u> "jiggle" them into combinations of the four limits you know.

3. <u>The Pinching Theorem.</u>   The two major limits of this section,

$$\lim_{h \to 0} \frac{\sin h}{h} = 1 \qquad \text{and} \qquad \lim_{h \to 0} \frac{\cos h - 1}{h} = 0$$

were both established using the Pinching Theorem (Theorem 3.3.1). This theorem will occur

again and again; fortunately it is a very

intuitive result, as the picture to the right

shows.

The picture shows that, under the stated conditions, there simply is nowhere else for f(x) to

go except to L ; it is "pinched" to L .

When attempting to compute a limit $\lim_{x \to a} f(x)$, the Pinching Theorem is valuable when

appropriate functions g and h can be found <u>which are simpler than f</u>! In particular, it

---

The Pinching Theorem

If f(x) is "pinched" inbetween two functions

g(x) and h(x) for all x values near x = a ,

<u>and</u> if g(x) and h(x) both approach the

same limit L as x approaches a ,

then $\lim_{x \to a} f(x)$ exists and also equals L .

should be much easier to compute $\lim\limits_{x \to a} g(x)$ and $\lim\limits_{x \to a} h(x)$ than it would be to compute

$\lim\limits_{x \to a} f(x)$. Notice this pattern in our two major limits:

$$\lim_{h \to 0} \frac{\sin h}{h} = 1 \qquad \text{follows from} \quad \cos h < \frac{\sin h}{h} < 1 \qquad \text{for} \quad -\frac{\pi}{2} < h < \frac{\pi}{2}, h \neq 0$$

$$\lim_{h \to 0} \frac{1 - \cos h}{h} = 0 \quad \text{follows from} \quad 0 \leq \frac{1 - \cos h}{h} \leq \frac{1}{2} h \quad \text{for} \quad h > 0$$

$$\text{and} \quad \frac{1}{2} h \leq \frac{1 - \cos h}{h} \leq 0 \quad \text{for} \quad h < 0$$

Example D.   Suppose f is a function that satisfies

$$x^2 + x - 3 \leq f(x) \leq 2x^2 - 3x + 1$$

for all x . Does $\lim\limits_{x \to 2} f(x)$ exist? If so, compute it.

Solution.   Let $g(x) = x^2 + x - 3$ and $h(x) = 2x^2 - 3x + 1$. Then

$$\lim_{x \to 2} g(x) = 4 + 2 - 3 = 3$$

$$\lim_{x \to 2} h(x) = 8 - 6 + 1 = 3$$

Ahh..., these limits are the same, so the Pinching Theorem applies to give

$$\lim_{x \to 2} f(x) = 3$$

If the limits of g and h had been <u>unequal</u> at x = 2 , then we could not have drawn any

conclusion about $\lim\limits_{x \to 2} f(x)$.   ☐

4.   <u>Now about this radian measure</u>....   We are now in a better position to explain why we insist

on using <u>radian measure</u> for angles instead of the more familar units of <u>degrees</u>.   It's very

simple: all our calculus formulas (limits, derivatives and, later on, integrals) would have

very unpleasant multiplicative factors if we had angles measured in degrees.   For example,

when   h   is measured in radians we have

$$\lim_{h \to 0} \frac{\sin h}{h} = 1 \qquad [\,h \text{ in radians }]$$

However, if   h   is measured in degrees, then

$$\lim_{h \to 0} \frac{\sin h}{h} = \frac{\pi}{180} \qquad [\,h \text{ in degrees }]$$

This, in turn, would take our pleasing derivative formula

$$\frac{d}{dx} [\,\sin x\,] = \cos x \qquad [\,x \text{ in radians }]$$

and change it into the not-so-pleasing

$$\frac{d}{dx} [\,\sin x\,] = \frac{\pi}{180} \cos x \qquad [\,x \text{ in degrees }]$$

Since these multiplicative factors would be a constant headache, the use of radian measure

is clearly to be preferred over degrees when dealing with the calculus of trigonometric functions.

Section 3.4: Δ-Notation; Differentials

1. <u>More notation (!?!): increments and differentials.</u> Suppose we are considering a particular

    point $P(x,y)$ on the graph of a function $y = f(x)$. Let's be clear as to what we mean by the

    <u>increments</u> Δx, Δy and the <u>differentials</u> dx, dy at $P(x,y)$:

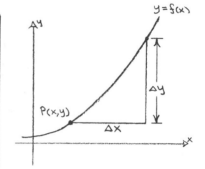

Increments Δx, Δy

Δx is any small change in the variable x.

Δy is the corresponding change in the

    variable y <u>along the curve</u>, i.e.,

$$\Delta y = f(x + \Delta x) - f(x)$$

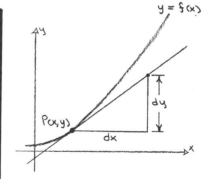

Differentials dx, dy

dx is any small change in the variable x,

dy is the corresponding change in the

    variable y <u>along the tangent line</u>

<u>at (x,y)</u>, i.e.,

$$dy = f'(x)\,dx$$

Setting Δx and dx equal to each other (as Anton generally

does) produces the picture to the right, showing the

relationship between Δy and dy.

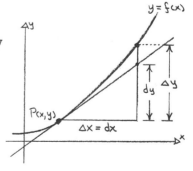

The formula $dy = f'(x)\,dx$ is valid for the following reason: By the definition of the differentials dx and dy, their quotient must equal the slope of the tangent line, which is $f'(x)$:

$$\frac{dy}{dx} = \left[\begin{array}{c}\text{slope of tangent}\\\text{line at } P(x,y)\end{array}\right] = f'(x)$$

Thus the <u>quotient of differentials</u> $\frac{dy}{dx}$ agrees with our earlier usage of $\frac{dy}{dx}$ as a <u>symbol for</u> <u>the derivative</u> ... it was, of course, planned that way!

The increments are related to the differentials in the following way. The <u>quotient of increments</u> is the slope of the secant line joining $P(x,y)$ to $Q(x + \Delta x, y + \Delta y)$. Thus, taking the limit as Δx approaches zero, our quotient approaches the slope of the tangent line at $P(x,y)$, i.e.,

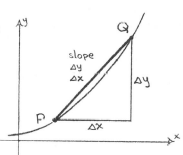

$$\boxed{\lim_{\Delta x \to 0} \frac{\Delta y}{\Delta x} = \underset{\substack{\text{slope of tangent}\\\text{line at } P(x,y)}}{f'(x)} = \frac{dy}{dx}}$$

Here are some simple examples involving increments and differentials:

<u>Example A.</u> Suppose $y = 2x^2 - x$.

    a.   Find Δy if $\Delta x = 2$ and the initial value of x is $x = 1$.

    b.   Find dy if $dx = 2$ and the initial value of x is $x = 1$.

<u>Solution.</u>   a.   To find Δy we use the defining formula for Δy, with $y = f(x) = 2x^2 - x$:

$$\Delta y = f(x + \Delta x) - f(x)$$
$$= f(3) - f(1) \qquad \text{since } x = 1,\ \Delta x = 2$$
$$= 15 - 1 = \boxed{14}$$

b.   To find   dy   we must first compute   f'(x) :

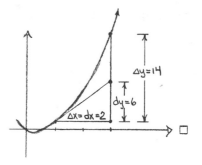

$$f'(x) = \frac{d}{dx} [\, 2x^2 - x \,] = 4x - 1$$

Thus, by the defining formula for   dy :

$$dy = f'(x)\, dx = f'(1) \cdot 2 = 3 \cdot 2 = \boxed{6}$$

Example B.   Suppose $y = \dfrac{\sin x}{1 + x^2}$.   Find   dy .

Solution.   When no specific values of   x   or   dx   are specified, then the answer for   dy   will involve the symbols   x   and   dx .   We must first compute   f'(x) :

$$f'(x) = \frac{d}{dx}\left[\frac{\sin x}{1 + x^2} \right] = \frac{(1 + x^2)\,\frac{d}{dx}[\,\sin x\,] - (\sin x)\,\frac{d}{dx}[\,1 + x^2\,]}{(1 + x^2)^2}$$

$$= \frac{(1 + x^2)\cos x - 2x \sin x}{(1 + x^2)^2}$$

Thus, using the defining formula for   dy   will give

$$dy = f'(x)\, dx = \boxed{\frac{(1 + x^2)\cos x - 2x \sin x}{(1 + x^2)^2}\, dx}$$

As you can see, the computation of a differential is really nothing but the computation of a derivative!

2. <u>Tangent line approximations.</u>   Anton shows that

the equation for the tangent line to the graph

of $y = f(x)$ at the point $(x_0, f(x_0))$ is

given by

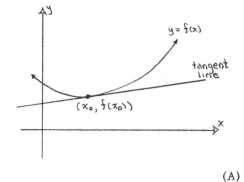

Tangent Line to $y = f(x)$ at x_0	$$y = f(x_0) + f'(x_0)(x - x_0)$$	(A)

However, if $\Delta x = x - x_0$ is small, then

notice (in the diagram to the right) how close the

y-value   corresponding to   x   on the <u>graph</u> is

to the   y-value   corresponding to   x   on the

<u>tangent line</u> (i. e. , the vertical closeness of Q_1

and Q_2). The   y-value   on the graph is

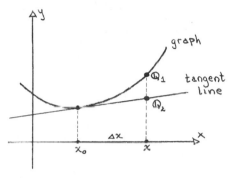

merely $f(x)$; the   y-value on the tangent line is given by the tangent line equation   (A).

Hence

$$f(x) \approx f(x_0) + f'(x_0)(x - x_0)$$

when $\Delta x = x - x_0$ is small .

Substituting $x = x_0 + \Delta x$ we obtain

Linear Approximation of $f(x)$ near x_0	$$f(x_0 + \Delta x) \approx f(x_0) + f'(x_0)\Delta x$$ when Δx is small	(B)

This is Anton's Equation (11).

Prior to Example 3 Anton describes how this is used: if you need an approximate value for $f(x)$, but $f(x)$ is difficult to compute directly, then find some x_0 close to x for which $f(x_0)$ and $f'(x_0)$ are easy to compute. Then $f(x) = f(x_0 + \Delta x)$ can be approximated by (B) using $\Delta x = x - x_0$. Anton illustrates this procedure in Examples 3 and 4.

3. <u>Error Propagation.</u> Suppose a quantity x is <u>measured</u> to within error limits of $\pm \Delta x$. Further suppose that a quantity y is <u>calculated</u> from x by a function formula $y = f(x)$. What are the possible error limits $\pm \Delta y$ for y that result from the possible errors in x?

Well, if in our increment/differentials picture we take the independent variables Δx and dx to be equal $(\Delta x = dx)$ and <u>very small</u>, then Δy and dy will closely approximate each other, i.e.,

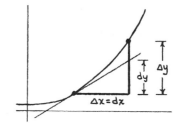

$$\Delta y \approx dy \quad \text{if} \quad \Delta x = dx \quad \text{is small, i.e.,}$$

Error Propagation	$\Delta y \approx f'(x) \Delta x$ if $\Delta x = dx$ is small.

Thus the possible error bounds on y are approximately the error bounds on x multiplied by the derivative of f at x.

<u>Example C.</u> According to the ideal gas law, the pressure P exerted by a confined gas is related to its volume V and its temperature T (measured in degrees Kelvin) by

$$PV = kT$$

where k is a constant depending on the amount of gas present. Suppose $T = 300^{\circ}$, and V is measured to be 20 m^3 with a maximum error of $.2 \text{ m}^3$.

a. Use differentials to estimate the maximum error in the calculated pressure.

b. Estimate the maximum relative errors in the volume and pressure.

Solution. a. The example asks us to estimate the maximum value of ΔP given $.2$ as a maximum value for ΔV. Thus we need to express P as a function of V :

$$P = f(V) = \frac{kT}{V} = \frac{300\,k}{V}$$

We now can find ΔP by our error propagation formula:

$$\Delta P \approx dP = f'(V)\,\Delta V$$

$$= \left(- \frac{300\,k}{V^2}\,\Delta V \right) \quad \text{since} \quad f'(V) = \frac{d}{dV}(300\,k\,V^{-1})$$

$$= (-1)(300\,k)\,V^{-2} - \frac{300\,k}{V^2}$$

$$= \left(- \frac{300\,k}{400} \right)(.2) \quad \text{since} \quad V = 20,\ \Delta V = .2$$

$$= -.15\,k$$

Thus the error in the calculated pressure should be at most $\pm.15\,k$, a value that depends on k.

b. If an error of Δx is made in a quantity x, then the relative error is $\frac{\Delta x}{x}$. (When expressed as a percentage, then $\Delta x/x$ is called the percentage error.) The maximum percentage error in our volume V is given by

$$\left| \frac{\Delta V}{V} \right| = \frac{.2}{20} = .01 = 1\%$$

The maximum percentage error in the pressure P is given by

$$\left| \frac{\Delta P}{P} \right| \approx \left| \frac{dP}{P} \right| = \frac{.15\,k}{P} \quad \text{since} \quad dP = -.15\,k \quad \text{from part (a)},$$

$$= \frac{.15\,k}{15\,k} \quad \text{since} \quad P = \frac{300\,k}{V} = \frac{300\,k}{20} = 15\,k$$

$$= .01 = 1\% \qquad \square$$

Percentage error problems are more common than the ordinary (absolute) error estimations... and a bit harder! Here's one that's trickier than most, but which illustrates a technique that is important in applications:

Example D. The electrical resistance R of a certain wire is given by $R = k/r^2$ where k is a constant and r is the radius of the wire. Estimate the maximum permissible percentage error in the measured value of r if the percentage error in R must be kept within $\pm 3\%$.

Solution.   Translating the last sentence into variables we have

"Estimate the maximum value of $\left|\dfrac{\Delta r}{r}\right|$ if $\left|\dfrac{\Delta R}{R}\right|$ must be less than or equal to $.03$ "

Our plan will be to express this inequality in terms of r and see if we can extract a bound for $\left|\dfrac{\Delta r}{r}\right|$ out of the result.   We proceed as follows:

Since $R = k r^{-2}$, then (using our error propagation equation $\Delta R \approx f'(r)\Delta r$)

$$\Delta R \approx -2 k r^{-3}\,\Delta r$$

Thus
$$\left|\frac{\Delta R}{R}\right| \approx \left|\frac{-2 k r^{-3}\,\Delta r}{k r^{-2}}\right| = \left|\frac{2\,\Delta r}{r}\right| = 2\left|\frac{\Delta r}{r}\right|$$

Therefore, in order to ensure that $\left|\dfrac{\Delta R}{R}\right| \le .03$, we must have

$$2\left|\frac{\Delta r}{r}\right| \le .03 , \qquad \text{or} \qquad \left|\frac{\Delta r}{r}\right| \le .015$$

Thus the maximum permissible percentage error in the measured value of r is 1.5%. □

4. Other applications of differentials.   The theory of differentials led to our developement of both tangent line approximations and error propagation equations. However, the most important use of differentials will not come until §5.3 when we encounter the method of u-substitution for integration. The time spent now on differentials will help later with u-substitution.

Section 3.5 :   The Chain Rule

   In §3.2 we learned how to compute the derivatives of sums, differences, products and quotients of differentiable functions. In this section we learn how to compute the derivative of a <u>composition</u> of differentiable functions by use of the Chain Rule.   This rule is the single most important law of differentiation.

1. <u>The Chain Rule: the dy/dx version.</u>   The Chain Rule is Theorem 3.5.2 :

The Chain
Rule;
dy/dx version

> Suppose g is differentiable at x and f is differentiable at u = g(x) .
>
> Then the composition f ∘ g is differentiable at x . Moreover,
>
> $$\quad\quad\text{if}\quad\quad y = f(u)\quad\text{and}\quad u = g(x)$$
>
> $$(\text{so that}\quad y = f(g(x))) ,$$
>
> then
>
> $$\frac{dy}{dx} = \frac{dy}{du} \cdot \frac{du}{dx}$$

Thus, the derivative of the composition $y = f(g(x))$ is the product of two intermediate derivatives. In the dy/dx notation the Chain Rule looks almost "obvious" : just "cancel" the du's in the numerator and denominator. Although far from being a proof, this makes the Chain Rule easy to remember!

   An extremely good <u>intuitive</u> justification for the Chain Rule is provided through the rates of change interpretation. As Anton points out in his initial discussion, if y changes at 4 times the rate of u $\left(\frac{dy}{du} = 4 \right)$, and u changes at 2 times the rate of x $\left(\frac{du}{dx} = 2 \right)$, then y must change at $4 \times 2 = 8$ times the rate of x , i.e.,

$$\frac{dy}{dx} = \frac{dy}{du} \cdot \frac{du}{dx} = 4 \times 2 = 8$$

The first hurdle to overcome in using the Chain Rule is learning to recognize when a function is the composition of two simpler functions. * We'll focus on this in the next few examples.

Example A.   Find $f'(x)$ if $f(x) = 3 \sin(x^3 + 1)$.

Solution.   How do we see that f is a composition? Well, if you compute $f(x)$ for a specific value of x ,   what do you do? First you cube x and add   1,   and then take   3 times the sine of this result... ahh, a second operation on the output of the first operation. Thus we can set

$$u = x^3 + 1 \qquad \text{(the first operation)}$$

$$y = 3 \sin u \qquad \text{(the second operation)}$$

so that $y = 3 \sin u = 3 \sin(x^3 + 1)$ is our original function now seen as a composition.

Thus $$\frac{dy}{dx} = \frac{dy}{du} \cdot \frac{du}{dx}$$

$$= \frac{d}{du}[\,3 \sin u\,] \cdot \frac{d}{dx}[\,x^3 + 1\,]$$

$$= (3 \cos u)(3x^2)$$

$$= 9x^2 \cos u$$

$$= 9x^2 \cos(\underbrace{x^3 + 1})$$

> notice that in our final answer we eliminate   u by substituting $u = x^3 + 1$. Don't forget this step!

☐

---

* This was discussed at length in §2.2.3 of <u>The Companion</u>. You should read this subsection over again, especially Examples F , G and H !

<u>Example B.</u>   Find   ds/dt   if $s = 3 \sin^3 t + 1$.

<u>Solution.</u>   Given a value of   t ,   our first operation is to compute

$$u = \sin t \qquad \text{(the first operation)}$$

and then take that result, cube it, multiply by   3 ,   and add   1 :

$$s = 3 u^3 + 1 \qquad \text{(the second operation)}$$

Thus $\dfrac{ds}{dt} = \dfrac{ds}{du} \cdot \dfrac{du}{dt}$

$$= \frac{d}{du} [3 u^3 + 1] \cdot \frac{d}{dt} [\sin t]$$

$$= (9 u^2)(\cos t)$$

$$= 9 \sin^2 t \cos t \qquad \text{(Don't forget to eliminate   u   by}$$
$$\text{substituting   u} = \sin t .)$$

There are two important comments to be made about these examples.   First, notice how the use of   t   and   s   instead of   x   and   y   in Example B causes no problem in using the Chain Rule; we just change the variables in the Chain Rule accordingly (also see Anton's Example 2).   Second, in both examples <u>we had a choice</u> of how to decompose the given function:

<u>Example A</u> : $y = 3 \sin (x^3 + 1)$

we chose: $y = 3 \sin u$ and $u = x^3 + 1$

another choice: $y = 3 \sin (u + 1)$ and $u = x^3$

<u>Example B</u> : $s = 3 \sin^3 t + 1$

we chose: $s = 3 u^3 + 1$ and $u = \sin t$

another choice: $s = 3 u + 1$ and $u = \sin^3 t$

still another choice: $s = u + 1$ and $u = 3 \sin^3 t$

Why did we make the choices as we did? In each example <u>we chose the decomposition which provided the two easiest functions to differentiate!</u> Had we chosen a different decomposition it would not have been a disaster --- but we would have needed to use the Chain Rule for a second time on one of the two "subfunctions, " certainly a waste of time.

<u>The Moral:</u> Choose your decomposition so that the two subfunctions are easiest to differentiate.

2.   <u>The Chain Rule: the (f ∘ g)' (x) version.</u>   In Equation (5) Anton gives another notational way to state the Chain Rule which is extremely useful in computations:

The Chain Rule: (f ∘ g)' version

> Suppose g is differentiable at x and f is differentiable at u = g(x) .
>
> Then the composition f ∘ g is differentiable at x . Moreover,
>
> $$(f \circ g)'(x) = f'(g(x)) \cdot g'(x)$$

Yes, this really is the same rule as before! To see this, consider the earlier notation,

$$y = (f \circ g)(x) = f(g(x)) = f(u) , \quad \text{where} \quad u = g(x) \quad .$$

Then the original Chain Rule,

$$\frac{dy}{dx} = \frac{dy}{du} \cdot \frac{du}{dx}$$

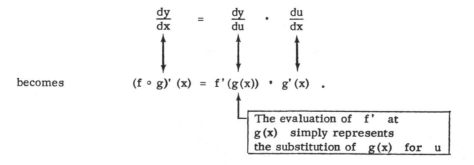

becomes

$$(f \circ g)'(x) = f'(g(x)) \cdot g'(x) \quad .$$

The evaluation of f' at g(x) simply represents the substitution of g(x) for u

The new version of the Chain Rule may not look as memorable or intuitive as the original version, but when set into words, <u>it is precisely the way that the Chain Rule is used in practice</u>:

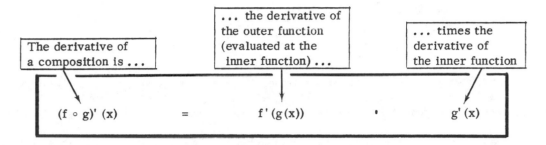

Memorize this sentence!! It will give you an <u>extremely efficient</u> method of using the Chain Rule, as we will now illustrate.

<u>Example C.</u>   Find $\dfrac{d}{dx}$ [2 cos $(x^3 - x)$] .

<u>Solution.</u>   We could proceed as in the previous examples, introducing the variables $u = x^3 - x$ and $y = 2 \cos u \ldots$, etc. However, <u>our aim is to eliminate the need for such additional variables!</u>

We start by observing that $2 \cos (x^3 - x)$ is a composition of an <u>inner function</u> $u = x^3 - x$ with an <u>outer function</u> $y = 2 \cos u$ (we merely observe this; we do not write it down unless it is absolutely necessary). We then proceed as follows:

$\dfrac{d}{dx}$ [2 cos $(x^3 - x)$] = ...   "The derivative of
                                                          a composition is ...

= - 2 sin (...   "... the derivative of
                                                          the outer function ...

= - 2 sin $(x^3 - x)$ ...   "... (evaluated at the
                                                          inner function) ...

= - 2 sin $(x^3 - x)$ • $(3x^2 - 1)$ "... times the derivative
                                                          of the inner function. "

So what actually appears on your paper? Nothing but the answer which you have written down in one step (!): $\dfrac{d}{dx} [\, 2 \cos (x^3 - x)\,] = -2 \sin (x^3 - x) \cdot (3x^2 - 1)$

You might wish, of course, to reorder the result:

$$\dfrac{d'}{dx} [\, 2 \cos (x^3 - x)\,] = -2\,(3x^2 - 1) \sin (x^3 - x) = (2 - 6x^2) \sin (x^3 - x)$$

but this is just "tidying up." We still have essentially a one-line procedure. ☐

Obviously your success with this method will depend on how well you are able to "visually decompose" a given function into a composition. This simply takes practice, so do lots and lots of Anton's exercises. We cannot over-emphasize how important the Chain Rule is! Here are three more examples, each illustrating some common procedure involving the Chain Rule.

<u>Example D</u>. Find $\dfrac{dw}{dz}$ where $w = z^2 (z^3 + 1)^3$.

<u>Solution</u>. In attempting to differentiate w with respect to z we first encounter a <u>product</u> of z^2 with $(z^3 + 1)^3$. Thus we start off with the product rule:

$$\frac{dw}{dz} = \frac{d}{dz} [\, z^2\,(z^3 + 1)^3\,] = z^2\,\frac{d}{dz} [\,(z^3 + 1)^3\,] + (z^3 + 1)^3\,\frac{d}{dz} [\, z^2\,]$$

Hmmm.... We'll need the <u>Chain Rule</u> for this derivative...

... The inner function is $u = z^3 + 1$ while the outer function is u^3.

Thus the derivative is $3(z^3 + 1)^2 \cdot (3z^2)$

$$= z^2 [\, 3(z^3 + 1)^2 \cdot (3z^2)\,] + (z^3 + 1)^3\,(2z) \qquad \text{Here's the answer.}$$

$$= 9z^4\,(z^3 + 1)^2 + 2z\,(z^3 + 1)^3$$

$$= z(z^3 + 1)^2\,[\, 9z^3 + 2z^3 + 2\,]$$

$$= z(z^3 + 1)^2\,(11z^3 + 2)$$

This is just "tidying up." ☐

The use of the Chain Rule in the middle of a differentiation is quite common.

Example E.   Find $f'(x)$ where $f(x) = (x + \tan^3 x)^4$.

Solution.   We quickly see that the 4-th power is the outermost operation, so we start off our

differentiation with that:

$$\frac{d}{dx} [(x + \tan^3 x)^4] = \dots$$ "The derivative of
                                                                           a composition is ...

$$= 4 (\dots)^3 \dots$$ "... the derivative of
                                                  the outer function ...

$$= 4 (x + \tan^3 x)^3 \dots$$ "... (evaluated at the
                                                           inner function) ...

$$= 4 (x + \tan^3 x)^3 \cdot \frac{d}{dx} [x + \tan^3 x]$$ "... times the derivative
                                                                                       of the inner function."

$$= 4 (x + \tan^3 x)^3 \cdot [1 + \frac{d}{dx} (\tan^3 x)]$$ (using Theorem 3.2.4)

Oh ...   Part of our inner function requires
the Chain Rule itself!

$$\frac{d}{dx} (\tan^3 x) - 3 (\dots)^2 \dots$$

$$= 3 (\tan x)^2 \dots$$

$$- 3 (\tan x)^2 \sec^2 x$$

$$= 4 (x + \tan^3 x)^3 \cdot [1 + 3 \tan^2 x \sec^2 x]$$

So what would actually appear on your paper? Only

$$\frac{d}{dx} [(x + \tan^3 x)^4] = 4 (x + \tan^3 x)^3 \cdot [1 + \frac{d}{dx} (\tan^3 x)]$$

$$= 4 (x + \tan^3 x)^3 \cdot [1 + 3 \tan^2 x \sec^2 x]$$

and when you become skilled at it you can even dispense with the middle expression!   □

<u>Example F.</u>   Evaluate $\dfrac{dy}{dt}\Big|_{t=2}$, given that $y = x^3$, $x = 2u - u^3 - 2$, and $u = t^2 - 2t + 1$.

<u>Solution.</u>   Using the $\dfrac{dy}{dx}$ version of the Chain Rule we see

$$\frac{dy}{dt} = \frac{dy}{dx} \cdot \frac{dx}{dt}$$

However, since u is an intermediate variable between x and t, we apply the Chain Rule again to dx/dt (i. e., $\dfrac{dx}{dt} = \dfrac{dx}{du} \cdot \dfrac{du}{dt}$) so that

$$\boxed{\frac{dy}{dt} = \frac{dy}{dx} \cdot \frac{dx}{du} \cdot \frac{du}{dt}}$$

<u>Chain Rules of any number of variables are easy to generate in this way!</u> We want dy/dt when $t = 2$. Thus, for later reference, we observe that

$$t = 2 \quad \text{implies} \quad u = 4 - 4 + 1 = 1$$
$$\text{and} \quad x = 2(1) - 1 - 2 = -1$$

We now compute our three derivatives, and evaluate them at the values $t = 2$, $u = 1$ and $x = -1$:

$$\frac{dy}{dx} = \frac{d}{dx}[x^3] = 3x^2, \qquad \text{so} \quad \frac{dy}{dx}\Big|_{x=-1} = 3$$

$$\frac{dx}{du} = \frac{d}{du}[2u - u^3 - 2] = 2 - 3u^2, \qquad \text{so} \quad \frac{dx}{du}\Big|_{u=1} = -1$$

$$\frac{du}{dt} = \frac{d}{dt}[t^2 - 2t + 1] = 2t - 2, \qquad \text{so} \quad \frac{du}{dt}\Big|_{t=2} = 2$$

Hence
$$\frac{dy}{dt}\Big|_{t=2} = (3)(-1)(2) = \boxed{-6}$$

Notice that plugging in the specific values of t, u and x (<u>after</u> the differentiation of course!) avoided a great deal of unpleasant computation.   □

Section 3.6.   Implicit Differentiation.

§1.   Differentiating expressions in x and y.   Before examining the full process of implicit

differentiation, let's consider the key step: differentiating expressions such as y^3 or

sin (xy) with respect to x , where y is a function of x .

Example A.   Assuming y to be a differentiable function of x , compute the following derivatives:

    a. $\dfrac{d}{dx}$ [3y]

    b. $\dfrac{d}{dx}$ [y^2]

    c. $\dfrac{d}{dx}$ [xy]

    d. $\dfrac{d}{dx}$ [sin (xy)]

Solution.   a. $\dfrac{d}{dx}$ [3y] = ? No, the answer is NOT 3. The (multiplicative) constant 3

moves past the derivative sign to give

$$\frac{d}{dx} [3y] = 3 \frac{dy}{dx}$$

Unless more is known about how y depends on x , this is as complete an answer as we can

give.

b. $\dfrac{d}{dx}$ [y^2] = ? In this case you must recognize the need for the Chain Rule: the inner

function (of x) is y , while the outer function is $(\dots)^2$. Therefore:

$$\frac{d}{dx} [y^2] = 2 (\dots)^1 \dots \qquad \text{derivative of outer function} \dots$$

$$= 2y \dots \qquad \dots \text{(evaluated at inner function)} \dots$$

$$= 2y \frac{dy}{dx} \qquad \dots \text{times derivative of inner function.}$$

3.6.2

Thus

$$\frac{d}{dx} [\, y^2 \,] = 2y \, \frac{dy}{dx}$$

A common mistake here is:

$$\frac{d}{dx} [\, y^2 \,] = ? = 2y \qquad \underline{\text{WRONG!}} \text{ Do not forget } \frac{dy}{dx} \,!$$

c. $\frac{d}{dx} [\, xy \,] = ?$ Here you must recognize the need for the Product Rule since we have a product of two functions of x:

$$\frac{d}{dx} [\, xy \,] = x \frac{d}{dx} [\, y \,] + y \frac{d}{dx} [\, x \,]$$

$$= x \frac{dy}{dx} + y \cdot 1$$

Thus

$$\frac{d}{dx} [\, xy \,] = x \frac{dy}{dx} + y$$

d. $\frac{d}{dx} [\, \sin (xy^2) \,] = ?$ Here is a more complex expression, typical of implicit differentiation problems. Since we have a composition (the sine of xy^2), we start with the Chain Rule:

$$\frac{d}{dx} [\, \sin (xy^2) \,] = \cos (\ldots) \ldots \qquad \text{derivative of outer function} \ldots$$

$$= \cos (xy^2) \ldots \qquad \ldots \text{(evaluated at inner function)} \ldots$$

$$= \cos (xy^2) \frac{d}{dx} (xy^2) \ldots \qquad \ldots \text{times derivative of inner function.}$$

The derivative of xy^2 requires the Product Rule and the Chain Rule:

$$\frac{d}{dx}(xy^2) = x\frac{d}{dx}[y^2] + y^2\frac{d}{dx}[x] \qquad \text{(Product Rule)}$$

$$= x\left(2y\frac{dy}{dx}\right) + y^2 \cdot 1 \qquad \text{(Chain Rule, as in (b))}$$

$$= 2xy\frac{dy}{dx} + y^2$$

Thus

$$\boxed{\frac{d}{dx}[\sin(xy^2)] = \cos(xy^2)\left(2xy\frac{dy}{dx} + y^2\right)} \qquad \square$$

In all four parts of Example A the key assumption was

$$\boxed{\quad y \quad \text{is a function of} \quad x \quad}$$

This is the thought which should be uppermost in your mind when computing such derivatives.

§2.   The mechanics of implicit differentiation.   Suppose y is a function of x that satisfies an equation such as

$$x^? + y^3 = 1 \quad \text{or} \quad x\sin y + y\cos x = \sqrt{2}\ \pi/4$$

Then there are two ways we could try to compute dy/dx : (explicit) differentiation * or implicit differentiation.

Example B.   Compute dy/dx if $x^2 + y^3 = 1$.

_____

* "Explicit differentiation" is what has been called plain old "differentiation" up to now. We are adding the word "explicit" in this section simply to emphasize the difference between it and implicit differentiation.

(Explicit) Differentiation	Implicit Differentiation

| 1. __Solve__ for y | 1. __Differentiate__ with respect to x |

$$y^3 = 1 - x^2$$

$$y = (1 - x^2)^{1/3}$$

$$\frac{d}{dx}(x^2 + y^3) = \frac{d}{dx}(1)$$

$$\frac{d}{dx}(x^2) + \frac{d}{dx}(y^3) = 0$$

$$2x + 3y^2 \frac{dy}{dx} = 0$$

| since y is a function of x |

| 2. __Differentiate__ with respect to x | 2. __Solve__ for dy/dx |

$$\frac{dy}{dx} = \frac{1}{3}(1 - x^2)^{-2/3}(-2x)$$

$$\frac{dy}{dx} = -\frac{2x}{3(1 - x^2)^{2/3}}$$

$$\frac{dy}{dx} = -\frac{2x}{3y^2}$$

The two answers we have obtained look different:

$$\frac{dy}{dx} = -\frac{2x}{3(1 - x^2)^{2/3}}$$

$$\frac{dy}{dx} = -\frac{2x}{3y^2}$$

However, substituting $y = (1 - x^2)^{1/3}$ into the second answer for dy/dx will yield the first, so there are no inconsistencies in our work. □

The two methods used in Example B are summarized as follows:

Explicit Differentiation: First solve (for y) , then differentiate.

Implicit Differentiation: First differentiate, then solve (for dy/dx) .

Let's attempt to use both methods on a more complicated equation:

Example C.   Compute dy/dx if x sin y + y cos x = $\sqrt{2}$ $\pi/4$.

Explicit Differentiation.   Solve for y ... ? Well, ... we can't solve for y as a function of

x because the equation is too "tangled. " Thus we can't use explicit differentiation in this

example!

Implicit Differentiation.   We first differentiate

$$\frac{d}{dx} [\text{ x sin y } + \text{ y cos x }] = \frac{d}{dx} [\sqrt{2} \ \pi/4] = 0$$

$$\frac{d}{dx} [\text{ x sin y }] + \frac{d}{dx} [\text{ y cos x }] = 0$$

The Product Rule is now used twice:

$$x \frac{d}{dx} [\sin y] + (\sin y) \frac{d}{dx} [x] + y \frac{d}{dx} [\cos x] + (\cos x) \frac{d}{dx} [y] = 0$$

$$x \cos y \ \frac{dy}{dx} + \sin y \ - \ y \sin x \ + \ \cos x \ \frac{dy}{dx} = 0$$

since y is a function of x

Finally we solve for dy/dx :

$$(x \cos y + \cos x) \frac{dy}{dx} + \sin y \ - \ y \sin x = 0$$

$$(x \cos y + \cos x) \frac{dy}{dx} = y \sin x \ - \ \sin y$$

$$\frac{dy}{dx} = \frac{y \sin x \ - \ \sin y}{x \cos y + \cos x}$$ $\square$

3.6.6

As Example C illustrates, it is frequently the case that an equation cannot be solved explicitly for y in terms of x. In such a situation we have only one method for computing dy/dx: implicit differentiation.

Example D. Find the slope of the tangent line to the curve

$$\tan (x + y^2) = xy + 1 \quad \text{at} \quad \left(0, \frac{\sqrt{\pi}}{2} \right)$$

Solution. We wish to compute dy/dx at the values $x = 0$ and $y = \sqrt{\pi}/2$. To find dy/dx, first we differentiate both sides of our equation with respect to x:

$$\frac{d}{dx} [\tan (x + y^2)] = \frac{d}{dx} [xy + 1]$$

$$\sec^2 (x + y^2) \frac{d}{dx} (x + y^2) = \frac{d}{dx} [xy] \quad \text{by the Chain Rule,}$$

$$\sec^2 (x + y^2) (1 + 2y \frac{dy}{dx}) = x \frac{dy}{dx} + y \quad \text{as in Example A}$$

since y is a function of x

Since we have finished the differentiation, we may now plug in the values $x = 0$ and $y = \sqrt{\pi}/2$:

$$\sec^2 \left(\frac{\pi}{4} \right) \left(1 + \sqrt{\pi} \frac{dy}{dx} \right) = \frac{\sqrt{\pi}}{2}$$

However,

$$\sec^2 \left(\frac{\pi}{4} \right) = \frac{1}{\cos^2 \left(\frac{\pi}{4} \right)} = \frac{1}{\left(\frac{\sqrt{2}}{2} \right)^2} = 2$$

Thus

$$2 \left(1 + \sqrt{\pi} \frac{dy}{dx} \right) = \frac{\sqrt{\pi}}{2}$$

Now we can solve for   dy/dx :

$$1 + \sqrt{\pi}\, \frac{dy}{dx} = \frac{\sqrt{\pi}}{4}$$

$$\sqrt{\pi}\, \frac{dy}{dx} = \frac{\sqrt{\pi}}{4} - 1$$

$$\boxed{\frac{dy}{dx} = \frac{1}{4} - \frac{1}{\sqrt{\pi}} \approx -.3142}$$

□

Notice that in Example D we plugged in the specified x and y values immediately after completing the differentiation. Plugging in the specified x and y values before the differentiation is complete is a grave (but all-too-common) mistake! Waiting until the very end of the problem to plug in these values is legal, but it generally entails needless computational details (and thus leads to more careless errors!). If, however, you were to follow this route and solve for dy/dx , you would obtain

$$\frac{dy}{dx} = \frac{y - \sec^2(x + y^2)}{2y \sec^2(x + y^2) - x}$$

Then plugging in $x = 0$ and $y = \sqrt{\pi}/2$ will still yield dy/dx $\approx -.3142$.

3. Implicit differentiation and higher derivatives. In Example 3 Anton uses implicit differentiation as his first step in obtaining a second derivative $d^2 y / dx^2$ when $4x^2 - 2y^2 = 9$. The steps in his solution form a reasonable procedure in the general case:

Step 1.   Differentiate the given equation implicitly ....

$$8x - 4y\, \frac{dy}{dx} = 0$$

<u>Step 2.</u>   Solve for $\dfrac{dy}{dx}$ ...

$$\frac{dy}{dx} = \frac{2x}{y}$$

<u>Step 3.</u>   Differentiate the expression in Step 2 to obtain $d^2 y / dx^2$ ....

$$\frac{d^2 y}{dx^2} = \frac{d}{dx}\left[\frac{dy}{dx}\right] = \frac{d}{dx}\left[\frac{2x}{y}\right]$$

$$= \frac{y \dfrac{d}{dx}[2x] - 2x \dfrac{d}{dx}[y]}{y^2}$$

— using the Quotient Rule (Theorem 3. 2. 6)
and remembering that <u>y   is a function of   x</u>

$$= \frac{2y - 2x \dfrac{dy}{dx}}{y^2}$$

<u>Step 4.</u>   Substitute the expression for   dy/dx   (Step 2) into the expression for $d^2 y / dx^2$

(Step 3)....

$$\frac{d^2 y}{dx^2} = \frac{2y - 2x\left[\dfrac{2x}{y}\right]}{y^2} = \frac{2y - \dfrac{4x^2}{y}}{y^2}$$

$$= \frac{2y^2 - 4x^2}{y^3}$$

The critical observation to make concerning this procedure is that the expression for

dy/dx   obtained in Step 2 will very often contain both   x   and   y.   Thus the differentiation

in Step 3 will have to be carefully done, using the fact that   <u>y   is a function of   x</u>.

Example E.   Find y" by implicit differentiation if

$$\sin y + x = 1$$

Solution.   Step 1.   We implicitly differentiate the equation as it stands:

$$\frac{d}{dx}[\sin y + x] = \frac{d}{dx}[1] = 0$$

$$(\cos y)\, y' + 1 = 0$$

REMEMBER: y   IS A FUNCTION of   x

Step 2.   Solving this equation for y' yields

$$y' = -\frac{1}{\cos y} = -\sec y$$

Step 3.   Differentiating this to get y" yields

$$y" = -(\sec y \tan y)\, y'$$

REMEMBER: y   IS A FUNCTION of   x

Step 4.   Substituting for y' yields

$$y" = -(\sec y \tan y)(-\sec y) \qquad \text{since} \qquad y' = -\sec y$$

Thus $y" - \sec^2 y \tan y$.   □

4.   The Power Rule for rational exponents.   In §3.2 we established the Power Rule for any

integer power   n   (Theorem 3.2.7):

$$\frac{d}{dx}[x^n] - n x^{n-1}$$

Using implicit differentiation Anton shows in this section that the rule is also true for any rational* power $r = m/n$:

$$\frac{d}{dx} [x^r] = r x^{r-1}$$

The method is to set $y = x^r = x^{m/n}$,   so that

$$y^n = x^m$$

and then to implicitly differentiate this equation.   Anton shows this leads to the desired result.

Example F.   Find dy/dx if $y = \sqrt{1 - x^2}$.

Solution. $y = (1 - x^2)^{1/2}$,   so that

$$\frac{dy}{dx} = \frac{1}{2} (...)^{(1/2)-1} ...$$ Derivative of outer function ...

$$= \frac{1}{2} (1 - x^2)^{-1/2} ...$$ ... (evaluated at inner function) ...

$$= \frac{1}{2} (1 - x^2)^{-1/2} (-2x)$$ ... times derivative of inner function.

Thus $y = -x/\sqrt{1 - x^2}$.   Notice that this derivative is defined only when $-1 < x < 1$,   while the original function is defined for $-1 \le x \le 1$.   □

Anton's derivation of the power rule for rational exponents does leave one gap: he assumes without proof that the function $y = x^r$ is differentiable! We will return to this question in §8.1 of The Companion.

5.   Implicit functions.   Given an equation in x and y ,   then essentially any function

---

* In Theorem 7.4.6 we will actually extend the rule to cover any real number exponent r .

y = f(x) which satisfies the equation is said to be <u>implicitly defined</u> by the equation. *
Determining whether a given equation does have implicitly defined functions, and whether these
functions are differentiable, can be difficult. We will leave these questions to a course in
advanced calculus.

However, <u>once we know that an equation has implicitly defined functions which are</u>
<u>differentiable, then implicit differentiation will always give us the derivative!</u> From this point
of view, implicit differentiation is simply a technique used to compute derivatives for implicitly
defined functions. This is what we did in Examples B, C, D and E, and what Anton does in
Examples 1, 2, 3 and 7.

Section 3.7.   Continuity.

The concept of <u>continuity</u> is important for a number of reasons. First, most phenomena
in the physical world can be modeled by functions which are continuous. Second, a great many
of our subsequent theorems and definitions will require the hypothesis of continuity; these results
could not even be stated, let alone used, without continuity.

There are many new definitions and theorems in this section, so we will try to organize
them in a useful and memorable way.

---

* There are some necessary restrictions on the function f.
Take any point (x_0, y_0) on the graph of f. Then there
must exist a rectangle with center (x_0, y_0) whose
intersection with the <u>graph of the equation</u> contains nothing
but points on the <u>graph of f</u>.

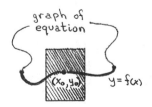

1. <u>The intuitive concept of continuity.</u>   The intuitive idea is simple: a function $y = f(x)$ is

<u>continuous at a point $x = c$</u>   if there is no "hole" or "gap" in the graph at $y = f(x)$ over

the point $x = c$; it is <u>discontinuous at $x = c$</u>   if there is a "hole" or "gap" over $x = c$.

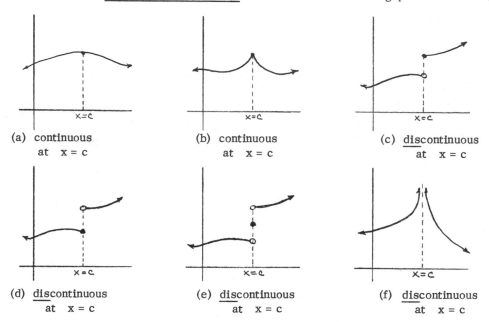

(a) continuous at $x = c$	(b) continuous at $x = c$	(c) <u>discontinuous</u> at $x = c$
(d) <u>discontinuous</u> at $x = c$	(e) <u>discontinuous</u> at $x = c$	(f) <u>discontinuous</u> at $x = c$

We can refine these intuitive concepts even further. In diagram (c) we have a function

which is discontinuous at $x = c$; however, the curve does not appear to have a gap or hole

when approached <u>only from the right at $x = c$</u> . Thus we will say that in diagram (c) we have

a function which is <u>continuous from the right at $x = c$</u> . Similarly, in diagram (d) we have a

function which is <u>continuous from the left at $x = c$</u> . Diagrams (e) and (f) show functions

which are not continuous from either direction at $x = c$. The function in diagram (f) is

particularly ill-behaved at $x = c$: it isn't even defined at that point, so no form of continuity

could possibly hold.

It is wise to clarify immediately the relationship between <u>continuity and differentiability</u>.

If a function $y = f(x)$ is differentiable

at $x = c$, then its graph has a tangent

line at $x = c$ whose slope is $f'(c)$.

In such a case, how could the graph

possibly have a "hole" or a "gap" at

$x = c$?? The answer is: it can't! Thus,

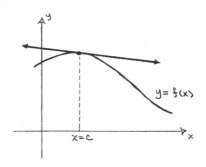

if $y = f(x)$ is differentiable at $x = c$, it must also be continuous at $x = c$. (This is proved

in Theorem 3.7.5.) However, continuity does not necessarily imply differentiability, as shown

in diagram (b) above. In diagram (b) we see what can go wrong: there is no "hole" or "gap"

at $x = c$, so f is continuous at that point, but there is a sharp corner (hence no tangent line),

showing that f is not differentiable at $x = c$.

Our six diagrams can now be given more detailed labels:

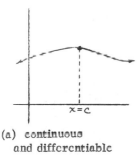

(a) continuous
    and differentiable
    at $x = c$

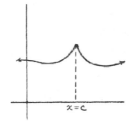

(b) continuous but
    not differentiable
    at $x = c$

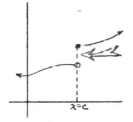

(c) continuous
    from the right
    at $x = c$

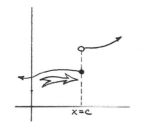

(d) continuous
    from the left
    at $x = c$

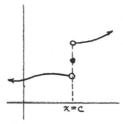

(e) discontinuous
    at $x = c$, but
    $f(c)$ exists

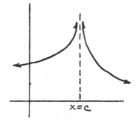

(f) discontinuous
    at $x = c$, and
    $f(c)$ does not exist

Remembering these six cases should help you quite a bit in dealing with continuity.

2.   <u>The definition of continuity.</u>   To say there is no "hole" or "gap" in the graph of $y = f(x)$

over the point $x = c$ is to assert that <u>$f(c)$ exists and as x approaches c , the function</u>

<u>values $f(x)$ approach $f(c)$</u> (i. e., $f(c)$ is "where it should be" on the graph to fill any

possible "hole"). This leads to the definition of continuity (and similarly to the definition of

continuity from the right and left):

| Definition of Continuity |

$y = f(x)$ is $\left\{\begin{array}{l}\text{continuous} \\ \text{continuous from the right} \\ \text{continuous from the left}\end{array}\right\}$ at $x = c$ if

$f(c)$ and $\left\{\begin{array}{l}\lim\limits_{x \to c} f(x) \\ \lim\limits_{x \to c^+} f(x) \\ \lim\limits_{x \to c^-} f(x)\end{array}\right\}$ both exist and are equal, i. e.,

$\left\{\begin{array}{ll}\lim\limits_{x \to c} f(x) = f(c) & \text{(continuity at } x = c) \\ \lim\limits_{x \to c^+} f(x) = f(c) & \text{(right continuity at } x = c) \\ \lim\limits_{x \to c^-} f(x) = f(c) & \text{(left continuity at } x = c)\end{array}\right\}$

Generally we are interested in continuity not just at one point but at a whole collection of points:

> **Continuity on various sets**
>
> $y = f(x)$ is <u>continuous on the open interval   (a , b)</u>
>
>   if it is continuous at every point in   (a , b)
>
> $y = f(x)$ is <u>continuous on the closed interval   [a , b]</u>
>
>   if it is   (i)   continuous on the open interval   (a , b) ,
>
>   (ii)   continuous from the right at   x = a ,
>
>   (iii)   continuous from the left at   x = b .
>
> $y = f(x)$ is <u>continuous</u>
>
>   if it is continuous at every point   x = c   in the domain of   f .

We know this list of definitions looks horribly long, but they are all the definitions Anton

gives in this section, collected together for <u>convenient reference</u> and for <u>comparison</u>. Here's

our first illustration of these concepts:

Example A.   Let   f   be the function whose graph is shown

to the right. On which of the following intervals, if any,

is   f   continuous?   a.   (1, 2)

                          b.   [1 , 2]

                          c.   [2 , 3]

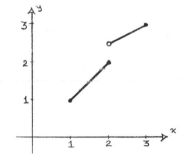

Solution.   (a) (1, 2) . There is no gap or hole in the graph of   f   above any point   x   in the

open interval   (1, 2) . Hence   f   is <u>continuous on   (1, 2)</u> .

(b)   [1, 2] . Since   f   is not defined for   x < 1 ,   then   f   cannot be continuous at   x = 1 .

Moreover, there is a gap in the graph of   y = f(x)   above   x = 2 ,   and hence   f   is not

continuous at   x = 2 . However, for   f   to be continuous on the <u>closed</u> interval   [1 , 2]   we only need continuity at the interior points $(1, 2)$ and the appropriate one-sided continuity at the endpoints:

(i) f is continuous on the open interval $(1, 2)$ from (a) ,

(ii) f is seen to be continuous from the right at $x = 1$ since

$$\lim_{x \to 1^+} f(x) = 1 = f(1)$$

(iii) f is seen to be continous from the left at $x = 2$ since

$$\lim_{x \to 2^-} f(x) = 2 = f(2)$$

Thus f is continuous on $[1, 2]$

(c) $[2, 3]$. As in (b) we check the interior and the two endpoints:

(i) There are no gaps or holes in the graph of f above the open interval $(2, 3)$,
so f is continuous on $(2, 3)$.

(ii) f is not, however, continuous from the right at $x = 2$ since

$$\lim_{x \to 2^+} f(x) = 2\frac{1}{2} \neq 2 = f(2)$$

(iii) f is continuous from the left at $x = 3$ since

$$\lim_{x \to 3^-} f(x) = 3 = f(3)$$

Thus f is not continuous on $[2, 3]$ since it is not continuous from the right at $x = 2$.   □

3.   <u>General results on continuity.</u>   Anton verifies the continuity at   x = c   for three very general classes of functions:

<table><tr><td>

General
classes
of
continuous
functions

</td><td>

(1)   <u>algebraic combinations</u>   (e.g., $f + g$, $f - g$, $f \cdot g$,   and f/g when $g \neq 0$)   of functions which are themselves continuous at   x = c   (Theorem 3. 7. 3),

(2)   <u>differentiable functions</u> at   x = c   (Theorem 3. 7. 5),

(3)   <u>compositions</u> $f \circ g$ where   g   is continuous at   c   and   f   is continuous at   g(c)   (a corollary of Theorem 3. 7. 6).

</td></tr></table>

These results allow us, in some instances, to verify continuity without having to resort to the definition in terms of limits.   In particular, noting that differentiability implies continuity, we see that the following specific classes of <u>differentiable</u> functions are all continuous:

<table><tr><td>

Specific
classes
of
continuous
functions

</td><td>

<u>polynomials</u>

<u>rational functions</u>
   (whenever the denominator is non-zero)

<u>trigonometric functions</u>
   (whenever they are defined)

<u>rational exponent functions</u> $y = x^r$
   (whenever they are defined, with perhaps the exception of   x = 0).

</td></tr></table>

<u>Example B.</u>   Determine where the following function is continuous:

$$y = 1/\sqrt{1 + x^2 + \sin^2 x} \quad .$$

Solution.   Our list tells us that the polynomial $1 + x^2$ and the trigonometric function $\sin x$ are both continuous everywhere.   Then the product $\sin^2 x = (\sin x)(\sin x)$ is continuous (an algebraic combination) and the sum $1 + x^2 + \sin^2 x$ is continuous (another algebraic combination).   This quantity,

$$g(x) = 1 + x^2 + \sin^2 x$$

is always greater than or equal to one and is composed with the rational exponent function $f(u) = 1/\sqrt{u} = u^{-1/2}$,   which is continuous for all $u > 0$.   Hence, by our result on compositions, the composition function $(f \circ g)(x) = \sqrt{1 + x^2 + \sin^2 x}$ is continuous everywhere.   □

<u>Example C.</u>   Find the points of discontinuity for $y = \cot x$.

Solution.   We know from our list that the trigonometric function $\cot x$ is continuous wherever it is defined.   Thus it is discontinuous only where it is undefined, i.e., at the points $x = 2\pi n$ for n any integer.   □

4.   <u>Determining continuity from the definition.</u>   When asked to check the continuity of a specific function, you often cannot totally avoid using the definition of continuity in terms of limits.   In these cases you will need to make heavy use of   Theorem 2.5.1   (algebraic properties of limits) and   Theorem 3.7.6   (limits of composition functions):

Theorem 3.7.6	$\lim f(g(x)) = f(\lim g(x))$ whenever f is continuous at the point $\lim g(x)$

In words, <u>the limit symbol can be moved through a continuous function.</u>   Here are some examples:

Example D.   Determine where the following function is continuous:

$$f(x) = \sqrt{1 - x^2}$$

Solution.   The function is defined only for $-1 \leq x \leq 1$. Since it is easy to check that f is differentiable for $-1 < x < 1$ (just compute the derivative $f'(x) = -x/\sqrt{1 - x^2}$ and note that it exists at these values), then it must be continuous for $-1 < x < 1$.

What about the endpoints $x = \pm 1$? Since f is not defined for $x < -1$ nor for $x > 1$, then we cannot have continuity at the endpoints. However, we could have one-sided continuity. Let's check:

$$\lim_{x \to -1^+} f(x) = \lim_{x \to -1^+} \sqrt{1 - x^2} = 0 = f(-1)$$

$$\lim_{x \to +1^-} f(x) = \lim_{x \to +1^-} \sqrt{1 - x^2} = 0 = f(1)$$

Thus f is continuous from the right at $x = -1$ and continuous from the left at $x = 1$. This proves that f is continuous on the closed interval $[-1, 1]$.   □

Example E.   Determine whether the following function is continuous at $x = -1$:

$$f(x) = \begin{cases} (x^2 - 1)/(x + 1) & \text{for} \quad x < -1 \\ -2 & \text{for} \quad x = -1 \\ -\sqrt{2 - x^2} & \text{for} \quad x > -1 \end{cases}$$

Solution.   When given a function f which is defined differently on either side of $x = c$, determining whether it is continuous at $x = c$ requires computing the one-sided limits and

3. 7. 10

comparing them with the value of $f(c)$ itself.

(i) $\quad \lim\limits_{x \to -1^-} f(x) = \lim\limits_{x \to -1^-} \dfrac{x^2 - 1}{x + 1} = \lim\limits_{x \to -1^-} \dfrac{(x - 1)(x + 1)}{x + 1} = \lim\limits_{x \to -1^-} (x - 1) = \boxed{-2}$

(ii) $\quad f(-1) = \boxed{-2}$

(iii) $\quad \lim\limits_{x \to -1^+} f(x) = \lim\limits_{x \to -1^+} -\sqrt{2 - x^2} = -\sqrt{\lim\limits_{x \to -1^+} (2 - x^2)} \qquad$ by Theorem 3. 7. 6

$$= -\sqrt{2 - 1} = \boxed{-1}$$

Hence $\quad \lim\limits_{x \to -1^-} f(x) = f(-1) \neq \lim\limits_{x \to -1^+} f(x)$, proving that f is <u>continuous from the left</u>

at $x = -1$ (but <u>not</u> continuous at $x = -1$ since the left limit and the right limit are not equal). $\quad\square$

<u>Example F.</u> Find a value for the constant k that will make the following function continuous:

$$f(x) = \begin{cases} 3x^2 - x, & x \leq 1 \\[2mm] kx + 1, & x > 1 \end{cases}$$

<u>Solution.</u> Since $3x^2 - x$ and $kx + 1$ are both polynomials, then $f(x) = 3x^2 - x$ is continuous for $x < 1$ and $f(x) = kx + 1$ is continuous for $x > 1$. Thus the only place where f could be discontinuous is at $x = 1$. We thus compute $f(1)$ and the one-sided limits:

(i) $\quad \lim\limits_{x \to 1^-} f(x) = \lim\limits_{x \to 1^-} (3x^2 - x) = 3 - 1 = \boxed{2}$

(ii) $\quad f(1) = 3(1)^2 - 1 = \boxed{2}$

(iii) $\lim\limits_{x \to 1^+} f(x) = \lim\limits_{x \to 1^+} (kx + 1) = \boxed{k + 1}$

Hence, in order to make these three numbers equal, we have only to set $k + 1 = 2$ and find

$\boxed{k = 1}$.   Then   f   will be continuous everywhere.   □

5.   <u>The Intermediate Value Theorem.</u>   Continuous functions on closed intervals $[a, b]$ have

many extremely useful properties which will be developed and exploited in subsequent chapters.

The first such property is encountered in   Theorem 3. 7. 8 :

<table>
<tr><td>The Intermediate
Value Theorem</td><td>Suppose   f   is continuous on the <u>closed</u> interval $[a, b]$

    and   C   is any number strictly between $f(a)$ and $f(b)$.

Then there is at least one number   x   in $[a, b]$ such that

$$f(x) = C$$</td></tr>
</table>

From a picture this result is "obvious. "

However, be sure to understand how

crucial the assumption of continuity is

to the theorem.   If   f   is not continuous

on $[a, b]$,   then the result can be false,

as is shown to the right: the horizontal line

$y = C$ passes right through a "gap"   in

the graph of $y = f(x)$,   and hence there

is no corresponding   x-value.

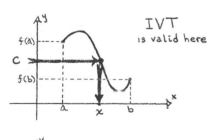

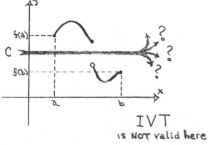

3.7.12

Although very important in more advanced work, we will encounter the Intermediate Value Theorem only occasionally in beginning calculus. For example, it will be needed in the proof of Theorem 5.10.1 (the Mean-Value Theorem for Integrals) and in the development of Newton's Method in §4.9. In Newton's Method the Intermediate Value Theorem appears via a corollary which Anton gives as Theorem 3.7.9.

Chapter 4:   Applications of Differentiation

Section 4.1.   Related Rates.

1.   Word problems... ! In §3.1 we observed that the single most important interpretation of

the derivative is as a rate of change: a measurement of how fast one quantity is changing with

respect to another quantity.   It is natural, therefore, that our first application of differentiation

centers on rates of change, in particular, on related rates problems:

> find the rate of change of a quantity
>
> by relating it to the known rates of change
>
> of other quantities.

Assuming you have adequately mastered the differentiation techniques of the previous

chapter, there is only one serious difficulty with related rates problems: they are word

problems! To many people the very thought of a word problem is enough to bring on a cold

sweat. This should not be so, for word problems are not impossible if you approach them in a

careful, disciplined and organized fashion.

Related rates comprise just the first of many groups of word problems that you will

encounter in calculus. For that reason there is a detailed supplement on word problems

(Appendix H) which is specifically geared to calculus-type applications. No matter how good

you are with word problems, we strongly advise that you read at least part of Appendix H before

starting on related rates! If word problems are the bane of your existence, then read all of

Appendix H. Your time will be well-spent!

2.   How to tackle a related rates problem ... and live to tell about it! Anton gives a four step

method for handling related rates problems. We'll reproduce his list here, but with some additional suggestions and warnings. To make the reading less abstract we will interweave the list with the solution to Example A :

Example A.   Superman is in level flight 6 miles above the ground. His flight path passes directly over Wobnoid College.   How fast is he flying when the distance between him and Wobnoid College is 10 miles and this distance is increasing at a rate of 5 miles per minute?

Solution.

Step 1.   Label the quantities that vary.

- First draw a picture if appropriate (it usually is!) and assign labels (x, y, h, t, etc.)   to any quantities which vary.

- In the picture put a numerical value on any quantity which never changes throughout the problem.   DO NOT PUT IN SPECIFIC VALUES OF VARYING QUANTITIES! To do so will almost surely lead to disaster.

- Carefully note the specific values and rates which you are given, and carefully identify the rate you are being asked to find.

The picture is as shown to the right. Notice how we have labeled the two varying distances by   x   and   h , and have noted the constant altitude of 6 miles. We do not put into the picture the specific value 10 for the varying quantity   h !! Instead, we write down the given values and

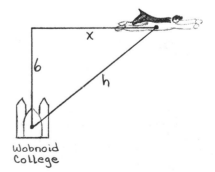

Wobnoid College

rates, along with the desired rate, in the box shown

to the right. The desired rate is $\frac{dx}{dt}$ = ? ("How

fast is he flying...?") when h = 10 ("... when

the distance between him and Wobnoid College is

$$\boxed{\begin{aligned} \frac{dx}{dt} &= ? \\ \text{when } h &= 10 \\ \text{and } \frac{dh}{dt} &= 5 \end{aligned}}$$

10 miles ... ") and $\frac{dh}{dt}$ = 5 ("... and this distance is increasing at a rate of 5 miles per

minute?"). Notice that <u>we have taken every piece of information from the problem and have</u>

<u>placed it either in the diagram or in the boxed equations!</u>

<u>Step 2.</u>　　　Find an equation relating the quantity with the unknown rate of

change to quantities whose rates of change are known.

- Generally these equations will come from known formulas

  such as those for areas and volumes, or by application of

  the <u>Pythagorean Theorem</u> or <u>similar triangles.</u>

- Sometimes more than one equation, using a number of variables,

  will be needed. Then you must simplify down to one equation by

  elimination of variables.

We need an equation which relates x (the quantity with the unknown rate of change)

to h (the quantity with the known rate of change). So we stare at the diagram...

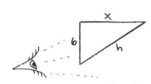 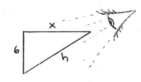 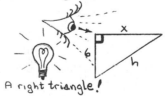

... until we realize that it is a right triangle, and hence the Pythagorean Theorem applies:

$$36 + x^2 = h^2$$

There's our equation -- it couldn't have been easier!

Step 3.   Differentiate both sides of this equation and solve for the derivative that will give the unknown rate of change.

- This differentiation is generally (but not always) with respect to time t .

Since we are considering rates of change with respect to time, we differentiate with respect to time t . (All five of Anton's examples involve differentiations with respect to time.)

$$\frac{d}{dt}(36 + x^2) = \frac{d}{dt}(h^2)$$

$$0 + 2x\frac{dx}{dt} = 2h\frac{dh}{dt}$$

(where we have used the Chain Rule twice, since both x and h are functions of t)

$$\frac{dx}{dt} = \left(\frac{h}{x}\right)\frac{dh}{dt}$$

Step 4.   Evaluate this derivative at the appropriate point.

- Here's where you plug in the specific values for varying quantities. You'll be glad you waited!

We now wish to evaluate $\dfrac{dx}{dt}$ when $h = 10$ and $\dfrac{dh}{dt} = 5$ ... but wait a minute! Our equation for $\dfrac{dx}{dt}$ also has an x in it. What is x?

No problem! Since x and h are related by $36 + x^2 = h^2$, then $h = 10$ implies $x = \sqrt{100 - 36} = \sqrt{64} = 8$. Thus

$$\frac{dx}{dt} = \left(\frac{h}{x}\right)\frac{dh}{dt} = \left(\frac{10}{8}\right)5 = \boxed{\frac{25}{4}} \text{ miles per minute}$$ □

Here is an example similar to Anton's Example 5. We chose it to illustrate what happens when there is more than one equation in Step 2. . . .

Example B.   A reservoir has the shape of an inverted cone whose cross-section is an equilateral triangle. If water is being pumped out of the reservoir at the rate of $2\,m^3/sec$, at what rate is the height of the water changing when the height is $40\,m$?

Solution.   Step 1.   A cross-section of the reservoir is as shown to the right. We have labeled three quantities: the volume and height of the water were mentioned in the problem itself, while the radius of the circular surface of the water is almost certain to play a role in the solution. We do not put into the picture the specific value 40 for the varying quantity y! Translating the rest of the problem into mathematics, we have $\dfrac{dV}{dt} = -2$ (". . . water is being pumped out of the reservoir at the rate of $2\,m^3/sec$, . . .") and we want $\dfrac{dy}{dt}$ (" . . . at what rate is the height of the water changing . . .") when $y = 40\,m$ (" . . . when

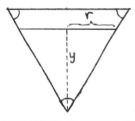

V = volume (m^3/sec) at time t

y = height of water (m) at time t

r = radius of surface (m) at time t

$\dfrac{dV}{dt} = -2$

$\dfrac{dy}{dt} = ?$

when $y = 40$

the height is 40 m ?").

Step 2.   The formula for the volume of a cone is given by

$$V = \frac{1}{3} \pi r^2 y$$

The problem here is the presence of the unwanted variable r . We clearly need another

equation to eliminate r , so we turn to our diagram for

such a relationship. Since the cross-section of the cone

is an equilateral triangle, then the hypotenuse of the right

triangle shown to the right is 2 r . The Pythagorean

Theorem then yields

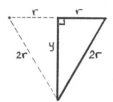

$$y^2 + r^2 = 4r^2 , \qquad \text{or} \qquad r^2 = \frac{1}{3} y^2$$

We can use this to eliminate r from our volume equation so that it becomes

$$V = \frac{1}{9} \pi y^3$$

Step 3.   Differentiating with respect to t we obtain

$$\frac{dV}{dt} = \frac{1}{3} \pi y^2 \frac{dy}{dt}$$

(remember that both V and y
are function of t !!)

which can be solved for dy/dt to yield

$$\frac{dy}{dt} = \frac{3}{\pi y^2} \frac{dV}{dt}$$

Step 4.   In this case   dy/dt   is expressed totally in terms of known quantities:

$$\frac{dy}{dt} = \frac{3}{\pi (40)^2} (-2) \approx \boxed{-.00119 \text{ m/sec.}}$$ □

Example C.   A lighthouse has its light mounted 200 feet above the ground. A ball is dropped from

this same height (by "Butterfingers" Superman who just happens to be flying by) at a distance of

30 feet from the lighthouse. Assuming the ball falls $16t^2$ feet during the first   t   seconds,

how fast is its shadow moving along the ground when the ball is 56 feet from the ground?

Solution.   Step 1.   The diagram is shown to the

right, where we have labeled the distance fallen

at time   t   as   s   and the distance of the shadow

from the base of the lighthouse at time   t   as   x .

Translating the rest of the problem into symbols

we obtain the boxed equations.

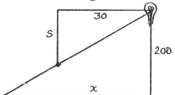

$$\begin{array}{l} s = 16t^2 \\[4pt] \dfrac{dx}{dt} = ? \\[4pt] \text{when} \quad s = 200 - 56 \\[4pt] \qquad\quad = 144 . \end{array}$$

Step 2.   We wish to have an equation relating   s   and   x . To get this we examine

our diagram ...

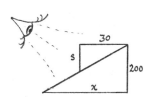   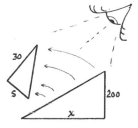   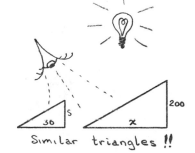

Similar   triangles !!

... and notice a pair of similar triangles! Thus

$$\boxed{\dfrac{30}{s} = \dfrac{x}{200}}$$

__Step 3.__   Differentiating with respect to t yields

$$-\dfrac{30}{s^2}\dfrac{ds}{dt} = \dfrac{1}{200}\dfrac{dx}{dt}$$ (remember: s and x are both functions of t)

so that

$$\boxed{\dfrac{dx}{dt} = -\dfrac{6000}{s^2}\dfrac{ds}{dt}}$$

__Step 4.__   To evaluate this quantity at $s = 144$ we must first compute $\dfrac{ds}{dt}$ at $s = 144$.

Since $s = 16t^2$ we obtain

$$\dfrac{ds}{dt} = 32t$$

so we need the t-value corresponding to $s = 144$:

$$s = 16t^2$$
$$144 = 16t^2$$
$$9 = t^2$$
$$3 = t$$

Thus $\dfrac{ds}{dt}\bigg|_{t=3} = 32(3) = 96$, which yields

$$\dfrac{dx}{dt}\bigg|_{s=144} = -\dfrac{6000}{(144)^2}(96) = \boxed{-27\dfrac{7}{9}\ \text{ft/sec}} \qquad \square$$

This last example illustrates the use of similar triangles. Many people overlook similar triangles in problems, so keep a sharp eye out ⤚ for them!

Section 4. 2.   Intervals of Increase and Decrease; Concavity

      In this section we learn how to determine where a function is <u>increasing or decreasing</u> and where it is <u>concave up or concave down</u>. In addition to uses with <u>max. -min. problems</u> which will be encountered in §4.3 , these two concepts form the basis for <u>curve sketching</u>, a procedure which is introduced in this section, but which is taken up in earnest in §§4. 4 and 4. 5 .

1.   <u>The intuitive concepts.</u>   A function f is <u>increasing</u> if the y-values increase as the x-values increase; it is <u>decreasing</u> if the y-values decrease as the x-values increase (see the pictures to the right).

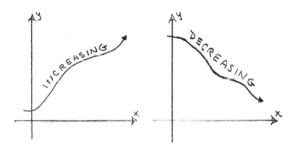

      Concavity is just as intuitive. As Anton describes it, a curve that is concave up " holds water;" a curve that is concave down "spills water." You can also remember concave up as a "smile," with concave down as a "frown. "

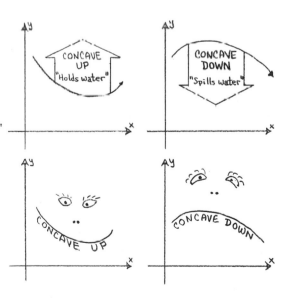

More rigorously described, concavity is a statement about the direction in which the curve $y = f(x)$ is turning. Place Howard Ant on the graph as in Chapter 2 and let him move in the direction of increasing x; if he turns to his left (i.e., counterclockwise), then the function is concave up, while if he turns to his right (i.e., clockwise) then the function is concave down.

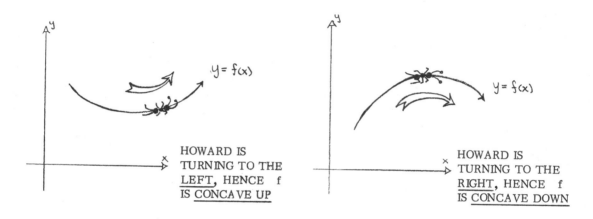

One point should be emphasized: any of the four combinations of increasing/decreasing with concave up/concave down are possible, as the following pictures show.

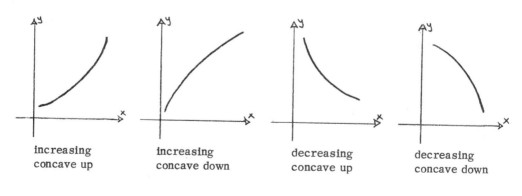

increasing   increasing   decreasing   decreasing
concave up   concave down   concave up   concave down

*   *   *   *   *

If a function changes concavity at a point

x_0 in its domain (i. e. , if it changes from concave

down to concave up), then   f   is said to have an

<u>inflection point</u> at x_0 .

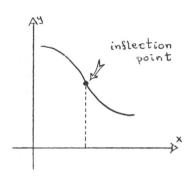

2.   <u>Determining increasing/decreasing and concavity.</u>   We know that if the derivative   f'   of

a function   f   is positive at x_0 ,   then the slope of the tangent

line to the graph of   f   at x_0 is positive.   Hence this

line "slopes upward, " and   f   is seen to be increasing

near x_0 as shown to the right.   Thus <u>a positive</u>

<u>derivative   f'   implies that   f   is an increasing</u>

<u>function.</u>   In a similar fashion, a negative derivative

f'   implies that   f   is a decreasing function.

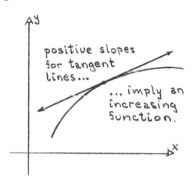

Now consider the second derivative   f"   of   f , i. e. , the derivative of the derivative.

From the underlined sentence above applied to   f'   instead of   f, we see that a positive second

derivative   f"   implies that the derivative   f'   is

an increasing function.   But this means that the

tangent line slopes are increasing which, as can

be seen to the right, implies that   f   is concave up.

<u>Thus a positive second derivative   f"   implies that   f</u>

<u>is concave up.</u>   Similarly a negative second derivative

f"   implies that   f   is concave down.

When stated precisely (in Theorems 4. 2. 2   and   4. 2. 4) these results apply to functions

over <u>open intervals</u>   (a , b) :

> Consider a function f on an open interval (a, b).
>
> If $f'(x) > 0$, then f is increasing
>
> If $f'(x) < 0$, then f is decreasing
>
> If $f''(x) > 0$, then f is concave up
>
> If $f''(x) < 0$, then f is concave down

<u>Example A.</u> Consider the function $f(x) = x^2 + x - 1 - x^{-1}$. Determine the open intervals on which f is increasing, decreasing, concave up and concave down. Also determine all points of inflection.

<u>Solution.</u> We have only to determine where f' and f'' are positive and negative.

$$f(x) = x^2 + x - 1 - x^{-1} = \frac{x^3 + x^2 - x - 1}{x}$$

$$f'(x) = 2x + 1 + x^{-2} = \frac{2x^3 + x^2 + 1}{x^2}$$

$$f''(x) = 2 - 2x^{-3} = \frac{2(x^3 - 1)}{x^3}$$

1. f will be increasing when

$$f'(x) = \frac{2x^3 + x^2 + 1}{x^2} > 0$$

To determine the x-values which satisfy this inequality, we use the interval method as described in §1.1 of <u>The Companion.</u> We see that $f'(x)$ is not defined for $\boxed{x = 0}$. Moreover, $f'(x) = 0$ whenever

$$2x^3 + x^2 + 1 = 0$$

Checking for possible rational roots (see the "Rational Root Test" in Appendix D. 6) ,

we find that $\boxed{x = -1}$ is a root, and thus $x + 1$ is a factor.   Long division

(Appendix D. 2) gives

$$(x + 1) (2 x^2 - x + 1) = 0$$

where the second factor cannot be factored further.   Thus $x = -1$ is the only root.

  Thus, by the interval method, we need only check the sign of $f'(x)$ on the three

intervals determined by $x = -1$ and $x = 0$.   This is done by checking the sign of

each factor of $f'(x)$:

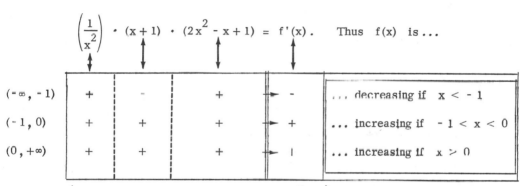

$$\left(\frac{1}{x^2}\right) \cdot (x + 1) \cdot (2 x^2 - x + 1) = f'(x) . \qquad \text{Thus} \quad f(x) \quad \text{is} \ldots$$

$(-\infty, -1)$	$+$	$-$	$+$	$-$	$\ldots$ decreasing if $x < -1$
$(-1, 0)$	$+$	$+$	$+$	$+$	$\ldots$ increasing if $-1 < x < 0$
$(0, +\infty)$	$+$	$+$	$+$	$+$	$\ldots$ increasing if $x > 0$

$\left(\begin{array}{l}\text{The } + \text{ and } - \text{ signs are the signs of the}\\ \text{respective factors on the indicated intervals.}\end{array}\right)$

2. f will be concave up when

$$f''(x) = \frac{2 (x^3 - 1)}{x^3} > 0$$

We use the interval method again.   First we note that $f''(x)$ is not defined at $\boxed{x = 0}$.

Moreover, $f''(x) = 0$ whenever

$$x^3 - 1 = 0 , \qquad \text{i. e.,} \qquad \boxed{x = 1}$$

Thus we need to check the sign of $f''(x)$ on the three intervals determined by $x = 0$ and $x = 1$:

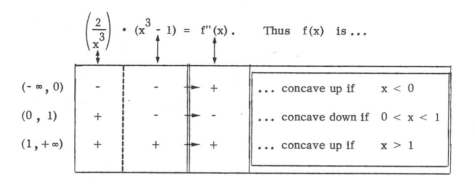

$$\left(\frac{2}{x^3}\right) \cdot (x^3 - 1) = f''(x) . \qquad \text{Thus} \quad f(x) \text{ is} \ldots$$

$(-\infty, 0)$	−	−	+	... concave up if $x < 0$
$(0, 1)$	+	−	−	... concave down if $0 < x < 1$
$(1, +\infty)$	+	+	+	... concave up if $x > 1$

3.   Once we know where f is concave up and concave down, then we simply read off the inflection points:

    f changes from concave up to concave down at $x = 0$;

        however, f is not defined at $x = 0$, and

        thus $x = 0$ is not an inflection point.

    f changes from concave down to concave up at $x = 1$;

        since f is defined at $x = 1$, then

        $\boxed{x = 1}$ is an inflection point.

(Note: The graph of f is sketched in Example C of §4.4.) □

3. Curve sketching: an introduction.   Knowing where a function is increasing or decreasing and concave up or concave down obviously gives a great deal of information about its graph. Here is an example that uses such information to obtain the sketch of the graph of a function. It is like Anton's Exercises 16 through 19 , and it anticipates the curve sketching procedures to be studied

in Sections 4.4 and 4.5 :

<u>Example B.</u>   Sketch a continuous curve $y = f(x)$ with the properties $f(2) = 4$, $f'(2) = 1$,

$f''(x) > 0$ if $x < 2$,   and $f''(x) < 0$ if $x > 2$.

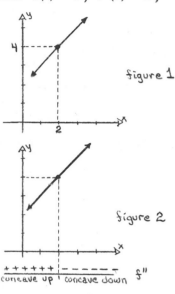

figure 1

<u>Solution.</u>   The information $f(2) = 4$ and $f'(2) = 1$

(i. e. , the slope of the tangent line at $x = 2$ equals   1)

can be placed directly on the graph as shown in

figure (1).   It is then convenient to record the second

derivative information (i. e. , the concavity) on a second

x-axis   below the original one, as shown in   figure (2).

figure 2

Then, "following the instructions" given on this second

x-axis ,   we sketch in a picture for the graph of   f   in

concave up ' concave down f''

figure (3).   Be sure not to read too much into this

sketch! For example, as   x   tends toward $+\infty$ or

$-\infty$,   we cannot say from the information given whether

y   tends towards $+\infty$, $-\infty$ or towards some finite limit.

figure 3

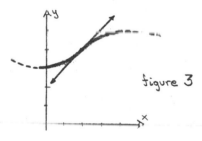

4.   <u>Some common misconceptions and errors ... BEWARE!</u>

<u>WARNING 1.</u>   All inflection points x_0 occur where $f''(x_0)$ is either zero or undefined.

However, the opposite is <u>not</u> the case, i. e. ,

> just because $f''(x_0)$ equals zero or is undefined
>
> does <u>not</u> mean that x_0 must be an inflection point !!

We can illustrate this as follows:

Example C.   Is $x = 0$ an inflection point for $f(x) = x^4$?

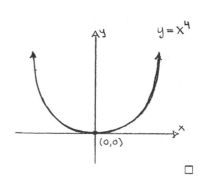

$y = x^4$

Solution.   Since $f''(x) = 12x^2$ we have $f''(0) = 0$.
But f'' is positive whenever $x \neq 0$, so f is
concave up on $(-\infty, 0)$ and $(0, +\infty)$. Thus $x = 0$
is not an inflection point (even though $f''(0) = 0$)
because the concavity of f does not change at $x = 0$!

$(0,0)$

WARNING 2.   If f is increasing on each of two abutting open intervals, (a, b) and (b, c) ,
that does not mean that f is increasing on the full interval (a, c) !

Example D.   Is $f(x) = -1/x$ increasing on $(-\infty, \infty)$?

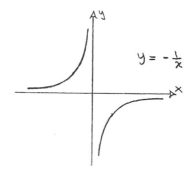

$y = -\frac{1}{x}$

Solution.   Since $f'(x) = 1/x^2$ is positive whenever
$x \neq 0$, then f is increasing on the two open intervals
$(-\infty, 0)$ and $(0, \infty)$. However, as shown to the right,
f is clearly not increasing on the interval $(-\infty, \infty)$!

WARNING 3.   As Example A shows, it is crucial that you be able to solve inequalities. Errors in
solving inequalities are the most common type of mistake with this material. If necessary, we
advise you to go back and reread the material on inequalities in §1.1 of The Companion.

Section 4.3.   Relative Extrema; First and Second Derivative Tests

1.   What are "relative extrema?"   The definitions of a relative maximum and relative minimum

for a function f are quite straightforward (Definitions 4.3.1 and 4.3.2). Stated in English:

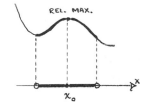

> f has a <u>relative maximum</u> at x_0 if
>
> $f(x_0)$ is the highest value of f near x_0
>
>   (i.e., on some open interval containing x_0).

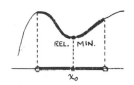

> f has a <u>relative minimum</u> at x_0 if
>
> $f(x_0)$ is the lowest value of f near x_0
>
>   (i.e., on some open interval containing x_0).

A relative maximum or relative minimum is called a <u>relative extremum</u> (the plural is relative

extrema).

If f has a relative extremum at x_0, then the definitions above require that f must

be defined on some open interval containing x_0. In particular, f must be defined <u>on both</u>

<u>sides</u> of x_0, and hence x_0 <u>cannot be an endpoint for the domain of f</u>:

> Relative extrema for a function f cannot
>
> occur at endpoints of the domain of f.

Said in compact form,

> Endpoints cannot give relative extrema.

Thus, even though $x_0 = 0$ gives the smallest value for $f(x) = \sqrt{x}$, $x_0 = 0$ is NOT a relative minimum for $f(x) = \sqrt{x}$ since it is an endpoint of $[0, +\infty)$, which is the domain of f.

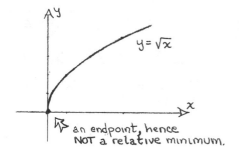

an endpoint, hence NOT a relative minimum.

$y = \sqrt{x}$

Keep in mind that a relative maximum point x_0 need not be an (absolute) maximum point, i.e., $f(x_0)$ may be the largest value of $f(x)$ for all x near x_0, but it does not have to be the largest value of f for all x (the toughest guy on the block might not be the toughest guy in the whole town!). The same comment applies to relative minimum points.

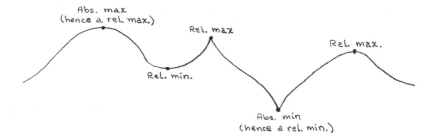

Abs. max (hence a rel. max.) Rel. max Rel. max. Rel. min. Abs. min (hence a rel. min.)

Determining when a relative extremum is an absolute extremum is an important and sometimes tricky question! We will return to it in §4.6. However, we first must answer a more basic question...

2. How do we find the relative extrema? The answer will boil down to: examine the first and second derivatives of the function. We'll give an intuitive development of these techniques to complement Anton's more formal approach.

Suppose f has a relative extremum at x_0.

In the case in which f has a tangent line at x_0

(i. e. , $f'(x_0)$ exists), then the picture to the

right should convince you that the tangent line must

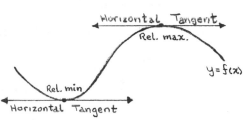

be horizontal (i. e. , $f'(x_0) = 0$). Hence, if f has

a relative extremum at x_0, then either f has no

tangent line (i. e. , $f'(x_0)$ does not exist) or f(x) has a

horizontal tangent line (i. e. , $f'(x_0)$ must equal zero). Points such as x_0 are given a name:

<table>
<tr><td>Critical
Points</td><td>x_0 is a <u>critical point</u> of f if $f'(x_0) = 0$ or $f'(x_0)$ is undefined</td></tr>
</table>

Using this terminology, the most basic result of this section seems almost obvious:

<table>
<tr><td>Theorem
4.3.4</td><td>Relative extrema can occur
only at critical points, i. e. ,
points where the derivative
is zero or undefined.</td></tr>
</table>

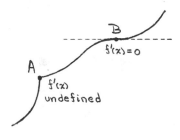

The converse of this result is not true however:

<u>every critical point does not have to give a relative</u>

<u>extremum</u> as is shown in the picture to the right!

A   and   B   are both critical points for f , but

neither is a relative extremum.   This naturally

leads us to a question:

"<u>When does a critical point give a relative extremum?</u>"

This is the burning question of the section! It is easy to compute the critical points for a function f : we first note those values of x for which f'(x) does not exist and then solve the equation f'(x) = 0 for x. (See Example A below.) But how do we test a critical point to determine whether it does give a relative extremum for f? Anton gives two answers for this problem, the First and Second Derivative Tests.

The First Derivative Test is built around plain common sense. Suppose f is continuous at a critical point x_0 , and that the derivative f' exists on both sides of x_0 (but not necessarily at x_0 itself). Then

<table>
<tr><td>

First
Derivative
Test

</td><td>

i.   the sign of f' near x_0 determines whether f is increasing or decreasing on either side of x_0 ; and

ii. knowing whether f is increasing or decreasing on both sides of x_0 determines whether x_0 is a relative maximum, a relative minimum, or neither.

</td></tr>
</table>

For example, suppose f' is positive immediately to the left of x_0 and negative immediately to the right of x_0 . Then f increases on the left of x_0 and decreases on the right of x_0 . So what must f have at x_0 ? A relative maximum, of course!

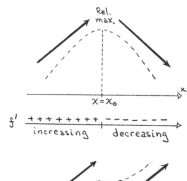

What if f' is positive on both sides of x_0 ? Well, then f increases on the left and on the right of x_0 , which means that f has a "nothing" at x_0 , i. e., f does not have a relative extremum at x_0 .

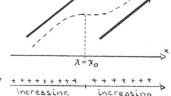

Anton states the First Derivative Test (Theorem 4.3.6) by spelling out the various possibilities for the sign of f' to the left and to the right of x_0; as demonstrated above, all these possibilities can be deduced from

$$f' > 0 \quad \text{means} \quad f \quad \text{increasing}$$
$$f' < 0 \quad \text{means} \quad f \quad \text{decreasing.}$$

Example A.   Find the relative extrema of

$$f(x) = \frac{3}{8} x^{8/3} + \frac{3}{5} x^{5/3} - 3x^{2/3} \ .$$

Solution.   We first observe that f is defined for all values of x. Then we determine the critical points by examining the first derivative:

$$f'(x) = x^{5/3} + x^{2/3} - 2x^{-1/3}$$
$$= \frac{1}{x^{1/3}} (x^2 + x - 2)$$
$$= \frac{1}{x^{1/3}} (x + 2)(x - 1)$$

Hence $f'(x)$ is undefined at $x = 0$ and equals zero at $x = -2$ and 1. These are the three critical points for f.

To use the First Derivative Test to classify each critical point we merely use the Interval Method (§1.1.2 of The Companion) to determine the sign of f' on the four intervals determined by -2, 0 and 1:

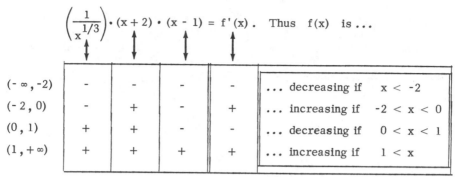

$$\left(\frac{1}{x^{1/3}}\right) \cdot (x + 2) \cdot (x - 1) = f'(x). \quad \text{Thus} \quad f(x) \quad \text{is} \ldots$$

$(-\infty, -2)$	$-$	$-$	$-$	$-$	$\ldots$ decreasing if $\quad x < -2$
$(-2, 0)$	$-$	$+$	$-$	$+$	$\ldots$ increasing if $\quad -2 < x < 0$
$(0, 1)$	$+$	$+$	$-$	$-$	$\ldots$ decreasing if $\quad 0 < x < 1$
$(1, +\infty)$	$+$	$+$	$+$	$+$	$\ldots$ increasing if $\quad 1 < x$

$$\left(\begin{array}{l} \text{The } + \text{ and } - \text{ signs are the signs of the} \\ \text{respective factors on the indicated intervals.} \end{array} \right)$$

As the increasing/decreasing diagram to the right indicates, the First Derivative Test shows that f has a relative maximum at $x = 0$ and two relative minima at $x = -2$ and 1.

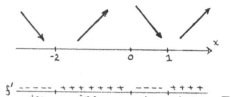

* * * * *

The other standard method for testing a critical point to determine whether or not it gives a relative extremum is the Second Derivative Test.

Second Derivative Test

Suppose x_0 is a stationary point for f and $f''(x_0)$ exists. Then

    (a) $f''(x_0) > 0$ implies f has a relative minimum at x_0

    (b) $f''(x_0) < 0$ implies f has a relative maximum at x_0

    (c) $f''(x_0) = 0$ implies no information at all about x_0 , i.e., the Second Derivative Test is inconclusive.

The Second Derivative Test is intuitively believable when you recall that $f''(x_0) > 0$ means f is concave up at x_0 and $f''(x_0) < 0$ means f is concave down at x_0. Then the two conclusive parts of the Test can be reworded in a "geometrically obvious" fashion:

(a)   If at x_0 the graph has a horizontal tangent line and is concave up, then a relative minimum must occur at x_0.

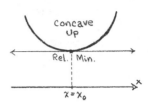

(b)   If at x_0 the graph has a horizontal tangent line and is concave down, then a relative maximum must occur at x_0.

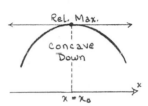

Also take notice of case (c) in the statement of the Test: if $f''(x_0)$ equals zero, then there is nothing we can conclude about x_0. It might give a relative extremum or it might not; the test simply does not say. When the Second Derivative Test is inconclusive, then you must use the First Derivative Test!

Similarly, if the second derivative $f''(x_0)$ is undefined, then the Second Derivative Test is inapplicable to the stationary point x_0. When the Second Derivative Test is inapplicable, then you must use the First Derivative Test!

Example B.   Use the Second Derivative Test to find the relative extrema of

$$f(x) = \frac{3}{8} x^{8/3} + \frac{3}{5} x^{5/3} - 3 x^{2/3}$$

Solution.   This is the same function as in Example A. It is defined everywhere,

$$f'(x) = x^{5/3} + x^{2/3} - 2 x^{-1/3}$$

and the critical points are $x = -2, 0, 1$, as computed earlier.

Right away we see there is a problem with $x = 0$: it is a critical point but $f''(0)$ is undefined since $f'(0)$ is undefined. Hence the Second Derivative Test cannot be used on $x = 0$, and (as done in Example A) we have to resort to the First Derivative Test to show $x = 0$ gives a relative maximum.

For $x = -2$ and $x = 1$ the situation is better:

$$f''(x) = \frac{5}{3} x^{2/3} + \frac{2}{3} x^{-1/3} + \frac{2}{3} x^{-4/3}$$

$$= \frac{1}{3 x^{4/3}} (5 x^2 + 2x + 2)$$

Thus

$$f''(-2) = \frac{1}{3 \cdot 2^{4/3}} (20 - 4 + 2) > 0$$

$$f''(1) = \frac{1}{1} (5 + 2 + 2) > 0$$

proving that f has relative minima at both $x = -2$ and 1. <u>Notice that we did not have to evaluate $f''(-2)$ and $f''(1)$ precisely; we only had to determine their signs.</u>   □

*   *   *   *   *

So which test is better, the First or the Second? Well, as discussed earlier and as seen in Example B, there are some situations in which we cannot use the Second Derivative Test: if $f''(x_0)$ doesn't exist, or if $f''(x_0) = 0$, then the Second Derivative Test fails. However, <u>if</u> the second derivative f'' exists (and is easy to compute, which is not always the case), and <u>if</u> $f''(x_0) \neq 0$, then the Second Derivative Test requires only the computation of the sign of one number, $f''(x_0)$. The First Derivative Test requires you to analyze $f'(x)$ for all the values of x near x_0.

For these reasons we recommend trying the 2nd Derivative Test first. If it works, you will have saved some time. And if it doesn't work, you'll know that quickly too, and you'll have lost very little time. Then you can use the 1st Derivative Test.

Section 4.4.   Sketching Graphs of Polynomials and Rational Functions

From §4.2 we know how to determine where a function f is <u>increasing and decreasing</u> and where it is <u>concave up and concave down.</u> From this we can formalize a curve sketching procedure that applies to almost any function you will encounter in applications.

In this section Anton discusses curve sketching for polynomials and rational functions. <u>It is the polynomial procedure that is the most basic and most important!</u> We will show how a few small modifications can make this procedure apply to a wide range of functions, including rational functions. This will give a simpler alternative to Anton's rational function method.

1.   <u>Graphing polynomials: the basic method.</u>   We will elaborate on Anton's four-step procedure in the following example:

<u>Example A.</u>   Sketch the graph of $y = f(x) = x^4 + 4x^3 - 2x^2 - 12x$

> <u>Step 1.</u>   Calculate $f'(x)$ and $f''(x)$.

$$f'(x) = 4x^3 + 12x^2 - 4x - 12$$
$$f''(x) = 12x^2 + 24x - 4$$

> <u>Step 2.</u>   Determine the x-values for which $f'(x) = 0$ (called the
>
> STATIONARY POINTS), and then the intervals where f is increasing
>
> and decreasing, i.e., where $f'(x) > 0$ and $f'(x) < 0$.

We must solve $0 = f'(x) = 4x^3 + 12x^2 - 4x - 12$

i.e., $0 = x^3 + 3x^2 - x - 3$

From the Rational Root Test in Appendix D.6 the only possible rational roots for this equation are ± 1 and ± 3; testing each value shows that 1, -1 and -3 are all roots. Thus

$$f'(x) = 4(x^3 + 3x^2 - x - 3) = 4(x-1)(x+1)(x+3)$$

By the interval method (§1.1.2 in The Companion) we must check the sign of $f'(x)$ on the four intervals determined by $x = 3, 1$ and -1:

$4(x-1) \cdot (x+1) \cdot (x+3) = f'(x)$.   Thus $f(x)$ is ...

$(-\infty, -3)$	$-$	$-$	$-$	$-$	... decreasing if $x < -3$
$(-3, -1)$	$-$	$-$	$+$	$+$	... increasing if $-3 < x < -1$
$(-1, 1)$	$-$	$+$	$+$	$-$	... decreasing if $-1 < x < 1$
$(1, +\infty)$	$+$	$+$	$+$	$+$	... increasing if $x > 1$

For convenience we place this information on an "auxiliary" x-axis, drawn below the xy-plane that will eventually contain the graph of f.

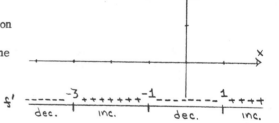

Step 3.   Determine the x-values for which $f''(x) = 0$, and then the intervals where f is concave up and concave down, i.e., where $f''(x) > 0$ and $f''(x) < 0$. Determine the inflection points.

We must solve $0 = f''(x) = 12x^2 + 24x - 4$

i.e., $0 = 3x^2 + 6x - 1$

The quadratic formula (Appendix D.5) yields

$$x = -1 \pm \frac{2}{3}\sqrt{3} \approx .1547 \text{ and } -2.1547$$

which shows that

$$f''(x) = 4(3x^2 + 6x - 1) \approx 12(x - .1547)(x + 2.1547)$$

By the interval method we must check the sign of $f''(x)$ on the three intervals determined by $x = .1547$ and -2.1547:

$12(x - .1547) \cdot (x + 2.1547) \approx f''(x).$ Thus $f(x)$ is ...

$(-\infty, -2.1547)$	-	-	+	... concave up if $x < -2.1547$
$(-2.1547, .1547)$	-	+	-	... concave down if $-2.1547 < x < .1547$
$(.1547, +\infty)$	+	+	+	... concave up if $x > .1547$

As with the first derivative information, we use a second "auxiliary" x-axis to record our facts.

Since the concavity changes at $x \approx -2.1547$ and $.1547$, and since both are in the domain of f, then these are both inflection points.

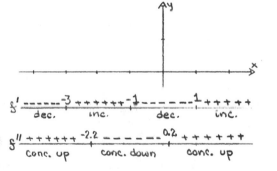

Step 4.   Plot a few well-chosen points (including the stationary points and inflection points) and sketch in the graph.

Computing $f(x)$ at the stationary and inflection points we find

x	f(x)
-3.0000	-9.0000
-2.1547	-1.8888
-1.0000	7.0000
.1547	-1.8888
1.0000	-9.0000

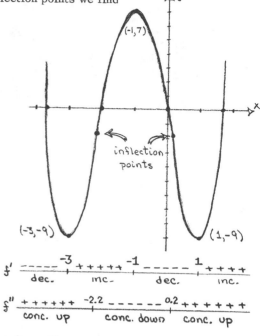

We also look for the roots of $f(x)$. It is
trivial to see that $\underline{x = 0}$ is one root;
the Rational Root Test will then yield
$\underline{x = -2}$ as a second root. Thus, factoring
f by $x(x+2)$ gives

$$f(x) = x(x + 2)(x^2 + 2x - 6)$$

and the quadratic formula applied to $x^2 + 2x - 6$ will then yield $x = -1 \pm \sqrt{7} \approx 1.65$ and
-3.65 as the final two roots. Plotting all of these points and then "following the instructions"
on the two auxiliary x-axes   produces the sketch shown above. □

2.   <u>Modifying the polynomial graphing procedure.</u> The polynomial graphing procedure actually
applies to <u>any</u> function f defined on an open interval (a, b) <u>provided that both f' and f"
are defined and continuous on the open interval</u>! The interval may be either bounded (i.e.,
both endpoints are finite numbers) or unbounded (i.e., one or both endpoints is infinite). We
have only to modify Step 4 in the following simple way:

Step 4. Plot a few well-chosen points (including the stationary points and inflection points), <u>compute the limits</u>

$$\lim_{x \to a^+} f(x) \quad \text{and} \quad \lim_{x \to b^-} f(x) \quad \text{if they exist}$$

and sketch the graph.

The two limits simply tell us what happens with the two ends of the graph of f. When either $a = +\infty$ or $b = -\infty$, we learn the behavior of f as x tends towards $+\infty$ or $-\infty$; when a or b is finite, we learn the behavior of f as x approaches a from the right or as x approaches b from the left.

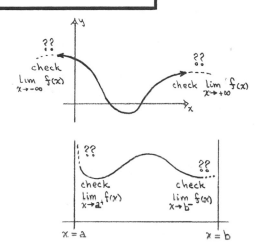

Everything else in this section and the next section is essentially concerned with analyzing these limits!

Here is an example which sketches part of the graph of a rational function:

<u>Example B.</u> Sketch the graph of $y = f(x) = \dfrac{x^2 - 4x + 4}{x^2}$ on $(0, \infty)$

<u>Step 1.</u> The derivatives. Since $f(x) = 1 - 4x^{-1} + 4x^{-2}$

then $f'(x) = 4x^{-2} - 8x^{-3} = \dfrac{4}{x^3}(x - 2)$

and $f''(x) = -8x^{-3} + 24x^{-4} = -\dfrac{8}{x^4}(x - 3)$

On the open interval $(0, \infty)$ both of these derivatives are defined and continuous. Hence our (slightly modified) polynomial graphing procedure will apply.

Step 2. <u>Increasing and decreasing.</u> Setting $f'(x)$ equal to zero gives

$$0 = \frac{4}{x^3} (x - 2)$$

Hence $x = 2$ is the only stationary point, and it does fall in our interval $(0, \infty)$. Thus

$$\left(\frac{4}{x^3}\right) \cdot (x - 2) = f'(x). \quad \text{Thus} \quad f(x) \text{ is} \ldots$$

$(0, 2)$	+	-	-	... decreasing if $0 < x < 2$
$(2, \infty)$	+	+	+	... increasing if $x > 2$

Step 3. <u>Concavity.</u> Setting $f''(x)$ equal to zero gives

$$0 = - \frac{8}{x^4} (x - 3)$$

Hence $x = 3$ is the only solution, and it does fall in our interval $(0, \infty)$. Thus

$$\left(- \frac{8}{x^4}\right) \cdot (x - 3) = f''(x). \quad \text{Thus} \quad f(x) \text{ is} \ldots$$

$(0, 3)$	-	-	+	... concave up if $0 < x < 3$
$(3, \infty)$	-	+	-	... concave down if $x > 3$

Since the concavity changes at $x = 3$, and $x = 3$ is in the domain of f, then this is an inflection point.

The
Modified
Step!

Step 4.   Endpoint limits and graph.   We must compute the two limits $\lim\limits_{x \to 0^+} f(x)$ and

$\lim\limits_{x \to +\infty} f(x)$, if they exist:

$$\lim_{x \to 0^+} f(x) = \lim_{x \to 0^+} \frac{x^2 - 4x + 4}{x^2} = \lim_{x \to 0^+} \left(1 + \frac{4 - 4x}{x^2} \right)$$

$$= 1 + \lim_{x \to 0^+} \left(\frac{4 - 4x}{x^2} \right)$$

Thus, as x tends toward zero (from either direction), the numerator tends toward 4 while the denominator tends toward 0 , but always remains positive. Hence this limit is $+\infty$

Thus

$$\boxed{\lim_{x \to 0^+} f(x) = +\infty}$$

$$\lim_{x \to +\infty} f(x) = \lim_{x \to +\infty} \frac{x^2 - 4x + 4}{x^2} = \lim_{x \to +\infty} 1 - \frac{4}{x} + \frac{4}{x^2}$$

$$= 1 - 0 + 0 = 1 \ .$$

Thus

$$\boxed{\lim_{x \to +\infty} f(x) = 1}$$

Before sketching the graph we compute f at the stationary point and the inflection point: $f(2) = 0$ and $f(3) = 1/9$. It is a good idea to plot another point or two; in this case we use $f(1) = 1$. We now can sketch our curve. Plot the points we have computed, and then put the f' and f'' information on two auxiliary x-axes. Then sketch the curve "following the instructions" on the auxiliary axes. For example, between the two points $(2, 0)$ and $(3, 1/9)$

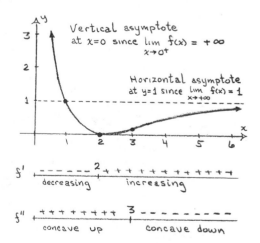

4. 4. 8

we know the curve is

increasing and

concave up. That leaves

little choice as to the shape

of the part of the curve which joins the two points, as shown to the left. The two limits we

computed show that f has a vertical asymptote x = 0 and a horizontal asymptote y = 1

(see Anton's Definition 4. 4. 1).   □

3.   Graphing rational functions.   A rational function f is the quotient of two polynomials, *

$$f(x) = P(x)/Q(x)$$

and it is defined for all x where the denominator Q(x) is non-zero. However, the denom-

inator can be zero only at a finite number of points $a_1, a_2, a_3, \ldots, a_m$ (the roots of Q(x)),

which means that the domain of f is divided into a finite number of open intervals:

$$(-\infty, a_1), (a_1, a_2), \ldots, (a_m, +\infty)$$

On each of these open intervals the rational function f has continuous first and second

derivatives (because the only possible points of discontinuity are the roots of Q(x)). Thus

on each open interval in its domain, a

rational function can be graphed by the

"modified polynomial method" of the previous section.

So this is how to graph a rational function:

---

* To avoid unnecessary complications we assume that P(x) and Q(x) have no common
factors, i.e., that P(x)/Q(x) is expressed "in lowest terms."

> Determine the points $a_1, a_2, \ldots, a_m$ where the denominator is zero, and then apply the "modified polynomial graphing method" on each open interval.

Although we could apply our four-step method to each open interval separately, it saves considerable time to handle all the intervals at once, as our next example shows.

Example C. Sketch the graph of $y = f(x) = \dfrac{x^3 + x^2 - x - 1}{x}$.

Solution. This is the function we analyzed in Example A of §4.2. We will use the results of that analysis at the appropriate times.

Since the denominator of f is $Q(x) = x$, then $x = 0$ is the only point where f is undefined. Thus the domain of f is the union of two intervals,

$$(-\infty, 0) \quad \text{and} \quad (0, +\infty)$$

Step 1. The derivatives. $\quad f'(x) = \dfrac{1}{x^2}(2x^3 + x^2 + 1)$

$$f''(x) = \dfrac{2}{x^3}(x^3 - 1)$$

As expected, these derivatives are defined and continuous on both intervals $(-\infty, 0)$ and $(0, +\infty)$.

Step 2. Increasing and decreasing. Analyzing $f'(x)$ in Example 4.2 A showed that f is:

$$\text{decreasing if} \quad x < -1$$
$$\text{increasing if} \quad -1 < x < 0$$
$$\text{increasing if} \quad x > 0$$

The only stationary point is $x = -1$.

Step 3.   Concavity.   Analyzing f"(x) in Example 4.2 A showed that f is:

concave up if   x < 0   or   x > 0 ,   and concave down if   0 < x < 1

Thus the only inflection point is   x = 1.

Step 4.   Endpoint limits and graph.   Since there are two intervals $(-\infty, 0)$ and $(0, +\infty)$

we have four limits to compute:

$$\lim_{x \to -\infty} f(x) = \lim_{x \to -\infty} x(x+1) - 1 - \left(\frac{1}{x}\right) = "(-\infty)(-\infty) - 1 - 0" = +\infty$$

$$\lim_{x \to 0^-} f(x) = \lim_{x \to 0^-} x^2 + x - 1 - \left(\frac{1}{x}\right) = 0 + 0 - 1 + \infty = +\infty$$

$$\lim_{x \to 0^+} f(x) = \lim_{x \to 0^+} x^2 + x - 1 - \left(\frac{1}{x}\right) = 0 + 0 - 1 - \infty = -\infty$$

$$\lim_{x \to +\infty} f(x) = \lim_{x \to +\infty} x(x+1) - 1 - \left(\frac{1}{x}\right) = "(+\infty)(+\infty) - 1 - 0" = +\infty$$

We now compute some specific points on the graph. The stationary and inflection

points are   x = -1   and   1 ,   so we naturally have one point in each of the two intervals

$(-\infty, 0)$ and $(0, \infty)$. Let's also compute   f   at   x = ±2   and   ±1/2 :

x	-2	-1	-1/2	1/2	1	2
f(x)	1.5	0	.75	-2.25	0	4.5

To sketch the curve we place the
points we have computed on the graph, and
then draw in the auxiliary f' and f"
axes. "Following the instructions" on
these axes, plus using the four limits above,
produces the picture shown to the right. As
you can see, the line   x = 0   is a <u>vertical</u>
<u>asymptote</u> for the graph.

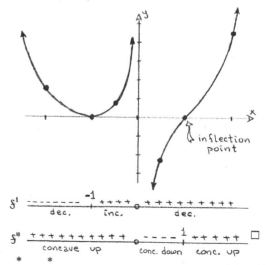

inflection
point

$\mathcal{f}'$ --------- -1 ++++ ++++++++++++
      dec.   inc.   dec.

$\mathcal{f}"$ +++++++++++ ----1++++++ □
      concave up   conc. down conc. up

*   *   *   *   *

When dealing with a rational function, behavior at the endpoints of the open intervals (where $Q(x) = 0$ 'or $x \to \pm\infty$) is easy to describe:

(1)   At a finite endpoint $x = a$ (i.e., where $Q(x) = 0$), there will always

be a <u>vertical asymptote.</u> This was illustrated in Examples B and C.

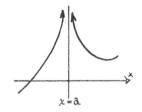 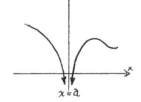 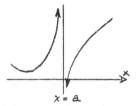

(2)   When x tends towards $\pm\infty$, there are two possibilities:

(a) $\lim\limits_{x \to \infty} f(x)$ or $\lim\limits_{x \to -\infty} f(x)$ is a finite number L.

In this case the line $y = L$ is a <u>horizontal asymptote.</u> This was

illustrated in Example B.

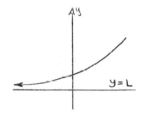 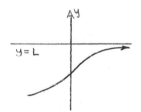

(b) $\lim\limits_{x \to \infty} f(x)$ or $\lim\limits_{x \to -\infty} f(x)$ is $+\infty$ or $-\infty$.

In this case one or both of the ends of the function will head off

to $+\infty$ or $-\infty$. This was illustrated in Example C.

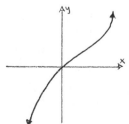 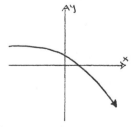 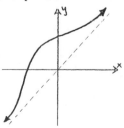

4. <u>The uses of symmetry.</u>   Be sure to take advantage of any symmetry properties which a curve

possesses.   If you need it, review the Symmetry Tests (Anton's Theorem 1.3.2).   Simply

put, symmetry can cut our graphing work almost in half, as the following example (and

Example A in the next section) will show.

<u>Example D.</u>   Sketch the graph of $y = \dfrac{x}{1 - x^2}$.

<u>Solution.</u>   We first look for symmetries: replacing x by - x will change the equation, as

will replacing y with - y.   However, replacing x <u>and</u> y with - x <u>and</u> - y will not

change the equation:

$$(- y) = \frac{(- x)}{1 - (- x)^2} = - \frac{x}{1 - x^2}$$

$$\text{so} \qquad y = \frac{x}{1 - x^2} \quad , \quad \text{the original equation}$$

Hence, by our symmetry tests (Theorem 1.3.2), the graph will be symmetric about

the origin (but not about the x- and y-axes).   We thus have only to plot half of the graph

of f; we choose $x \geq 0$.

Going through our curve sketching procedures
(the details of which we will leave to you) produces the
curves shown in the two solid lines to the right.   The
dotted curve is the rest of the graph obtained simply by
reflection through the origin.   □

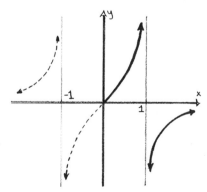

Section 4.5.   Other Graphing Problems

1.   The general graphing procedure.   Almost any function   f   encountered in applications will

have a domain consisting primarily of open intervals on each of which the function has continuous

first and second derivatives.   For such a function the graphing procedure of the previous section

carries over with hardly any modification; the only additional complications center on behavior

at the endpoints of the open intervals.

Very often the function will be defined at some of the interval endpoints, while its first

or second derivative is not defined at those points.   A simple example is $f(x) = \sqrt{x} = x^{1/2}$.

It is easy to see that   f   is defined for all $x \geq 0$; however, $f'(x) = \frac{1}{2} x^{-1/2} = \frac{1}{2\sqrt{x}}$,

which is defined only for $x > 0$ and not at $x = 0$.

For this reason we normally begin a general graphing problem as follows:

> Determine the domain of $y = f(x)$ and divide this
>
> domain up into intervals by the points where
>
> f , f'   or   f"   are undefined. *

The behavior of the graph at those interval endpoints where   f   is defined (but   f'   or   f"   is

not) must be determined in   Step 4   of our graphing procedure.   "Vertical tangent lines" or

"cusps" are common at such points, so this is what Anton focuses his attention on in this section.

2.   Vertical tangent lines and cusps.   Suppose   f   is a function which is continuous at $x = x_0$.

We say that   f   has a vertical tangent line at $x = x_0$ if the derivative of   f   behaves as

follows:

---
*
    This is exactly how we handled rational functions except that we did not have to check   f'   and
    f"   for undefined points.   The reason?   For a rational function the derivatives are defined when-
    ever the function itself is defined!

$$\lim_{x \to x_0^+} f'(x) = +\infty \quad \text{or} \quad -\infty$$

and

$$\lim_{x \to x_0^-} f'(x) = +\infty \quad \text{or} \quad -\infty$$

In English, as x approaches x_0 from either direction, the slopes of the tangent lines to the graph tend to $+\infty$ or $-\infty$.

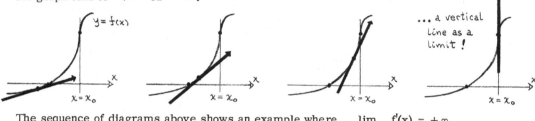

The sequence of diagrams above shows an example where $\lim_{x \to x_0^-} f'(x) = +\infty$.

As you can see, the tangent lines get steeper and steeper as x approaches x_0 from the left, until finally (i.e., in the limit) we obtain a "vertical tangent line" at $x = x_0$.

Since there are two possibilities $(\pm \infty)$ for each of the limits $\lim_{x \to x_0^-} f'(x)$ and $\lim_{x \to x_0^+} f'(x)$, there are four possibilities for having a vertical tangent line at $x = x_0$:

$\lim_{x \to x_0^-} f'(x)$	$+\infty$	$-\infty$	$+\infty$	$-\infty$
$\lim_{x \to x_0^+} f'(x)$	$+\infty$	$-\infty$	$-\infty$	$+\infty$

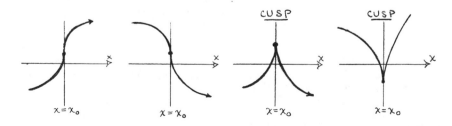

As the pictures show, when the two one-sided limits differ, we obtain a sharp point on the graph at $y = f(x)$, and say that f has a cusp at $x = x_0$.

How do we detect the presence of a vertical tangent line or a cusp in our graphing procedure? Well, if f behaves in this way at $x = x_0$, then f is defined at x_0 but f' will not be defined at x_0. Hence x_0 will be designated as an interval endpoint at the very beginning of our graphing procedure! This will alert us to the possibility of a vertical tangent line or a cusp at $x = x_0$, and we will check this possibility in Step 4 by computing

$$\lim_{x \to x_0^+} f'(x) \qquad \text{and} \qquad \lim_{x \to x_0^-} f'(x)$$

3. The curve sketching procedure; the final version.   For your reference we will write out below the full graphing procedure as it has evolved to this point:

To graph $y = f(x)$, first determine the domain of f and check for useful symmetry properties.

Step 1.   The derivatives.   Compute f' and f'', and check for points in the domain of f where f' and f'' are undefined.   Divide the domain up into intervals by the points where f, f', or f'' are undefined.

Step 2.   Increasing and decreasing.   Use the interval method ($\S 1.1$) to determine where $f'(x) > 0$ and where $f'(x) < 0$.   You must start by finding the stationary points, $f'(x) = 0$.

Step 3.   Concavity.   Use the interval method to determine where $f''(x) > 0$ and where $f''(x) < 0$.   You must start by finding the points where $f''(x) = 0$.   Inflection points occur where the concavity changes.

4.5.4

Step 4.   Endpoint limits and graph.

Compute the one-sided limits of $f(x)$ at the interval endpoints.

Compute the one-sided limits of $f'(x)$ at possible vertical tangent line

or cusp points x_0 (f continuous at x_0, but $f'(x_0)$ not defined).

Compute function values at all important x-values.

Using auxiliary f' and f'' axes and all the limit information ...

FINALLY SKETCH THE GRAPH!

Here is an example (at last!):

Example A.   Sketch the graph of the function $f(x) = x^2 - 3x^{2/3}$.

Solution.   The function is defined for every value of x and hence its domain in the whole

x-axis. Moreover, f is symmetric about the y-axis since substituting $-x$ for x

will not change the equation. Hence we have only to concern ourselves with $x \geq 0$.

Step 1.   The derivatives. $f'(x) = 2x - 2x^{-1/3} = \dfrac{2}{x^{1/3}} (x^{4/3} - 1)$

$$f''(x) = 2 + \frac{2}{3} x^{-4/3} = \frac{2}{3 x^{4/3}} (3 x^{4/3} + 1)$$

Before going on to Step 2, we check the domains of f' and f''. AHA! Neither is defined

at $x = 0$! Thus $x = 0$ must be considered an interval endpoint, and we keep in mind that

f could have a vertical tangent line or a cusp at $x = 0$!

Step 2.   Increasing and decreasing. $f'(x)$ equals zero only when $x = \pm 1$, so these are

the only stationary points. Therefore we check the sign of f' on the intervals $(0, 1)$ and

$(1, +\infty)$ [by our earlier symmetry observation we need not check $(-\infty, -1)$ and $(-1, 0)$]:

$$\left(\frac{2}{x^{1/3}}\right)\cdot\left(x^{4/3} - 1\right) = f'(x). \quad \text{Thus} \quad f(x) \quad \text{is} \ldots$$

(0, 1)	+	-	-	... decreasing if $0 < x < 1$
(1, +∞)	+	1	+	... increasing if $x > 1$

Step 3. <u>Concavity.</u> Setting $f''(x)$ equal to zero gives

$$3x^{4/3} + 1 = 0$$

$$\text{or} \quad (x^{1/3})^4 = -1/3 \qquad \text{(Huh?)}$$

Since the fourth power of a number cannot be negative, there are no solutions to this equation, and hence no points where $f''(x)$ equals zero in the interval $(0, \infty)$. In fact, on this interval $f''(x)$ is clearly positive so that f is concave up.

Step 4. <u>Endpoint limits and graph.</u> Since by symmetry we have only to graph the function on the interval $[0, \infty)$, then we have only two endpoint limits to consider:

$$\lim_{x \to 0^+} f(x) = \lim_{x \to 0^+} (x^2 - 3x^{2/3}) = 0 - 0 = 0$$

$$\lim_{x \to \infty} f(x) = \lim_{x \to \infty} x^{2/3}(x^{4/3} - 3) = +\infty$$

<u>But remember that</u> f' was undefined at $x = 0$! Thus, we must check for a vertical tangent line or a cusp at $x = 0$. By symmetry we have only to compute one limit:

$$\lim_{x \to 0^+} f'(x) = \lim_{x \to 0^+} \left(\frac{2}{x^{1/3}}\right)\left(x^{4/3} - 1\right) = -\infty$$

$$\begin{array}{ccc} \uparrow & \uparrow & \uparrow \\ \text{tends} & \text{tends} & \\ \text{toward} & \text{toward} & \ldots \quad \text{thus} \\ +\infty & -1 & \end{array}$$

The symmetry about the y-axis will thus give a <u>cusp</u> at $x = 0$.

We now compute some points on the graph:

x	0	1	2
f(x)	0	- 2	- .76

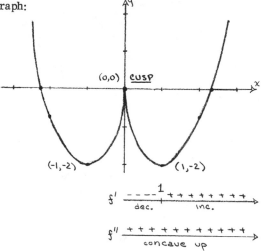

Moreover, $f(x) = 0$ when

$$x^2 - 3x^{2/3} = 0$$

$$x = 0 \quad \text{or} \quad x^{4/3} - 3 = 0$$

$$x = 0 \quad \text{or} \quad x = 3^{3/4} \simeq 2.28$$

Plotting all these points, adding in the auxiliary f' and f" axes, and "following the instructions" produces the graph as shown. The left-hand side of the graph was obtained by reflecting the right-hand side over the y-axis. □

*   *   *   *   *

There is a lot to consider when graphing a function, but keep in mind that most of the techniques we have discussed are just "bells and whistles" added to our basic four-step method for graphing polynomials !! Keep these four simple steps as your guide and don't get bogged down in the details.

Section 4.6.   Maximum and Minimum Values of a Function

1.   Existence of extreme values.   The definitions of the maximum and minimum values (which

are called extreme values) of a function   f   on an interval   I   are straightforward:

Maximum
Value

The value $f(x_0)$ is the (absolute) maximum value

for the function   f   on the interval   I   if

$f(x_0) \geqq f(x)$ for all   x   in   I .

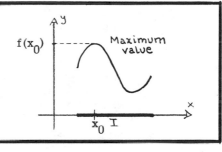

Minimum
Value

The value $f(x_0)$ is the (absolute) minimum value

for the function   f   on the interval   I   if

$f(x_0) \leqq f(x)$ for all   x   in   I .

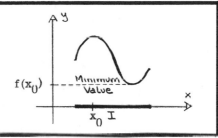

Note the differences between (absolute) maximum values and relative maximum values:

1.   The (absolute) maximum value $f(x_1)$

must be the largest value of   f   on

the entire interval   I.

A relative maximum value $f(x_2)$

need only be the largest value of   f

over a small open interval in   I.

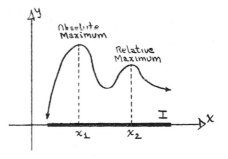

2.   The <u>(absolute) maximum value</u> $f(x_0)$

can occur at an endpoint of I.

A <u>relative maximum value</u> cannot

occur at an endpoint of I.

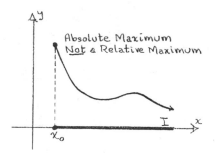

Similar statements can be made about minimum values.

Anton gives numerous examples in this section to illustrate the existence and <u>non-existence</u> of extreme values. This latter situation is important to understand: there are many instances in which extreme values for a function do not exist on a given interval.

Fortunately Theorem 4.6.4 guarantees the existence of extreme values in one important case:

Extreme- Value Theorem	If f is <u>continuous</u> on the <u>closed, bounded</u>* interval [a, b], then f has both a maximum and a minimum value on [a, b].

This theorem would not be true if any of the three hypotheses "continuous function," "closed interval" or "bounded interval" were dropped. In fact, Anton gives examples in which each one of these conditions is omitted:

---

* We have included the word "bounded" to emphasize that the interval has <u>finite</u> values for endpoints.

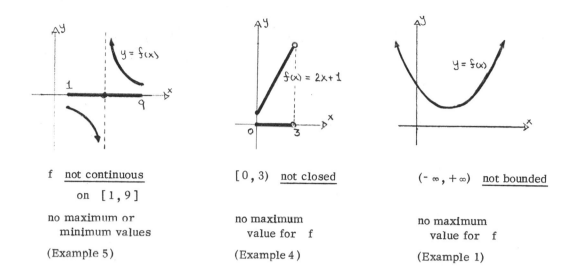

f <u>not continuous</u>   [0 , 3) <u>not closed</u>   (- ∞ , + ∞) <u>not bounded</u>

on [1 , 9]

no maximum or   no maximum   no maximum
minimum values   value for f   value for f

(Example 5)   (Example 4)   (Example 1)

The Extreme Value Theorem should make you wonder about two further question:

1.   What can we do about continuous functions * on intervals which are <u>not closed and bounded?</u>

Answer: we'll consider this question in Subsection 3 below.

2.   Given a continuous function f on a closed, bounded interval [a , b], how do we

<u>compute</u> the maximum and minimum values? Answer: we'll consider this in the very

next subsection . . .

2.   <u>Computation of extreme values on a closed, bounded interval.</u>   Assume that f is a continuous

function on the closed, bounded interval [a , b]. The Extreme Value Theorem tells us that f

has a maximum and minimum value on   [a , b]; unfortunately it does not tell us how to compute

these values!

However, it's not hard to do! First, we compile a short list of x-values   which are

---

* We will not worry about discontinuous functions. Most functions occurring in applications are
continuous, and no decent extreme value theory is possible for discontinous functions.

the only x-values at which f can achieve its maximum and minimum values. This list is obtained as follows:

1. The maximum and minimum values of f on [a, b] could occur at the endpoints of the interval. So start the list with the ENDPOINTS $x = a$ and $x = b$.

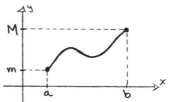

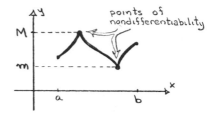

2. The maximum and minimum values of f on [a, b] could occur at points in the open interval (a, b). If this happens, the absolute extreme value will also be a relative extreme value. By Theorem 4.3.4, relative extrema can occur only at critical points, so add all CRITICAL POINTS of f to the list.

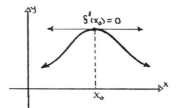

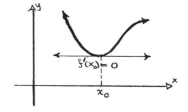

Once we have our list $\{x_1, x_2, \ldots, x_n\}$ of the only x-values at which f can achieve its maximum and minimum values, then finding the maximum and minimum values themselves is easy: simply compute the function values $\{f(x_1), f(x_2), \ldots, f(x_n)\}$ and choose the largest and smallest!

This discussion motivates the following three-step procedure:

To find the extreme values of a continuous function f on a closed, bounded interval

[a, b]:

Step 1.   Find the critical points of f , i. e. , the stationary points and

the points of non-differentiability.

Find
Critical
Points

- To do this, we compute f'(x) , note the x-values for which

f'(x) does not exist, and then solve the equation f'(x) = 0

for x .

Step 2.   Evaluate f at the critical points and the endpoints a and b .

Evaluate
Function

- Here we simply plug all our compiled x-values into the

function.

Step 3.   The largest of the values in Step 2 is the maximum and the

Pick
Extrema

smallest is the minimum.

We emphasize that this procedure is valid only for continuous functions on closed,

bounded intervals [a, b]. More general situations will be considered in the next subsection.

Example A.   Find the extreme values for the function

$$f(x) = x^{3/5} - 3x^{1/5} \quad \text{on} \quad [-32, 2^{5/3}]$$

Solution.   Step 1: Critical points.   We start by computing the derivative of f :

$$f'(x) = \left(\frac{3}{5}\right)x^{-2/5} - \left(\frac{3}{5}\right)x^{-4/5}$$

The best way to handle a sum or difference of (fractional) powers is to factor out the lowest power; in this case the lowest power is $x^{-4/5}$.

$$f'(x) = \left(\frac{3}{5}\right)x^{-4/5}\left[x^{-2/5 + 4/5} - 1\right]$$

$$= \left(\frac{3}{5}\right)x^{-4/5}\left[x^{2/5} - 1\right]$$

Notice that because of the <u>negative</u> power $-4/5$, the derivative $f'(x)$ is not defined when $x = 0$. Since $x = 0$ is in the interval under consideration, we include it as a point of non-differentiability.

Now we look for the stationary points by solving $f'(x) = 0$. Since a product can equal zero only if one of the factors is zero, then we see that $f'(x) = 0$ only when

$$x^{-4/5} = 0 \qquad \text{(which can never happen!)}$$

<u>or</u>

$$x^{2/5} - 1 = 0$$
$$x^{2/5} = 1$$
$$x = \pm 1$$

Since these x-values are in the open interval $(-32, 2^{5/3})$, they are both stationary points for f on $[-32, 2^{5/3}]$.

Thus the list of critical points for f on $[-32, 2^{5/3}]$ is

$$\{-1, 0, 1\}$$

<u>Step 2: Evaluate.</u> The x-values we must test are $\{-32, -1, 0, 1, 2^{5/3}\}$, the critical points plus the endpoints.

x	$f(x) = x^{3/5} - 3x^{1/5}$
- 32	$-8 - 3(-2) = -8 + 6 = -2$
- 1	$-1 - 3(-1) = -1 + 3 = 2$
0	$0 - 3(0) = 0 - 0 = 0$
1	$1 - 3(1) = 1 - 3 = -2$
$2^{5/3}$	$2 - 3(2^{1/3}) \approx 2 - 3.78 = -1.78$

Step 3: Pick maximum and minimum.   From our list we see that the maximum value

of f on $[-32, 2^{5/3}]$ is 2 , which occurs when x = -1 , and the minimum value of f

on $[-32, 2^{5/3}]$ is -2 , which occurs when x = -32 and when x = 1 .   □

*   *   *   *   *

Example B.   Prove or disprove: $\tan x \geq x$ for all x in the interval $[0, \pi/4]$.

Solution.   By the methods of this section it is easy to solve this problem.   We let

$$f(x) = \tan x - x$$

The question now reads:

prove or disprove: $f(x) \geq 0$ for all   x   in $[0, \pi/4]$.

To do this we compute the minimum value for $f(x)$ on $[0, \pi/4]$. If the minimum value is

less than zero, then we have disproved our conjecture.   If it is greater than or equal to zero,

then we have proved our conjecture.

Step 1: Critical points. $f'(x) = \sec^2 x - 1$.

Since this derivative is defined for all x in the desired interval, there are no points of non-

differentiability.   To determine the stationary points we set $f'(x) = 0$, i. e.,

$$\sec^2 x = 1$$

Since   sec x = 1/cos x ,   then

$$\cos^2 x = 1$$

$$\cos x = \pm 1$$

However, this equality is never true on the <u>open</u> interval $(0, \pi/4)$. Thus there are no stationary points for f on $[0, \pi/4]$.

Step 2: Evaluate. Since there are no critical points for f on $[0, \pi/4]$, we have only to evaluate f at the endpoints x = 0 and x = $\pi/4$:

x	f(x) = tan x - x
0	0 - 0 = 0
$\pi/4$	$1 - \frac{\pi}{4} \approx .2146$

Step 3: Pick maximum and minimum. The minimum value for tan x - x on $[0, \pi/4]$ is 0 , which occurs when x = 0 . Thus we do have tan x $\geq$ x for all x in $[0, \pi/4]$. In fact, we have actually shown that tan x > x for all x in $(0, \pi/4]$. The reason? Because f(x) = tan x - x achieves its minimum value on $[0, \pi/4]$ <u>only</u> when x = 0!   □

(If you are adventuresome you might try to generalize Example B as follows: prove that tan x $\geq$ x + x^3/3 for all x in the interval $[0, \pi/4]$. This is a bit trickier than the inequality in Example B.)

*   *   *   *   *

Example C.   Find the extreme values for the function

$$f(x) = \left| x^2 - 4x - 5 \right| + x \quad \text{on} \quad [-2, 5.5]$$

Solution.   Step 1: Critical points.   The presence of the absolute value makes this problem a little more challenging. We need to consider where the term inside the absolute value is positive and where it is negative. Since

$$x^2 - 4x - 5 = (x + 1)(x - 5)$$

then the term equals zero when $x = -1$ or $x = 5$. Then we find that

$$x^2 - 4x - 5 \quad \text{is positive when} \quad x < -1 \quad \text{or} \quad x > 5$$

$$\text{and negative when} \quad -1 < x < 5$$

We can therefore rewrite the function $f(x)$ as

$$f(x) = \begin{cases} (x^2 - 4x - 5) + x = \quad x^2 - 3x - 5 \quad \text{when} \quad x \leq -1 \quad \text{or} \quad x \geq 5 \\ -(x^2 - 4x - 5) + x = -x^2 + 5x + 5 \quad \text{when} \quad -1 \leq x \leq 5 \end{cases}$$

This expresses f without the absolute value signs. However, f is given by two different formulas on different intervals. Since there are no points of non-differentiability other than possibly the change-over points $x = -1$ and $x = 5$, the only question is whether these change-over points are points of non-differentiability. However:

> we may assume that f is non-differentiable
> at the change-over points $x = -1$ and $x = 5$.

If this assumption is false, then it will simply cause us to have "too many" points in our list of critical points for f (and hence on our list of candidates for the maximum and minimum values); we would certainly not lose our maximum and minimum values!

Continuing, we find the stationary points by setting the derivative equal to zero:

$$0 = f'(x) = \begin{cases} 2x - 3 & \text{when} & x < -1 \text{ or } x > 5 \\ -2x + 5 & \text{when} & -1 < x < 5 \end{cases}$$

Then the stationary points are $x = 1.5$ when $x < -1$ or $x > 5$ (which is not possible) or $x = 2.5$ when $-1 < x < 5$ (which is possible). Thus $x = 2.5$ is the only stationary point.

Step 2: Evaluate. Evaluating at the 3 critical points and 2 endpoints gives:

x	$f(x) = \lvert x^2 - 4x - 5 \rvert + x$
-2	5
-1	-1
2.5	11.25
5	5
5.5	8.75

Step 3: Pick maximum and minimum. The maximum value for $f(x)$ on $[-2, 5.5]$ is 11.25, which occurs when $x = 2.5$. The minimum value is -1, which occurs when $x = -1$. $\qquad \square$

3. Computation of extreme values on a general interval. In the previous subsection, we learned that a continuous function f on a closed, bounded interval $[a, b]$ must have a maximum and a minimum value (Theorem 4.6.4, The Extreme Value Theorem). However, when the interval in question is not closed or not bounded, the situation is much more complicated. First of all, extreme values might not even exist! (See §4.6.1 of The Companion or Anton's Examples 1 and 3.) And even if they do exist, the method we used in the previous subsection to find these values must be modified.

When finding extreme values on a general interval it is often helpful to <u>sketch the graph</u>, using all or part of the sketching procedures of §4.5.3 of <u>The Companion</u>. In this way you can "see" if there are any maximum or minimum values, and determine approximately where they are located. This use of sketching is illustrated in Anton's Example 8, as well as in Examples D and E below.

No matter how you proceed in searching for extreme values, invariably you encounter the following problem:

> How do we determine when a <u>relative</u>
>
> extreme value is an (<u>absolute</u>) extreme value?

Alas, there is no one, simple answer to this question since there are so many possible combinations of relative and absolute extreme values. Anton does give one useful result, however (Theorem 4.6.6):

Single
Relative
Extremum

> Suppose f is continuous and has <u>only one</u>
>
> relative extremum on an interval.
>
> Then the relative extremum must be an
>
> absolute extremum on the interval.

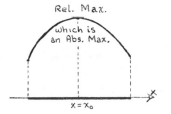

Rel. Max.
which is
an Abs. Max.

$x = x_0$

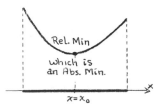

Rel. Min
which is
an Abs. Min.

$x = x_0$

Example D. Find the extreme values (if they exist) for

$$f(x) = \frac{x^2 + 3x}{x - 1}$$

on the interval $[-4, 1)$.

Solution. The function f is defined everywhere but at $x = 1$. Computing f' we obtain

$$f'(x) = \frac{(x - 1)(2x + 3) - (x^2 + 3x)}{(x - 1)^2} = \frac{x^2 - 2x - 3}{(x - 1)^2}$$

$$= \frac{(x - 3)(x + 1)}{(x - 1)^2}$$

Thus $f'(x) = 0$ on $[-4, 1]$ only when $x = -1$ (the value $x = 3$ is not in the interval) and $f'(x)$ is never undefined on $[-4, 1)$. Thus $x = -1$ is the only critical point. Hence, by Theorem 4.6.6, if $x = -1$ can be shown to be a relative extremum, then it must be an absolute extremum.

Since f'' looks unpleasant to compute, we use the 1st Derivative Test to test the critical point $x = -1$. Examining the first derivative near $x = -1$ we find:

$$\left(\frac{1}{(x - 1)^2}\right) \cdot (x - 3) \cdot (x + 1) = f'(x). \quad \text{Thus} \quad f(x) \text{ is} \ldots$$

$(-4, -1)$	$+$	$-$	$-$	$+$	... increasing if $-4 < x < -1$
$(-1, 1)$	$+$	$-$	$+$	$-$	... decreasing if $-1 < x < 1$

Hence the First Derivative Test tells us that $x = -1$ gives a relative maximum. However, since $x = -1$ gives the only relative extremum on the interval $[-4, 1)$,

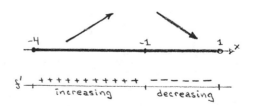

then it must give an absolute extremum. Thus $f(-1) = 1$ is the absolute maximum value of f on $[-4, 1)$.

What about minimum values? Our increasing/decreasing diagram for f given above shows that the only possible location for an absolute minimum is at the left endpoint $x = -4$ ($x = 1$ is not in the domain of f so the absolute minimum could not occur at this right end-point). However, checking what happens as x approaches 1 from the left we find

$$\lim_{x \to 1^-} f(x) = \lim_{x \to 1^-} \frac{x(x+3)}{x-1} = -\infty$$

since the numerator tends toward 4 while the denominator increases toward zero through negative values. As our rough sketch to the right indicates, f does not have an absolute minimum on $[-4, 1)$. □

Example E.   Find the extreme values (if they exist) for

$$f(x) = \frac{x^2 + 3x}{x - 1}$$

on the interval $[-4, 0)$.

Solution.   This is the same function as in Example D, but with a smaller domain. The same reasoning as used in Example D shows that f has an absolute maximum at $x = -1$.

What about minimum values? The increasing/decreasing diagram for f given to the right shows that the only possible location for an absolute minimum is at the left endpoint $x = -4$. (Once again, the right endpoint $x = 0$

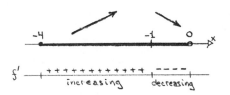

is not in the interval). At $x = -4$ we find $f(-4) = -4/5$. Checking the behavior of f   near

0   we find

$$\lim_{x \to 0^-} f(x) = \lim_{x \to 0^-} \frac{x(x + 3)}{x - 1} = 0$$

We can thus make the rough sketch of

f   shown to the right, which shows that

$x = -4$ does indeed give an absolute

minimum value for f . That value is $-4/5$.   □

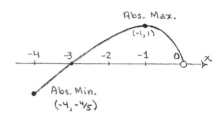

Abs. Max.
(-1, 1)

Abs. Min.
(-4, -⁴/₅)

Examples D and E illustrate two useful observations to be made about the absolute

extrema for a continuous function f :

1.   When they exist, absolute extrema

occur either at relative extrema or

at endpoints of intervals in the domain

of f. (Example E)

2.   When absolute extrema do not

exist, the reason is some unusual

behavior at an endpoint of an interval

in the domain of f. (Example D)

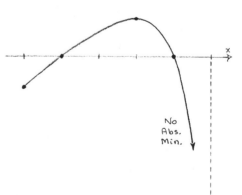

Abs. Max. at
a Rel. Max.

Abs. Min. at
an endpoint

No
Abs.
Min.

4. <u>A summary diagram.</u>   The relationships between the various types of points discussed so

far - critical points, relative and absolute extrema, and endpoints - is conveniently

summarized as follows:

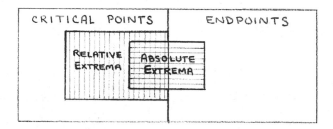

Thus a relative extremum is always a critical point; an absolute extremum is either a relative

extremum or an endpoint, etc.

<u>Section 4.7.   Applied Maximum and Minimum Problems.</u>

1. <u>Word problems again</u>...! As with related rates, the difficulty with applied maximum and

minimum problems is that they are <u>word problems</u>! If you have not yet looked at Appendix H

on word problems, we suggest that you do so now! Some people tend to find "max.-min."

problems even more difficult than related rates.

        The common thread which holds all of the problems in this section together is the

following:

> Each solution requires finding the
>
> maximum and/or minimum value of
>
> a <u>continuous function</u> f on a
>
> <u>closed, bounded* interval</u> [a, b].

The difficult part in any specific problem is usually the determination of the function f and the interval [a, b], i.e., the translating of the problem into mathematics. After that, the straightforward techniques of the previous section will produce the answer.

Before giving some additional illustrations of Anton's solution procedure, we should note that not all applied max-min problems translate into problems involving continuous functions on <u>closed, bounded</u> intervals. In practice many other types of intervals can occur, and their analysis needs more delicate techniques than those presented here. We will return to this question in 4.8.

2. <u>How to tackle a max.-min. problem ... and live to tell about it!</u> Anton summarizes his procedure for solving max.-min. problems in a 5-step list just before Example 2. We will elaborate on this list, giving additional suggestions and warnings, as we work the following examples:

<u>Example A.</u> A manufacturer wishes to produce cylindrical aluminum cans each with a volume of 10 inches^3 and with a ratio of the radius to the height of between $1/4$ and 2. There is no waste in the production of the sides of the cans; however, the tops and bottoms of the cans are disks cut from square pieces of

_____

* As in 4.6 we have included the word "bounded" to emphasize that the interval has <u>finite</u> values for endpoints.

aluminum, and the scraps are discarded.   What dimensions for the cans will require the least

amount of aluminum?

Solution.

Step 1.   Label the quantities relevant to the problem.

- First draw a picture if appropriate (it usually is!) and assign labels

  $(x, y, h, t, \text{etc.})$ to any quantity that seems relevant to the

  problem.

- Carefully translate all of the given information into equation form

  and identify the quantity (variable) to be maximized or minimized!

The picture shown here indicates all the important relationships in the problem.   Notice how we have labeled the radius r of the top and bottom of the can as well as its height h . Translating the rest of the problem naturally introduces two more quantities: the volume of the can, $V = 10$, and the amount A of aluminum needed to produce one can.   We wish to minimize A .

V = volume of can

   = 10

$$\frac{1}{4} \le \frac{r}{h} \le 2$$

A = amount of aluminum needed to produce one can

minimum value of A = ?

4.7.4

Step 2.　　　Find a formula for the quantity to be maximized or minimized.

- Do not hesitate to use as many variables as necessary in

your formula!

The variable A , the amount of aluminum needed to produce one can, is the sum of

the areas of a rectangle (side of can) and two squares (from which the top and bottom are

obtained). Thus

$$A = h(2\pi r) + (2r)^2 + (2r)^2$$

$$A = 2\pi rh + 8r^2$$

Step 3.　　　Using the conditions stated in the problem to eliminate variables,

express the quantity to be maximized or minimized as a function

of one variable.

- If your formula uses 2 "independent" variables, then you need

one relationship between the two variables to eliminate one of

them. If your formula uses 3 "independent" variables, then

you need two relationships, etc.

Our formula for A uses two "independent" variables, r and h , and we are not

explicitly given an equation relating r and h . However, we have not used the constant

volume V = 10. Since the volume of the can is given by $V = \pi r^2 h$ (formula 2(iii) in

Appendix A), then

$$\boxed{\pi r^2 h = 10}$$

is the relationship we want. Solving for h gives

$$h = \frac{10}{\pi r^2}$$

so that

$$A = 2\pi r \left(\frac{10}{\pi r^2}\right) + 8r^2$$

$$\boxed{A = \frac{20}{r} + 8r^2}$$

Step 4.   Find the interval of possible values for this variable

from the physical restrictions in the problem.

We have the restrictions

$$\frac{1}{4} \le \frac{r}{h} \le 2$$

which in view of the fact that $h = \frac{10}{\pi r^2}$ (from Step 3) translates into

$$\frac{1}{4} \le r / \left(\frac{10}{\pi r^2}\right) \le 2$$

$$\frac{5}{2\pi} \le r^3 \le \frac{20}{\pi}$$

$$\left(\frac{5}{2\pi}\right)^{1/3} \le r \le \left(\frac{20}{\pi}\right)^{1/3}$$

Thus the variable r must satisfy

$$\boxed{.9267 \leq r \leq 1.8534}$$

Step 5.   If applicable, use the techniques of the previous section to obtain

the maximum or minimum.

- Remember: these techniques are applicable if you are maximizing

or minimizing a <u>continuous function</u> on a <u>closed, bounded interval</u>!

We wish to minimize $A = \frac{20}{r} + 8r^2$ on the interval $[.9267, 1.8534]$. This function
is continuous and the interval is closed and bounded, so we can proceed as in the previous
section:

<u>Find critical points:</u> $\dfrac{dA}{dr} = -\dfrac{20}{r^2} + 16r$

This is defined for all $r \neq 0$ so that there are no points of non-differentiability
in the interval under consideration.

Setting $\dfrac{dA}{dr} = 0$ we find $-20 + 16r^3 = 0$ or

$$r = \left(\frac{5}{4}\right)^{1/3} \approx 1.0772$$

This does fall in the desired interval. It is our only stationary point and only
critical point.

<u>Evaluate:</u>

	r	$A = \frac{20}{r} + 8r^2$
(endpoint)	.9267	28.4521
	1.0772	27.8495
(endpoint)	1.8534	38.2717

<u>Pick minimum value:</u> $r = 1.0772$ gives the minimum value of $A = 27.8495$. Thus the dimensions of the can requiring the least amount of aluminum are

$$r \approx 1.0772 \quad \text{and} \quad h = \frac{10}{\pi r^2} \approx 2.7432$$ □

*   *   *   *   *

Summarizing Anton's five step method in a few words we have:

<u>Step 1.</u>	Diagram and labels
<u>Step 2.</u>	Formula
<u>Step 3.</u>	Formula in one variable
<u>Step 4.</u>	Interval
<u>Step 5.</u>	Max.-min. techniques

<u>Example B.</u> Assume that the operating cost of a certain truck (excluding the driver's wages) is $10 + x/2$ cents per mile when the truck travels at x mph. If the driver earns $12.50 per hour, what is the most economical speed at which to operate the truck on a turnpike where the minimum speed is 40 mph and the maximum speed is 55 mph?

<u>Solution.</u> <u>Step 1: Diagram and labels.</u> There seems to be no useful diagram in this type of problem. However, there are variables to be identified and labeled. First, we wish to find that value of

$$x = \text{the speed (miles per hour)}$$

in the interval

$$40 \leq x \leq 55$$

which minimizes

$$\boxed{C = \text{the total cost (dollars)}}$$

of any trip made on the turnpike. We are given:

operating cost of the truck (excluding driver's wages)

$$= 10 + x/2 \quad \text{cents per mile}$$

$$= \frac{1}{100} (10 + x/2) \quad \text{dollars per mile}$$

and

driver's wages = 12.50 dollars per hour

What other variables would seem important in calculating C ? Well, the total length of any trip and the total time spent:

$$D = \text{length of trip (miles)} ,$$

$$t = \text{total time (hours) of trip}$$

Step 2: Formula. It is clear that

$$\begin{bmatrix} \text{total cost} \\ C \end{bmatrix} = \begin{bmatrix} \text{operating} \\ \text{cost} \end{bmatrix} + \begin{bmatrix} \text{driver's} \\ \text{wages} \end{bmatrix}$$

$$= \begin{bmatrix} \text{operating cost} \\ \text{per mile} \end{bmatrix} \cdot \begin{bmatrix} \text{number} \\ \text{of miles} \end{bmatrix} + \begin{bmatrix} \text{driver's wages} \\ \text{per hour} \end{bmatrix} \cdot \begin{bmatrix} \text{number} \\ \text{of hours} \end{bmatrix}$$

Thus
$$\boxed{C = \left[\frac{1}{100} \left(10 + \frac{x}{2} \right) \right] \cdot D + [12.5] \cdot t}$$

where the units are

$$[\text{ dollars }] = \left[\frac{\text{dollars}}{\text{mile}} \right] \cdot \text{ miles } + \left[\frac{\text{dollars}}{\text{hour}} \right] \cdot \text{ hours}$$

Step 3: Formula in one variable.   Our formula for C is written in terms of the quantities

x , D and t . However, only two of these (x and t) are variables. The third quantity,

the distance D of the trip, is a constant for any given trip, and cannot be eliminated.

Thus we look for one relationship to eliminate t from our equation for C . The

relationship is the familiar

$$D = x t \qquad (\text{distance} = \text{rate} \times \text{time})$$

Thus t = D/x , giving

$$C = \frac{1}{100} \left(10 + \frac{x}{2} \right) D + 12.5 \left(\frac{D}{x} \right)$$

Step 4: Interval.   The range of allowable x-values was given to us:

$$40 \le x \le 55 .$$

Step 5: Max.-min. techniques.   The function for C in terms of x found in Step 3 is

continuous on the closed, bounded interval [40 , 55]. Thus the procedure of the previous

section does apply:

Find critical points: $\frac{dC}{dx} = \frac{1}{100} \left(\frac{1}{2} \right) D - 12.5 \left(\frac{D}{x^2} \right) = \frac{D}{200 x^2} [x^2 - 2500]$

(Remember: D is just a constant)

This is defined for all $x \neq 0$, so there are no points of non-differentiability in the interval $(40, 55)$.

To find the stationary points we set $dC/dx = 0$:

$$0 = \frac{D}{200 x^2} [x^2 - 2500]$$

Thus $x = 50$ is the only stationary point and the only critical point. (Note that $x = -50$ is not in the interval $[40, 55]$; it also makes no sense to talk of a negative speed!)

Evaluate:

x	$C = \frac{1}{100} \left(10 + \frac{x}{2}\right) D + 12.5\left(\frac{D}{x}\right)$
(endpoint) 40	.6125 D
50	.6000 D
(endpoint) 55	.6023 D

Pick minimum value: $x = 50$ mph gives the minimum value of $C = .6D$. Thus 50 mph is the most economical speed at which to drive the truck no matter what the distance of the trip. □

*   *   *   *   *

Max.-min. problems can be hard -- no question about it. But if you practice at them, and carefully follow the 5-step procedure, you will become much more adept at solving them. As a first step, we suggest that you cover up the solutions to Anton's six examples and try to do them yourself using the 5-step procedure. Don't peek unless you absolutely have to!

Section 4. 8.   More Applied Maximum and Minimum Problems

The problems in this section are essentially just like those in the previous section:

> Each solution requires us to find
>
> the maximum and/or minimum
>
> value of a continuous function f on
>
> an interval I .

The difference is that in §4. 7 the interval I was always a closed, bounded interval [a , b] ; in this section any type of interval might appear, e. g., (a , b] , (- ∞, b) , [a , + ∞) , etc.

The five-step procedure of §4. 7 still applies:

Step 1:   Diagram and labels

Step 2:   Formula

Step 3:   Formula in one variable

Step 4:   Interval

Step 5:   Max. - min. techniques.

However, in Step 5 you will need to employ the techniques of §4. 6 which apply to a general interval rather than just a closed, bounded interval. To be specific, you will:

Step 5

> First, determine the critical points for the function of Step 3 on the
>   interval of Step 4 ;
>
> Second, determine which of the critical points are relative extrema by
>   using the First or Second Derivative Tests; and
>
> Finally, determine the (absolute) maximum or minimum by examining
>   the relative extrema and the behavior of the function near the
>   endpoints of the interval.   A graph might be useful.

It is very common to obtain only one relative extremum on the interval; in that case, Theorem 4. 6. 6 will tell you that the relative extremum is actually an absolute extremum!

*   *   *   *   *

As in §4. 7 the difficulty of these problems does not lie in the max. -min. techniques themselves, but in the translation from words into mathematics. If your translation is in error, then no matter how good the mathematics is which follows, you will get a wrong answer. As we stressed in §§4. 1 and 4. 7, you must approach this translation in a careful, disciplined and organized fashion.

Anton's Examples 1 and 2 both illustrate the types of problems you may encounter and the solution procedure. Here is one more.

Example A. You wish to fence in a rectangular pasture of 30,000 square feet, with one side on a neighbor's lot. If the neighbor agrees to pay for half the cost of the fence that abuts his lot, what dimensions for the fence will minimize your cost?

Step 1: Diagram and labels. A sketch of the pasture is shown to the right, where we have labeled the side lengths as x and y. We need to minimize your total costs C of the fencing materials; to do this we will certainly need to consider the cost per foot of the fencing material, which we'll label as c.

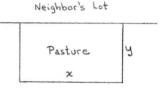

Neighbor's Lot

Pasture y

x

Area = 30,000 ft^3

x = ? and y = ?

to minimize

C = total cost to you of the fence.

Step 2: Formula. In terms of x, y and c, the total cost C

for the fence is

$$C = xc + \frac{1}{2} xc + yc + yc \qquad (\frac{1}{2} xc \text{ because you pay for}$$
$$\text{only half of the abutting fence. })$$

or

$$C = \left(\frac{3}{2} x + 2y \right) c$$

Step 3: Formula in one variable. The formula for C involves the two variables x and y

(c, the cost per foot, is a constant), so we must eliminate one of the quantities x and y

by finding a relationship between them. That's easy:

$$30,000 \text{ ft}^2 = \text{area} - xy$$

Thus $y = 30,000/x$, which gives

$$C = \left(\frac{3}{2} x + \frac{60,000}{x} \right) c$$

Step 4: Interval. Given any x > 0 we can find a y > 0 such that xy = 30,000. Thus,

in the absence of any given restrictions on the dimensions of the pasture, we must let

$$0 < x < + \infty$$

Step 5: Max. - min. techniques. Since we are minimizing a function on an open and unbounded

interval we must use the techniques of §4. 6 which apply to a general interval:

Find critical points: $$\frac{dC}{dx} = \left(\frac{3}{2} - \frac{60,000}{x^2} \right) c$$

This is defined for all $x \neq 0$, so there are no points of non-differentiability in the interval $(0, \infty)$.

Setting $\dfrac{dC}{dx}$ equal to zero we find (since $c \neq 0$)

$$\frac{60,000}{x^2} = \frac{3}{2}, \qquad \text{or} \qquad x^2 = \frac{120,000}{3} = 40,000$$

$$x = 200$$

Thus $\boxed{x = 200}$ is the only critical point in the interval.

Find relative extrema: Using the Second Derivative Test we find

$$\frac{d^2C}{dx^2} = + \frac{120,000}{x^3}$$

so that

$$\left. \frac{d^2C}{dx^2} \right|_{x = 200} = \frac{120,000}{8,000,000} > 0$$

Hence C has a relative minimum at $x = 200$.

Find absolute minimum: Since $x = 200$ gives the only relative extremum on the interval $(0, +\infty)$, then Theorem 4.6.6 guarantees that this gives an absolute minimum. Hence your cost for the fence is minimized when

$$\boxed{\begin{aligned} x &= 200 \text{ ft} \\ y &= 150 \text{ ft.} \end{aligned}}$$

y being determined from the equation $y = 30,000/x$. $\square$

Section 4.9.   Newton's Method.

1.   Newton's Method: a succession of tangent lines

Howard Ant is at the point x_1 on the x-axis,
and he wishes to get to $x = r$, a zero (root) of
the equation $y = f(x)$. Fortunately, Howard
has taken a course in calculus, and has learned
Newton's Method... which he will now demonstrate
for us:

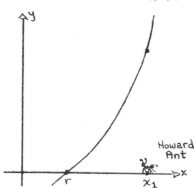

   First Howard crawls straight up*
   until he reaches the point
   $(x_1, f(x_1))$ on the graph of $y = f(x)$ ...

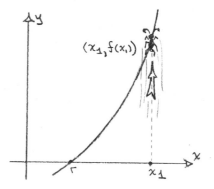

   Then Howard slides down the
   tangent line at $(x_1, f(x_1))$
   until he hits the x-axis at a
   point we'll call x_2.

   Given a little luck, Howard's new
   value x_2 should be closer to r
   than his initial value x_1.

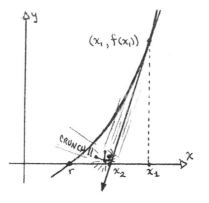

It is easy to compute x_2 in terms of x_1! As Anton shows, the equation for the tangent line
that Howard slides down is

_____

* Howard will crawl down if $f(x_1) < 0$.

$$y - f(x_1) = f'(x_1)(x - x_1)$$

Then x_2 is merely the x-value we get from this equation when we take $y = 0$. Taking $y = 0$ and solving for $x = x_2$ yields

$$\boxed{x_2 = x_1 - \frac{f(x_1)}{f'(x_1)}}$$

But suppose Howard Ant is not satisfied with x_2 as an approximation of r. Well... he can repeat the previous steps, this time starting at $x = x_2$:

Howard crawls straight up from $x = x_2$ to $(x_2, f(x_2))$ ...

... and then slides down the tangent line until he hits x_3.

As above, we compute x_3 from x_2 by

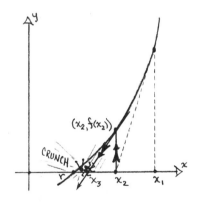

$$\boxed{x_3 = x_2 - \frac{f(x_2)}{f'(x_2)}}$$

Continuing in this way Howard can generate a string of approximations for r:

$$x_1, \, x_2, \, x_3, \, \ldots, \, x_n, \, x_{n+1}, \, \ldots$$

and the equation used to compute any approximation x_{n+1} from the previous x_n is simply

Newton's
Method

$$\boxed{x_{n+1} = x_n - \frac{f(x_n)}{f'(x_n)} \qquad n = 1, 2, 3, \ldots}$$

This is <u>Newton's Method</u> for approximating the solutions for $f(x) = 0$.

<u>Warning 1.</u> If it happens that $f'(x_n)$ equals zero for some n , then Newton's Method has

failed (because you can't divide by zero, and hence you can't determine x_{n+1} !). You should

then go back and choose a different initial value $x_1 \ldots$ and hope for better luck.

<u>Warning 2.</u> Occasionally Newton's Method will generate a string of values $x_1, x_2, x_3, \ldots ,$

that do NOT get closer to the desired $x = r$. As in Warning 1 , you should just go back and

choose a different initial value x_1 . (See Example B below.)

<u>Example A.</u> Use Newton's Method to approximate a root of $f(x) = x^2 - 10$.

  <u>Solution.</u>   We have

$$f(x) = x^2 - 10$$

$$f'(x) = 2x$$

Thus the Newton's Method equation becomes

$$x_{n+1} - x_n - \frac{f(x_n)}{f'(x_n)} = x_n - \frac{x_n^2 - 10}{2x_n}$$

which simplifies to

$$\boxed{x_{n+1} = \frac{x_n^2 + 10}{2x_n}}$$

Now we must <u>guess</u> an initial value x_1 (this is a crucial step!). Since

$f(3) = 9 - 10 = -1 < 0$ and $f(4) = 16 - 10 = 6 > 0$, then one root of f must lie between

3 and 4 (by Theorem 3.7.8, the Intermediate Value Theorem). We'll start with a guess

of $x_1 = 3 \ldots$ and now we're ready to go:

4.9.4

Calculate x_2 from x_1:

$$x_2 = \frac{x_1^2 + 10}{2x_1} = \frac{3^2 + 10}{2 \cdot 3} \cong 3.166666667$$

Calculate x_3 from x_2:

$$x_3 = \frac{x_2^2 + 10}{2x_2} \cong \frac{(3.166666667)^2 + 10}{2(3.166666667)} \cong 3.162280702$$

Calculate x_4 from x_3:

$$x_4 = \frac{x_3^2 + 10}{2x_3} \cong \frac{(3.162280702)^2 + 10}{2(3.162280702)} \cong 3.162277660$$

Calculate x_5 from x_4:

$$x_5 = \frac{x_4^2 + 10}{2x_4} \cong \frac{(3.162277660)^2 + 10}{2(3.162277660)} \cong 3.162277660$$

Ah ha! Since x_5 equals x_4, then we have reached the limits of accuracy of the calculator we are using. Thus

$$x_4 \cong 3.162277660$$

is the best approximation of r obtainable by Newton's Method using this calculator. ☐

2. Choose the initial approximation x_1 with care. Here's an example which shows how a "perfectly good" choice for the initial approximation x_1 can sometimes lead to undesirable results:

Example B.   Use Newton's Method to approximate the root of $f(x) = 2x^3 - 4x + 1$ which lies

between   0   and   1.

Solution.   Since $f(0) = 1 > 0$, while

$$f(1) = 2 - 4 + 1 = -1 < 0$$

then there is certainly a root of   f   which lies between   0   and   1   (by   Theorem 3.7.8,

the Intermediate Value Theorem).   We'll start with a guess of $x_1 = 1$...   that seems

reasonable, doesn't it?

We have $f(x) = 2x^3 - 4x + 1$.

$$f'(x) = 6x^2 - 4$$

so that the Newton's Method equation becomes

$$x_{n+1} = x_n - \frac{f(x_n)}{f'(x_n)} = x_n - \frac{2x_n^3 - 4x_n + 1}{6x_n^2 - 4}$$

which simplifies to

$$x_{n+1} = \frac{4x_n^3 - 1}{6x_n^2 - 4}$$

So starting with $x_1 = 1$ we obtain

$$x_2 = \frac{4x_1^3 - 1}{6x_1^2 - 4} = \frac{4(1)^3 - 1}{6(1)^2 - 4} = 1.5$$

$$x_3 = \frac{4x_2^3 - 1}{6x_2^2 - 4} = \frac{4(1.5)^3 - 1}{6(1.5)^2 - 4} \cong 1.315789474$$

$$x_4 = \frac{4x_3^3 - 1}{6x_3^2 - 4} \cong \frac{4(1.315789474)^3 - 1}{6(1.315789474)^2 - 4} \cong 1.315789474$$

Since x_4 equals x_3, we're not going to get any different answers by continuing this process... but we did not get a root which lies between 0 and 1! Why?

A sketch of the graph $y = f(x)$ indicates the problem:

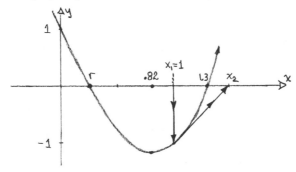

As we can see, the initial guess of $x_1 = 1$ and the desired root r are separated by a relative minimum value for $y = f(x)$... and this causes our tangent line to move in the direction of the root 1.315789474, NOT in the direction of r.

So what value should we choose for x_1? Well, practically any value that's substantially less than .8 should work; let's try $x_1 = .5$. With this initial value, the successive approximations we generate are

$$x_2 = .2$$
$$x_3 = .257446808$$
$$x_4 = .258651398$$
$$x_5 = .258652022$$
$$x_6 = .258652022$$

Since $x_6 = x_5$, then we have reached the limits of accuracy of the calculator we are using. Thus the approximation for our desired root is

$$r \approx .258652022$$ □

The Moral: Choose the initial approximation x_1 carefully.

Frequently a graph of the function under

consideration will prove helpful in making this choice.

3.   Solving equations by Newton's Method.   As Anton docs in Example 2 , Newton's Method can

be used to solve equations in   x   by a simple (but often overlooked) trick:

The solutions of the equation

$g(x) = h(x)$ are merely ...

... the zeroes of the equation

$f(x) = 0$,   where $f(x) = g(x) - h(x)$.

And Newton's Method can solve this problem!

Example C.   Approximate the coordinates of the point of intersection of the graphs of $y = 1/x$

and $y = x^2 - 4x$.

Solution.   The   x-coordinate of the point of intersection is found by setting $x^2 - 4x$ equal

to $1/x$:

$$x^2 - 4x = 1/x$$

$$x^2 - 4x - 1/x = 0$$

$$(x^3 - 4x^2 - 1)/x = 0$$

So the   x-coordinate   we are seeking will be a root of the equation

$$f(x) = x^3 - 4x^2 - 1 = 0$$

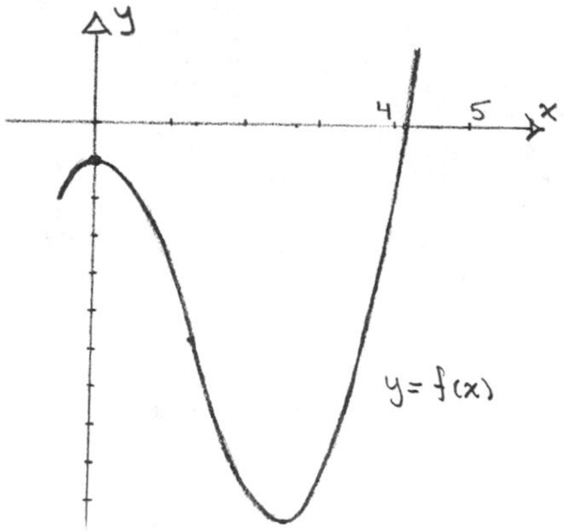

Using the curve sketching procedures of §4.4 will produce the graph to the right. As can be seen there is only one root for this equation, and it lies between $x = 4$ and $x = 5$. The value $x_1 = 4$ appears to be a good initial approximation.

We have

$$f(x) = x^3 - 4x^2 - 1$$

$$f'(x) = 3x^2 - 8x$$

so that the Newton's Method equation becomes

$$x_{n+1} = x_n - \frac{f(x_n)}{f'(x_n)} = x_n - \frac{x_n^3 - 4x_n^2 - 1}{3x_n^2 - 8x_n}$$

which simplifies to

$$x_{n+1} = \frac{2x_n^3 - 4x_n^2 + 1}{3x_n^2 - 8x_n}$$

Starting with $x_1 = 4$, this will generate the following approximations:

$$x_2 = 4.0625$$

$$x_3 = 4.060648680$$

$$x_4 = 4.060647029$$

$$x_5 = 4.060647029$$

Since x_5 equals x_4 , then our approximation for the x-coordinate of the point of inter-

section is $x \approx 4.060647029$. The y-coordinate is therefore

$$y = 1/x \approx .246266172$$ □

Section 4.10   Rolle's Theorem; Mean Value Theorem

At first reading Rolle's Theorem and the Mean Value Theorem (MVT) may appear to

be very technical and of little importance to calculus. Well, such is not the case! Rolle's

Theorem is a special case of the MVT which is needed to prove the MVT , and as Anton

states, the MVT has "... so many major consequences ... that it is regarded as one of the

most fundamental results in calculus. "

Later in this section and in §4.11 we will give some of the consequences of the MVT

and discuss in general why the theorem is so useful. We begin, however, by carefully

discussing each theorem and showing how to apply them in specific situations.

1.   Rolle's Theorem.   The statement of the theorem is best made with three clearly labeled

hypotheses.

Rolle's
Theorem

> Suppose   (i) f   is differentiable on the <u>open</u> interval (a , b) ,
>
> (ii) f   is continuous on the <u>closed</u> interval $[a , b]$,
>
> (iii) $f(a) = f(b) = 0$.
>
> Then there exists at least one point   c   in (a , b) such that
>
> $$f'(c) = 0$$

Geometrically we are saying that if a

"well-behaved" curve intersects the

x-axis at the points   a   and   b , then

somewhere between   a   and   b   <u>there is at</u>

<u>least one point on the curve that has a horizontal tangent line.</u>

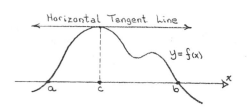

We emphasize that all three hypotheses must be valid in order to know that the number

c   exists.   To convince you of this, here are functions demonstrating what can go wrong when

any one of the three hypotheses fails:

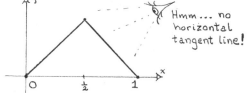

(1) $f(x) = \begin{cases} x & \text{if } 0 \le x \le \frac{1}{2} \\ 1 - x & \text{if } \frac{1}{2} \le x \le 1 \end{cases}$

OOPS! ⟶ (i) f is <u>NOT</u> differentiable on (0, 1)   since $f'(\frac{1}{2})$ is undefined!

(ii) f is continuous on   [0, 1]

(iii) $f(0) = f(1) = 0$.

Since (i) fails, it is not surprising that Rolle's Theorem fails, i. e. , there is

no horizontal tangent line between   0   and   1 .

(2) $f(x) = \begin{cases} x & \text{if } 0 \le x < 1 \\ 0 & \text{if } x = 1 \end{cases}$

(i) f is differentiable on (0, 1)

OOPS! ⟶ (ii) f is <u>NOT</u> continuous on   [0, 1]   since it is not continuous from

the left at   x = 1 !

(iii) $f(0) - f(1) = 0$.

Since (ii) fails, we again have no horizontal tangent line, and hence Rolle's

Theorem fails again.

(3) $f(x) = x$ if $0 \leq x \leq 1$

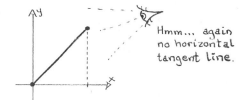

Hmm... again no horizontal tangent line.

   (i)   f is differentiable on $(0, 1)$

   (ii) f is continuous on $[0, 1]$

OOPS! ⟩⟩(iii) $f(0) = 0$ but $f(1)$ does <u>NOT</u> equal 0.

   Since (iii) fails, Rolle's Theorem fails once more.

Thus, when applying Rolle's Theorem to a particular function, you must verify all three

hypotheses!!

<u>Example A.</u>   Determine if all the hypotheses of Rolle's Theorem are satisfied for $f(x) = 4x^{2/3} - x^2$

on the interval $[-1, 1]$. If so, find all values of c that satisfy the conclusion of the theorem.

<u>Solution.</u>   We attempt to verify the three hypotheses of Rolle's Theorem:

   (i) $f'(x) = 4\left(\frac{2}{3}\right) x^{-1/3} - 2x$ ... Oops! That's the end of the game! This derivative

      is not defined at $x - 0$ (even though f is defined at $x = 0$). Since $x = 0$

      falls in the open interval $(-1, 1)$, then f is <u>NOT</u> differentiable on $(-1, 1)$.   □

When the hypotheses of Rolle's Theorem fail, we simply cannot say whether there exist values of

c   for which $f'(c) = 0$; they might or might not exist.

For the function f in Example A , it turns out that

such values do not exist, as can be seen from the graph

of f given to the right.

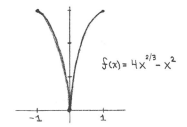

$f(x) = 4x^{2/3} - x^2$

<u>Example B.</u>   Determine if all the hypotheses of Rolle's Theorem are satisfied for $f(x) = (x - x^2)^{1/2}$

on the interval $[0, 1]$. If so, find all values of c that satisfy the conclusion of the theorem.

4.10.4

Solution.　(i)　$f'(x) = (\frac{1}{2})(x - x^2)^{-1/2}(1 - 2x)$,　so that

$$f'(x) = \frac{1 - 2x}{2\sqrt{x - x^2}}\quad.$$

Although　$f'(x)$　is not defined at　$x = 0$　or　$x = 1$,　that is not important.　We only require

f　to be differentiable on the open interval　$(0, 1)$,　and　f　is differentiable on that open

interval.

　　　(ii)　Since　f　is differentiable on the open interval　$(0, 1)$,　it must also be continuous

on　$(0, 1)$.　(This is　Theorem 3.7.5.)

　　　Hence we have only to verify that　f　is continuous from the right at　$x = 0$　and

continuous from the left at　$x = 1$:

$$\lim_{x \to 0^+} f(x) = \lim_{x \to 0^+} \sqrt{x - x^2} = \sqrt{\lim_{x \to 0^+}(x - x^2)} = 0 = f(0)$$

> valid since
> $x - x^2 > 0$ whenever
> $0 < x < 1$

$$\lim_{x \to 1^-} f(x) = \lim_{x \to 1^-} \sqrt{x - x^2} = \sqrt{\lim_{x \to 1^-}(x - x^2)} = 0 = f(1)\quad.$$

Hence　f　is continuous on the closed interval　$[0, 1]$.

　　　(iii)　$f(0) = f(1) = 0$,　as we saw above.

　　　Since all the hypotheses of Rolle's Theorem are valid, the conclusion must also be valid,

i.e., there is a　c　in　$(0, 1)$　with　$f'(c) = 0$.　We thus have only to solve　$f'(c) = 0$　for　c:

$$\frac{1 - 2c}{2\sqrt{c - c^2}} = 0$$

$1 - 2c = 0$ since $\sqrt{c - c^2} \neq 0$ when $0 < c < 1$

$\boxed{c = 1/2}$ □

A useful technique: Notice how verifying the differentiability of f   on   (a , b)   also verifies

continuity on   (a, b) by Theorem 3. 7. 5. Hence to prove continuity on [a , b] you

only have to show that   f   is continuous from the right (respectively, left) at   x = a

(respectively,   x = b).

2.   The Mean Value Theorem (MVT).   The statement of the   MVT   has one less hypothesis than

Rolle's Theorem and a slightly altered conclusion:

<table>
<tr><td>Mean
Value
Theorem</td><td>Suppose (i) f is differentiable on the open interval (a , b) ,

(ii) f is continuous on the closed interval [a , b].

Then there exists at least one point   c   in   (a , b)   such that

$$f'(c) = \frac{f(b) - f(a)}{b - a}$$</td></tr>
</table>

Geometrically we are saying that if a

"well-behaved" curve   y = f(x)   contains

the points   (a , f(a))   and   (b , f(b)) , then

somewhere between   a   and   b   there is at

least one point   c   for which the tangent line

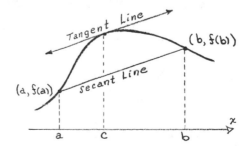

at (c , f(c))   is parallel to the secant line joining   (a , f(a))   to   (b , f(b)). (Remember: the slope of the secant line is $\frac{f(b) - f(a)}{b - a}$).

Notice that if $f(a)$ and $f(b)$ both equal zero, then the MVT becomes Rolle's Theorem. Thus Rolle's Theorem is the special case of the MVT when $f(a) = f(b) = 0$. As with Rolle's Theorem, you must verify all the hypotheses before applying the MVT to any particular function.

Example C. Determine if all the hypotheses of the MVT are satisfied for $f(x) = \frac{\sqrt{x}}{\sin x}$ on the interval $[0, 1]$. If so, find all values of c that satisfy the conclusion of the theorem.

Solution. (i) $f'(x) = \dfrac{(\sin x)(\frac{1}{2} x^{-1/2}) - \sqrt{x} \cos x}{\sin^2 x}$, which is defined for all $0 < x < 1$.

Hence f is differentiable on $(0, 1)$.

(ii) Wait a minute .... f is not even defined at $x = 0$, and hence it certainly is not continuous from the right at $x = 0$. Thus f is not continuous on the closed interval $[0, 1]$, showing that the hypotheses of the MVT do not hold for this function. $\square$

Anton's Example 2 illustrates how to proceed when the MVT is valid for a function f.

3. More unusual uses of Rolle's Theorem and the MVT. Before talking about the important theoretical value of our two theorems, we'll illustrate some practical uses that might surprise you.

Example D. Use Rolle's Theorem to prove that

$$f(x) = x^5 + 4x^3 + 2x + 3 = 0$$

does not have more than one real root.

Solution. At first sight this hardly looks like a problem that Rolle's Theorem will help with!

However, using Rolle's Theorem the problem becomes very easy. For suppose f has two

roots x = a and x = b , i.e., f(a) = f(b) = 0 . Since f is a polynomial, it is differ-

entiable on (a, b) and continuous on [a, b] . Hence f satisfies the three hypotheses of

Rolle's Theorem, and the conclusion must also follow:

there exists at least one value   c   between

a   and   b   such that   f'(c) = 0 .

However, $f'(x) = 5x^4 + 12x^2 + 2$, which is <u>always positive</u>! Hence, for any   c ,   the

derivative   f'(c)   must always be positive, proving that the   c   of Rolle's Theorem cannot exist

and that our assumption of two real roots   a   and   b   must be false! Thus   f   has at most one

real root.   □

<u>Example E.</u>   Use the   MVT   to show that

$$| \cos x - \cos y | < | x - y |$$

holds for all real values of   x   and   y .

<u>Solution.</u>   Once again we'll have a surprising application of the   MVT ! Writing the desired

inequality in the form

$$\left| \frac{\cos x - \cos y}{x - y} \right| \le 1 \qquad \text{when} \qquad x \neq y$$

might make you suspect how to use the   MVT: since   f(x) = cos x   is continuous and differ-

entiable everywhere, it satisfies the hypotheses of the   MVT   on any interval   [x, y] . Thus

there exists a value   c   between   x   and   y   such that

$$\frac{\cos x - \cos y}{x - y} = \frac{f(x) - f(y)}{x - y} = f'(c) = - \sin c$$

Thus $\left| \dfrac{\cos x - \cos y}{x - y} \right| = \left| \sin c \right| \leq 1 ,$ as desired. $\square$

4.   <u>Consequences of the Mean Value Theorem.</u>   The essential reason for the importance of the MVT   is this:

> Values of a function may be expressed in terms of
>
> values of the derivative; then questions about the
>
> function can be transformed into questions about
>
> the derivative which may be easier to answer!

You might protest by saying we've been doing this procedure all throughout Chapter 4 without having the   MVT :

(1)   the question "Where is a function increasing?" was transformed into "Where is the <u>derivative</u> positive?";

(2)   the question "When does a function have a maximum value at a critical point?" was transformed into questions about the signs of <u>derivatives</u> in the First and Second Derivative Tests.

The   MVT   wasn't needed in these cases, was it?

Yes it was!! In the next (optional) section it is shown that the rigorous justification of these results rests on the   MVT   (we just haven't given rigorous justifications before now).

Two other very innocent looking consequences of the   MVT   are given at the end of this section:

Theorem 4. 10. 3

> If $f'(x) = 0$ for all x in an interval,
>
> then f is constant on the interval.

| Corollary 4. 10. 4 |

> If $f'(x) = g'(x)$ for all x in an interval,
>
> then f and g differ by at most a
>
> constant on the interval.

Do not be deceived by the simplicity of these results: they are of critical importance in our forthcoming study of "integration," the other major operation of calculus.

To make this clear, let's look ahead a bit. "Indefinite integration" is the reverse process of differentiation:

> to integrate a function h means to find
>
> all the functions f whose derivatives f'
>
> equal h, i.e., given h, find all f
>
> such that $f' = h$.

We can now write Theorem 4. 10. 3 in terms of "integration":

| Theorem 4. 10. 3; Integration version |

> Integrating the zero function $f = 0$ on an interval
>
> produces only the constant functions $f = c$.

We now know how "to integrate" the most basic function in existence: the zero function. The important part of the theorem is that there are no functions beside the constants produced when integrating $h = 0$ (we already knew that constants differentiate to zero; now we know there are no other functions which differentiate to zero, at least on an interval).

The Corollary can also be written in terms of "integration":

Corollary 4. 10. 4; Integration version	Two functions f and g produced by integrating a given function h on an interval can differ by at most a constant!

Ahh, ... when integrating a function h on an interval, you really have to find only one solution f ; then any other solution g is merely f with a constant tacked on: $g(x) = f(x) + c$. This result is critical in integration theory; in fact, it is restated in the integration form as Theorem 5.2.2 in Chapter 5.

The moral of this story? Although you may not often use the MVT directly, a huge number of our standard calculus techniques are based on the MVT and its consequences.

Section 4.11. Proofs of Key Results Using the Mean Value Theorem. (Optional).

In this section Anton demonstrates the fundamental importance of the MVT by showing how it is used in the proofs of certain key theorems of Chapter 4 .

1. Increasing functions. Suppose we have a function f whose derivative f' is positive on an open interval (a , b). Taking any two points $x_1 < x_2$ in (a , b) we see that

$$\frac{f(x_2) - f(x_1)}{x_2 - x_1} = f'(c) > 0$$

for some c
by MVT

Since $x_2 - x_1$ is positive, this shows

$$f(x_2) - f(x_1) > 0$$

which proves

$$f(x_2) > f(x_1) \qquad \text{whenever} \qquad a < x_1 < x_2 < b \quad .$$

Thus f is increasing on (a, b) whenever the derivative f' is positive on (a, b), proving Theorem 4. 2. 2(a). The rest of the theorem can be verified in a similar fashion.

Notice how the MVT takes information about the derivative ("f' is positive") and transforms it into information about the function itself ("f is increasing"). Thus a hard question about the function ("Is f increasing?") can be handled by answering an easier question about the derivative ("Is f' positive?")

2.   The First and Second Derivative Tests.   Suppose we have a function f whose derivative f' is positive on (a, x_0), and negative on (x_0, b). Then taking $a < x < x_0$ we obtain

$$\frac{f(x_0) - f(x)}{x_0 - x} = f'(c) > 0 \qquad \text{since} \qquad a < x < c < x_0$$

$$\boxed{\text{for some } x < c < x_0 \atop \text{by the MVT}}$$

Hence

$$f(x_0) - f(x) > 0$$

proving

$$\boxed{f(x_0) > f(x) \qquad \text{for} \qquad a < x < x_0}$$

Reasoning in a similar fashion for $x_0 < x < b$ will show

$$\boxed{f(x_0) > f(x) \qquad \text{for} \qquad x_0 < x < b}$$

Taken together, the two results prove that f has a relative maximum at $x = x_0$ whenever f' is positive to the left of x_0 and negative to the right of x_0. This verifies part (a) of the First Derivative Test (Theorem 4.3.6); the rest of the theorem can be proved in a similar fashion.

Notice again how the MVT takes information about the derivative (f' is positive on (a, x_0) and negative on (x_0, b)) and transforms it into information about the function itself (f has a relative maximum at x_0).

The proof of the Second Derivative Test does not explicitly use the MVT. However, it relies on the First Derivative Test which does use the MVT. Thus the Second Derivative Test ultimately needs the MVT for its derivation.

3. Hints on some of the exercises.   Exercise Set 4.11 asks you to prove a number of general results, some of which are pretty tricky. As an aid we'll give you a few hints and suggestions. You should first try the problems without our help, and look at the hints only if you get into trouble. To save space we will not reproduce the exercise statements.

Exercise 6: This one's really cute! Try applying Theorem 4.2.2 to the function $h = g - f$.

Exercise 7 (a): You know that f is increasing on (a, x_0) and (x_0, b) (why?), and so you will be done if you can show

(i) $f(x) < f(x_0)$ for all $a < x < x_0$, and

(ii) $f(x_0) < f(x)$ for all $x_0 < x < b$.

To prove (i) and (ii), notice that the MVT applies to the interval $[x, x_0]$ in (i) and to the interval $[x_0, x]$ in (ii) ....

Exercise 8(a):   The following general result on continuous functions should prove useful:

if $h(x_0) > 0$ and h is continuous at x_0, then there is an open interval containing

x_0 on which h is positive. Although very

believable, it is a tricky result to prove.

Here's the proof: We have

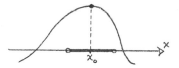

$$\lim_{x \to x_0} h(x) = h(x_0) \quad \text{since h is continuous at } x_0.$$

Thus (see §2.6) given $\varepsilon = h(x_0) > 0$, there exists $\delta > 0$ such that

$$\left| h(x) - h(x_0) \right| < \varepsilon = h(x_0) \quad \text{whenever } \left| x - x_0 \right| < \delta.$$

Thus $-h(x_0) < h(x) - h(x_0) < h(x_0)$ whenever $-\delta < x - x_0 < \delta$

or $0 < h(x)$ whenever $x_0 - \delta < x < x_0 + \delta$.

Ahh...!! That's our result: h is positive on the open interval $(x_0 - \delta, x_0 + \delta)$.

Exercise 9(a):   Remember that $f'' - (f')'$, i.e., f'' is the derivative of the function f'.

Exercise 10:   Since $f'(a) f'(b) < 0$, then $f'(a)$ and $f'(b)$ have opposite sign. Thus

assume $f'(a) > 0$ and $f'(b) < 0$ (the other case can be treated in a similar way).

Now f must have a maximum value on $[a, b]$ (why?). Assume this maximum value

occurs at $x = c$ in $[a, b]$. If c is not equal to either a or b, then we are

done (why?). Exercise 8 can now be used to show that c cannot equal a or b!

4. A preview of Taylor's Theorem. Using x in place of b and rearranging the final

equation, the MVT can be rewritten as follows:

Suppose f is differentiable on (a , x) and continuous on [a , x].

Then there is at least one point c in (a , x) such that

$$f(x) \; = \; f(a) \; + \; f'(c)\,(x - a)$$

Written in this form we see that $f(x)$ can be approximated by $f(a)$ with an error of

$f'(c)\,(x - a)$. Suppose, however, that f is known to have a first and second derivative. Can

we use f" to get an even better approximation to $f(x)$? The following result gives the answer:

Suppose f' is differentiable on (a , x) and continuous on [a , x].

Then there is at least one point c in (a , x) such that

$$f(x) \; = \; f(a) \; + \; f'(a)\,(x - a) \; + \; \frac{f''(c)}{2}\,(x - a)^2$$

Hence $f(x)$ can be approximated by $f(a) + f'(a)\,(x - a)$ with an error of $\frac{1}{2}\,f''(c)\,(x - a)^2$.

In general, if f has $n + 1$ derivatives, then we have the following:

Suppose $f^{(n)}$ is differentiable on (a , x) and continuous on [a , x].

Then there is at least one point c in (a , x) such that

$$f(x) = f(a) + f'(a)\,(x - a) + \cdots + \underbrace{\frac{f^{(n)}(a)}{n!}\,(x - a)^n} + \underbrace{\frac{f^{(n+1)}(c)}{(n + 1)!}\,(x - a)^{n+1}}.$$

This is called the   "n-th Taylor
Polynomial for f   about   a"   " Lagrange's form
of the remainder"

This result is known as Taylor's Theorem , and appears as Theorem 11. 10. 1 in

Chapter 11. Its value lies in approximating arbitrary functions by polynomials (the Taylor

polynomials) and in giving a useful formula for the approximation error (Lagrange's form of the

remainder). The importance of this generalization of the MVT for computations and applications

cannot be over emphasized! It will be extensively studied in the latter part of Chapter 11.

Chapter 5:   Integration

Section 5.1.   Introduction

1.   The importance of the area problem

At first, the "second major problem of calculus" looks like a serious misnomer.   After

all, what's so important about being able to find the area of a region like this

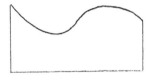

where three of the sides are straight lines? And why does such a specialized problem deserve

the grandiose title "The Area Problem?"

One answer is that being able to calculate areas of regions with three straight sides

is the crucial step in being able to calculate the areas of nearly all regions.

To see this, suppose that we want to calculate the area of a general region (an ink-blot)

such as

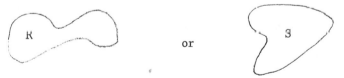

or

Roughly speaking it can be done like this: For convenience, locate the region in the first quadrant

of a coordinate system (certainly, the area of a region is the same wherever it is located).

Then

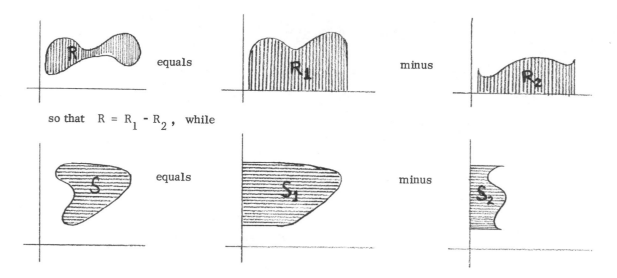

so that $R = R_1 - R_2$, while

so that $S = S_1 - S_2$. (Note that S_1 and S_2 have the y-axis, and not the x-axis, as a boundary, but they are still regions with three straight sides.) The point is that <u>if we know how to calculate areas of regions with three straight sides</u> (such as R_1, R_2, S_1 and S_2), <u>we can calculate areas of almost all reasonable regions</u> (such as R and S).

But why is so much fuss made about calculating areas? After all, how often do you need to find areas? Probably not very often. The truth is that if the solution to the area problem applied only to calculating areas, it would not be very interesting. However, it turns out that <u>many "real-world" applications of calculus which do not appear to be area problems can be reduced to calculating areas of regions with three straight sides.</u> We won't go into detail at this point, but look ahead to Sections 5.4 , 5.8 and 6.2 - 6.7 if you want to see some examples. This is why the Area Problem is of central concern in calculus.

2. <u>Special conditions in the area problem.</u>   In the statement of The Area Problem, two words should not be overlooked.   One is " CONTINUOUS" and the other is "NONNEGATIVE. "

Neither should be surprising. After all, if f were not continuous it might look like the

following function over [a, b]:

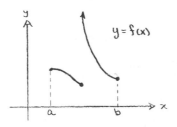

Then we would have a hard time even <u>defining</u> the area in question because it "spills out" from

under the curve. Since 99% of the "real world" applications involve regions with continuous

boundaries, assuming continuity in The Area Problem makes good sense. Likewise, the presence

of the word "nonnegative" in the statement of The Area Problem makes life easier without

reducing the generality of our analysis. For, as we remarked earlier, a region has the same

area regardless of where it is located so, for convenience, we may locate all regions in the

first quadrant, thus making the upper-boundary curve f nonnegative.

3. <u>Areas from derivatives.</u> Suppose A(x) is the area

of a region under the continuous curve y = f(x) and

over the interval [a, x]. As Anton points out,

the key to finding A(x) (i. e., to solving the

Area Problem) is the following fact:

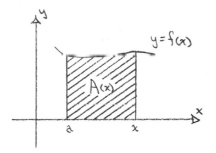

> The derivative of A(x) is f(x), i. e.,
>
> $$A'(x) = f(x)$$

5. 1. 4

This result, known as the <u>First Fundamental Theorem of Calculus,</u> will be proved in §5. 8. (The proof requires a rigorous definition for the concept of <u>area</u> which will be developed in §5. 6 and § 5. 7.)

The First Fundamental Theorem of Calculus transforms The Area Problem into a problem of finding functions $A(x)$ whose derivative is a given function $f(x)$, i.e.,

$$\boxed{\begin{array}{c} \text{Given} \quad f(x), \quad \text{find} \quad A(x) \quad \text{such that} \\ A'(x) \ = \ f(x) \end{array}}$$

The process of finding $A(x)$ such that $A'(x) = f(x)$ is known as <u>antidifferentiation</u> or <u>(indefinite) integration</u> ... and it will be the focus of our attention in §§5. 2 - 5.4.

<u>Example A.</u> Find an antiderivative $A(x)$ for the function

$$f(x) \ = \ 4x^3 + \cos x$$

<u>Solution.</u> The term $4x^3$ should remind you of the power rule $\frac{d}{dx}(x^n) = nx^{n-1}$. Hmmm. If $n = 4$, then we get $\frac{d}{dx}(x^4) = 4x^3$, so part of our antiderivative $A(x)$ should be x^4.

What about the $\cos x$ term? Since the derivative of $\sin x$ is $\cos x$, then $\sin x$ should also be part of $A(x)$. Thus a good guess for $A(x)$ will be $x^4 + \sin x$. Let's check it:

$$A'(x) \ = \ \frac{d}{dx}[x^4 + \sin x] = 4x^3 + \cos x = f(x) \qquad \square$$

Section 5. 2.   Antiderivatives; The Indefinite Integral

1.   Antiderivatives and Indefinite Integrals: two names for the same thing

The method for calculating areas described in §5. 1. 3 depends on the ability to obtain

a function $F(x)$ whose derivative is a given function $f(x)$. Any function $F(x)$ such that

$$F'(x) = f(x)$$

is called an antiderivative of $f(x)$. [The prefix "anti" is used in the same way in "antithesis"

(the direct opposite) and "antitoxin" (counteracting poison).] The antiderivative is the opposite

of the derivative since finding it counteracts (undoes, reverses) the process of differentiation.

An antiderivative of $f(x)$ is also called an (indefinite) integral of $f(x)$. You'will just have

to get used to using the terms "antiderivative" and "indefinite integral" interchangeably. Sorry!

THE MAJOR GOAL OF THIS SECTION (AND OF MANY SECTIONS TO FOLLOW) IS TO

TAKE SOME OF THE GUESSWORK OUT OF FINDING ANTIDERIVATIVES.

2.   The Key Result: one antiderivative gives them all.   At first glance, Theorem 5. 2. 2 seems

almost too good to be true. It has two parts:

(i)   If $F(x)$ is an antiderivative of $f(x)$, so is $F(x) + C$ where C is a constant.

(That is, adding a constant to an antiderivative gives another antiderivative.)

(ii) Any antiderivative of $f(x)$ on an interval has the form $F(x) + C$ where $F(x)$

is one fixed antiderivative. (That is, on an interval one antiderivative gives them all

(by adding constants.)) *

---

*
    This is an important consequence of the Mean Value Theorem (see §4. 10. 4 of The Companion).

For example, we know that x^4 is an antiderivative of $4x^3$ (because $\frac{d}{dx}(x^4) = 4x^3$).
Part (i) of Theorem 5.2.2 says that $x^4 + 7$, $x^4 - 113 = x^4 + (-113)$ and $x^4 + \pi$ are
also antiderivatives of $4x^3$. Part (ii) of the theorem is the remarkable part. It says that
<u>any</u> antiderivative of $4x^3$ has the form $x^4 + C$ where C is a constant.

We can put part (ii) of Theorem 5.2.2 more colorfully like this:

> any antiderivative consists of a "main part" plus a constant.

For example, any antiderivative of $4x^3$ has the form $x^4 + C$; here x^4 is the "main
part" and C is the constant. The same "main part" will work for all antiderivatives of a
given function *; once you have found it, you have essentially found <u>all</u> antiderivatives of the
function. Here are some other examples:

$f(x)$	the general antiderivative of $f(x)$	"main part"
$2x^2 + x$	$\frac{2}{3}x^3 + \frac{x^2}{2} + C$	$\frac{2}{3}x^3 + \frac{x^2}{2}$
$2\sin x + 3$	$-2\cos x + 3x + C$	$-2\cos x + 3x$
$\sqrt{x} - \pi$	$\frac{2}{3}x^{2/3} - \pi x + C$	$\frac{2}{3}x^{2/3} - \pi x$

Note that the "main part" is the general antiderivative with $C = 0$. Also note that in
every case it can be verified that the antiderivative is correct (by differentiating the antiderivative and checking that the result is $f(x)$).

---

* At times there can be a number of choices for the "main part" of an antiderivative. However,
any two "main parts" of an antiderivative must differ by a constant. For example, either
$\frac{1}{2}\sin^2 x$ or $-\frac{1}{2}\cos^2 x$ can be the "main part" of the antiderivative of $f(x) = \sin x \cos x$.
(Differentiate each of them to check the accuracy of this statement.) However,

$$\frac{1}{2}\sin^2 x - (-\frac{1}{2}\cos^2 x) = \frac{1}{2}(\sin^2 x + \cos^2 x) = \frac{1}{2}$$

so that the two do differ by a constant.

3. How to find a "main part" $F(x)$; it's not always easy. Theorem 5. 2. 2 says that to find the

(general) antiderivaitve of a function f, all you have to do is find one antiderivative, what

we have called the "main part" of the antiderivative. However, finding the "main part" is not

always easy. It involves a combination of

+-----------------------------+
| (i)   Memorization |
| (ii)   Trickery |
| (iii)   Guesswork |
+-----------------------------+

i)   Memorization. You must memorize certain basic formulas, such as those in Table

5. 2. 1, which are derived from differentiation formulas. Learning those formulas

should not be much of a chore since the integration formulas in the right-hand column

are simply the differentiation formulas in the left-hand column read backwards. And

you know (or you should!) those differentiation formulas.

Example A.   Evaluate $\displaystyle\int \sec^2 x \, dx$.

Solution.   You should remember that $\sec^2 x$ is the derivative of $\tan x$. Hence, turning

the process around, $\tan x$ is an antiderivative of $\sec^2 x$. Using $\tan x$ as our "main

part" for the antiderivative of $\sec^2 x$ yields

$$\int \sec^2 x \, dx \; = \; \tan x \; + \; C$$

This is simply Formula 6 in Table 5. 2. 1.   □

ii) <u>Trickery.</u> You must learn <u>tricks</u> (techniques) to break down or transform unfamiliar

integrals into ones you know. This section discusses two very useful ones (Theorem

5. 2. 3):

(a) $\displaystyle\int c\,f(x)\,dx = c\int f(x)\,dx$ (multiplicative constants "move through" integral signs)

(b) $\displaystyle\int [f(x) \pm g(x)]\,dx = \int f(x)\,dx \pm \int g(x)\,dx$ (the integral sign can be "distributed" over a sum or difference of functions)

<u>Example B.</u> Evaluate $\displaystyle\int (2x^3 + 3x^2 - 6)\,dx$.

<u>Solution.</u> $\displaystyle\int (2x^3 + 3x^2 - 6)\,dx = \int 2x^3\,dx + \int 3x^2\,dx - \int 6\,dx$

"distribute" the $\displaystyle\int (\ \ )\,dx$ by (b)

$\displaystyle = 2\int x^3\,dx + 3\int x^2\,dx - 6\int dx$

"move the constants through" by (a)

Now the integrals have been reduced to known integrals (Formulas 1 and 2 in Table

5. 2. 1). The computation is completed as follows:

$$2\int x^3\,dx + 3\int x^2\,dx - 6\int dx = 2\left(\frac{x^4}{4} + C_1\right) + 3\left(\frac{x^3}{3} + C_2\right) - 6\left(x + C_3\right)$$

$$= \frac{x^4}{2} + x^3 - 6x + 2C_1 + 3C_2 - 6C_3$$

$$= \boxed{\frac{x^4}{2} + x^3 - 6x + C}$$

where $C = 2C_1 + 3C_2 - 6C_3$ □

Here are some other simple "tricks" which can be used to transform difficult integrals into friendly ones:

$$\boxed{\text{"Square out" quantities in the integral}}$$

Example C.   Evaluate $\displaystyle\int (3 - t^3)^2\,dt$.

Solution.   First we "square out" the expression $(3 - t^3)^2$:

$$\int (3 - t^3)^2\,dt = \int (9 - 6t^3 + t^6)\,dt$$

Now we can use Theorem 5.2.3 :

$$= \int 9\,dt - \int 6t^3\,dt + \int t^6\,dt \qquad \text{"distribute"}$$

$$= 9\int dt - 6\int t^3\,dt + \int t^6\,dt \qquad \text{"move the constants through"}$$

$$= (9t + C_1) - 6\left(\frac{t^4}{4} + C_2\right) + \left(\frac{t^7}{7} + C_3\right) \qquad \text{Table 5.2.1}$$

$$= \boxed{9t - \frac{3}{2}t^4 + \frac{1}{7}t^7 + C}$$

where $C = 9C_1 - 6C_2 + C_3$ □

5.2.6

Example D.   Evaluate $\displaystyle\int \frac{3y^5 + 2y^3 + 1}{y^3}\, dy$.

Solution.   First we divide the denominator y^3 into the numerator:

$$\int \frac{3y^5 + 2y^3 + 1}{y^3}\, dy = \int (3y^2 + 2 + y^{-3})\, dy$$

Now we can use Theorem 5.2.3:

$$= \int 3y^2\, dy + \int 2\, dy + \int y^{-3}\, dy \qquad \text{"distribute"}$$

$$= 3\int y^2\, dy + 2\int dy + \int y^{-3}\, dy \qquad \text{"move the constants through"}$$

$$= 3\left(\frac{y^3}{3} + C_1\right) + 2(y + C_2) + \left(\frac{y^{-2}}{-2} + C_3\right) \qquad \text{Table 5.2.1}$$

$$= \boxed{\; y^3 + 2y - \frac{1}{2y^2} + C \;}$$

where $C = 3C_1 + 2C_2 + C_3$ □

Example E.   Evaluate $\displaystyle\int \frac{\sin 2x}{\sin x}\, dx$.

Solution.   First we use the trigonometric identity $\sin 2x = 2\sin x \cos x$ (Equation (17a)) in Anton's Appendix 1):

$$\int \frac{\sin 2x}{\sin x} \, dx = \int \frac{2 \sin x \cos x}{\sin x} \, dx$$

$$= \int 2 \cos x \, dx$$

Now we can use Theorem 5. 2. 3 :

$$= 2 \int \cos x \, dx \qquad \text{"move the constant through"}$$

$$= 2 \left[\sin x + C_1 \right] \qquad \text{Table 5. 2. 1}$$

$$= \boxed{2 \sin x + C}$$

$$\text{where} \quad C = 2C_1 \qquad\qquad \square$$

We have just given a few of the "tricks" that can be applied to integration problems. Others will be discussed (at length) in the rest of this chapter and in succeeding chapters. In fact, Chapter 9 is devoted entirely to the study of techniques of integration.

iii)   Guesswork.   In order to use these techniques, you will frequently have to guess (guess substitutions, tricks, etc.).   And you may guess wrong.   Even people with years of experience make wrong guesses and false starts.   You should not become discouraged when the same thing happens to you (and it will!).   Simply remember:

| Antiderivatives can always be checked by differentiation. |

That is, any guess you make can be verified, or proven to be wrong, instantly.   And sometimes this checking procedure indicates how to correct a wrong answer, as the next example shows.

<u>Example F.</u>   Evaluate $\displaystyle\int \csc 2x \cot 2x \, dx$.

<u>Solution.</u>   Suppose your calculations produce $F(x) = 4 \csc 2x$ as an antiderivative of

$f(x) = \csc 2x \cot 2x$.   Checking by differentiation, you find $F'(x) = -8 \csc 2x \cot 2x$,

indicating that your answer is wrong.   However, the fact that you are off by a factor of -8

points to a revised answer $G(x) = -\dfrac{1}{8} F(x) = -\dfrac{1}{2} \csc 2x$.   Checking this by differentiation,

you find $G'(x) = \csc 2x \cot 2x$,   as desired, so $G(x) = -\dfrac{1}{2} \csc 2x$ is a correct anti-

derivative ("main part").   The general antiderivative is thus

$$\int \csc 2x \cot 2x \, dx \;=\; -\frac{1}{2} \csc 2x \;+\; C$$

$\square$

4.   <u>Notational problems: watch out or you'll be in trouble!</u>   The important equation

$$\frac{d}{dx}\left[\int f(x) \, dx \right] \;=\; f(x)$$

should help you understand the indefinite integral.   In English, it says that <u>the derivative of the</u>

<u>antiderivative (indefinite integral) of a function $f(x)$ is the function $f(x)$.</u>   That is, if you

begin with a function $f(x)$,   integrate it (obtaining $\displaystyle\int f(x) \, dx$)   and then differentiate what you

get, the result is $f(x)$ again.   Or, more simply put,

differentiation "undoes" integration.

<u>There are two important notational errors to avoid like the plague:</u>

1.   The notation $\displaystyle\int [\quad] dx$ must be treated as a sandwich: <u>whenever you write</u>

$\displaystyle\int$,   <u>you must also write   dx</u>.   The "sandwich" is incomplete without both pieces of "bread." It is a common mistake to write expressions such as $\displaystyle\int x^2$ or

$\displaystyle\int [3 \sin x + 7x]$, dropping the dx.

$$\boxed{\text{Never write } \int \text{ without } dx\,!}$$

2.   The symbol $\displaystyle\int f(x)\,dx$ is always equal to a function $F(x)$ (the "main part")

plus an arbitrary constant C:

$$\int f(x)\,dx \quad = \quad F(x) \quad + \quad C$$

$\uparrow$	$\uparrow$	$\uparrow$
antiderivative or indefinite integral of $f(x)$	"main part"	arbitrary constant

(This is Theorem 5.2.2). Make writing the "$+ C$" a reflex action:

$$\boxed{\begin{array}{l}\text{An answer to an indefinite integral } \displaystyle\int f(x)\,dx \\[2mm] \text{is incomplete without the } \text{"}+ C\text{"}\end{array}}$$

As we shall see below, it is perfectly all right to forget about the arbitrary

constant until the end and tack it on as the last step. But don't forget it!

Its importance will become clear in later sections.

Section 5.3.   Integration By u-Substitution

1.   u-substitution: simple in concept, not-so-simple in practice

You have now learned (we hope) several integration formulas, the ones in Table 5.2.1.

But it is apparent that there are lots of integrals such as $\int x \sqrt{x^2 + 2}\ dx$ which are not on

this list. (This will always be true, no matter how long our list of integral formulas grows,

so it is not simply that our list is not long enough.) However, as we mentioned in the last

section, there are tricks ("techniques" is a nicer word) which can be used to transform a

given integral into a familiar one.

One of the most important techniques is a simple one called u-substitution. At

least it is simple in concept. But it is decidedly not simple in practice because it frequently

involves a trial-and-error approach with several false starts.

The basic goal of u-substitution   (and by the way, there is nothing sacred about the

letter u - any other letter works just as well) is as follows: by grouping the terms of a given

integral in a certain way, the integral becomes one we know. Nothing in the integral is changed;

its terms are simply rearranged.

Question: Where does the u come in?

Answer: It is simply a visual aid. That is, it serves as a label for a group of terms

in the original integral so that it is easier to regard that grouping as a block. When it has served

its purpose, it is discarded (i. e., replaced with the original grouping).

Sound simple? Wrong! The difficulty lies in grouping the terms correctly. Not all

groupings will convert the integral to one you know; in fact, most of the time, only one grouping

will work. That means that you must be clever in picking your grouping, and being clever is not always easy.   You'll get better with practice (that's a key word in this section), but it will

sometimes require several false starts.

2.   Practice, practice, practice.

Example A.   Evaluate $\displaystyle\int x \sqrt{x^2 + 2}\ dx$.

Solution.   If we are looking for terms to group together, the expression $x^2 + 2$ under the

radical seems like a "natural" grouping.   To make it easier to remember that we want to look

at those terms as a group, we christen this expression   "u" .   That is, we let $\boxed{u = x^2 + 2}$

We need to get rid of all the   x's   in our integral and replace them with   u's.   But we

also have a   "dx"   to get rid of; how do we do that? Differentiate the function   u   with respect

to   x,   that's how! We get $\dfrac{du}{dx} = 2x$ or $du = 2x\ dx$ or $\boxed{\dfrac{1}{2}\,du = x\ dx}$.   (Note that we

are manipulating the differentials   du   and   dx   as if they were real numbers and thus we are

regarding $\dfrac{du}{dx}$ as a fraction.   If you are not comfortable with this, review the discussion of

differentials in   Section 3.4 .)

Now we can see a way to find $\displaystyle\int x\sqrt{x^2 + 2}\ dx$ by rearranging terms.   The rearrange-

ment is simply to reorder the integrand $x\sqrt{x^2 + 2}$,   obtaining $\displaystyle\int \sqrt{x^2 + 2}\ (x\ dx)$.   Replacing

$x^2 + 2$ by   u   and   x dx   by $\dfrac{1}{2}\,du$ from above, this becomes $\displaystyle\int \sqrt{u}\ \left(\dfrac{1}{2}\,du\right)$.   And this

integral is one we know! For $\displaystyle\int \sqrt{u}\ \left(\dfrac{1}{2}\,du\right) = \dfrac{1}{2}\int \sqrt{u}\ du = \dfrac{1}{2}\int u^{1/2}\ du = \dfrac{1}{2}\,\dfrac{u^{3/2}}{3/2} + C =$

$= \dfrac{1}{3}\,u^{3/2} + C$ by   Theorem 5.2.3   and   Formula 2   in   Table 5.2.1.   (Note that we didn't

forget the   "+ C".)

But $\dfrac{1}{3}\,u^{3/2} + C$ is not an entirely satisfactory answer, for the original problem was stated in terms of the variable x and the answer is in terms of the variable u. The solution?

Convert the answer to one in terms of x by replacing u by $x^2 + 2$. Thus we get the

answer $\frac{1}{3}\left(x^2 + 2\right)^{3/2} + C$ (i.e., $\int x \sqrt{x^2 + 2}\ dx = \frac{1}{3}\left(x^2 + 2\right)^{3/2} + C$). If we differentiate

the right-hand side of this equality, we get $x \sqrt{x^2 + 2}$, validating our answer.

Of course, we can shorten the explanation of our solution considerably:

$$\int x \sqrt{x^2 + 2}\ dx = \int \sqrt{u}\ \frac{1}{2}\ du = \frac{1}{2} \int u^{1/2}\ du$$

$$u = x^2 + 2$$

$$\frac{du}{dx} = 2x$$

$$\frac{1}{2}\ du = x\ dx$$

$$= \frac{1}{2}\ \frac{u^{3/2}}{(3/2)} + C$$

$$= \frac{1}{3}\ u^{3/2} + C$$

$$= \frac{1}{3}\left(x^2 + 2\right)^{3/2} + C$$

Note: We recommend this format for presenting a solution. It is essentially the same as the

method given in the blue box preceding Example 1 of Anton, but it is easier to follow. To

make sure the point is clear, here again is the solution, but now with Anton's 5 steps noted:

$$\int x \sqrt{x^2 + 2}\ dx = \int \sqrt{u}\ \frac{1}{2}\ du = \frac{1}{2} \int u^{1/2}\ du$$

Step 1 $u = x^2 + 2$

Step 2 $\frac{du}{dx} = 2x$ Step 3

Extra Step $\frac{1}{2}\ du = x\ dx$

Step 4

$$= \frac{1}{2}\ \frac{u^{3/2}}{(3/2)} + C$$

$$= \frac{1}{3}\ u^{3/2} + C$$

$$= \frac{1}{3}\left(x^2 + 2\right)^{3/2} + C$$

Step 5

Note that we have added one small step. Inside the box, we took the extra step of solving for

x dx in terms of du (x dx = $\frac{1}{2}$ du). The reason for doing this is to make the substitution

easier. We will be substituting for the expression x dx and so we determine explicitly what

x dx is in terms of u. Then we simply "plug in" for x dx in the integral. We recommend

taking this extra step; it makes it much less likely that you will make a substitution mistake.

We happened to solve Example A with the first try. But we might not have been so

clever (or lucky). For instance, in the solution of $\int x \sqrt{x^2 + 2}\ dx$ we might have proceeded

as follows:

$$\int x\sqrt{x^2 + 2}\ dx = \int \sqrt{u + 2}\ \frac{1}{2}\ du = \frac{1}{2}\int \sqrt{u + 2}\ du$$

$$= \text{TILT!}$$

$$\boxed{\begin{array}{l} u = x^2 \\[4pt] \dfrac{du}{dx} = 2x \\[4pt] \frac{1}{2}\ du = x\ dx \end{array}}$$

Extra step ⟶

By TILT!, we mean that this is not one of the integrals we know how to evaluate. At this

point, we have two choices: 1) to go ahead and see if we can evaluate the integral

$\frac{1}{2}\int \sqrt{u + 2}\ du$ or   2) to bail out entirely and try a new substitution in the original integral

$\int x\sqrt{x^2 + 2}\ dx$.

Either approach will lead to a correct answer in this particular example (provided, of

course, that we are clever enough). For instance, we can use a w-substitution to evaluate

$\frac{1}{2}\int \sqrt{u + 2}\ du$ as follows:

$$\frac{1}{2} \int \sqrt{u + 2} \; du = \frac{1}{2} \int \sqrt{w} \; dw = \frac{1}{2} \int w^{1/2} \, dw$$

$$\boxed{\begin{array}{l} w = u + 2 \\[2mm] \dfrac{dw}{du} = 1 \\[2mm] dw = du \end{array}}$$

Extra step $\longrightarrow$

$$= \frac{1}{2} \frac{w^{3/2}}{(3/2)} + C$$

$$= \frac{1}{2} w^{3/2} + C = \frac{1}{3} (u + 2)^{3/2} + C = \frac{1}{3} \left(x^2 + 2 \right) + C$$

Note that once again we re-substitute (in this case $w = u + 2$ and $u = x^2$) to leave the answer in terms of the same variable used in the original problem.

However, if one reaches TILT! , it may be necessary to go back and attack the original problem again, to "go back to the drawing board," as it were. As we have said before, if this happens to you, don't despair. You have had what is euphemistically called "a learning experience." (You have learned that your approach doesn't work.) Too many such "learning experiences" will discourage anyone, but in doing u-substitutions you have to steel yourself and be prepared for them. Remember that Thomas Edison went through about one thousand filaments before finding one that worked in his light bulb. ☐

Let's work through a non-trivial example in a stream-of-consciousness fashion

Example B (Exercise Set 5.3 , Number 30). Evaluate $\int x^2 \sqrt{2 - x} \; dx$. [Even Anton says this is tricky!]

Solution. Casey Calcwhiz began his solution as follows:

1-st Attempt

$$\int x^2 \sqrt{2 - x} \; dx = \int u \sqrt{2 - x} \; \frac{1}{2x} \; du$$

$$\boxed{\begin{array}{l} u = x^2 \\[2mm] \dfrac{du}{dx} = 2x \\[2mm] \dfrac{1}{2x} \, du = dx \end{array}}$$

= TILT ! (because both u's and x's are left). NO GOOD since ALL x's in the integral must be eliminated.

So... back to the drawing board. Casey then tried this:

2-nd Attempt

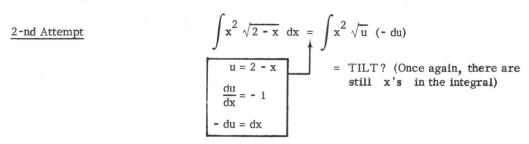

$$\int x^2 \sqrt{2-x} \ dx = \int x^2 \sqrt{u} \ (-\,du)$$

u = 2 - x

$\dfrac{du}{dx} = -\,1$

- du = dx

= TILT? (Once again, there are
   still x's in the integral)

So... back to the drawing board again. In desperation, Casey made this third try:

3-rd Attempt

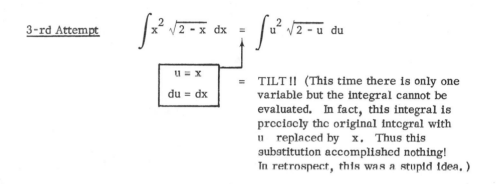

$$\int x^2 \sqrt{2-x} \ dx \ = \ \int u^2 \sqrt{2-u} \ du$$

u = x

du = dx

= TILT!! (This time there is only one
   variable but the integral cannot be
   evaluated. In fact, this integral is
   precisely the original integral with
   u replaced by x. Thus this
   substitution accomplished nothing!
   In retrospect, this was a stupid idea.)

So... back to the drawing board once more. But now Casey is (temporarily!) out of ideas.

This problem occurs in an exercise set after a section on u-substitution , so any fool knows

it has to be done with a u-substitution. But Casey has tried all the likely substitutions. So

what does he do? Quit? Well, ... maybe go on to other problems and come back to this one

later, but before doing this let's look over Casey's three attempts (disregarding the 3-rd

"stupid" attempt) to see if any of them can be patched up and made to work.

In the first attempt, the problem was that not all the x's were eliminated.

However, ... since $u = x^2$, then (at least for positive values of x), $x = \sqrt{u}$. Ah ha!

Now Casey can get rid of all those pesky x's:

1-st Attempt Revisited

$$\int x^2 \sqrt{2-x}\ dx \ = \ \int u \sqrt{2-x}\ \frac{1}{2x}\ du$$

$u = x^2$
$\dfrac{du}{dx} = 2x$
$\dfrac{1}{2x}\ du = dx$
$x = \sqrt{u}$

$$= \int u \sqrt{2 - \sqrt{u}}\ \frac{1}{2\sqrt{u}}\ du \qquad \text{using the extra}$$
$$\text{step}\quad x = \sqrt{u}$$

$$= \frac{1}{2} \int \sqrt{u}\ \sqrt{2 - \sqrt{u}}\ du$$

$$= \text{TILT!} \ \ \text{(Yuk! This is a much worse}$$
$$\text{integral than we started with!)}$$

So Casey's elation was short-lived. The substitution $u = x^2$, when pushed as far as it will

go, leads only to a very ugly integral. Thus $u = x^2$ is a definite TILT! and should be

discarded.

So we come to the second attempt. Here Casey was left with the integral $\int x^2 \sqrt{u}\ (-du)$.

Once again, the x must be replaced by an expression involving only u. Since the substitution

was $u = 2 - x$, then $x = 2 - u$, and $\int x^2 \sqrt{u}\ (-du)$ becomes $\int (2-u)^2 \sqrt{u}\ (-du)$, an

integral written entirely in terms of u. And all of a sudden, Casey sees the way out! For

$$\int (2-u)^2 \sqrt{u}\ (-du) \ = \ -\int (4 - 4u + u^2) u^{1/2}\ du \ = \ \int (-4u^{1/2} + 4u^{3/2} - u^{5/2})\, du$$

and we know how to integrate that! Joy has come to Mudville because mighty Casey has finally

stopped striking out!

What Casey has discovered, then, is the following solution to the problem:

2-nd Attempt Revisited (BINGO!)

$$\int x^2 \sqrt{2-x}\; dx \;=\; \int (2-u)^2 \sqrt{u}\; (-du)$$

$$\boxed{\begin{array}{l} u = 2 - x \\[4pt] \dfrac{du}{dx} = -1 \\[6pt] -\,du = dx \\[4pt] x = 2 - u \\[4pt] x^2 = (2-u)^2 \end{array}}$$

$$= -\int (4 - 4u + u^2)\, u^{1/2}\, du$$

$$= -4 \int u^{1/2}\, du + 4 \int u^{3/2}\, du - \int u^{5/2}\, du$$

$$= -4\, \frac{u^{3/2}}{(3/2)} + 4\, \frac{u^{5/2}}{(5/2)} - \frac{u^{7/2}}{(7/2)} + C$$

$$= -\frac{8}{3} u^{3/2} + \frac{8}{5} u^{5/2} - \frac{2}{7} u^{7/2} + C$$

$$= -\frac{8}{3} (2-x)^{3/2} + \frac{8}{5} (2-x)^{5/2} - \frac{2}{7} (2-x)^{7/2} + C$$

Differentiating the right-hand side yields $x^2 \sqrt{2-x}$ (try it!) so Casey knows he is correct.
Written out like this it looks deceptively simple (well, ...) and masks the extra work put
into all those false starts. (For one more solution to Example B see Subsection 4 below.)   □

   The correct solutions to Examples A and B and Anton's Examples 1 , 4 , 7 , 9 and 10
suggest that in   u-substitutions

> letting u   be an expression in parentheses or under
>
> a radical can be fruitful, even if its derivative is not
>
> apparent in the integral

You will do well to keep this point in mind. Here is one more example:

Example C.   Evaluate $x^3 \sqrt{2 - x^2}\; dx$.

Solution.   The observation above leads us to try the substitution $u = 2 - x^2$. Then, however,

$\dfrac{du}{dx} = -2x$ or $-\dfrac{1}{2} du = x\, dx$, so we "break off" one of the x's in x^3 , i.e., we write

$x^3 = x^2 x$ as follows:

$$\int x^3 \sqrt{2 - x^2}\; dx = \int x^2 \sqrt{2 - x^2}\; x\, dx$$

$$= \int (2 - u)\sqrt{u}\left(-\frac{1}{2}\, du\right)$$

$$= -\frac{1}{2}\int\left(2u^{1/2} - u^{3/2}\right) du$$

$$= -\frac{1}{2}\left(\frac{2u^{3/2}}{3/2} - \frac{u^{5/2}}{5/2}\right) + C$$

$$= -\frac{2}{3} u^{3/2} + \frac{1}{5} u^{5/2} + C$$

$$= -\frac{2}{3}\left(2 - x^2\right)^{3/2} + \frac{1}{5}\left(2 - x^2\right)^{5/2} + C$$

Box:

$u = 2 - x^2$

$\dfrac{du}{dx} = -2x$

$-\dfrac{1}{2} du = x\, dx$

$x^2 = 2 - u$

Differentiating the right-hand side validates our answer.   □

3. <u>Common mistakes and stumbling blocks in u-substitution.</u>   Beware of the following:

<u>Mistake I.</u>   Simply replacing "dx" with "du". (You can't do this unless $\dfrac{d}{dx}(u) = 1$.)

<u>Example:</u>   The following is incorrect:

$$\int (2x - 3)^6\, dx \;\overset{\boxed{\text{FALSE!}}}{=}\; \int u^6\, du = \frac{u^7}{7} + C = \frac{(2x - 3)^7}{7} + C$$

$\boxed{u = 2x - 3}$

The format we have recommended will help you avoid this common mistake.

$$\int (2x-3)^6 \, dx \;=\; \int u^6 \left(\tfrac{1}{2}\, du\right) \;=\; \tfrac{1}{2} \int u^6 \, du \;=\; \frac{u^7}{14} + C \;=\; \frac{(2x-3)^7}{14} + C$$

$$u = 2x - 3$$

$$\frac{du}{dx} = 2$$

$$\tfrac{1}{2}\, du = dx$$

Mistake II.   Stopping with one unsuccessful guess for u . Example B shows what

perseverance will do for you.

Mistake III.   Failing to replace __all__ the x's with u's .

Example.   The following is incorrect:

$$\int x^2 \sqrt{2-x} \; dx \;=\; \int x^2 \sqrt{u}\,(-du) \;=\; -\int x^2 \, u^{1/2} \, du$$

$$u = 2 - x$$

$$\frac{du}{dx} = -1$$

$$-du = dx$$

$$\boxed{\text{FALSE!}}$$

$$= \quad -x^2 \, \frac{u^{3/2}}{(3/2)} + C$$

$$= \quad -\tfrac{2}{3} x^2 (2-x)^{3/2} + C$$

Differentiating the answer will convince you that it is wrong. A correct solution is given in

Example B above.

Mistake IV.   Carelessly writing   "u"   where you mean   "x"   (this might be called "over-

zealous substitution").

Example.   The following is incorrect:

$$\int x^2 \sqrt{2-x} \; dx \quad \boxed{\text{FALSE!}} \;=\; \int u^2 \sqrt{u}\,(-du) \;=\; -\int u^{5/2} \, du$$

$$u = 2 - x$$

$$\frac{du}{dx} = -1$$

$$-du = dx$$

hmmm... $= \dfrac{-u^{7/2}}{(7/2)} + C$

$$= \quad -\tfrac{2}{7} (2-x)^{7/2} + C$$

Once again, differentiating the answer will convince you that it is wrong.

Compare this incorrect solution with the correct one in Example B above.

4. <u>Luck</u>. Finally, it should be remarked that in u-substitutions you occasionally get lucky. Sometimes the wildest substitutions work. As an example, let us return to the integral

$\int x^2 \sqrt{2-x} \, dx$ of Example B:

$$\int x^2 \sqrt{2-x} \, dx = \int \left(2 - u^2\right)^2 u \, (-2u \, du)$$

$$u = \sqrt{2-x}$$

$$\frac{du}{dx} = \frac{1}{2}(2-x)^{-1/2}(-1) = \frac{-1}{2u}$$

$$-2u \, du = dx$$

$$u^2 = 2 - x$$

$$x = 2 - u^2$$

$$x^2 = (2 - u^2)^2$$

$$= -2 \int u^2 \left(4 - 4u^2 + u^4\right) du$$

$$= \int \left(-8u^2 + 8u^4 - 2u^6\right) du$$

$$= -\frac{8}{3} u^3 + \frac{8}{5} u^5 - \frac{2}{7} u^7 + C$$

$$= -\frac{8}{3} \left(\sqrt{2-x}\right)^3 + \frac{8}{5} \left(\sqrt{2-x}\right)^5 - \frac{2}{7} \left(\sqrt{2-x}\right)^7 + C$$

$$= -\frac{8}{3} (2-x)^{3/2} + \frac{8}{5} (2-x)^{5/2} - \frac{2}{7} (2-x)^{7/2} + C$$

This is the (correct) answer we obtained above! And we got it by a substitution that probably did not occur to most of you.

<u>The Moral</u>: <u>When all else fails, any substitution, no matter how wild, is worth trying,</u> particularly where radicals are involved! For another example, see Exercise 35 in Exercise Set 5.3.

The several correct and incorrect solutions in Subsections 2, 3 and 4, together with Anton's solutions to his ten examples, should give you ample "reference material" on u-substitutions. Now it's up to <u>you</u>. As we said, ... PRACTICE, PRACTICE, PRACTICE!

5. <u>An alternate derivation of the u-substitution method</u>. (Optional) Anton explains the theory

behind u-substitution in the beginning of this section. Here is another derivation of

u-substitution, one which makes clear its connection with the Chain Rule:

Suppose we are given an integral $\int h(x)\,dx$ to evaluate. Further suppose that a

substitution $u = g(x)$ transforms the integral as follows:

$$\int h(x)\,dx \quad = \quad \int f(u)\,du$$

$$\boxed{\begin{array}{l} u = g(x) \\ du = g'(x)\,dx \end{array}}$$

where f is a "good" function, i.e., it has a <u>known</u> antiderivative F. (Guessing a sub-

stitution $u = g(x)$ which produces a "good" function is the <u>art</u> of u-substitution, as we

have seen in the examples above.)

The Key Theoretical Step (KTS) is next:

$$\int h(x)\,dx \; - \; \int f(u)\,du \; = \; F(u) \; + \; C \; = \; F(g(x)) \; + \; C$$

$$\boxed{\text{substituting } u} \qquad \boxed{\text{KTS}} \qquad \boxed{\text{substituting for } u}$$

In actual evaluations, the KTS is done almost without thought - study the examples in this

section! However, look closely at what is happening in $\int f(u)\,du = F(u) + C$. We are

treating the <u>function</u> $u = g(x)$ just as though it were a <u>variable</u> x. This requires justification:

$$\int f(u)\ du\ =\ \int f(g(x))\ g'(x)\ dx \qquad\qquad \text{since} \quad u = g(x)\ ,\ du = g'(x)\ dx$$

$$=\ \int F'(g(x))\ g'(x)\ dx \qquad\qquad \text{since} \quad F' = f$$

$$=\ \int (F \circ g)'(x)\ dx \qquad\qquad \text{by the } \underline{\text{Chain Rule}}!$$

$$=\ (F \circ g)(x)\ +\ C \qquad\qquad \text{since integration undoes differentiation}$$

$$=\ F(g(x))\ +\ C \qquad\qquad \text{by definition of composition}$$

$$=\ F(u)\ +\ C \qquad\qquad \text{since} \quad u = g(x)$$

This proves the KTS! The heart of the derivation is the Chain Rule. Loosely, one can say that the method of u-substitution is the inverse operation to the Chain Rule.

Section 5.4. Rectilinear Motion (An Application of The Indefinite Integral)

Although motion along a straight line may seem to be a concept of very limited applicability, you need to master it before going on to more general motion problems (which are discussed in Section 14.5).

1. Position, Velocity and Acceleration

When we talk about the straight-line motion of a particle, we are interested in four things:

The <u>position</u>   s(t)   of the particle at time   t

The <u>velocity</u>   v(t)   at time   t,   i.e.,   how fast and in
    which direction the particle is moving at time   t

The <u>speed</u> |v(t)| at time   t,   i.e.,   how fast the
    particle is moving at time   t

The <u>acceleration</u>   a(t)   at time   t,   i.e.,   how fast and in
    what direction the velocity is changing at time   t

Note that all of these are functions of   t   (time).   That is, each may vary according to where

the particle is on its path, which in turn depends on how long the particle has been moving.

    The following diagram is helpful in remembering the relationships among the various

motion quantities.   For reference purposes, we'll call it the   Four-Box Diagram:

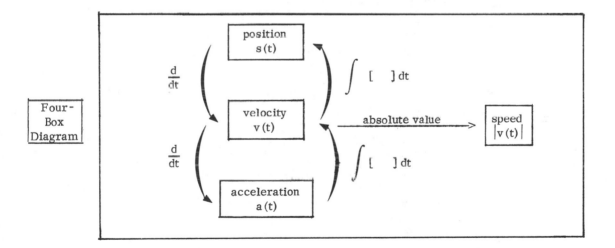

2.   <u>The meanings of positive and negative</u>

    It is very important to cement in your mind the various meanings of <u>positive</u> and its

opposite, <u>negative</u>.   Here is a summary of their meanings when they are used to describe the

position, velocity and acceleration, respectively, of a particle:

position
- positive means the particle is located to the right of the origin (x-axis) or up from the origin (y-axis)
- negative means the particle is located to the left of the origin (x-axis) or down from the origin (y-axis)

velocity
- positive means the motion is to the right (x-axis) or up (y-axis)
- negative means the motion is to the left (x-axis) or down (y-axis)

acceleration
- positive means the velocity of the particle is increasing
- negative means the velocity of the particle is decreasing

speed is always positive since it is an absolute value.

In the table we have interpreted the sign of acceleration in terms of its effect on velocity:

If $a(t)$ is positive, then $v(t)$ is increasing.

If $a(t)$ is negative, then $v(t)$ is decreasing.

What about the effect on speed? (Knowing this effect is important for the correct set-up of many motion problems.) Since speed is the absolute value $|v(t)|$ of velocity, we need to consider when $v(t)$ is positive and when it is negative:

If velocity $v(t)$ is positive, then speed equals velocity, and the sign of acceleration affects speed in the same way that it effects velocity.

> If velocity $v(t)$ is negative, then speed equals the negative
> of velocity, and the sign of acceleration affects speed
> in the opposite way that it effects velocity.

Thus whether speed is increasing or decreasing depends on the signs of both velocity and

acceleration:

	$v(t)$ positive	$v(t)$ negative
$a(t)$ positive	speed increasing	speed decreasing
$a(t)$ negative	speed decreasing	speed increasing

3.   Two types of motion problems.   The motion problems of this section are of two types:

A.   Motion under gravity.   Let $g = 32$ ft/sec^2 be the gravitation constant (a number which

has been determined experimentally).   Then the equations of motion for an object which is

subject to the force of gravity are:

$$
\begin{aligned}
s(t) &= -\tfrac{1}{2} g t^2 + v_0 t + s_0 \qquad &(v_0 - \text{ initial velocity}) \\
v(t) &= -g t + v_0 \qquad &(s_0 = \text{ initial position}) \\
a(t) &= -g
\end{aligned}
$$

These equations can be memorized or you can memorize one of them and use the Four-Box

Diagram to derive the other two.   [For example, start with $a(t) = -g$ in the Four-Box

Diagram.   Integrate once to get $v(t) = -gt + C$.   Find $C = v_0$ by using $v(0) = v_0$.

Integrate again to get $s(t) = -\tfrac{1}{2} g t^2 + v_0 t + C$.   Then find $C = s_0$ by using $s(0) = s_0$.]

5.4.5

In any case, it is important to know that the motion equations under the force of gravity are intially derived from the Four-Box Diagram.

B. <u>Motion under arbitrary acceleration.</u> Here you <u>must</u> use the Four-Box Diagram.

We will illustrate both types of motion problems with four examples, all about Superman. (The solutions are organized according to the word problem procedures described in Appendix H.)

<u>Example A</u>. Superman, standing at the base of a skyscraper which is 875 feet tall, sees Lois Lane dangling precariously off the edge of the top of the building. If he leaps upward with an acceleration of

$$a(t) = 12t^2 + 20 \ \text{ft/sec}^2$$

what is Superman's speed when he saves Lois? [Hint: $875 = 5^4 + 10 \cdot 5^2$]

Step 1: Translate the problem into mathematics

<u>Solution.</u> This is a Type B (non-gravity) problem because the acceleration is not always $-g$. The position function $s(t)$ gives Superman's vertical distance from the ground at time t (see the picture). At time $t = 0$, he is standing (i.e., $v(0) = 0$) on the ground (i.e., $s(0) = 0$). If t_1 is the time at which he reaches Lois, a distance of 875 feet (i.e., $s(t_1) = 875$), then we want to find his speed at time t_1 (i.e., $|v(t_1)|$).

Thus we can translate the problem as follows:

Superman, standing at
base of skyscraper, $\longrightarrow$ $\Big\{$

$$s(0) = 0$$
$$v(0) = 0$$

accelerates with given acceleration. $\longrightarrow$ $\quad a(t) = 12\, t^2 + 20$

Find speed when he reaches Lois. $\longrightarrow$ $\quad |v(t_1)| = ?$ when $s(t_1) = 875$

Step 2:
Devise
a plan

We need to find t_1 and then plug it into $|v(t_1)|$. This means we will also need to find $v(t)$. And to find t_1, we will need to find $s(t)$ (so that we can solve $s(t_1) = 875$ for t_1). The Four-Box Diagram reminds us how to get $v(t)$ and $s(t)$ from $a(t)$. Thus we will

    i)   integrate twice to go from $a(t)$ to $v(t)$ to $s(t)$

    ii)   find t_1 such that $s(t_1) = 875$

    iii) compute $|v(t_1)|$.

Step 3:
Execute
the
plan

i) $a(t) = 12\, t^2 + 20$ so

$$v(t) = \int a(t)\, dt = \int (12\, t^2 + 20)\, dt = 4\, t^3 + 20t + C$$

Use $v(0) = 0$ to evaluate C :

$$0 = v(0) = 4(0)^3 + 20(0) + C \qquad \text{so} \qquad C = 0$$

Thus $\qquad\qquad\qquad\qquad\qquad v(t) = 4\, t^3 + 20t$

Then

$$s(t) = \int v(t)\, dt = \int (4\, t^3 + 20t)\, dt = t^4 + 10\, t^2 + C$$

Use $s(0) = 0$ to evaluate C :

$$0 = s(0) = (0)^4 + 10(0)^2 + C \qquad \text{so} \qquad C = 0$$

Thus
$$s(t) = t^4 + 10 t^2$$

ii)   Set $t_1^4 + 10 t_1^2 = s(t_1) = 875$ and, using the hint in the problem, find $t_1 = 5$ sec.

iii)   Thus $|v(t_1)| = |v(5)| = |4(5)^3 + 20(5)| = 600$ ft/sec. $\qquad\qquad\square$

<u>Example B.</u>   Superman is at the top of a cliff when he spots Howard Anton drowning in a river 160 ft

below. He throws himself straight downward with an initial velocity of 48 ft/sec. What is

Superman's speed when he saves Howard?

Step 1: Translate the problem

Solution.   This is a Type A (gravity) problem so we will use the equations

$$a(t) = -g$$

$$v(t) = -gt + v_0$$

$$s(t) = -\frac{1}{2} g t^2 + v_0 t + s_0$$

We need to determine v_0 and s_0.

The position function $s(t)$ gives Superman's

vertical distance <u>measured upwards from the river.</u>

(See the Note at the end of the problem.) At time

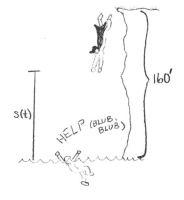

$t = 0$ he is at the top of the cliff (i. e., $s_0 = s(0) = 160$)

and he throws himself down with initial velocity 48 ft/sec (i. e., $v_0 = v(0) = -48$, <u>minus</u>

because it is downward). Thus the gravity equations giving Superman's acceleration, velocity

and position are

$$a(t) = -g$$

$$v(t) = -gt - 48$$

$$s(t) = -\frac{1}{2}gt^2 - 48t + 160$$

If t_1 is the time at which Superman reaches Howard in the river (i.e., $s(t_1) = 0$), we want to find his speed at time t_1 (i.e., we want $|v(t_1)|$). Thus we can translate the problem into mathematics as follows

Superman, at top of 160 ft. cliff, $\longrightarrow$ | $s(0) = s_0 = 160$

leaps downward with initial velocity 48 ft/sec. $\longrightarrow$ | $v(0) = v_0 = -48$

Find speed when he reaches Howard. $\longrightarrow$ | $|v(t_1)| - ?$ when $s(t_1) = 0$

Step 2: Devise a plan

We will (i) set $s(t_1)$ equal to zero, (ii) solve for t_1 and (iii) plug it in $|v(t_1)|$. (Note that we have already used the $s_0 = 160$ and $v_0 = -48$ conditions to determine the precise gravity equations.)

Step 3: Execute the plan

1)   Set $0 = s(t_1) = -\frac{1}{2}gt_1^2 - 48t_1 + 160 = -16t_1^2 - 48t_1 + 160$.

ii)   Dividing by -16 and factoring, we get

$$0 = t_1^2 + 3t_1 - 10 = (t_1 + 5)(t_1 - 2)$$

$$\text{or} \quad t_1 = -5 \text{ or } 2$$

We eliminate $t_1 = -5$ on physical grounds (even Superman can't make time negative) and are left with $t_1 = 2$ secs.

iii)   Then $|v(t_1)| = |v(2)| = |-g(2) - 48| = |-64 - 48| = 112$ ft/sec. $\square$

5. 4. 9

Note: In this example, we set "ground zero," i.e., $s = 0$, to be the river. Thus Superman's

original position is $s_0 = 160$. It is always important to state clearly what your "ground zero"

is. Otherwise, it might "wander around" during the solution and thus cause you to wander off

to confusion. . .)

Example C. Superman is standing next to a straight railroad track when a train passes him at a

constant speed of 2 miles/minute. After waiting one minute to tie his shoe, Superman dashes

off after the train, running with a constant acceleration of $a = 3$ miles/min^2 . How soon does

he catch the train?

Solution. This is a Type B (non-gravity) problem and so we are going to have to use the

Four-Box Diagram.

Step 1: Trans- late the problem	Let t be the time in minutes (it is important

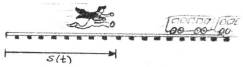

Let t be the time in minutes (it is important

to say clearly what the units of measurement are)

measured from when Superman starts to run (it is also important to say clearly when your

clock starts!). We want to find the time t_1 when he catches the train (i.e., when Superman

and the train are at exactly the same position).

Turning our attention to the train, it moves at a constant speed (which is a signal to

use the familiar formula $d = rt$)* of 2 miles/minute and it has a 1 minute head start on our

hero (so that when the train and Superman are at the same position, it has been $t_1 + 1$

minutes since it passed him). Hence the train has traveled a distance of $d = rt = 2(t_1 + 1)$

$= 2t_1 + 2$ miles when Superman catches it. Thus Superman must have traveled a distance

of $2t_1 + 2$ miles as well. That is, if s(t) is Superman's position function (see the

picture), $s(t_1) = 2t_1 + 2$.

---
*
  distance = rate x time

At this point, think ahead a bit. We are going to use the Four-Box Diagram and that will produce constants of integration. By now, we know that those constants are closely related to (in fact, they frequently are) the initial velocity and initial position. So before we finish this translation step, let's read the problem one more time to see if there is any information given, explicitly or implicitly, about initial velocity and initial position. Well, ... there is certainly no explicit information. But, because the problem does not say what Superman's initial velocity is, the logical assumption to make is that there is none. That is, the problem is telling us implicitly that the initial velocity is zero (i.e., $v(0) = 0$). Moreover, the initial position is zero (i.e., $s(0) = 0$) as well because of our "ground zero" assumption. That is, we have chosen to measure all distances from the point where the train passes Superman; that point is "ground zero," i.e., where $s = 0$.

So after several readings (whew!), we are ready to translate our problem as follows

Train passes a standing
  Superman, who
chases it with a constant acceleration.
Find time it takes him to catch it
if train has constant velocity of 2 mi/min.
and Superman gives it a 1 minute
head start.

$$s(0) = 0$$
$$v(0) = 0$$
$$a(t) = 3$$
$$t_1 = ? \quad \text{when} \quad s(t_1) = 2t_1 + 2$$

Step 2:
Devise a plan

We will i) use the Four-Box Diagram to integrate twice to go from $a(t)$ to $v(t)$ to $s(t)$, using $s(0) = 0$ and $v(0) = 0$ to determine the constants of integration, and then

ii)   solve $s(t_1) = 2t_1 + 2$ for t_1.

Step 3:
Execute the plan

i) $a(t) = 3$ so $v(t) = \int a(t)\, dt = \int 3\, dt = 3t + C$

Use $v(0) = 0$ to evaluate C:

$$0 = v(0) = 3(0) + C \qquad \text{so} \qquad C = 0$$

Thus $v(t) = 3t$.

Then $s(t) = \displaystyle\int v(t)\,dt = \int 3t\,dt = \frac{3}{2}t^2 + C$

Use $s(0) = 0$ to evaluate C:

$$0 = s(0) = \frac{3}{2}(0)^2 + C \qquad \text{so} \qquad C = 0$$

Thus $s(t) = \frac{3}{2}t^2$.

ii)   Thus $s(t_1) = 2t_1 + 2$ is $\dfrac{3t_1^2}{2} = 2t_1 + 2$. We solve this as follows:

$$\frac{3t_1^2}{2} = 2t_1 + 2$$

$$3t_1^2 - 4t_1 - 4 = 0$$

$$(3t_1 + 2)(t_1 - 2) = 0$$

$$t_1 = -\frac{2}{3} \quad \text{and} \quad t_1 = 2$$

We are not interested in negative time (before the train reaches Superman), so we discard the solution $t_1 = -\frac{2}{3}$. Thus Superman catches the train after $t_1 = 2$ minutes.   □

A Note About Constants of Integration.   In the previous example, note the necessity of using the given information about particular values of velocity and position correctly to calculate the constants of integration. Sometimes this information is given explicitly and sometimes you have to dig it out. But you always need it to use the methods of the Four-Box Diagram. The most common mistakes in straight-line motion problems involve improper treatment of the constants of integration. And frequently, the major part of solving a problem is the evaluation

of the constants of integration. For example, note the critical role of the constant of integration

and the way it is handled in the following problem.

Example D.   Just for fun, Superman rolls Howard Anton into a ball and, after taking a little run

down a giant ski-jump to gather momentum, bowls him down the remainder of the ski-jump (an

inclined plane). If Howard's velocity t seconds after Superman releases him is $\sqrt{t}$, and

if he has rolled to a point 5 miles down from the top of the jump after 3 seconds, where did

Superman release him? (i. e. , with time measured from the time of release, where were

Howard and Superman at time $t = 0$?)

Solution.   This is a Type B (non-gravity) problem with time measured from the time of release.

Let $s(t)$ be Howard's position at

Step 1: Tans-late

time t measured from the top of the

jump (i. e. , the top of the jump is "ground

zero" - see picture). We are given $s(3) = 5$

and we want to find $s(0)$. Thus the problem translates as follows:

Superman bowls Howard with given velocity. ──────▶	$v(t) = \sqrt{t}$
In 3 seconds Howard is 5 miles from the top. ────▶	$s(3) = 5$
Find the point of release. ──────────────▶	$s(0) = ?$

Step 2: Devise a plan

We will i) use the Four-Box Diagram to integrate $v(t)$ to $s(t)$, ii) use $s(3) = 5$

to evaluate the constant of integration and iii) find $s(0)$.

Step 3: Execute the plan

i) $v(t) = \sqrt{t} = t^{1/2}$ so

$$s(t) = \int v(t)\,dt = \int t^{1/2}\,dt = \frac{2}{3} t^{3/2} + C$$

ii)   Then $5 = s(3) = \frac{2}{3}(3)^{3/2} + C = 2\sqrt{3} + C$ (since $3^{3/2} = (\sqrt{3})^3 = (\sqrt{3})^2(\sqrt{3}) =$

$= 3\sqrt{3}$)   so $C = 5 - 2\sqrt{3}$.   Hence $s(t) = \frac{2}{3} t^{3/2} + 5 - 2\sqrt{3}$.

iii)   Thus $s(0) = \frac{2}{3}(0)^{3/2} + 5 - 2\sqrt{3} = 5 - 2\sqrt{3}$ miles from the top of the jump. $\square$

Section 5.5. Sigma Notation.

The important objective to set for yourself in this section is to become comfortable enough with $\sum$ notation to be able to use it without much deep thought. Mathematical notation is simply a form of shorthand and, as with all shorthand, practice is essential to learning it. However, all shorthand is relatively easy to master and the $\sum$ notation is no exception. Thus in many ways this section is an oasis in a desert of difficult concepts.

1.   Basic Laws of Summations.   The key theorem of the section is Theorem 5.5.1, which we may restate in English (as opposed to in Mathematics) as follows

---

THEOREM 5.5.1.

a)   Sigma of a sum is the sum of the sigmas.

b)   Sigma of a difference is the difference of the Sigmas.

c)   Multiplicative constants can be moved through Sigma signs.

Of course, such abbreviated statements leave out many important details, but they capture the essence of the theorem and they are easy to remember.

It is important to note the similarity between the laws of summation (Theorem 5.5.1) and the laws of integration (Theorem 5.2.3). Comparing the two, we see that the $\sum$ symbol and the $\int$ [ ] dx symbol behave in the same way with respect to sums (and differences) and multiplication by constants. For instance, constants "move through" both. This similarity is no accident: both the "elongated S" and the Greek letter $\sum$ are abbreviations for "sum." (That discussion will be pursued in earnest in Section 5.7.)

2. Special summations. Another result which is exceedingly useful in particular situations is Theorem 5.5.2. You should memorize (learn) at least parts a) and b) of it. In English, the theorem says

---

THEOREM 5.5.2

a) The sum of the first n positive integers is $\dfrac{n(n + 1)}{2}$

b) The sum of the first n squares is $\dfrac{n(n + 1)(2n + 1)}{6}$

c) The sum of the first n cubes is $\left[\dfrac{n(n + 1)}{2}\right]^2$

You will be amazed at how often this information will prove useful! For instance, note Anton's Examples 3 and 4. Here is one more example:

Example A.   Evaluate $\displaystyle\sum_{k=3}^{50} (3k + 2).$

5.5.3

<u>Solution.</u> We will want to use the summation formulas of Theorem 5.5.2. However, these formulas start the summations at $k = 1$, while our summation starts at $k = 3$. To remedy this situation we will start our summation at $k = 1$ ... and then <u>subtract</u> away the two terms corresponding to $k = 1$ and $k = 2$:

$$\sum_{k=3}^{50} (3k + 2) = \sum_{k=1}^{50} (3k + 2) - \sum_{k=1}^{2} (3k + 2)$$

$$= \left[3 \sum_{k=1}^{50} k + \sum_{k=1}^{50} 2 \right] - \left[(3(1) + 2) + (3(2) + 2) \right] \qquad \text{Theorem 5.5.1}$$

$$= \left[3 \cdot \frac{50(50 + 1)}{2} + \underbrace{(2 + 2 + \cdots + 2)}_{50 \text{ repetitions}} \right] - [5 + 8] \qquad \text{Theorem 5.5.2(a)}$$

$$= [\, 3825 + 100 \,] - 13$$

$$= \boxed{3912} \qquad\qquad\qquad \Box$$

A comment about the proofs of Theorem 5.5.2: as you read them, they seem almost like magic, in that they seem to require just the right observations or tricks at just the right times. However, this is true only because Anton is trying to avoid using the concept of mathematical induction to prove them. The principle of mathematical induction is discussed in Appendix I and parts (a) and (b) of Theorem 5.5.2 are proved there using that principle. Once the principle is understood, those proofs are much less mysterious than the proofs in Anton.

3. <u>Common Mistakes.</u> Problems 38 and 39 in Exercise Set 5.5 are important because they focus on some <u>common mistakes.</u> We have said (Theorem 5.5.1) that "Sigma of a sum (difference) is the sum (difference) of the Sigmas" so it is natural to want to say

---

FALSE THEOREMS

    a)   Sigma of a product is the product of the Sigmas; that is,

FALSE $\sum_{i=1}^{n} a_i b_i = \left(\sum_{i=1}^{n} a_i \right) \left(\sum_{i=1}^{n} b_i \right)$

    b)   Sigma of a quotient is the quotient of the Sigmas; that is,

FALSE $\sum_{i=1}^{n} \frac{a_i}{b_i} = \left(\sum_{i=1}^{n} a_i \right) \left(\sum_{i=1}^{n} b_i \right)$

    c)   Sigma of a power is the power of the Sigma; that is, $\sum_{i=1}^{n} a_i^k = \left(\sum_{i=1}^{n} a_i \right)^k$

FALSE

        where k is an integer.

However, <u>none of these statements is true</u>, a fact which we can establish by giving a counter-example to each. (Remember that to show a statement is not always true it suffices to give just <u>one</u> instance where it is false.)

<u>Counterexample to "Sigma of a product equals the product of the Sigmas"</u>

$$\sum_{k=1}^{2} a_k b_k = a_1 b_1 + a_2 b_2 \longleftarrow \boxed{\text{CLEARLY UNEQUAL}}$$

$$\left(\sum_{k=1}^{2} a_k \right) \left(\sum_{k=1}^{2} b_k \right) = (a_1 + a_2)(b_1 + b_2) = a_1 b_1 + a_1 b_2 + a_2 b_1 + a_2 b_2$$

5. 5. 5

Counterexample to "Sigma of a quotient equals the quotient of the Sigmas"

$$\sum_{k=1}^{2} \frac{a_k}{b_k} = \frac{a_1}{b_1} + \frac{a_2}{b_2}$$

CLEARLY UNEQUAL

$$\frac{\sum_{k=1}^{2} a_k}{\sum_{k=1}^{2} b_k} = \frac{a_1 + a_2}{b_1 + b_2}$$

Counterexample to "Sigma of a power equlas the power of the Sigma"

$$\sum_{k=1}^{2} a_k^2 = a_1^2 + a_2^2$$

CLEARLY UNEQUAL

$$\left(\sum_{k=1}^{2} a_k \right)^2 = (a_1 + a_2)^2 = a_1^2 + 2a_1 a_2 + a_2^2$$

Example B.　Express the following in sigma notation:

$$-1 + \pi/2! - \pi^2/3! + \pi^3/4! - \pi^4/5!$$

Solution.　Since there are five terms, we will use a summation from $k = 1$ to 5. Set up a table for comparison:

k	1	2	3	4	5
k-th term	-1	$\pi/2!$	$-\pi^2/3!$	$-\pi^3/4!$	$\pi^4/5!$

Except for sign the k-th term is seen to be $\pi^{k-1}/k!$. Since the sign alternates from term-to-term, and the $k = 1$ term is negative, we must multiply by $(-1)^k$ (if the $k = 1$ term were positive we would multiply by $(-1)^{k+1}$). Thus the series equals

$$\sum_{k=1}^{5} (-1)^k \pi^{k-1} / k!$$

□

Section 5.6.   Areas As Limits

1.   "Areas as limits" versus "Areas as antiderivatives"

In this section we return to the Area Problem as first discussed in §5. 1; however, this time around we treat the concept of "area" in a more rigorous fashion. Indeed, it would be best (just for this section and the following one) simply to forget that you ever saw §5. 1, or that you ever heard of the computation of areas by the use of "antiderivatives. " Instead, we develop a method using "limits of rectangles. "

If you have read Anton's section, you are probably perplexed: "You mean this 'limit of rectangles' method for calculating areas is better than antiderivatives? You gotta be nuts!" The response to this question is basic to an understanding of both the theory and application of integration. While the details will have to be developed over the next few sections, at least the broad outline can be stated now. A rigorous definition of area cannot be made without "limits of rectangles. " Moreover, areas (and other quantities which behave like "area") arise in real world applications through "limits of rectangles. " Thus the approach to area given in this section is important for both theoretical and practical reasons. However ... any fool, after having looked at the examples of this section, will see that the calculation of areas by "limits of rectangles" is at best a gigantic bore and at worst an algebraic and notational headache. From the First Fundamental Theorem of Calculus (§5. 8) we will see that area, when defined by "limits of rectangles, " can still be computed by "antiderivatives" as in §5. 1. Thus

> areas arise in applications through
> "limits of rectangles, " but they are
> computed via "antiderivatives. "

Thus we will be able to calculate areas almost as routinely as we are able to calculate derivatives. To see how this will work, peek ahead to Example 2 of Anton's §5.8. Even if you can't understand the symbols yet, you'll have to admit that it looks pretty easy!

2. <u>A basic overview of "areas as limits.</u>" In the "limit of rectangles" approach, we take the area under a curve $y = f(x)$ above the interval $[a, b]$ and approximate it by areas we know. Specifically, the area under $y = f(x)$ above $[a, b]$ is approximated by a collection of <u>inscribed</u> or <u>circumscribed</u> rectangles in such a way that the more rectangles used, the better the approximation. Then the number of rectangles is increased without limit and, bingo, we get the area!

This concept is straightforward... but the notation is somewhat formidable and the algebra can be messy and tricky.

3. <u>Bookkeeping with a fixed n</u>. Anton has a number of detailed examples written out in Section 5.6. Here we give an additional example, first using a <u>fixed</u> number of subintervals and then going to the limit with an increasing number of subintervals.

<u>Example A</u>. Approximate the area under $y = 2x - 3$ above the interval $[a, b] = [2, 5]$ by dividing $[2, 5]$ into $n = 4$ subintervals of equal length and computing a) the sum of the areas of the inscribed rectangles and b) the sum of the areas of the circumscribed rectangles (cf. Problems 1 - 4 in Exercise Set 5.6).

Solution.   First draw a rough graph of the

function over the interval in question, i. e. ,

$f(x) = 2x - 3$ over $[2, 5]$.

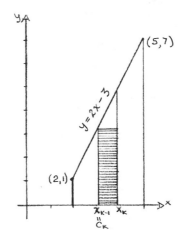

Inscribed rectangles.   We can use the

following step-by-step approach to

compute the areas of the inscribed

rectangles:

Step 1.   Compute Δx,   the length of each subinterval. For this we use the formula

$$\Delta x = \frac{b - a}{n} = \frac{5 - 2}{4} = \frac{3}{4}$$

Step 2.   Compute x_1, x_2, x_3,   the points of subdivision

     for $[2, 5]$:

$x_1 = 2 + \Delta x = 2 + 3/4 = 11/4$

| since x_1 is a distance |
| of Δx to the right of 2 |

$x_2 = 2 + 2\Delta x = 2 + 2(3/4) = 7/2$

| since x_2 is a distance |
| of $2\Delta x$ to the right of 2 |

$x_3 = 2 + 3\Delta x = 2 + 3(3/4) = 17/4$

| since x_3 is a distance |
| of $3\Delta x$ to the right of 2 |

5. 6. 4

<u>Step 3.</u> Compute c_1, c_2, c_3, c_4, the points where f

assumes minimum values in the four subintervals

$[a, x_1], [x_1, x_2], [x_2, x_3]$ and $[x_3, b]$.

Since f is increasing, the minimum value for

$f(x)$ on each subinterval occurs at the left endpoint,

as shown in the diagram to the right. Thus

on $[a, x_1]$,　　$c_1 = a = 2$

on $[x_1, x_2]$,　　$c_2 = x_1 = 11/4$

on $[x_2, x_3]$,　　$c_3 = x_2 = 7/2$

on $[x_3, b]$,　　$c_4 = x_3 = 17/4$

<u>Step 4.</u> Compute $f(c_k) \Delta x$ $(k = 1, 2, 3, 4)$,

the areas of the inscribed rectangles:

$$f(c_1) \Delta x = (2c_1 - 3) \Delta x$$
$$= (2(2) - 3)(\frac{3}{4}) \quad = \frac{3}{4}$$
$$f(c_2) \Delta x = (2c_2 - 3) \Delta x$$
$$= (2(\frac{11}{4}) - 3)(\frac{3}{4}) = \frac{15}{8}$$
$$f(c_3) \Delta x = (2c_3 - 3) \Delta x$$
$$= (2(\frac{7}{2}) - 3)(\frac{3}{4}) = 3$$
$$f(c_4) \Delta x = (2c_4 - 3) \Delta x$$
$$= (2(\frac{17}{4}) - 3)(\frac{3}{4}) = \frac{33}{8}$$

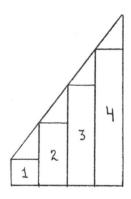

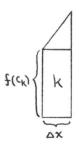

<u>Step 5.</u>   Compute $\displaystyle\sum_{k=1}^{4} f(c_k) \Delta x$,   the sum of the areas of the inscribed rectangles. This

simply means adding up the areas computed in Step 4:

$$\sum_{k=1}^{4} f(c_k) \Delta x - \frac{3}{4} + \frac{15}{8} + 3 + \frac{33}{8} = \frac{39}{4} = \boxed{9.75}$$

<u>Circumscribed rectangles.</u>   The procedure here is the same as for inscribed rectangles

except that maximum values for $f(x)$ over the subintervals are used in place of minimum

values.

<u>Step 1.</u> $\Delta x = \dfrac{b - a}{n} = \dfrac{5 - 2}{4} = \dfrac{3}{4}$

<u>Step 2.</u> $x_1 = 2 + \Delta x = 2 + 3/4 = 11/4$

$x_2 = 2 + 2\Delta x = 2 + 2(3/4) = 7/2$

$x_3 = 2 + 3\Delta x = 2 + 3(3/4) = 17/4$

<u>Step 3.</u>   Compute d_1, d_2, d_3, d_4,   the points in

our four subintervals where   f   assumes maximum

values. Since   f   is increasing, the maximum value

tor   f(x)   on each subinterval occurs at tho right

endpoint, as shown in the diagram to the right. Thus

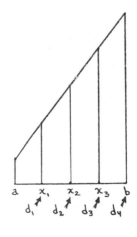

on $[a, x_1]$, $d_1 = x_1 = 11/4$

on $[x_1, x_2]$, $d_2 = x_2 = 7/2$

on $[x_2, x_3]$, $d_3 = x_3 = 17/4$

on $[x_3, b]$, $d_4 = b = 5$

5. 6. 6

<u>Step 4.</u> Compute $f(d_k) \Delta x$ $(k = 1, 2, 3, 4)$,

the areas of the circumscribed rectangles:

$$f(d_1) \Delta x = (2d_1 - 3) \Delta x$$
$$= (2(\tfrac{11}{4}) - 3)(\tfrac{3}{4}) = \tfrac{15}{8}$$

$$f(d_2) \Delta x = (2d_2 - 3) \Delta x$$
$$= (2(\tfrac{7}{2}) - 3)(\tfrac{3}{4}) = 3$$

$$f(d_3) \Delta x = (2d_3 - 3) \Delta x$$
$$= (2(\tfrac{17}{4}) - 3)(\tfrac{3}{4}) = \tfrac{33}{8}$$

$$f(d_4) \Delta x = (2d_4 - 3) \Delta x$$
$$= (2(5) - 3)(\tfrac{3}{4}) = \tfrac{21}{4}$$

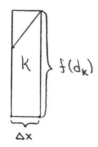

<u>Step 5.</u> Compute $\displaystyle\sum_{k=1}^{4} f(d_k) \Delta x$, the sum of the areas of the

circumscribed rectangles:

$$\sum_{k=1}^{4} f(d_k) \Delta x = \tfrac{15}{8} + 3 + \tfrac{33}{8} + \tfrac{21}{4} = \tfrac{57}{4} = \boxed{14.25}$$

Let A be the actual area under $y = 2x - 3$ above the interval $[2, 5]$. This area must

be greater than the area of the inscribed rectangles (9.75), while less than the area of the

circumscribed rectangles (14.25). Thus

$$9.75 < A < 14.25$$

[Note that in this case the exact area A is easily computed to

be $A = A_1 + A_2 = 9 + 3 = 12$ (see the figure) using the

basic area formulas $A_1 = \tfrac{1}{2} bh = \tfrac{1}{2}(3)(6) = 9$ and $A_2 = lw = (3)(1) = 3.$] $\square$

4. <u>Bookkeeping with an increasing number of subintervals.</u>   If there is an elusive part of the

concept of approximating area by rectangles, it is the notion that increasing the number of

rectangles without limit gives a better and better approximation of the area.   To see that this

is so, let us return to Example A.

<u>Example A (revisited).</u>   Approximate the area under $y = 2x - 3$ above the interval $[2, 5]$ by

dividing $[2, 5]$ into (i) 6 and (ii) 12 subintervals of equal length and computing the

sums of the areas of the inscribed and circumscribed rectangles.

<u>Solution.</u>

<u>Inscribed rectangles with</u> $n = 6$.   We follow the same

procedures as in Example A :

<u>Step 1.</u> $\Delta x = \dfrac{b - a}{n} = \dfrac{5 - 2}{6} = \dfrac{1}{2}$

<u>Step 2.</u> $x_1 = 2 + \Delta x = 2 + (1/2) = 5/2$

$x_2 = 2 + 2\Delta x = 2 + 2(1/2) = 3$

$x_3 = 2 + 3\Delta x = 2 + 3(1/2) = 7/2$

$x_4 = 2 + 4\Delta x = 2 + 4(1/2) = 4$

$x_5 = 2 + 5\Delta x = 2 + 5(1/2) = 9/2$

<u>Step 3.</u>   Since f is increasing we have:

$$c_1 = a = 2 \qquad\qquad c_4 = x_3 = 7/2$$

$$c_2 = x_1 = 5/2 \qquad\qquad c_5 = x_4 = 4$$

$$c_3 = x_2 = 3 \qquad\qquad c_6 = x_5 = 9/2$$

5.6.8

Step 4.
$$f(c_1)\Delta x = (2c_1 - 3)\Delta x = (4 - 3)(1/2) = 1/2$$
$$f(c_2)\Delta x = (2c_2 - 3)\Delta x = (5 - 3)(1/2) = 1$$
$$f(c_3)\Delta x = (2c_3 - 3)\Delta x = (6 - 3)(1/2) = 3/2$$
$$f(c_4)\Delta x = (2c_4 - 3)\Delta x = (7 - 3)(1/2) = 2$$
$$f(c_5)\Delta x = (2c_5 - 3)\Delta x = (8 - 3)(1/2) = 5/2$$
$$f(c_6)\Delta x = (2c_6 - 3)\Delta x = (9 - 3)(1/2) = 3$$

Step 5.
$$\sum_{k=1}^{6} f(c_k)\Delta x = \frac{1}{2} + 1 + \frac{3}{2} + 2 + \frac{5}{2} + 3 = \frac{21}{2} = \boxed{10.5}$$

By now these mechanical computations must be boring you. We'll spare you the details of the remaining three calculations and just give the answers:

Circumscribed rectangles with n = 6.   Area = $\boxed{13.5}$

Inscribed rectangles with n = 12.   Area = $\boxed{11.25}$

Circumscribed rectangles with n = 12.   Area = $\boxed{12.75}$ □

If A is the actual area under y = 2x - 3 above the interval [2, 5], we have obtained the following approximations of A in our three passes at Example A :

number of subintervals	approximation of A
4	9.75 < A < 14.25
6	10.5 < A < 13.5
12	11.25 < A < 12.75

[Recall that we computed A = 12 at the end of our first solution to Example A .] Observe that as the number of subintervals increases, the value of the inscribed sum increases and the value of the circumscribed sum decreases, so that each is successively a better approx-

imation of A. Put another way, it is always true that

$$\text{inscribed sum} \le A \le \text{circumscribed sum}$$

It should not be too hard to believe that if we continue this process indefinitely (that is, if we

increase the number of subintervals without limit), our approximations will approach

perfection. That is, the inscribed sum and the circumscribed sum will approach a common

value, and that common value will be the true value of A. We can actually <u>prove</u> that this

happens.

<u>Example A (re-revisited)</u>. Use circumscribed rectangles to find the area under the curve

y = 2x - 3 over the interval [2, 5].

<u>Solution.</u> For each value of n, we must divide [2, 5] into n equal pieces and then

compute $\displaystyle\sum_{k=1}^{n} f(d_k) \Delta x$. To do this in an organized fashion we can proceed just as we did

for n = 4 and n = 6. However, we will not deal with each subinterval separately, but will

just deal with a "general" subinterval $[x_{k-1}, x_k]$. This is done below. Remember: <u>this is the</u>

<u>same procedure that we used earlier (with one additional step), but in a more abstract setting.</u>

We first draw a rough graph of the

function f(x) = 2x - 3 over the interval

[2, 5]. The major piece of information to

note is that f is increasing on this interval;

hence, over any subinterval, f will assume

its maximum value at the right endpoint.

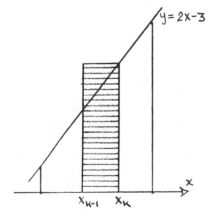

Step 1. Compute Δx, the length of each subinterval:

$$\Delta x = \frac{b - a}{n} = \frac{5 - 2}{n} = \frac{3}{n}$$

Step 2. Compute $x_1, x_2, \dots, x_{n-1}$, the points

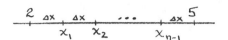

of subdivision for $[2, 5]$. Our formula

will be for the general "k-th" subdivision point:

$$x_k = a + k\Delta x = 2 + 3k/n$$

> since x_k is a distance
> of $k\Delta x$ to the right of $a = 2$

Step 3. Compute $d_1, d_2, \dots, d_n$, the points where f assumes

maximum values in the n subintervals

$$[a, x_1], [x_1, x_2], \dots [x_{k-1}, x_k], \dots [x_{n-1}, b]$$

As we observed before Step 1, since f is increasing it will assume maximum

values at right endpoints of intervals. Hence,

$$\text{on } [x_{k-1}, x_k], \quad d_k = x_k = 2 + 3k/n$$

Step 4. Compute $f(d_k)\Delta x$ $(k = 1, \dots, n)$,

the area of the circumscribed rectangle over

the k-th subinterval $[x_{k-1}, x_k]$:

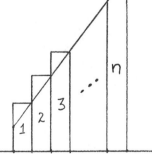

$$f(d_k)\Delta x = [\quad 2d_k \quad - 3]\Delta x$$

$$= [2(2 + 3k/n) - 3](3/n)$$

$$= [1 + 6k/n](3/n)$$

$$= \frac{3}{n} + \left(\frac{18}{n^2}\right)k$$

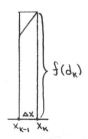

Step 5. Compute $\displaystyle\sum_{k=1}^{n} f(d_k)\,\Delta x$, the sum of the areas of the circumscribed rectangles.

This means adding up the areas computed in Step 4.

$$\sum_{k=1}^{n} f(d_k)\,\Delta x = \sum_{k=1}^{n} \left[\frac{3}{n} + \left(\frac{18}{n^2}\right) k \right] = ?$$

Now things become more interesting ... and you see why we studied the $\displaystyle\sum$ notation in the previous section! We proceed as follows:

$$\sum_{k=1}^{n} \left[\frac{3}{n} + \left(\frac{18}{n^2}\right) k \right] = \sum_{k=1}^{n} \left(\frac{3}{n}\right) + \sum_{k=1}^{n} \left(\frac{18}{n^2}\right) k$$

We will evaluate each of the two sums which appear on the right separately. DO NOT CONFUSE THE ROLES OF n AND k IN THESE SUMMATIONS: the n is a fixed positive integer, while k varies from 1 to n.

The first sum:

$$\sum_{k=1}^{n} \left(\frac{3}{n}\right) = \frac{3}{n} + \frac{3}{n} + \cdots + \frac{3}{n} \quad \text{(n times)}$$

$$= n\left(\frac{3}{n}\right) = 3$$

The second sum:

$$\sum_{k=1}^{n} \left(\frac{18}{n^2}\right) k = \left(\frac{18}{n^2}\right) \sum_{k=1}^{n} k$$

—— by Theorem 5. 5. 1 c: multiplicative constants move through summations. (Remember: n is constant while k varies!)

$$= \left(\frac{18}{n^2}\right)\left(\frac{n(n+1)}{2}\right) \qquad \text{Theorem 5. 5. 2 a}$$

$$= 9\left(1 + \frac{1}{n}\right)$$

Thus, adding our two sums together yields

$$\sum_{k=1}^{n} f(d_k) \Delta x = 3 + 9 \left(1 + \frac{1}{n}\right)$$

$$= \boxed{12 + \frac{9}{n}}$$

Step 6. Compute the limit as n goes to infinity of the answer obtained in Step 5. This answer will be the area A under the curve $y = 2x - 3$ over the interval $[2, 5]$.

$$A = \lim_{n \to \infty} \sum_{k=1}^{n} f(d_k) \Delta x$$

$$= \lim_{n \to \infty} \left(12 + \frac{9}{n}\right) \qquad \text{from Step 5}$$

$$= 12 + 0 = \boxed{12}$$

As you can see, the limit of circumscribed rectangles does give the actual area of 12. If we use inscribed rectangles the following is what would result (try it yourself!):

Step 5.
$$\sum_{k=1}^{n} f(c_k) \Delta x = \sum_{k=1}^{n} \left[\left(\frac{3}{n} - \frac{18}{n^2}\right) + \left(\frac{18}{n^2}\right) k \right]$$

$$= 12 - 9/n$$

Step 6.
$$\lim_{n \to \infty} \sum_{k=1}^{n} f(c_k) \Delta x = \lim_{n \to \infty} \left(12 - \frac{9}{n}\right) = \boxed{12}$$

As expected, the limit of inscribed rectangles also gives the actual area of 12. □

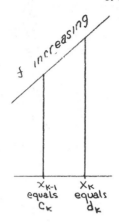

<u>Warning</u>: Since $f(x) - 2x - 3$ is an increasing function,

in Example A we had an easy time picking the points

c_k and d_k (the values in $[x_{k-1}, x_k]$ which

give the minimum and maximum values for f ,

respectively): the c_k were the left endpoints

and the d_k were the right endpoints, i. e. ,

f increasing	$c_k = x_{k-1}$ and $d_k = x_k$

THIS IS NOT ALWAYS THE CASE!! Suppose f is a

decreasing function; then the c_k will be the

right endpoints while the d_k will be the

left endpoints, i. e. ,

f decreasing	$c_k = x_k$ and $d_k = x_{k-1}$

Worse yet, if f is neither increasing nor decreasing

on $[x_{k-1}, x_k]$, then c_k and d_k could be

<u>anywhere</u> in the interval!

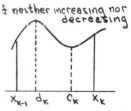

Fortunately Anton has taken pity on us: his exercises

use only functions which are strictly increasing or strictly decreasing on the appropriate

intervals. As we pointed out above, however, <u>you do need to distinguish between the increasing</u>

<u>and decreasing cases!</u>

Section 5.7.   The Definite Integral

This is perhaps the most theoretical section of the entire book. And it is one of the most important, for this is the section in which the second of the two basic calculus operations, (definite) integration, is defined.

1.   Riemann Sums: Discussion and Examples.   Riemann Sums are obviously quite similar to the inscribed and circumscribed sums of Section 5.6. However, if we compare

$$\sum_{k=1}^{n} f(c_k) \, \Delta x \qquad \text{and} \qquad \sum_{k=1}^{n} f(d_k) \, \Delta x \qquad \text{to} \qquad \sum_{k=1}^{n} f(x_k^*) \, \Delta x_k$$

(inscribed sum)   (circumscribed sum)   (Riemann sum)

there are two significant differences:

Significant Difference I.   In the inscribed and circumscribed sums, $[a, b]$ is subdivided into n subintervals of equal width, but in the Riemann Sum, that requirement is dropped. That is, the n subintervals are not necessarily of equal width in the Riemann Sum. This explains the change from "Δx" to "Δx_k" in the notation: Δx represents the (common) width of all subintervals in the inscribed/circumscribed sums while $\Delta x_1, \Delta x_2, \ldots, \Delta x_n$ denote the (possibly different) widths of the 1-st, 2-nd, ... , n-th subintervals in the Riemann Sums. *

Significant Difference II.   In the inscribed and circumscribed sums, specific points c_k and d_k in each subinterval are used. (They were chosen so that $f(c_k)$ and $f(d_k)$ are minimum

---

* It also explains why the symbol $\displaystyle \lim_{\max \Delta x_k \to 0}$ must be used instead of the simpler

$\displaystyle \lim_{\Delta x \to 0}$. In order to make the widths of all the subintervals approach zero, we require that

the width of the largest one $(\max \Delta x_k)$ approach zero.

and maximum values of f, respectively, on the subinterval.) In the Riemann Sum, <u>an</u>

<u>arbitrary point</u> x_k^* <u>is chosen in each subinterval.</u> This freedom to choose x_k^* makes

the computation of specific Riemann Sums easier than the computation of inscribed or

circumscribed sums. Here is an example:

<u>Example A.</u>   Find the value of the Riemann Sum

$$\sum_{k=1}^{n} f(x_k^*) \Delta x_k$$

for $f(x) = x^3 - 3x + 2$ on the interval $[-2, 2]$ using the partition points $x_k = -1, -1/2,$

$1/2, 3/2$ and the intermediate evaluation points $x_k^* = -1, -9/10, 0, 5/6$.

<u>Solution.</u>   Although a bit messy, this is really an easy problem.

<u>Step 1.</u>   Compute the values Δx_k, $k = 1, \ldots, n$. The

quantity Δx_k is simply the length of the subinterval

$[x_{k-1}, x_k]$. Thus $\Delta x_k = x_k - x_{k-1}$, which gives us:

$$\Delta x_1 = x_1 - a = -1 \quad - (-2) \quad = 1$$

$$\Delta x_2 = x_2 - x_1 = -1/2 - (-1) \quad = 1/2$$

$$\Delta x_3 = x_3 - x_2 = \quad 1/2 - (-1/2) = 1$$

$$\Delta x_4 = x_4 - x_3 = \quad 3/2 - \quad 1/2 \quad = 1$$

$$\Delta x_5 = b - x_4 = \quad 2 - \quad 3/2 \quad = 1/2$$

Step 2. Compute the sum $\displaystyle\sum_{k=1}^{n} f(x_k^*)\,\Delta x_k$. We first compute the individual terms:

$$f(x_1^*)\,\Delta x_1 = [\,(x_1^*)^3 \quad - 3(x_1^*) \quad + 2\,]\Delta x_1$$
$$= [\,(-1)^3 \quad - 3(-1) \quad + 2\,](1) \quad = 4$$

$$f(x_2^*)\,\Delta x_2 = [\,(x_2^*)^3 \quad - 3(x_2^*) \quad + 2\,]\Delta x_2$$
$$= [\,(-9/10)^3 - 3(-9/10) + 2\,](1/2) = 1.9855$$

$$f(x_3^*)\,\Delta x_3 = [\,(x_3^*)^3 \quad - 3(x_3^*) \quad + 2\,]\Delta x_3$$
$$= [\,(0)^3 \quad - 3(0) \quad + 2\,](1) \quad = 2$$

$$f(x_4^*)\,\Delta x_4 = [\,(x_4^*)^3 \quad - 3(x_4^*) \quad + 2\,]\Delta x_4$$
$$= [\,(5/6)^3 \quad - 3(5/6) \quad + 2\,](1) \quad \approx .0787$$

$$f(x_5^*)\,\Delta x_5 = [\,(x_5^*)^3 \quad - 3(x_5^*) \quad + 2\,]\Delta x_5$$
$$= [\,(2)^3 \quad - 3(2) \quad + 2\,](1/2) = 2$$

Thus $\displaystyle\sum_{k=1}^{5} f(x_k^*)\,\Delta x_k \approx 4 \;+\; 1.9855 \;+\; 2 \;+\; .0787 \;+\; 2 \;=\; \boxed{10.0642}$ □

Comparing this example with the examples in Section 5.6 should convince you that computing Riemann Sums is generally easier than computing inscribed/circumscribed sums.

2. The Definite Integral: Definition. Here is a diagram summarizing the definition of the definite integral (Definition 5.7.2):

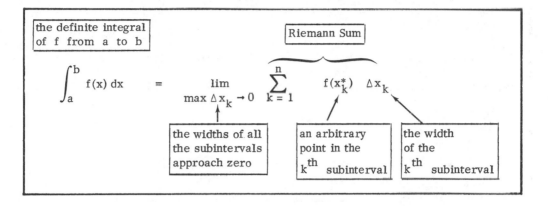

It is far more important to understand the concepts used in this definition than it is to memorize word-for-word Anton's three-step build-up to it. Here is a translation of those three steps into less technical language:

<u>Step 1.</u>   Break the interval $[a, b]$ up into n pieces (where n is some number). These pieces do not have to be the same size.

<u>Step 2.</u>   Approximate the area above each subinterval by picking a (any) point in the subinterval and computing the area of the rectangle whose height is the function value of that point and whose width is the width of the subinterval. Do this for each subinterval and add up the areas of all the approximating rectangles to get an approximation of the area above the interval $[a, b]$.

In Steps 1 and 2, a Riemann Sum $\displaystyle\sum_{k=1}^{n} f(x_k^*)\, \Delta x_k$ has been formed (using essentially the same two steps that were used in Example A) as an approximation to the actual area.

Step 3.   Make this approximation approach the actual area by taking the limit as

n (the number of subintervals) goes to $+\infty$ in such a way that the width of each sub-

interval approaches zero.   Then ask this question: do all the Riemann Sums formed in

Steps 1 and 2 have the same (finite) number as a limit? IF SO, then this number is

denoted by

$$\int_a^b f(x)\, dx \;=\; \lim_{\max \Delta x_k \to 0} \sum_{k=0}^{n} f(x_k^*)\, \Delta x_k$$

and is called the definite integral of f from a to b.

3.   The Definite Integral: Computation.   While calculating a specific Riemann Sum (Steps 1 and 2)

is rather easy, taking the limit of such sums (Step 3) just from the definition is essentially

impossible!

Example B.   Evaluate $\displaystyle\int_1^2 x^2\, dx$.

Frustrating non-solution (using only Definition 5. 7. 2).

The definition says

$$\int_1^2 x^2\, dx \;=\; \lim_{\max \Delta x_k \to 0} \sum_{k=1}^{n} (x_k^*)^2\, \Delta x_k \;=$$

$$=\; \lim_{\max \Delta x_k \to 0} \left[(x_1^*)^2\, \Delta x_1 + (x_2^*)^2\, \Delta x_2 + \cdots + (x_n^*)^2\, \Delta x_n \right]$$

What do we do with this? The only alternative we have at this point is to calculate like mad !

That is, we would choose a value of n , say n = 6 , 6 partition points $x_1, x_2, \ldots, x_6$

and 6 intermediate evaluation points $x_1^*, x_2^*, \ldots, x_6^*$, and compute the Riemann Sum

$\sum_{k=1}^{6} (x_k^*)^2 \, \Delta x_k.$ Then we would choose a larger value of n, say $n = 12$, and repeat

the process, computing $\sum_{k=1}^{12} (x_k)^2 \, \Delta x_k$. Then we would choose a still larger value of n

and compute the Riemann Sum, and so on. If, after ten or so of these calculations, it appears

that the values of these Riemann Sums are "closing in" on some number, we would guess that

that number is the value of $\displaystyle\int_1^2 x^2 \, dx$. However, if we are not certain as to whether the

Riemann Sum values are approaching $\frac{7}{3}$ as opposed to $\frac{48}{21}$, say, we would have to make

another 10 (or 110 or 1,010!) Riemann Sum calculations and then guess again. Even then

we can't be sure our guess is correct. And this is a very simple integral! Oh dear!!   □

THERE MUST BE A BETTER WAY TO COMPUTE DEFINITE INTEGRALS! Fortunately,

there is - by using antiderivatives - as you will see in the next section. However, until

you get to that section, you do not have any effective way to evaluate definite integrals!

4.   The Importance of Riemann Sums. All this leaves us with some nagging questions. If Definition

5.7.2 is not useful tor calculating the value of the definite integral, why make it? Why not use

a definition which is more useful for compuations?

        The answer to this question is vitally important: many quantities important in both

the physical and social sciences can be defined only as limits of Riemann Sums and hence as

definite integrals. This is the importance of our "uncomputable" Definition 5.7.2 for the

definite integral: it provides the link between "The Area Problem" and many seemingly

unrelated real-world problems.

In subsequent chapters we will develop many of these applications of integration. However, if you want to see such an application now, the next (optional) example describes the use of integration with the physics concept of work:

Example C.   Work (Optional)

Suppose Howard Anton exerts a constant force   F   on a concrete block and pushes it a distance   d   along the   x-axis ,   from   x = a   to   x = b.   The work   W   Howard has done is defined to be:

$$W \;=\; F \cdot d \qquad \text{(work equals force times distance)}* \tag{A}$$

[Common sense says that this definition is reasonable: the greater the weight of the block and/or the greater the distance pushed, the harder you "work."]

Now suppose that different parts of the   x-axis   offer differing resistance to the block (because there are oil slicks, patches of gravel, etc.).   Then at each   x-value   Howard will have to exert a different force, say $F(x)$ (i. e., the force will no longer be constant, but will vary with   x).   Now how much work does Howard do in moving the block from   x = a   to   x = b ?

We attack this problem as follows: Break up the interval   [a, b]   into   n   small pieces $[x_{k-1}, x_k]$ and approximate Howard's work ΔW_k on each subinterval by

---

* This definition assumes that the direction of the force is along the   x-axis.

$$\Delta W_k \approx F(x_k^*) \Delta x_k \qquad \text{(B)}$$

where Δx_k is the width of the subinterval $[x_{k-1}, x_k]$ and x_k^* is an arbitrary point in it. [Here we are approximating the work done by a <u>variable</u> force $F(x)$, x in $[x_{k-1}, x_k]$, by the work done by a <u>constant</u> force $F(x_k^*)$ over $[x_{k-1}, x_k]$. By (A), the latter is $F(x_k^*) \Delta x_k$. This approximation should be pretty good if the width Δx_k of $[x_{k-1}, x_k]$ is small.] Now Howard's hard work W going all the way from $x = a$ to $x = b$ is the sum of all the ΔW_k:

$$W = \sum_{k=1}^{n} \Delta W_k \approx \sum_{k=1}^{n} F(x_k^*) \cdot \Delta x_k \qquad \text{(C)}$$

(Ah ha! A Riemann Sum!) Taking smaller and smaller subintervals will make (C) a better and better approximation. In fact, if we make all the subinterval widths approach zero, (B) and hence (C) will become "perfect approximations," i. e.

$$W = \lim_{\max \Delta x_k \to 0} \sum_{k=1}^{n} F(x_k^*) \Delta x_k = \int_a^b F(x) \, dx$$

(the last equality is Definition 5. 7. 2). Oh ho! <u>Work done by a variable force over an interval is the definite integral of the force.</u> [Work will be discussed at length in Section 6. 6.]   □

Example C illustrates two important points:

(i) Integrals represent more than just areas.
(ii) Integrals arise in applications via Riemann Sums.

5. **Integrability.** As Anton points out following his definition of the definite integral, not every function is **integrable**, i.e., there exist many functions f for which the limit

$$\lim_{\max \Delta x_k \to 0} \sum_{k=1}^{n} f(x_k^*) \, \Delta x_k$$

does not exist and hence the definite integral for f from a to b

$$\int_a^b f(x) \, dx$$

does not exist.

A natural and important question is: "What functions are integrable?" Although the full answer to this question is too complex to deal with at this level of calculus, Anton does provide us with a big collection of functions which are known to be integrable:

> Continuous functions on [a, b] are integrable on [a, b]

(Theorem 5.7.3 a). Hence, whenever f is a continuous function on [a, b], we know that $\int_a^b f(x) \, dx$ exists. (This does NOT, however, guarantee that we can compute the value of the integral!)

Although the above result is generally sufficient for our needs, occasionally a function such as that shown to the right will crop up. It is continuous everywhere on the interval [a, b] except at the point $x = x_0$. Are such functions integrable? The answer is "yes" ... assuming the function is **bounded** on the interval [a, b]:

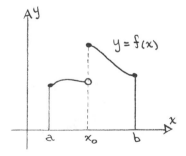

> Suppose f   has only finitely many points of
>
> discontinuity on the interval [a, b], and
>
> there is a number M   such that
>
> $\quad$ - M $\leq$ f(x) $\leq$ M   for all x in [a, b].
>
> Then f is integrable on [a, b].

(Theorem 5. 7. 3 b).   The existence of the number M is what makes f "bounded": its

values on the interval [a, b] cannot run off to

positive or negative infinity   Boundedness happens

to be a necessity for an integrable function; thus

the function shown to the right is <u>not</u> integrable

on [a, b].

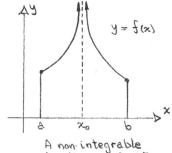

A non-integrable
function on [a, b]

6.   <u>Areas from integrals.</u>   Anton ends this section with a few remarks on how a general definite

integral is interpreted in terms of area:

$$\int_a^b f(x)\, dx \;=\; \left[\begin{array}{c} \text{area above} \\ \text{x-axis} \end{array}\right] \;-\; \left[\begin{array}{c} \text{area below} \\ \text{x-axis} \end{array}\right]$$

The "area above the x-axis" refers to

the area A_1 obtained when f is positive

(i. e. , above the x-axis). The "area below

the x-axis" refers to the area A_2 obtained

when f is negative (i. e., below the x-axis).

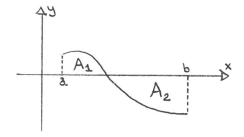

<u>Be careful of the minus sign that appears in the above formula!</u> The integral is a difference of

the areas above and below the x-axis, not a sum of those areas... and this can lead to errors

Example D. Find the area between the curve $y = x^3 - x^2 - 2x$ and the x-axis.

Solution. If we graph $y = x^3 - x^2 - 2x$, we see that the curve intersects the x-axis at $x = -1, 0$ and 2 (simply solve $0 = x^3 - x^2 - 2x = x(x+1)(x-2)$), and that the area we desire is $A_1 + A_2$. Now if we were careless, we might write

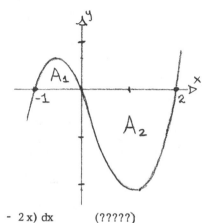

$$\begin{bmatrix} \text{area between} \\ \text{curve and} \\ \text{x-axis} \end{bmatrix} = \int_{-1}^{2} (x^3 - x^2 - 2x)\, dx \qquad (?????)$$

WRONG! This will be the difference of areas

$$A_1 - A_2$$

So how do we compute the sum $A_1 + A_2$? Simple! We compute the areas above and below the x-axis separately. That is, we compute A_1 and A_2 separately by the integrals

$$A_1 = \int_{-1}^{0} (x^3 - x^2 - 2x)\, dx \qquad \text{and} \qquad A_2 = -\int_{0}^{2} (x^3 - x^2 - 2x)\, dx$$

Why, the negative sign on the A_2 integral? Well, on $[0,2]$ the function $x^3 - x^2 - 2x$ is always negative, so there is no "area above x-axis." Thus

$$\int_{0}^{2} (x^3 - x^2 - 2x)\, dx = - \begin{bmatrix} \text{area below} \\ \text{x-axis} \end{bmatrix} = -A_2$$

Just multiply by -1 to get our formula for A_2.

Using techniques to be developed in the next section one can show

$$A_1 = \int_{-1}^{0} (x^3 - x^2 - 2x)\,dx = \frac{5}{12}$$

$$A_2 = -\int_{0}^{2} (x^3 - x^2 - 2x)\,dx = \frac{8}{3}$$

Thus the sum of our two areas $A_1 + A_2$ is

$$\begin{bmatrix} \text{area between} \\ \text{curve and} \\ \text{x-axis} \end{bmatrix} = A_1 + A_2 = \frac{5}{12} + \frac{8}{3} = \frac{37}{12} \qquad \square$$

Section 5.8.   The First Fundamental Theorem of Calculus

1.   Calculating definite integrals by the 1st FTC.   In the previous section we DEFINED the definite integral.   In this section we give you the tool necessary to COMPUTE definite integrals: the modestly-named First Fundamental Theorem of Calculus (1st FTC).

The
First
FTC

> If f is continuous on $[a,b]$, then
>
> $$\int_{a}^{b} f(x)\,dx = F(b) - F(a)$$
>
> where F is an antiderivative of f on $[a,b]$.

The 1st FTC gives an easy way of computing definite integrals (provided you can find antiderivatives).   This is one reason it is "fundamental."   It also establishes a profound relationship between the two basic calculus operations of definite integration and differentiation. This is another reason it is "fundamental."

Example A.   Evaluate $\displaystyle\int_0^{\pi/4}$ sin x cos x dx .

Solution 1.   Since $\dfrac{d}{dx}(\sin^2 x) = 2\sin x \cos x$, we know that $F(x) = \dfrac{1}{2}\sin^2 x$ is an anti-

derivative of $f(x) = \sin x \cos x$. Thus

$$\int_0^{\pi/4} \sin x \cos x \, dx = F\left(\frac{\pi}{4}\right) - F(0) \qquad (\text{1st FTC}*)$$

$$= \frac{1}{2}\sin^2\left(\frac{\pi}{4}\right) - \frac{1}{2}\sin^2(0)$$

$$= \frac{1}{2}\left(\frac{1}{\sqrt{2}}\right)^2 - \frac{1}{2}(0) = \frac{1}{4} \qquad \square$$

Notice that <u>any antiderivative of f will do in the 1st FTC</u>. We illustrate this by

returning to the example above:

Example A (revisited).   Evaluate $\displaystyle\int_0^{\pi/4}$ sin x cos x dx .

Solution 2.   If $G(x) = \dfrac{1}{2}\sin^2 x + 7$, then $G'(x) = \sin x \cos x$, so G is an anti-

derivative of $f(x) = \sin x \cos x$. Thus

$$\int_0^{\pi/4} \sin x \cos x \, dx = G\left(\frac{\pi}{4}\right) - G(0) \qquad (\text{1st FTC})$$

$$= \left(\frac{1}{2}\sin^2\left(\frac{\pi}{4}\right) + 7\right) - \left(\frac{1}{2}\sin^2(0) + 7\right)$$

$$= \frac{1}{2}\sin^2\left(\frac{\pi}{4}\right) - \frac{1}{2}\sin^2(0) = \frac{1}{4}$$

---

* Note that in the 1st FTC the subtraction is of the form

     F (<u>top</u> limit) - F (<u>bottom</u> limit)

<u>Solution 3.</u>   If $W(x) = -\dfrac{1}{2}\cos^2 x$,   then $W'(x) = \sin x \cos x$,   so W is an anti-derivative of f.   Thus

$$\int_0^{\pi/4} \sin x \cos x \, dx = W\left(\frac{\pi}{4}\right) - W(0) \qquad \text{(1st FTC)}$$

$$= -\frac{1}{2}\cos^2\left(\frac{\pi}{4}\right) + \frac{1}{2}\cos^2(0)$$

$$= -\frac{1}{2}\left(\frac{1}{\sqrt{2}}\right)^2 + \frac{1}{2}(1)^2 = -\frac{1}{4} + \frac{1}{2} = \frac{1}{4} \qquad \Box$$

There is no mystery about why <u>any</u> antiderivative works in the 1st FTC. Recall that <u>any two antiderivatives of   f   differ only by a constant</u> (Theorem 5.2.2).* Thus, if $F(x)$ and $G(x)$ are any two antiderivatives of f,   there exists a constant C such that $G(x) = F(x) + C$.   Hence

$$G(b) - G(a) = [F(b) + C] - [F(a) + C] = F(b) - F(a)$$

This shows that the right hand side of the equation in the 1st FTC is the same regardless of which antiderivative is used.

Since any antiderivative will work, you might as well choose one which is convenient. Usually, but not always, this is one with constant term zero (or with "no constant term," if you prefer that language). That is, in Example A, Solution 1 (or Solution 3, depending on which occurs to you first) is slightly shorter than Solution 2 because there is no constant term to contend with.

---

*
  That's certainly clear for the antiderivatives used in the first two solutions to Example A above; it applies to the third antiderivative as well if we use the trigonometric identity $\sin^2 x + \cos^2 x = 1$ to see

$$W(x) = -\frac{1}{2}\cos^2 x = \frac{1}{2}\sin^2 x - \frac{1}{2} = F(x) - \frac{1}{2}$$

2. <u>Some theoretical observations on the 1st FTC. (Optional)</u> The statement of the 1st FTC applies to any continuous function f <u>which has an antiderivative</u> F. Does every continuous function on [a , b] have an antiderivative? The answer is "yes" (assuming that f is continuous on some open interval I containing [a , b]), as we will show in §5.10 using the <u>2nd</u> FTC.

    As for the proof of the 1st FTC: it takes only a page in Anton's text... but it depends on some difficult results discussed previously:

    1.   The Mean Value Theorem   (Theorem 4.10.2)

    2.   The integrability of continuous functions   (Theorem 5.7.3)

    3.   The Extreme Value Theorem   (Theorem 4.6.4)
            (The MVT -- and hence the 1st FTC --
            depends on this result)

    In fact, the proofs of these last two results are not given in Anton because they are too advanced for an introductory calculus course. These observations should impress on you that the 1st FTC is a deep as well as powerful result.

3. <u>Algebraic properties of the definite integral.</u>   The 1st FTC allows us to carry over algebraic properties of indefinite integrals (antiderivatives) to definite integrals. The simplest of these are gathered in Theorem 5.8.2:

THEOREM 5.8.2	(a) multiplicative constants move through definite integrals
	(b) and (c) the definite integral may be distributed over addition and subtraction

These can be used to advantage when evaluating complicated definite integrals as in the following example.

Example B.   Evaluate $\displaystyle\int_0^{\pi} (3 \sin x - x^2)\, dx$.

Solution.

$$\int_0^{\pi} (3 \sin x - x^2)\, dx \;=\; \int_0^{\pi} 3 \sin x\, dx \;-\; \int_0^{\pi} x^2\, dx \qquad \text{(Theorem 5. 8. 2(c))}$$

$$= \; 3 \int_0^{\pi} \sin x\, dx \;-\; \int_0^{\pi} x^2\, dx \qquad \text{(Theorem 5. 8. 2(a))}$$

$$= \; 3 \left[-\cos (\pi) \;-\; (-\cos (0)) \right] \;-\; \left[\frac{(\pi)^3}{3} - \frac{(0)^3}{3} \right] \qquad \text{(1st FTC)}$$

$$= \; 3 \left[-(-1) \;-\; (-1) \right] \;-\; \frac{(\pi)^3}{3} \;=\; 6 - \frac{(\pi)^3}{3}$$

(The 1st FTC equation uses the facts that $-\cos x$ is an antiderivative of $\sin x$ and

that $\dfrac{x^3}{3}$ is an antiderivative of x^2 .)   □

There are also some important, but not surprising, results concerning inequalities

involving definite integrals:

THEOREM 5. 8. 3	

> (a) Suppose f is continuous and $f(x) \geq 0$ for all x in $[a, b]$.
>
> Then $\displaystyle\int_a^b f(x)\, dx \geq 0$
>
> (b) Suppose f and g are continuous, and $f(x) \geq g(x)$ for all x in $[a, b]$.
>
> Then $\displaystyle\int_a^b f(x)\, dx \geq \int_a^b g(x)\, dx$

When f is continuous and non-negative, then $\displaystyle\int_a^b f(x)\, dx$ represents the area under f

over [a, b] (Theorem 5. 7. 4); thus the first inequality in Theorem 5. 8. 3 merely observes that "if f is non-negative, the area under f over [a, b] is non-negative. " To interpret the second inequality in a similar way, it is easiest to assume both f and g are non-negative: in that case, the second inequality can be interpreted as "the higher the curve, the more area beneath it. "

These results can be extremely useful in estimating definite integrals, as the next example shows.

Example C.   Find upper and lower bounds for the definite integral

$$\int_0^{\pi/2} \cos^2 \sqrt{x} \ \sin x \ dx$$

Solution.   Don't bother looking for an antiderivative F for the function $f(x) = \cos^2 \sqrt{x} \ \sin x$ ... no "elementary" antiderivative exists for f. * Instead we observe that for x in the interval $[0, \pi/2]$ we have

$$0 \le \cos^2 \sqrt{x} \ \sin x \le \sin x$$

$$\llcorner \text{because} \ \cos^2 \sqrt{x} \le 1$$

Thus

$$0 \le \int_0^{\pi/2} \cos^2 \sqrt{x} \ \sin x \ dx \qquad \text{by Theorem 5. 8. 3(a)}$$

_____

* Antiderivatives for f do exist (this will be a consequence of the 2nd FTC in §5. 10); however, these antiderivatives cannot be written down in terms of functions we already know.

and $\displaystyle\int_0^{\pi/2} \cos^2 \sqrt{x}\ \sin x\ dx \leq \int_0^{\pi/2} \sin x\ dx$ by Theorem 5. 8. 3 (b)

$\qquad\qquad\qquad\qquad = [\ -\cos x\ ]_0^{\pi/2}$ by the 1st FTC since
$\qquad\qquad\qquad\qquad\qquad\qquad\qquad$ $F(x) = -\cos x$ is an
$\qquad\qquad\qquad\qquad\qquad\qquad\qquad$ antiderivative for $\sin x$

$\qquad\qquad\qquad\qquad = -\cos(\pi/2)\ -\ (-\cos 0)$

$\qquad\qquad\qquad\qquad = -0 + 1 = 1$

Summarizing,

$$0 \leq \int_0^{\pi/2} \cos^2 \sqrt{x}\ \sin x\ dx \leq 1 \qquad\qquad \square$$

Remark.   In later sections we will learn methods for approximating definite integrals to as high a degree of accuracy as we desire.

4. Rectilinear motion revisited: total distance traveled.   In Section 5.4 , we discussed the concepts of position, displacement, velocity, speed and acceleration in rectilinear motion (i. e. , motion on a straight line). This section adds the consideration of total distance traveled to this list.

The relationship between distance traveled and the other motion quantities can be remembered by the following diagram:

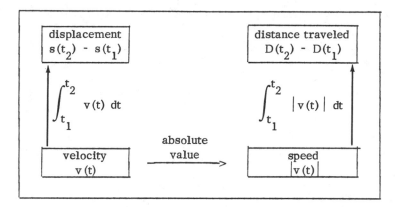

That is,

    - <u>displacement</u> from time t_1 to time t_2 is found by

        integrating the <u>velocity</u> function $v(t)$ from t_1 to t_2

    - <u>distance traveled</u> from time t_1 to time t_2 is found by

        integrating the <u>speed</u> function $|v(t)|$ from t_1 to t_2

Here are two examples:

<u>Example D.</u> Suppose $v(t) = \sqrt{2}\, \sin t - 1$ is the velocity (in meters/sec.) of a particle moving on a coordinate line. Find the displacement and the distance traveled by the particle during the time interval $\frac{\pi}{4} \le t \le \pi$.

<u>Solution.</u> The diagram reminds us that $\quad \text{displacement} = \displaystyle\int_{\pi/4}^{\pi} v(t)\, dt = \int_{\pi/4}^{\pi} (\sqrt{2}\, \sin t - 1)\, dt$,

and $\quad \text{distance traveled} = \displaystyle\int_{\pi/4}^{\pi} |v(t)|\, dt = \int_{\pi/4}^{\pi} |\sqrt{2}\, \sin t - 1|\, dt$. Evaluating the first of

these integrals, we get

$$\text{displacement} = \int_{\pi/4}^{\pi} (\sqrt{2} \sin t - 1) \, dt = (-\sqrt{2} \cos t - t) \Big|_{\pi/4}^{\pi}$$

$$= 1 + \sqrt{2} - \frac{3\pi}{4} = .15 \text{ meters}$$

To evaluate the second integral, we need to know where $v(t) = \sqrt{2} \sin t - 1$ is positive and

where it is negative, so we will use the graphing techniques of Chapter 4 to sketch its graph

on $[\frac{\pi}{4}, \pi]$. We obtain the following information

$$\left. \begin{cases} v(t) = \sqrt{2} \sin t - 1 \\ v'(t) = \sqrt{2} \cos t \\ v''(t) = -\sqrt{2} \sin t \end{cases} \right\}$$

so $v(t) = 0$ for $t = \frac{\pi}{4}$ and $t = \frac{3\pi}{4}$

$v'(t) = 0$ for $t = \frac{\pi}{2}$

$v'(t) > 0$ for $t < \frac{\pi}{2}$

$v'(t) < 0$ for $t > \frac{\pi}{2}$, and

$v''(\frac{\pi}{2}) < 0$

Hence the graph of $v(t) = \sqrt{2} \sin t - 1$, $\frac{\pi}{4} \leq t \leq \pi$, is

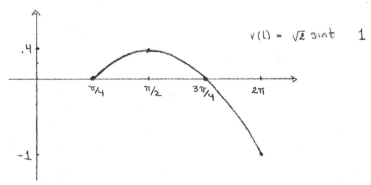

That is, $v(t) = \sqrt{2} \sin t - 1$ is positive for $\frac{\pi}{4} < t < \frac{3\pi}{4}$ and negative for $\frac{3\pi}{4} < t < \pi$.

Hence

5. 8. 10

$$\text{distance traveled} = \int_{\pi/4}^{\pi} \left| \sqrt{2} \ \sin t \ - \ 1 \right| dt$$

$$\stackrel{*}{=} \int_{\pi/4}^{\frac{3\pi}{4}} (\sqrt{2} \ \sin t \ - \ 1) \ dt \ + \ (- \ 1) \int_{\frac{3\pi}{4}}^{\pi} (\sqrt{2} \ \sin t \ - \ 1) \ dt$$

SEE FOOTNOTE

interval over which $v(t)$ is positive: $\left| v(t) \right| = v(t)$

interval over which $v(t)$ is negative: $\left| v(t) \right| = - \ v(t)$

$$= (- \sqrt{2} \ \cos t \ - \ t) \Big|_{\pi/4}^{\frac{3\pi}{4}} - (\sqrt{2} \ \cos t \ - \ t) \Big|_{\frac{3\pi}{4}}^{\pi}$$

$$= 1 + \sqrt{2} \ - \ \frac{\pi}{4} \ = \ 1.6 \ \text{meters}$$

These two answers say that the particle ends up only .15 meters   from where it started,
but that it has traveled a total of   1.6 meters   in the interim.   □

Example E.   Superman is jogging along a straight road. His position at time t is given by
$s(t) = t^2 - 12t$ where   s   is in miles and   t   is in seconds. How far away from his starting
point is Superman after he has jogged for 15 seconds? How far has he run during those
15 seconds?

Solution.   We can translate the problem into mathematics as follows:

---

*
The step $\int_{\pi/4}^{\pi} (\ldots) \ dt = \int_{\pi/4}^{3\pi/4} (\ldots) \ dt + \int_{3\pi/4}^{\pi} (\ldots) \ dt$ requires Theorem 5.9.2
of the next section. This Theorem is also used in our next example, and in Anton's
Example 5. In fact, rarely can you integrate the absolute value of a function without splitting
up the integral according to where the function is positive and where it is negative. See
Example 3 in §5.9.

Step 1: Translate the problem	

Superman's position function $\longrightarrow$ | $s(t) = t^2 - 12t$

... and his starting point $\longrightarrow$ | at $t = 0$, $s(0) = 0^2 - 12(0) = 0$

"How far away ... (after) 15 seconds ... ?" $\longrightarrow$ | what is displacement
    from $t = 0$ to $t = 15$?

"... how far has he run ... (in) 15 seconds?" $\longrightarrow$ | what is distance traveled
    from $t = 0$ to $t = 15$?

Step 2: Devise a plan	

We will i) compute $v(t)$ using $v(t) = s'(t)$ and then

        ii) use the diagram above to find displacement and distance traveled

Step 3: Execute the plan	

i)   Now $\qquad v(t) = s'(t) = \dfrac{d}{dt}(t^2 - 12t) = 2t - 12$

so from the diagram

ii) $\qquad \begin{bmatrix} \text{displacement from} \\ t = 0 \ \text{to} \ t = 15 \end{bmatrix} = \displaystyle\int_0^{15} v(t)\,dt = \int_0^{15} (2t - 12)\,dt$

$$= \left[t^2 - 12t \right]_0^{15} = 45$$

$\begin{bmatrix} \text{distance traveled from} \\ t = 0 \ \text{to} \ t = 15 \end{bmatrix} = \displaystyle\int_0^{15} |v(t)|\,dt = \int_0^{15} |2t - 12|\,dt$

Theorem 5. 9. 2	$\xrightarrow{\ =\ }$	$\displaystyle\int_0^6 |2t - 12|\,dt + \int_6^{15} |2t - 12|\,dt$

	interval over which $v(t) = 2t - 12$ is negative	interval over which $v(t) = 2t - 12$ is positive

$$= \int_0^6 (-2t + 12)\,dt + \int_6^{15} (2t - 12)\,dt$$

5. 8. 12

$$= \left[- t^2 + 12t \right]_0^6 + \left[t^2 - 12t \right]_6^{15}$$

$$= 117$$

Thus at the end of 15 seconds Superman is 45 miles from his starting point and has traveled 117 miles. Here is a "graph" of $s(t) = t^2 - 12t$ which shows that Superman jogs to the left for the first six seconds and then turns around and goes back to the right:

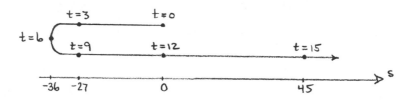

The following exercises may be done in addition to those of Exercise Set 5. 8 for additional practice.

Supplementary Exercises 5. 8

1.   Superman wants to show off for Lois Lane. He marks off a long straight line and asks her to sit on it. He starts at a position 2 miles to the right of Lois and, on her signal, runs along the line with velocity $v(t) = (1 - t)(2 - t)$ miles per second. At the end of 4 seconds, Lois blows a whistle and Superman stops immediately. Where is he (in relation to Lois) and how far has he traveled during those 4 seconds?

2.   Not to be outdone, Lois sits Superman on the same straight line and, starting 2 meters to the left of him, runs along the line with velocity $v(t) = t(1 - t)$ meters per second. To tease her, Superman does not blow his whistle after 4 seconds, but waits 10. Lois is exhausted and angry. Where is she (in relation to Superman) after 10 seconds? How far did she run? How much farther did she run than she would have run in 4 seconds?

3.   Superman and his brother are picking on Howard Anton again. They are playing catch with him by throwing him back and forth in a straight line from one to the other. Howard is getting a headache and, in order to persuade them to stop, he makes a few calcuations. He determines that his velocity is $v(t) = (t - 1)/(t + 1)^3$ meters per second. Then he tells the brothers that his mother says he can play until he has traveled a total of 100 meters, but then she wants him home for supper. How long will it take for Howard to travel 100 meters?

4.   Lois Lane is teaching Superman how to drive, but he is having trouble learning how to shift gears. He goes forward in first gear, but then, instead of shifting into second, he throws the car into reverse, backs (straight) up and then shifts into first again and lurches forward. Lois is getting irritated, not to mention getting whiplash, and she determines that she can take this punishment for only 3 seconds at a time. She calculates that, with Superman driving, the velocity of the car for the first 3 seconds is $v(t) - 3t^2 - 15t + 18$ feet per second. Find a)   the portion of the three seconds in which the car is moving forward   b)   the portion in which it is moving backwards and   c)   the total distance the car travels in three seconds.

5.   Superman is jogging along the equator with an acceleration $a(t) = \cos t - \sin t$ miles per second per second. His initial velocity is one mile per second and, for variety, he stops occasionally and reverses direction. His workout lasts only π seconds (yes, $\pi \approx 3.14$ seconds!). How much of this time is he going in the positive direction and what is the total distance he travels?

[Answers.   (1) $7\frac{1}{3}$ miles to the right; $5\frac{2}{3}$ miles.   (2) $285\frac{1}{3}$ meters to the left; $283\frac{2}{3}$ meters; 137 meters.   (3)   approximately .095 sec.   (4)   forward for first 2, backward for third sec.; 14.5 ft.   (5)   first $\frac{3}{4}\pi$ sec ; $2\sqrt{2}$ miles .]

Section 5.9. Properties of the Definite Integral; Average Value

1. More algebraic properties of the definite integral

This section begins with three important properties of the definite integral. Two are stated as definitions and the third as a theorem without proof. The three are

(1) $$\int_a^a f(x)\,dx = 0$$ If f is positive, then $\int_a^a f(x)\,dx$ is the area under f

above [a, a], and the area of

a line segment is zero!

(2) $$\int_a^b f(x)\,dx = -\int_b^a f(x)\,dx$$ Or, in English, switching the limits of

integration changes the value of the integral by a negative sign. This definition

frees us from the constraint of "lower limit < upper limit."

(3) $$\int_a^b f(x)\,dx = \int_a^c f(x)\,dx + \int_c^b f(x)\,dx.$$ This gives a way of expressing

definite integrals as the sum of integrals (of the same function) over subintervals.

This theorem is easy to believe if a < c < b (see Anton's Figure 5.9.1 and the

accompanying explanation), but it is important to remember that it is true no

matter how a, b and c are ordered. (However, this is not easy to prove!)

These three properties are easy to remember and easy to believe; they are also very important.

For instance, note that Anton's Examples 2 and 3 of this section cannot be done without the third

property. The same is true of problems in which the total distance traveled in rectilinear motion is to be calculated (such as Example 5 of Section 5.8).

Here is an example illustrating how Properties 2 and 3 can be used:

Example A. If $\int_0^1 f(x)\,dx = 6$, $\int_0^2 f(x)\,dx = 4$ and $\int_2^5 f(x)\,dx = 1$, find $\int_5^1 f(x)\,dx$.

Solution.

$$\int_5^1 f(x)\,dx = \int_5^0 f(x)\,dx + \int_0^1 f(x)\,dx \qquad\qquad \text{Property 3}$$

$$= \left(\int_5^2 f(x)\,dx + \int_2^0 f(x)\,dx \right) + \int_0^1 f(x)\,dx \qquad \begin{array}{l}\text{Property 3}\\ \text{again}\end{array}$$

$$= -\int_2^5 f(x)\,dx - \int_0^2 f(x)\,dx + \int_0^1 f(x)\,dx \qquad \text{Property 2}$$

$$= -1 - 4 + 6 = 1 \qquad\qquad\qquad\qquad\qquad \square$$

2. Limits of integration under u-substitution. When evaluating a definite integral involving the use of a u-substitution, one must be careful to treat the limits of integration correctly. The basic point is that in $\int_a^b f(x)\,dx$ the integral is being evaluated from $x = a$ to $x = b$. If the variable is changed to u, say, you CANNOT simply keep the same limits of integration and evaluate from $u = a$ to $u = b$. There are two ways of handling the situation:

I. Leave the limits as values for x and wait until the antiderivative is converted back to x to plug in these limits. This approach works as follows:

$$\int_a^b f(g(x))\, g'(x)\, dx \;\underset{\begin{array}{|c|}\hline u = g(x)\\ du = g'(x)\,dx\\\hline\end{array}}{=}\; \int_{x=a}^{x=b} f(u)\, du = F(u)\Big|_{x=a}^{x=b} = F(g(x))\Big|_a^b$$

$$\boxed{\begin{array}{l}\text{We waited}\\ \text{to plug in!}\end{array}} \longrightarrow = F(g(b)) - F(g(a))$$

If you use this approach, you <u>must</u> use "x = a" and "x = b" instead of just "a"
and "b" to indicate that the limits in the second and third steps are values for x, <u>not u</u>!

    II. <u>Convert the limits to values for u</u>, as follows:

$$\int_a^b f(g(x))\, g'(x)\, dx \quad \underset{\uparrow}{=} \quad \int_{g(a)}^{g(b)} f(u)\, du \;=\; F(u)\Big|_{g(a)}^{g(b)} \;=\; F(g(b)) - F(g(a))$$

$$
\begin{aligned}
u &= g(u) \\
du &= g'(x)\, dx \\
u &= g(a) \ \text{when} \ x = a \\
u &= g(b) \ \text{when} \ x = b
\end{aligned}
$$

Notice that with this method
we do NOT return to the x
variable in the antiderivative

This change of limits is easy to remember in the following way: just say to yourself, "when
x = a , what does u equal? When x = b , what does u equal?"

    It is clear that either of these leads to the correct answer. What <u>NOT</u> to do is the
following:

$$\int_a^b f(g(x))\, g'(x)\, dx \quad \underset{\uparrow}{=} \quad \int_a^b f(u)\, du \;=\; F(u)\Big|_a^b \;=\; F(b) - F(a)$$

$$
\begin{aligned}
u &= g(x) \\
du &= g'(x)\, dx
\end{aligned}
$$

All messed up! The x - limits a and b
have been treated like u - limits!

Here is an example illustrating the two methods:

Example B.   Evaluate $\displaystyle\int_0^4 6x\left(\sqrt{x^2+9} - 10\right)dx$.

Correct Solution using I.

$$\int_0^4 6x\left(\sqrt{x^2+9} - 10\right)dx \qquad = \qquad 3\int_{x=0}^{x=4}\left(\sqrt{u} - 10\right)du$$

$$\boxed{\begin{array}{l} u = x^2 + 9 \\ du = 2x\,dx \\ 3\,du = 6x\,dx \end{array}}$$

$$= \left. 3\left(\frac{2}{3}u^{3/2} - 10u\right)\right|_{x=0}^{x=4}$$

$$- \left[2\left(x^2+9\right)^{3/2} - 30\left(x^2+9\right)\right]_0^4$$

$$= [\,2(25)^{3/2} - 30(25)\,]$$

$$- [\,2(9)^{3/2} - 30(9)\,]$$

$$= -284$$

Correct Solution using II.

$$\int_0^4 6x\left(\sqrt{x^2+9} - 10\right)dx \qquad = \qquad 3\int_9^{25}\left(\sqrt{u} - 10\right)du$$

$$\boxed{\begin{array}{l} u = x^2 + 9 \\ du = 2x\,dx \\ 3\,du = 6x\,dx \\ u = (0)^2 + 9 = 9 \ \text{ when } \ x = 0 \\ u = (4)^2 + 9 = 25 \ \text{ when } \ x = 4 \end{array}}$$

$$= \left. 3\left(\frac{2}{3}u^{3/2} - 10u\right)\right|_9^{25}$$

$$= [\,2(25)^{3/2} - 30(25)\,]$$

$$- [\,2(9)^{3/2} - 30(9)\,]$$

$$= -284$$

5.9.5

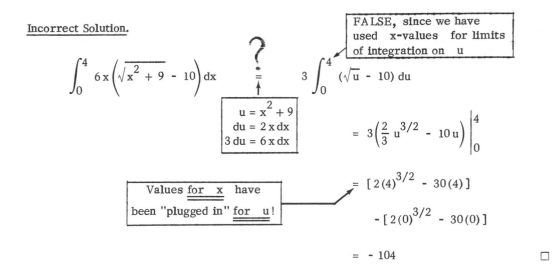

Incorrect Solution.

$$\int_0^4 6x\left(\sqrt{x^2+9}-10\right)dx \overset{?}{=} 3\int_0^4 (\sqrt{u}-10)\,du$$

FALSE, since we have used x-values for limits of integration on u

$$u = x^2+9$$
$$du = 2x\,dx$$
$$3\,du = 6x\,dx$$

$$= 3\left(\frac{2}{3}u^{3/2}-10u\right)\Big|_0^4$$

Values for x have been "plugged in" for u!

$$= [2(4)^{3/2}-30(4)]$$

$$-[2(0)^{3/2}-30(0)]$$

$$= -104 \qquad \square$$

3.   Average Value.   If you were writing the very first calculus book, one of the best parts would be making the definitions.   You could define things in any way you wanted - for instance, you could define what we call the derivative $f'(x)$ as $\displaystyle\lim_{h\to 6}\frac{h}{f(13h)-4}$ and call it a "whoop-de-doo." The test of your definitions would come later, of course; if they did not prove to be useful, they would just fade away, no matter how clever they were originally.

Of course good definitions arise naturally. They spring directly from the subject being considered. One good example is the definition of average value (or mean value) f_{ave} of a function f on an interval $[a,b]$ (Definition 5.9.5). After a little thought (of the sort Anton develops in the pages preceding Definition 5.9.5), one comes almost inevitably to the definition

Definition of the Average Value of f

$$f_{ave} = \frac{1}{b-a}\int_a^b f(x)\,dx$$

People frequently think mathematical definitions are obscure and mysterious, with little connection to reality. In calculus this is never the case, and the definition of f_{ave} is a classic example.

The definition of average value is very easy to use. Here is an example:

Example C.   Find the average value of $f(x) = \sin x$ on the interval $[0, \pi]$.

Solution.

$$f_{ave} = \frac{1}{b - a} \int_a^b f(x)\, dx = \frac{1}{\pi - 0} \int_0^\pi \sin x\, dx$$

$$= \frac{1}{\pi} \left[- \cos x \right]_0^\pi$$

$$= \frac{1}{\pi} \left[- \cos \pi + \cos 0 \right]$$

$$= \frac{1}{\pi} \left[- (- 1) + 1 \right]$$

$$= \frac{2}{\pi} \qquad\qquad \square$$

A physical interpretation of the results of Example C might be as follows: Suppose you build a fence on the interval $[0, \pi]$ whose height at any point x is $\sin x$. Then the surface area of this fence would be the same as a fence on $[0, \pi]$ with uniform height $2/\pi$. Put in another way, which fence would need more paint to cover it? Answer: neither! They both would require the same amount of paint.

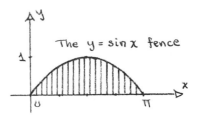

The $y = \sin x$ fence

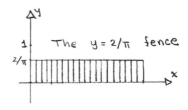

The $y = 2/\pi$ fence

Section 5. 10.   The Second Fundamental Theorem of Calculus

1.   The existence of antiderivatives.   Suppose f is a function which is continuous on an open interval I containing the closed interval [a, b]. At this point we know (Theorem 5. 8. 1):

1st FTC

$$\underline{\underline{\text{If}}} \quad f \text{ has an antiderivative } F \text{ on } [a,b], \quad \underline{\underline{\text{then}}}$$

$$\int_a^b f(x)\, dx = F(b) - F(a)$$

However, the 1st FTC leaves the following question unanswered:

" Does every function f which is continuous on an open interval have an antiderivative? "

Now we can give the answer. It is the first major consequence of the Second Fundamental Theorem of Calculus (2nd FTC):

Every function f which is continuous
on an open interval I has an antiderivative F on I, i.e., for all x in I

$$F'(x) = f(x)$$

Actually the 2nd FTC tells us even more, for it gives a formula for an antiderivative F :

2nd FTC

> If f is continuous on the open interval I,
>
> and a is any point in I, then an
>
> antiderivative F for f can be defined by
>
> $$F(x) = \int_a^x f(x)\, dx$$
>
> for every x in I.

Unfortunately, for reasons which Anton carefully outlines, the formula for F is not very useful if we want to express the antiderivative in terms of "familiar functions" (e. g., polynomials, rational functions or trigonometric functions). The major use of the formula for F given in the 2nd FTC is to define "new" functions. For instance, from Anton's Example 3 we know that the function F defined by

$$F(x) = \int_1^x \frac{1}{t}\, dt$$

is an antiderivative for $f(x) = 1/x$ on the interval $(0, +\infty)$. As Anton points out, F cannot be expressed as a simple combination of "familiar functions" ... and thus F is a "new" function. *

---

* This new function is so important that it is given a name - the "natural logarithm. " It is denoted by the symbol "ln, " i. e.,

$$F(x) = \ln x$$

Since $F'(x) = f(x) = 1/x$ for all x in $(0, +\infty)$, then

$$\frac{d}{dx}(\ln x) = \frac{1}{x} \text{ for all } x > 0$$

We will study the natural logarithm extensively in Chapter 7.

<u>Example A.</u>   Find an antiderivative of $f(x) = \cos^{1/3} x$.

<u>Solution.</u>   The function $f(x) = \cos^{1/3} x$ is continuous on $(-\infty, +\infty)$ so, by the 2nd FTC ,

an antiderivative of   f   on $(-\infty, +\infty)$ is

$$F(x) = \int_a^x f(t)\, dt = \int_a^x \cos^{1/3} t\, dt$$

where   a   is any constant. We cannot actually express this integral in terms of "familiar"

functions *, but, nonetheless, it is an antiderivative of   f .   □

<u>Example B.</u>   Define   F (x)   by

$$F(x) = \int_{-5}^x \left(t^4 + 3t^2 \right) dt$$

    (a)   Find   F'(x)   by first integrating and then differentiating.

    (b)   Find   F'(x)   by using the 2nd FTC .

    Which method is preferable?

<u>Solution.</u>   (a)   The indefinite integral (antiderivative)

$$\int \left(t^4 + 3t^2 \right) dt$$

is easy to compute using the techniques of §5. 2 :

---

*
  However once   a   is fixed, the value of   F (x)   for any   x   can be closely <u>approximated</u>
  using Riemann Sums or some techniques we will learn in §9. 9 .

$$\int \left(t^4 + 3t^2 \right) dt = \int t^4 \, dt + 3 \int t^2 \, dt$$

$$= \frac{1}{5} t^5 + 3 \left(\frac{1}{3} t^3 \right) + C$$

$$= \frac{1}{5} t^5 + t^3 + C$$

Thus

$$F(x) = \int_{-5}^{x} \left(t^4 + 3t^2 \right) dt$$

$$= \left[\frac{1}{5} t^5 + t^3 \right]_{-5}^{x} \qquad (\text{1st FTC. The constant } C \text{ is unnecessary when evaluating a definite integral.})$$

$$= \left[\frac{1}{5} x^5 + x^3 \right] - \left[\frac{1}{5} (-5)^5 + (-5)^3 \right]$$

$$= \frac{1}{5} x^5 + x^3 + 750$$

Thus

$$F'(x) = \frac{1}{5} (5x^4) + 3x^2 + 0 = \boxed{x^4 + 3x^2}$$

(b)   We can also find $F'(x)$ by the 2nd FTC. Since $f(x) = x^4 + 3x^2$ is continuous everywhere (it's a polynomial) then the 2nd FTC says that

$$F(x) = \int_{-5}^{x} \left(t^4 + 3t^2 \right) dt$$

is an antiderivative for $f(x)$, i. e.,

$$F'(x) = f(x) = \boxed{x^4 + 3x^2}$$

This agrees with the answer in (a)... and is certainly a lot easier to obtain!   □

Remark.   Notice that the variable   t   in   Example B, and in any application of the 2nd FTC

for that matter, is simply a "dummy" variable: if the integral $\int_a^x f(t)\,dt$ is evaluated,

we obtain a function of   x , not   t. The variable   t   plays the role of "something to

substitute for" and has been eliminated (i. e. , substituted for) by the time the final form of

F(x)   is obtained. In fact, any variable (other than   x)   can be used in place of   t, i. e. ,

all the following integrals are the same: $\int_a^x f(t)\,dt$, $\int_a^x f(\theta)\,d\theta$, $\int_a^x f(y)\,dy$.

2.   Intervals of validity for the 2nd FTC.   One must be very careful to check where the

formula for   F(x)   in the 2nd FTC   is valid. Here are two examples:

Example C.   Find an antiderivative for the function $f(x) = (x - 1)^{-1/3}$ on the interval   (1 , +∞).

Do the same for the interval   (- ∞, 1).

Solution.   First observe that   f   is a continuous function everywhere except at   x = 1.   Thus

f   is continuous on the two intervals   (1 , +∞)   and   (- ∞, 1),   and we can find antiderivatives

for   f   on each of these intervals by using the 2nd FTC. Will these two antiderivatives have

the same formula? NO:

> the lower limit of integration   "a"
>
> in the 2nd FTC   formula for
>
> an antiderivative on an interval   I
>
> MUST BE PICKED FROM   I !!

Thus, on $(1, +\infty)$ we can use

$$F_1(x) = \int_2^x (t-1)^{-1/3}\, dt \qquad \left[\begin{array}{l} \text{We have taken } a = 2\,; \\ \text{in fact, any value} \\ a > 1 \text{ will do} \end{array} \right]$$

On $(-\infty, 1)$ we can use

$$F_2(x) = \int_0^x (t-1)^{-1/3}\, dt \qquad \left[\begin{array}{l} \text{We have taken } a = 0\,; \\ \text{in fact, any value} \\ a < 1 \text{ will do} \end{array} \right]$$

The functions F_1 and F_2 are NOT the same. $F_1(x)$ is defined only for $x > 1$, while $F_2(x)$ is defined only for $x < 1$. Yes, they are both antiderivatives for f ... but over completely separate intervals!   □

Example D.   Over what open interval does the formula

$$F(x) = \int_{-1}^x \frac{dt}{t^3 - 4t}$$

represent an antiderivative of $f(x) = 1/(x^3 - 4x)$?

Solution.   We first need to determine the open intervals on which f is continuous. Since f is a rational function, it is continuous everywhere except where the denominator is zero: since $x^3 - 4x = x(x-2)(x+2)$, then the denominator is zero at the points $-2, 0, 2$. Thus f is continuous on the four open intervals

$$(-\infty, -2), \quad (-2, 0), \quad (0, 2), \quad (2, +\infty)$$

Now look at the formula for the antiderivative F. Its lower limit of integration is $a = -1$, which falls in the second open interval $(-2, 0)$. Thus F is an antiderivative for $f(x) = 1/(x^3 - 4x)$ on $I = (-2, 0)$.   □

3.   <u>Some theoretical observations on the 2nd FTC.   (Optional)</u>   The proof of the 2nd FTC
     takes only half a page in Anton's text ... but that hides the numerous difficult theorems
     upon which the proof depends:

     1.   The legality of splitting a definite integral up

            over two subintervals (Theorem 5.9.2).

            This is a surprisingly difficult result to prove!

     2.   The Mean Value Theorem for Integrals (Theorem 5.9.4).

            This is also a non-trivial result in that its

            proof depends on 3 and 4 below.

     3.   The Extreme Value Theorem (Theorem 4.5.4)

     4.   The Intermediate Value Theorem (Theorem 3.7.8)

It is interesting to notice that the proof of the 2nd FTC   does not make use of the 1st FTC.
They are quite separate results.

4.   <u>Functions as limits of integration. (Optional)</u>   Sometimes a function of x , and not just x
     itself, appears as one of the limits of integration, as in

$$\boxed{\begin{array}{c} g(x) , \quad \text{not} \\ \text{just} \quad x \end{array}} \longrightarrow \int_a^{g(x)} f(t)\, dt$$

where   a   is a constant. The derivative of this integral can be handled quite easily by letting

$$F(x) = \int_a^x f(t)\,dt$$

so that the original integral is the composite function

$$F(g(x)) = \int_a^{g(x)} f(t)\,dt$$

Differentiating $F(g(x))$ using the Chain Rule yields

$$\frac{d}{dx}\,F(g(x)) = F'(g(x))\,g'(x)$$

and then, since $F'(x) = f(x)$ by the 2nd FTC,

$$\frac{d}{dx}\,F(g(x)) = f(g(x))\,g'(x)$$

To summarize,

$$\boxed{\frac{d}{dx}\left[\int_a^{g(x)} f(t)\,dt\right] = F'(g(x))\,g'(x) = f(g(x))\,g'(x)}$$

Anton gives this useful formula as Exercise 12.

Example E.   Find $\dfrac{d}{dx}\displaystyle\int_3^{2x^2-4} \ln 2t\,dt$.

Solution.   This is $\displaystyle\int_a^{g(x)} f(t)\,dt$ with $a = 3$, $g(x) = 2x^2 - 4$ and $f(t) = \ln 2t$. Hence

$$\frac{d}{dx}\int_3^{2x^2-4} \ln 2t\,dt = f(g(x))\,g'(x)$$

$$= [\ln 2(2x^2 - 4)](4x)$$

$$= 4x\ln(4x^2 - 8) \qquad\qquad \square$$

If _both_ limits of integration are functions of x ,

$$\boxed{\begin{array}{c}\text{both limits}\\ \text{are functions}\\ \text{of} \ \ x\end{array}} \longrightarrow \begin{array}{c}\displaystyle\int_{h(x)}^{g(x)} f(t)\,dt\end{array}$$

we can use Theorem 5. 9. 2 and write

$$\int_{h(x)}^{g(x)} f(t)\,dt \ = \ \int_{h(x)}^{a} f(t)\,dt \ + \ \int_{a}^{g(x)} f(t)\,dt$$

whence, by Theorem 5. 9. 1 ,

$$\int_{h(x)}^{g(x)} f(t)\,dt \ = \ \int_{a}^{g(x)} f(t)\,dt \ - \ \int_{a}^{h(x)} f(t)\,dt$$

Thus the theorem in the box above can be used to obtain

$$\boxed{\ \frac{d}{dx} \int_{h(x)}^{g(x)} f(t)\,dt \ = \ f(g(x))\,g'(x) \ - \ f(h(x))\,h'(x)\ }$$

Anton gives this formula as Exercise 14 .

<u>Example F</u>.   Find $\displaystyle \frac{d}{dx} \int_{x^2-1}^{x^3+1} \frac{1}{t^2+1}\,dt$.

<u>Solution</u>.   This is $\displaystyle \int_{h(x)}^{g(x)} f(t)\,dt$ with $h(x) = x^2 - 1$, $g(x) = x^3 + 1$,   and $f(t) = 1/(t^2 + 1)$.

Hence

$$\frac{d}{dx} \int_{x^2-1}^{x^3+1} \frac{1}{t^2 + 1}\, dt = f(g(x))\, g'(x) \; - \; f(h(x))\, h'(x)$$

$$= \frac{1}{(x^3 + 1)^2 + 1}\,(3x^2)\; - \;\frac{1}{(x^2 - 1)^2 + 1}\,(2x)$$

$$= \frac{3x^2}{x^6 + 2x^3 + 2}\; - \;\frac{2x}{x^4 - 2x^2 + 2}\quad . \qquad\qquad \square$$

Chapter 6 :   Applications of the Definite Integral

Section 6. 1.   Area Between Two Curves

1.   The basic procedure.   As Anton shows at the start of this chapter,

the logical way to obtain the area   A   between

two continuous curves over an interval   [a , b]   is:

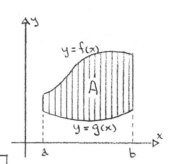

$$A = \int_a^b [f(x) - g(x)]\,dx$$

when $f(x) \geq g(x)$ for all   x   in   [a , b]

A good way both to memorize and to understand this formula is to interpret it

infinitesimally:

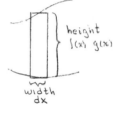

$[f(x) - g(x)]\,dx$ is the "area"

of a slice of our region with

infinitesimal width   dx ,

and

$$\int_a^b [f(x) - g(x)]\,dx \quad \text{is the "sum"}$$

of all the infinitesimal areas from

x = a   to   x = b.

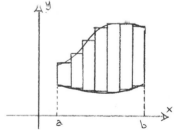

(This "infinitesimal approach" can be made rigorous by using the Riemann Sum definition for

the definite integral $\int_a^b [f(x) - g(x)]\,dx$.   This is discussed in §6. 8 of The Companion.)

<u>Example A.</u>   Compute the area enclosed by the curves $y = x^2 - 4x + 2$ and $y = -x^2 + 2x + 1$.

<u>Solution.</u>   Using $y = x^2 - 4x + 2 = (x - 2)^2 - 2$ and $y = -x^2 + 2x + 1 = -(x - 1)^2 + 2$

(see Appendix D §7 on completing the square), we obtain the following graph (with the aid of

the Translation Principles - see Appendix E §1):

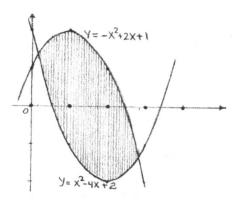

We now slice our region by a vertical line

segment at   x   as shown. This line segment

extends from $f(x) = -x^2 + 2x + 1$ on the top

curve to $g(x) = x^2 - 4x + 2$ on the bottom curve.

<u>The length of this line segment is the integrand in our area formula</u>:

$$f(x) - g(x) = (-x^2 + 2x + 1) - (x^2 - 4x + 2)$$

Hence

$$\boxed{f(x) - g(x) = -2x^2 + 6x - 1}$$

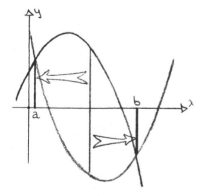

Shifting our line segment as far to the left

and as far to the right as possible will determine

the endpoints of the interval [a, b]. In this

example the endpoints a   and b   occur where

the two curves intersect; they can be found by equating

$$y = -x^2 + 2x + 1 \quad \text{with} \quad y = x^2 - 4x + 2$$

This yields $2x^2 - 6x + 1 = 0$, which by the quadratic formula (Appendix D §5) has roots $(3 \pm \sqrt{7})/2$. Thus

$$a = \frac{3 - \sqrt{7}}{2} \approx .177 \qquad \text{and} \qquad b = \frac{3 + \sqrt{7}}{2} \approx 2.823$$

Now we can plug everything into our area formula:

$$A = \int_a^b [\, f(x) - g(x) \,]\, dx$$

$$= \int_{.177}^{2.823} [\, -2x^2 + 6x - 1 \,]\, dx$$

$$= \left[-\frac{2x^3}{3} + 3x^2 - x \right]_{.177}^{2.823}$$

$$= \left[-\frac{2(2.823)^3}{3} + 3(2.823)^2 - 2.823 \right]$$

$$\qquad - \left[-\frac{2(.177)^3}{3} + 3(.177)^2 - .177 \right]$$

$$\approx 6.087 - (-.087) = \boxed{6.174} \qquad \qquad \square$$

We have used a four-step procedure in this problem:

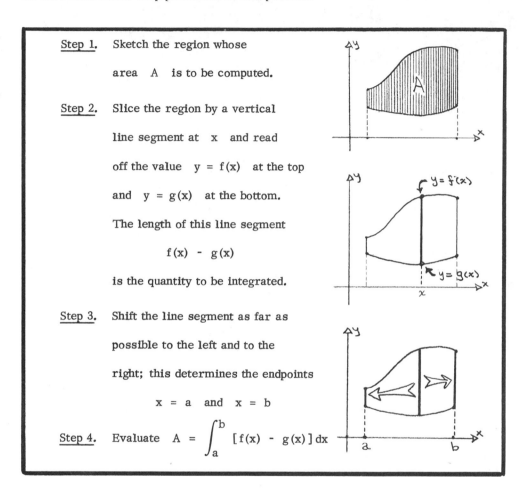

Calculating
Areas
Between
Curves

Step 1. Sketch the region whose

area A is to be computed.

Step 2. Slice the region by a vertical

line segment at x and read

off the value y = f(x) at the top

and y = g(x) at the bottom.

The length of this line segment

$$f(x) - g(x)$$

is the quantity to be integrated.

Step 3. Shift the line segment as far as

possible to the left and to the

right; this determines the endpoints

$$x = a \quad \text{and} \quad x = b$$

Step 4. Evaluate $A = \int_a^b [f(x) - g(x)] \, dx$

2. Exchange of Roles and Making Choices.   Anton's Definition 6. 1. 2 is simply Definition 6. 1. 1

with x and y interchanged. This sort of exchange of roles is common in mathematics;

for instance, most of the definitions in Chapter 6 will be restated in an exchange-of-roles

form. It is important to understand that the reason for stating a definition in two forms is to

give you a choice. You can pick the form which is more convenient and that can be a big

advantage.

For instance, Anton's Examples 3 and 4 make it clear that the area enclosed by the curves $y = \sqrt{x}$, $y = -x + 6$ and $y = 1$ can be computed by evaluating

$$\int_1^4 (\sqrt{x} - 1)\,dx \; + \; \int_4^5 [-x + 6]\,dx$$

or by evaluating

$$\int_1^2 [(6 - y) - y^2]\,dy$$

The choice is yours to make - the two approaches are equally good. You might regard the second as preferable because it involves evaluating only one integral instead of two, or you might prefer the first because the integrands are slightly less complicated. In this case, it is probably a toss-up. However, you will encounter situations in which a judicious choice of the roles for x and y can make life considerably easier. This is the case in Anton's Example 5 and also in the following example:

Example B.   Find the area between the curves $x = y^2$ and $x = 3 - 2y^2$.

Solution.   Step 1.   The region is as shown to the right, where we found the intersection points $(1, 1)$ and $1, -1)$ by equating y^2 with $3 - 2y^2$ to get $y = \pm 1$. Moreover, by symmetry we see that we have only to compute the area of the top half of our region (labeled R), and double it.

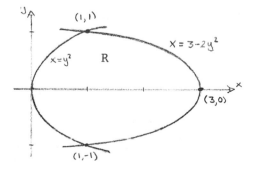

We now must choose between an **x-axis** approach (our original method) or a **y-axis** approach. We will do both, starting with the x-axis:

<u>Step 2.</u> (x-axis) From the diagram to the right we see that the y-value at the top of a vertical line segment will have a different formula depending on whether you are to the left or to the right of $x = 1$. (To the right we use $x = y^2$, which gives $y = \sqrt{x}$; to the left we use $x = 3 - 2y^2$, which gives $y = \sqrt{(3 - x)/2}$.) Hence we must split our area into two pieces, labeled A and B, and compute each separately. Since the bottom curve is the x-axis (i.e., the curve $y = 0$), the

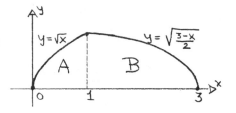
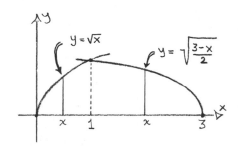

lengths of vertical line segments in A and B, respectively, will be

$$\sqrt{x} \quad \text{and} \quad \sqrt{(3 - x)/2}$$

<u>Step 3.</u> (x-axis) From our diagram above it is apparent that the limits of integration for A and B, respectively, are

$$0 \text{ and } 1 \ , \quad 1 \text{ and } 3$$

Step 4. (x-axis) Plugging into our formula will give

$$A = \int_0^1 \sqrt{x} \; dx = \left. \frac{2x^{3/2}}{3} \right|_0^1 = \frac{2}{3} - 0 = \boxed{\frac{2}{3}}$$

$$B = \int_1^3 \sqrt{\frac{3-x}{2}} \; dx = \int_1^0 \sqrt{u} \; (-2\,du) = (-2)\left. \left(\frac{2u^{3/2}}{3}\right) \right|_1^0 = \boxed{\frac{4}{3}}$$

$u = (3-x)/2$	$x = 1$ gives $u = 1$
$du = -\,dx/2$	$x = 3$ gives $u = 0$
$dx = -\,2\,du$	

Thus the total area of the (full) region is

$$2(A + B) = 2(2/3 + 4/3) = \boxed{4}$$

Now we'll use the y-axis approach:

Step 2. (y-axis) Cutting the region with a horizontal

line segment shows that the right-most x-value is

$w(y) = 3 - 2y^2$, while the left-most x-value is

$v(y) = y^2$. The length of this line segment is

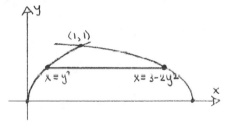

$$w(y) - v(y) = 3 - 2y^2 - y^2 = 3 - 3y^2$$

Step 3. (y-axis) Shifting our line segment down

as far as possible yields the lower endpoint to be

$c = 0$; shifting up as far as possible yields the

upper endpoint to be $d = 1$.

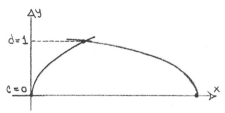

Step 4. (y-axis) . Using the formula in Definition 6.1.2 yields

$$R = \int_c^d [\, w(y) - v(y) \,] \, dy = \int_0^1 [\, 3 - 3y^2 \,] \, dy$$

$$= [\, 3y - y^3 \,]_0^1 = (3 - 1) - 0 = 2$$

Since R is only half of our original region, we see that our desired area is $\boxed{4}$. □

Although both the x-axis and y-axis approaches worked in Example B, we think you'll agree that the y-axis approach is preferable (just one easy integral).

Section 6.2 : Volumes By Slicing; Disks and Washers.

1. The method of slicing: use volumes of right cylinders. Everything in this section depends on the ability to calculate the volume of a right cylinder:

<table>
<tr><td>Definition of
a right
cylinder</td><td>a <u>right cylinder</u> is a solid that can

be generated by moving a plane region

along a line <u>perpendicular</u> to that region</td></tr>
</table>

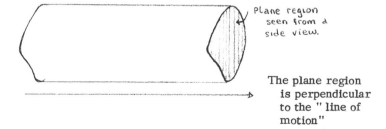

Plane region
seen from a
side view.

The plane region
is perpendicular
to the " line of
motion"

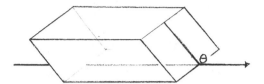

This is a right cylinder. Its triangular
face is perpendicular to the "line of
motion," i.e., angle θ is 90°.

This is not a right cylinder. Its square
face is not perpendicular to the "line of
motion," i.e., angle θ is not 90°.

Calculating the volume of a right cylinder is easy: the volume of a right cylinder is

the area of the base times the height. That is,

| The volume of a right cylinder |

$$V = A \cdot h$$

where A and h are as shown in the figure

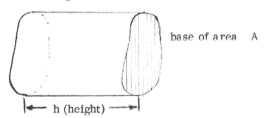

base of area   A

h (height)

Note that this right cylinder is on its side (which will prove to be the most convenient way for us

to look at right cylinders).

If you need some convincing that the $V = A \cdot h$ definition of the volume of a right

cylinder is a natural one, consider the right cylinder which is a 3-dimensional rectangle (i.e.,

a box)

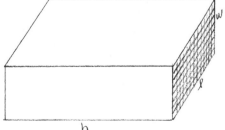

base of area   A = ℓw

You know (or you should!) that the volume of this box is the product $V = \ell wh$ of its dimensions. But of course the area of the base is $A = \ell w$ so $V = A \cdot h$, in accordance with the definition of the volume of a right cylinder.

Example A. Find the volume of the right cylinder whose base is an equilateral triangle of side length 2 inches and whose height is 6 inches.

Solution.

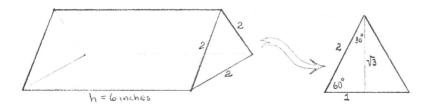

The altitude of the base triangle is $\sqrt{3}$ (from the standard $30^\circ - 60^\circ - 90^\circ$ triangle) so its area is $A = \dfrac{1}{2} bh = \dfrac{1}{2}(2)\sqrt{3} = \sqrt{3}$. Thus the volume V is

$$V = A \cdot h = \sqrt{3} \cdot 6 = 6\sqrt{3} \text{ in}^3 \qquad \qquad \square$$

The intuitive principle behind the method of slicing is simple: Take any solid and slice it into paper-thin slices (with a salami-slicer). Then figure out the volume of each slice and add those volumes together to get the volume of the whole solid.

The hard part is to "figure out" the volume of each slice. Since this may not be possible, we <u>approximate</u> the slice by a very thin right cylinder with (known) volume $A_k \Delta x_k$.

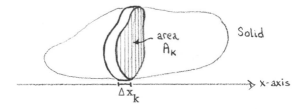

Then adding up these approximate volumes gives the approximate volume of the solid:

$$\text{Volume of solid} \approx \sum_{k=1}^{n} A_k \Delta x_k$$

If we let the widths Δx_k of the slices go to zero, we get a better and better approximation of the volume. Since the limit of these <u>Riemann Sums</u> is a definite integral, this motivates the following definition:

Volume of solid $= \displaystyle\int_a^b A(x)\, dx$

where $A(x)$ is the area of a

cross-sectional slice of the

solid by a plane <u>perpendicular</u>

to the x-axis at x .

A good way both to memorize and to understand this formula is to interpret it

<u>infinitesimally</u>: *

$A(x)\,dx$ is the "volume" of

the slice with infinitesimal

width dx, and

$\displaystyle\int_a^b A(x)$ is the "sum" of all

the infinitesimal volumes

from x = a to x = b .

_____

* This "infinitesimal approach" is merely a non-rigorous (and simpler) version of the volume formula using Riemann Sums. This is discussed in §6. 8 of <u>The Companion.</u>

6.2.5

A Summary of the Method of Slicing

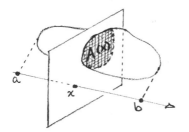

1. Axis. Introduce an x-axis running near or through the solid so that the cross-sectional areas A(x) of the solid <u>perpendicular</u> to the axis <u>are easy to compute.</u>

2. Area. Compute the cross-sectional area A(x) for each x.

3. Integral. "Add together" all the infinitesimal volumes A(x) dx, i.e., compute the integral

$$\text{Volume} = \int_a^b A(x)\, dx$$

where x = a and x = b are the smallest and largest allowable values of x.

Example B. The base of a certain solid is the region bounded by the curves $y = x^2$ and $y = 2x - x^2$. Cross-sections of the solid perpendicular to the x-axis are squares. Find the volume of the solid.

Solutions. First we need to sketch the region. Finding the points of intersection of the curves $y = x^2$ and $y = 2x - x^2$, we get

$$x^2 = 2x - x^2$$

$$0 = 2x(1 - x)$$

$$\text{so} \quad x = 0, 1$$

Thus $(0,0)$ and $(1,1)$ are the points of intersection,

and the graph of the region is as shown to the right.

The method of slicing now proceeds as follows:

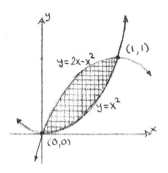

1. Axis.   The x-axis is given to be the slicing axis. From

the graph it is apparent that the allowable x-values run·from

$x = 0$ to $x = 1$.

2. Area.   A typical cross-section of the solid is a square

as shown to the right. The side length s of the square at

x is the difference of the y values, or

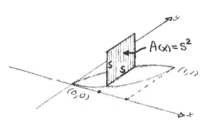

$s = (2x - x^2) - x^2 = 2x - 2x^2$. Thus the area of the cross-

section at x is $A(x) = s^2 = (2x - 2x^2)^2$.

3. Integral.   The volume of the solid is thus

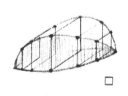

$$V = \int_a^b A(x)\,dx = \int_0^1 (2x - 2x^2)^2\,dx = \int_0^1 (4x^2 - 8x^3 + 4x^4)\,dx$$

$$= \left(\frac{4}{3}x^3 - 2x^4 + \frac{4}{5}x^5\right)\Bigg|_0^1 = \frac{4}{3} - 2 + \frac{4}{5} = \frac{2}{15} \qquad \square$$

2. The method of disks: a special case of the method of slicing.   A situation which is a natural for

the slicing method occurs when a solid is obtained by rotating the area under a curve in the

x-y plane about the axis which is adjacent to it:

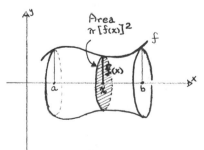

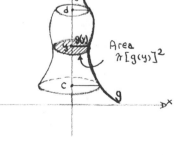

475
11 15

The solids pictured are obtained by rotating the region under $y = f(x)$, $a \leq x \leq b$ about the (adjacent) x-axis and the region under $x = g(y)$, $c \leq y \leq d$ about the (adjacent) y-axis. In each case the cross-section perpendicular to the axis of revolution is a circle (disk) with radius $f(x)$ or $g(y)$ so that the cross-sectional areas are $\pi[f(x)]^2$ and $\pi[g(y)]^2$. Thus the volumes of the solids on the left and right are, respectively,

The Method of Disks

$$V_1 = \int_a^b \pi[f(x)]^2 \, dx \text{ and } V_2 = \int_c^d \pi[g(y)]^2 \, dy$$

The first of these formulas, for rotating about the (adjacent) x-axis, probably should be memorized. [It's handy to have it on the tip of your tongue, but, if you forget, you can always derive the formula quickly from the method of slicing.] The second formula, for rotation about the (adjacent) y-axis, is the same as the first, except that y plays the role of x as the axis of revolution. (Compare this with the discussion of exchange of roles in §6. 1. 2 of The Companion.)

Example C. Find the volume of the solid of revolution obtained by rotating the curve $y = \dfrac{1}{x}$ for $1 \leq x \leq 36$ about the x-axis.

Solution. The solid is a long cornucopia (see the figure). By the method of disks, its volume V is

$$V = \int_1^{36} \pi\left[\frac{1}{x}\right]^2 dx = \pi \int_1^{36} x^{-2} dx = \pi\left[-\frac{1}{x}\right]_1^{36} = \pi\left[-\frac{1}{36} + 1\right] = \frac{35}{36}\pi \qquad \square$$

Example D.   Find the volume of the solid of revolution obtained by rotating the curve $y = x^2$

for $0 \leq x \leq 4$ about the   y-axis.

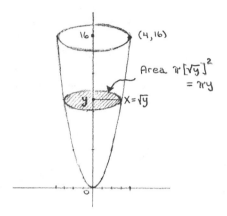

Solution.   The solid in a solid bowl. Because y

is the axis of revolution, we have to convert the

equation $y = x^2$, $0 \leq x \leq 4$, into $x = \sqrt{y}$,

$0 \leq y \leq 16$ (where y is the independent variable

and the limits are in terms of y).   Then by the

method of disks (with rotation about the   y-axis),

the volume   V   of the bowl is

$$ V = \int_0^{16} \pi [\sqrt{y}]^2 \, dy = \pi \int_0^{16} y \, dy = \pi [\frac{y^2}{2}] \Big|_0^{16} = \pi [\frac{256}{2} - 0] = 128\pi \qquad \square $$

3.   The method of washers: a special case of a special case.   We said that the method of disks is

a special case of the method of slicing. Well, the method of washers is a special case of the

method of disks, making it a special case of a special case!

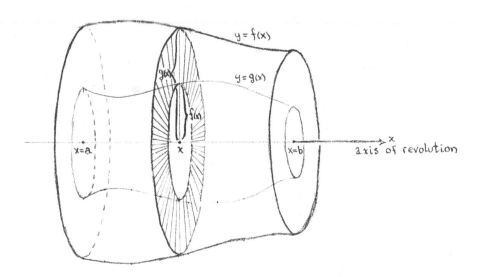

The volume of the solid of revolution obtained by revolving the region bounded by the curves $y = f(x)$ and $y = g(x)$ and the lines $x = a$ and $x = b$, where $g(x) \leq f(x)$ on $[a, b]$, about the x-axis is

$$V = \int_a^b \pi \left([f(x)]^2 - [g(x)]^2 \right) dx$$

$$= \int_a^b \pi [f(x)]^2 - \int_a^b \pi [g(x)]^2 \, dx$$

The Method of Washers

As our equations show,

the method of washers is nothing more than

a subtraction of two method-of-disks volumes.

The only important thing to keep in mind when using the method of washers is that <u>it is vital to know which curve is above the other</u> (with respect to the axis of revolution).

If you assume that the wrong curve is on top, your answer

will be off by a negative sign.

<u>Example E.</u> Find the volume of the solid of revolution obtained by rotating the region bounded by the parabolas $x = y - y^2 + 4$ and $x = y^2 + 1$ about the y-axis.

Solution.   By solving the two given equations simultaneously,

we find that the curves intersect at the points $(2, -1)$ and

$(\frac{13}{4}, \frac{3}{2})$, as shown in the figure to the right. Because the

curve $x = y - y^2 + 4$ is on top (from a   y-axis   point of

view) over the entire interval $-1 \le y \le \frac{3}{2}$, the volume   V

of the solid of revolution is obtained by the method of

washers using $x = f(y) = y - y^2 + 4$ (the "upper" function),

$x = g(y) = y^2 + 1$ (the "lower" function) and $[a, b] = [-1, 3/2]$:

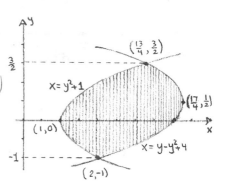

$$V = \int_{-1}^{3/2} \pi [y - y^2 + 4]^2 \, dy - \int_{-1}^{3/2} \pi [y^2 + 1]^2 \, dy$$

$$= \pi \int_{-1}^{3/2} [(y - y^2 + 4)^2 - (y^2 + 1)^2] \, dy$$

$$= \pi \int_{-1}^{3/2} (-2y^3 - 9y^2 + 8y + 15) \, dy$$

$$= \pi \left[-\frac{1}{2} y^4 - 3y^3 + 4y^2 + 15y \right]_{-1}^{3/2} = 875\pi/32 \qquad \square$$

Example F.   Find the volume of the solid of revolution obtained by revolving the area bounded by

the curve $y = x^2$, the   y-axis   and the lines $y = 1$ and $x = 2$ about the   x-axis.

Solution.   The region is shown to the right.

It is clear that the curve $y = x^2$ is below the line

$y = 1$ for $0 \le x \le 1$ and above it for $1 \le x \le 2$,

so we must consider the intervals $[0, 1]$ and $[1, 2]$

separately. That is, $V = V_1 + V_2$ where V_1 is the

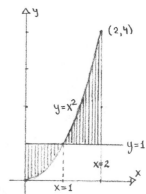

volume of the solid of revolution obtained by rotating the region bounded by $y = x^2$, $y = 1$, $x = 1$ and the y-axis about the x-axis, and V_2 is the volume of the solid of revolution obtained by rotating the region bounded by $y = x^2$, $y = 1$, $x = 1$ and $x = 2$ about the x-axis. Thus

$$V = V_1 + V_2 = \int_0^1 \pi [[1]^2 - [x^2]^2] \, dx + \int_1^2 \pi [[x^2]^2 - [1]^2] \, dx$$

$$= \pi \int_0^1 (1 - x^4) \, dx + \pi \int_1^2 (x^4 - 1) \, dx = \pi \left(x - \frac{x^5}{5} \right) \Bigg|_0^1 + \pi \left(\frac{x^5}{5} - 1 \right) \Bigg|_1^2$$

$$= \pi (1 - \frac{1}{5}) + \pi [(\frac{32}{5} - 1) - (\frac{1}{5} - 1)] = 7\pi \qquad \square$$

4. <u>The method of disks when the axis of revolution is not the x-axis or the y-axis.</u> (Optional)

There is a way of stating the method of disks so that it can be used when plane regions are rotated around <u>any</u> horizontal or vertical line.

First consider the case in which the axis of revolution is the horizontal line $y = y_0$.

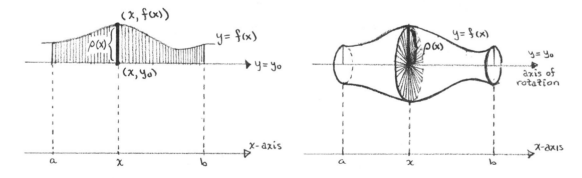

Consider the region between $y = f(x)$ and $y = y_0$, for $a \leq x \leq b$, shown in the left diagram above. The vertical distance between $f(x)$ and y_0 will be denoted by

$$\rho(x) - |f(x) - y_0|$$

If we rotate this region about the line $y = y_0$ (an adjacent axis of revolution), we obtain the

solid shown in the right diagram above. A cross-section of this solid perpendicular to the line

$y = y_0$ will be a disk of radius $\rho(x)$ and area

$$A(x) = \pi [\rho(x)]^2 = \pi (f(x) - y_0)^2$$

Therefore the volume of this solid can be computed by the Method of Slicing to be

Method of Disks about $y = y_0$	$$V = \int_a^b \pi [\rho(x)]^2 \, dx = \int_a^b \pi [f(x) - y_0]^2 \, dx$$

When $y_0 = 0$ we are rotating about the x-axis, and our new formula is simply the usual

Method of Disks formula.

<u>Example G.</u>   Find the volume of the solid obtained by rotating the region enclosed by the graphs of

$y = \sqrt{x}$, $y = 1$ and $x = 0$ about the line $y = 1$.

Solution.   A graph of the region is given to the right.

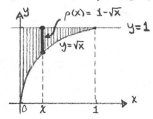

As can be seen we have the region between a curve and

a line $(y = \sqrt{x}$ and $y = 1)$ for $0 \le x \le 1$, and

we are revolving it about the (adjacent) line $y = 1$. The

radius of the cross-sectional disk at x is therefore

$$\rho(x) - |\sqrt{x} - 1|$$

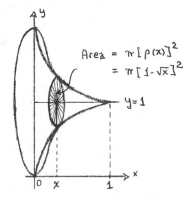

and the volume of the total solid will be

$$V = \int_0^1 \pi (\sqrt{x} - 1)^2 \, dx = \pi \int_0^1 (1 - 2\sqrt{x} + x) \, dx$$

$$= \pi \left[x - \frac{4}{3} x^{3/2} + \frac{1}{2} x^2 \right] \Bigg|_0^1 = \frac{\pi}{6} \qquad \square$$

Now consider a rotation in which the axis of revolution is the vertical line $x = x_0$.

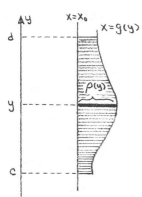
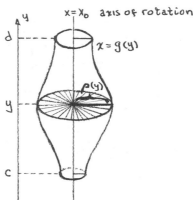

Rotating the region shown on the left about the vertical line $x = x_0$ will produce the solid shown on the right. Using the method of slicing as applied to rotations about (adjacent) horizontal lines, we find that the volume of this solid is

Method of Disks about $x = x_0$	$$V = \int_c^d \pi [\rho(y)]^2 \, dy = \int_c^d \pi [g(y) - x_0]^2 \, dy$$

Here $\rho(y)$ is the horizontal distance between $f(y)$ and x_0, i.e.,

$$\rho(y) = |g(y) - x_0|$$

Example H. Find the volume of the solid of revolution obtained by rotating the area bounded by the two parabolas $x = y - y^2$ and $x = y^2 - 3$ about the line $x = -4$.

<u>Solution.</u>   This is really a "Method of Washers" problem about the vertical axis $x = -4$.

Solving the equations simultaneously we

see that the parabolas $x = f(y) = y - y^2$

and $x = g(y) = y^2 - 3$ intersect at

$(-3/4, 3/2)$ and $(-2, -1)$. The desired

volume   V   may be determined by computing

$$V = V_1 - V_2$$

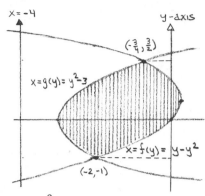

where V_1 is the volume obtained by rotating $x - f(y) = y - y^2$

about the line $x = -4$ (with $-1 \leq y \leq 3/2$),   and V_2 is the volume obtained by rotating

$x = g(y) = y^2 - 3$ about the line $x = -4$ (with $-1 \leq y \leq 3/2$). Thus

$$V = \int_{-1}^{3/2} \pi \, | f(y) + 4 |^2 \, dy \quad - \int_{-1}^{3/2} \pi \, [\, g(y) + 4 \,]^2 \, dy$$

$$= \int_{-1}^{3/2} \pi \, [y - y^2 + 4]^2 \, dy - \int_{-1}^{3/2} \pi \, [y^2 - 3]^2 \, dy$$

$$= \dots = 875\pi/32 \approx 85.9$$

(The details of the evaluation of these integrals may be found in Example E above.)   □

<u>Exercises.</u>   These are problems dealing with the optional material concerning rotations about

arbitrary horizontal and vertical lines.

Let   R   be the region bounded by the curve $y = x^2$ and the line $y = 4$. Find the

volume of the solid of revolution obtained by rotating   R   about

6.3.1

1.   the   y-axis

2.   the line   y = 4

3.   the   x-axis

4.   the line   y = -1

5.   the line   x = 2

Answers.   (1) 8π (2) $512\pi/15$ (3) $256\pi/5$ (4) $1088\pi/15$ (5) $128\pi/3$

Section 6.3.   Volumes by Cylindrical Shells

1.   Disks and Shells: similarities and differences.   The method of cylindrical shells and the method of disks are both ways of computing the volume of a solid of revolution (i. e., a solid obtained by revolving a plane region about an axis of revolution).   However, there are important differences between the two.   The fundamental difference in application centers around the axis of revolution used: In the disk method, the axis of revolution must be adjacent to the region being rotated and is the axis of the independent variable; in the method of cylindrical shells, the axis of revolution might be separated from the region being rotated and is the axis of the dependent variable.   That is,

Disks vs Shells		Axis of Revolution	
		geometric condition	analytic condition
	Disk Method	must be adjacent	must be axis of independent variable
	Cylindrical Shells Method	might be separated	must be axis of dependent variable

Hence if   R   is the region under $y = f(x)$ for $a \leq x \leq b$

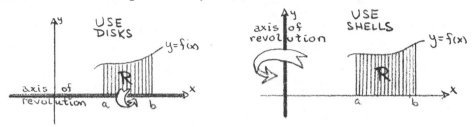

use disks to find the volume of revolution about the (adjacent)   x-axis   (since   x   is the

independent variable), and use shells to find the volume of revolution about the (separated)

y-axis   (since   y   is the dependent variable).

On the other hand, if   R   is the region "under" $x = g(y)$ for $c \leq y \leq d$

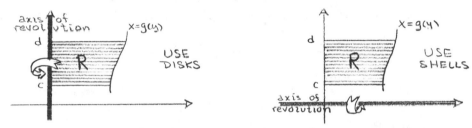

use disks to find the volume of revolution about the (adjacent)   y-axis   (since   y   is now the

independent variable), and use shells to find the volume of revolution about the (separated)

x-axis   (since   x   is now the dependent variable).

Note.   The preceding disks vs. shells discussion should be qualified in two respects:

i)   as we shall see below, frequently either disks or shells can be used to find a volume

of revolution, and

ii)   in the method of shells sometimes the "separated" axis of revolution is not actually

separated from the region, for example when the region under $y = f(x)$ for

$0 \leq x \leq b$

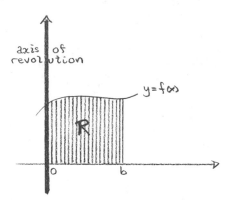

is rotated about the y-axis. However, in this case the axis of rotation is still the axis of the dependent variable.

With the basic difference in the use of the disks and shells methods in mind, we now focus on their basic difference in concept: Instead of dividing the solid into infinitesimal flat cross-sectional disks as in the method of disks, in the method of shells we divide the solid into infinitesimal curved cylindrical shells.

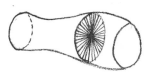

Method of Disks

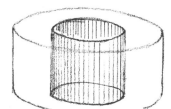

Method of Cylindrical Shells

Thus cylindrical shells cannot be a special case of the method of slicing -- we are not using flat cross-sections of the solid. The method of cylindrical shells is, however, another application of the Riemann Sum definition of the integral (Definition 5.7.2):

$$\int_a^b f(x)\,dx = \lim_{\max \Delta x_k \to 0} \sum_{k=1}^n f(x_k^*)\,\Delta x_k$$

Once again, that "abstract" definition proves to be just what is needed.

2. <u>Remembering the cylindrical shells formula: think infinitesimals.</u>* As with the method of

disks (see §6.2.1 of <u>The Companion</u>), if you

$$\boxed{\text{regard} \quad dx \quad \text{as an infinitesimal}}$$

there is an easy way to remember - and understand - the cylindrical shells formula. It

goes like this:

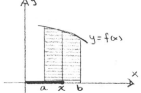

Suppose the region under $y = f(x)$ for $a \leq x \leq b$

is rotated about the   y-axis   to obtain...

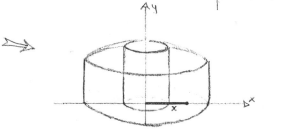

... a solid of revolution

Then:

i. $2\pi x$ = <u>circumference</u> of the

base circle with radius   x

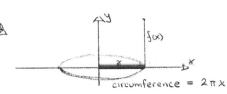

ii. $2\pi x\, f(x)$ = <u>surface area</u> of the cylindrical

sheet at   x   (since, if it is cut

open and rolled out flat, it is a

rectangle of length $2\pi x$ and

width $f(x)$)

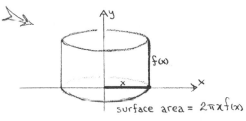

$$\boxed{\text{area} = 2\pi x\, f(x)} \quad f(x)$$

$2\pi x$

* The "infinitesimal approach" will be discussed in general in §6.8 of <u>The Companion</u>.

iii. $2\pi x\,f(x)\,dx$ = approximate <u>volume</u> of the "infinitesmial cylindrical shell" at x (since, if cut open and rolled out flat, it is a rectangular solid of length $2\pi\dot{x}$, width $f(x)$ and depth dx)

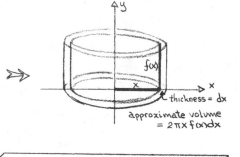

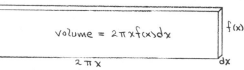

iv.   Then "add up" all the infinitesimal volumes, i.e., integrate over all possible x-values, $a \leqq x \leqq b$, to get the volume V of the solid:

Method of Cylindrical Shells

$$V = \int_{a}^{b} 2\pi x\,f(x)\,dx$$

In addition to being an aid to remembering and understanding the cylindrical shells formula, the infinitesimal approach can be helpful in specific problems:

<u>Example A.</u>   Find the volume of the solid of revolution obtained by rotating the region bounded by the curve $y = x^2$ and the lines x = 1 , x = 2 and y = 0 about the y-axis .

<u>Solution.</u>

The region is to be rotated about the y-axis to obtain ...

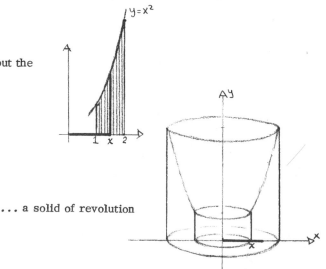

... a solid of revolution

Thus

i. $2\pi x$ = <u>circumference</u> of the

base circle with

radius x

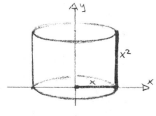

ii. $2\pi x\, f(x)$ = $2\pi x\,(x^2)$ = <u>surface area</u>

of the cylindrical sheet

at x

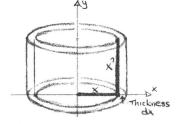

iii. $2\pi x\, f(x)\, dx$ = $2\pi x\,(x^2)\, dx$ = approximate <u>volume</u>

of the infinitesimal cylindrical

shell at x

iv. Then the volume of the solid is the sum of all the infinitesimal volumes over all x

values, $1 \leqq x \leqq 2$, i.e.

$$V = \int_1^2 2\pi x\,(x^2)\, dx = 2\pi \int_1^2 x^3\, dx = (\pi x^4/2)\Big|_1^2 = 15\pi/2 \qquad \square$$

Of course, you could solve Example A simply by plugging into the general Method

of Cylindrical Shells formula given above. However, using the "infinitesimal approach" will

help you avoid misapplying the general formula by constantly reminding you of <u>why</u> the formula

works. It also provides an additional degree of flexibility which the next example shows can be

very useful.

<u>Example B.</u>   Find the volume of the solid of revolution obtained by rotating the region enclosed

between $y = 2/x$ and $x + y = 3$ about the   y-axis.

<u>First Solution.</u>

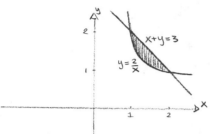

   The region is to be rotated

   about the   y-axis   to obtain ...

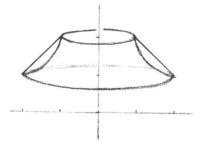

   ... a solid of revolution.   The

   volume V may be thought of as

   what remains when the volume V_2

   obtained by rotating $y = 2/x$,

   $1 \leqq x \leqq 2$,   is subtracted from

   the volume V_1 obtained by rotating

   $x + y = 3$, $1 \leqq x \leqq 2$.

Using this approach, we would obtain

$$V = V_1 - V_2 = \int_1^2 2\pi x (3 - x)\, dx - \int_1^2 2\pi x (2/x)\, dx$$

   (by two applications of the cylindrical shells formula)

$$= \ldots = 13\pi/3 - 4\pi = \pi/3$$

   | we'll leave the integration details to you |

However, the infinitesimal approach allows you to set up the problem in a natural way with just

one integral.   Here are the details:

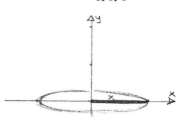

Second Solution.

i. $2\pi x$ = circumference of the base

circle with radius   x

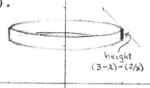

The height of our cylindrical shell is $f(x) - g(x) = (3 - x) - (2/x)$.
Thus...

ii. $2\pi x[f(x) - g(x)] = 2\pi x[(3 - x) - (2/x)] = $ surface

area of cylindrical sheet at   x

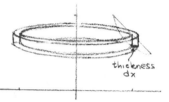

iii. $2\pi x[f(x) - g(x)]dx = 2\pi x[(3 - x) - (2/x)]dx = $ approximate

volume of infinitesimal cylindrical shell

at   x

iv.   Then the volume of the solid is the sum of all the infinitesimal volumes over all   x

values, $1 \le x \le 2$, i. e.,

$$V = \int_1^2 2\pi x[(3 - x) - (2/x)]dx = 2\pi \int_1^2 (3x - x^2 - 2)dx = 2\pi\left(\frac{3x^2}{2} - \frac{x^3}{3} - 2x\right)\Bigg|_1^2 = \pi/3$$

□

Reversing the roles of   x   and   y   allows us to use the method of cylindrical shells

with rotation about the   x-axis .   Then we obtain the formula

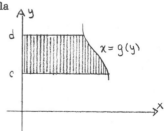

method of cylindrical shells about the x-axis	$$V = \int_c^d 2\pi y\, g(y)\, dy$$

for the volume of the solid of revolution obtained by rotating the region under   x = g(y)   for

c ≦ y ≦ d   about the   x-axis.   As with the   y-axis   version of this formula, we strongly

recommend using the infinitesimal approach in applications.   Here's an example:

Example C.   Use the Method of Cylindrical Shells to find the volume of the solid of revolution

obtained by rotating the area bounded by the curve $y = x^2$,   the line   y = 4   and the   y-axis

about the   x-axis.

Solution.

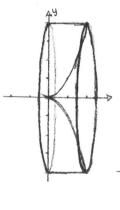

The region is to be rotated

about the   x-axis   to obtain...

... a solid of revolution

Thus

   i. $2\pi y$ = circumference of the

                base circle with radius   y

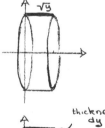

  ii. $2\pi y\, g(y)$ = $2\pi y\,(\sqrt{y})$ = surface area

                 of the cylindrical sheet at   y

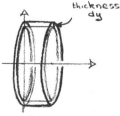

 iii. $2\pi y\, g(y)\, dy$ = $2\pi y\,(\sqrt{y})\, dy$ = approximate volume

                of the infinitesimal cylindrical shell

                at   y

iv.   Then the volume of the solid is the "sum" (i. e. , integral) of all the infinitesimal volumes

over all y values, $0 \leq y \leq 4$, i. e.,

$$V = \int_0^4 2\pi y (\sqrt{y})\, dy = 2\pi \int_0^4 y^{3/2}\, dy = (4\pi y^{5/2}/5)\Big|_0^4 = 128\pi/5 \qquad \square$$

3.   <u>Disks or Shells?</u> ; sometimes you have a choice.   It is frequently possible to use either the

method of disks (or washers) or the method of cylindrical shells to calculate the volume of a

solid of revolution.   In such a situation, one of the two methods will usually be more convenient

(fewer or dimpler integrals, less calculation, etc.).   As an illustration, let us solve Example C

by the Method of Disks:

<u>Example D.</u>   Use the method of disks to find the volume of the solid of revolution obtained by rotating

the area bounded by the curve $y = x^2$, the line $y = 4$ and the y-axis about the x-axis.

<u>Solution.</u>

The region is to be rotated about

the x-axis to obtain . . .

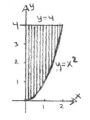

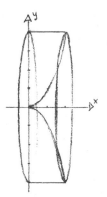

. . . a solid of revolution.

Since we wish to use the method of disks (acually washers) about the x-axis, we must use

the x-variable as our <u>independent</u> variable. You should see that the range of x-values on our region is $0 \leq x \leq 2$. Thus

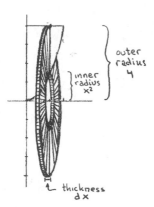

i. $\pi(4)^2 - \pi(x^2)^2 = \pi(16 - x^4)$

is the <u>area</u> of the washer at x.

ii. $\pi(16 - x^4) dx$ is the approximate <u>volume</u> of the infinitesimal washer at x

iii. Then the volume of the solid is the "sum" (i.e., integral) of the infinitesimal volumes over all x-values $0 \leq x \leq 2$, i.e.,

$$V = \int_0^2 \pi(16 - x^4) dx = \pi(16x - x^5/5)\Big|_0^2 = 128\pi/5 \qquad \square$$

<u>Even a third approach is possible in Example C-D</u>:

The volume V may be regarded as a difference $V = V_1 - V_2$ where V_1 is the volume of revolution of $y = 4$, $0 \leq x \leq 2$, about the x-axis and V_2 is the volume of revolution of $y = x^2$, $0 \leq x \leq 2$, about the x-axis.

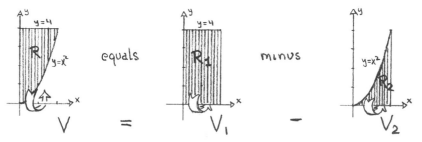

(This is like the first solution in Example B.) Both V_1 and V_2 can be computed easily by the method of disks, i.e.,

$$V_1 = \int_0^2 \pi (4)^2 \, dx = 32 \pi \quad \text{and} \quad V_2 = \int_0^2 \pi (x^2)^2 \, dx = \frac{32}{5} \pi, \text{ so } V = V_1 - V_2 = 128\pi/5$$

In this example all three solutions are of about the same degree of difficulty. This is not always the case: even when both disks and shells can be used in a volume problem, one method will usually be much easier than the other. The general rule is to use the method which produces the simpler integrand function.

4.   More on the "infinitesimal approach": cylindrical shells about any horizontal or vertical line

(optional).   Using the infinitesimal approach to finding volumes of revolution, it is not hard to extend the method of cylindrical shells to axes of revolution other than the coordinate axes. We start with an example.

Example E.   Find the volume of the solid generated when the region   R   enclosed by

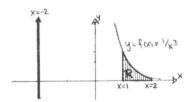

$$y = f(x) = 1/x^3, \quad x = 1, \quad x = 2, \quad y = 0$$

is rotated about the line $x = -2$.

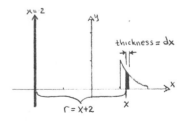

Solution.   Rotating the infinitesimal slice of   R   shown to the right about the line $x = -2$ will yield an infinitesimal cylindrical shell with approximate volume (found by cutting the shell open and rolling it out flat)

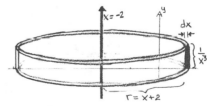

$$dV = 2\pi r \, f(x) \, dx$$
$$= 2\pi (x + 2) \left(\frac{1}{x^3} \right) dx$$

where $r = x - (-2) = x + 2$ is the distance from the slice at   x   to the axis of revolution

$x = -2$. Thus the total volume obtained by revolving R about the line $x = -2$ will be the "sum" of the infinitesimal volumes dV as x takes on all values in the interval $[1, 2]$, i. e.,

$$V = \int dV = \int_1^2 2\pi (x + 2) \left(\frac{1}{x^3}\right) dx$$

$$= 2\pi \int_1^2 x^{-2} dx + 4\pi \int_1^2 x^{-3} dx$$

$$= 2\pi (-x^{-1}) \Big|_1^2 + 2\pi (-x^{-2}) \Big|_1^2$$

$$= 5\pi/2 \qquad \square$$

The approach used in this example generalizes in the following way.

Suppose R is the region under a curve $y = f(x)$ above the x-interval $[a, b]$, and we wish to compute the volume obtained when R is rotated about a line $x = x_0$, with $x_0 \le a$. Rotating the infinitesimal slice of R shown to the right about the line $x = x_0$ will yield an infinitesimal cylindrical shell with approximate volume

$$dV = 2\pi r f(x) dx$$

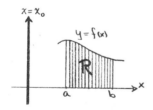

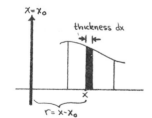

where $r = x - x_0$ is the distance from the slice at x to the axis of revolution $x = x_0$. Thus the total volume obtained by rotating R about the line $x = x_0$

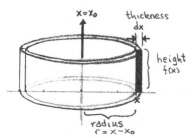

will be the "sum" of the infinitesimal volumes d V as x takes on all values in the interval

[a,b], i.e.,

Shells
about
x = x_0

$$V = \int dV = \int_a^b 2\pi\, r\, f(x)\, dx \qquad \text{where} \quad x_0 \le a\,, \text{ and}$$

$$r = x - x_0\,, \text{ the distance from } x \text{ to the axis of rotation } x = x_0\,.$$

It is by far more useful to learn the "infinitesimal approach" which leads to the formula

than to memorize the formula itself! However, the formula does highlight the following

important fact: in the method of cylindrical shells, we integrate $2\pi\, r\, f(x)$ where r is the

distance from an infinitesimal slice of the region at x to the axis of revolution.

In the derivation given above, the axis of revolution $x = x_0$ was to the left of the region

(i. e., $x_0 \le a$) and $r = x - x_0$, but the shells formula about $x = x_0$ is also valid when the

axis of revolution is to the right of the region (i. e., $b \le x_0$) and we change r to be

$r = x_0 - x$. This is illustrated in the next example.

Example F.   Find the volume of the solid generated when the region R   enclosed by

$$y = f(x) = x^3 - x + 1\,, \quad x = 0\,, \quad x = 1\,, \quad y = 0$$

is rotated about the line x = 2 .

Solution.   Rotating the infinitesimal slice of R

shown to the right about the line x = 2   will yield

an infinitesimal cylindrical shell with approximate

volume

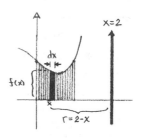

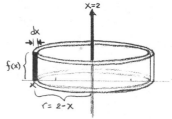

$$d V = 2\pi r f(x) dx = 2\pi (2 - x)(x^3 - x + 1) dx$$
$$= 2\pi (-x^4 + 2x^3 + x^2 - 3x + 2) dx \quad ,$$

where $r = 2 - x$ is the distance from the slice at x to the axis of rotation $x = 2$. (Notice that r equals $2 - x$ rather than $x - 2$ because x is less than 2 and <u>r is always chosen to be positive</u>.) Thus the total volume obtained by rotating R about the line $x = 2$ will be the "sum" of the infinitesimal volumes dV as x takes on all values in the interval $[0, 1]$, i.e.,

$$V = \int dV = \int_0^1 2\pi (-x^4 + 2x^3 + x^2 - 3x + 2) dx$$
$$= 2\pi \left(-\frac{x^5}{5} + \frac{x^4}{2} + \frac{x^3}{3} - \frac{3x^2}{2} + 2x \right) \Bigg|_0^1$$
$$= 2\pi \left(-\frac{1}{5} + \frac{1}{2} + \frac{1}{3} - \frac{3}{2} + 2 \right) = \frac{34}{15} \pi \qquad \square$$

<u>Example G.</u>   Find the volume of the solid generated when the region R enclosed by the curves

$$x = g(y) = y^2 \quad \text{and} \quad x = h(y) = 3y - 2y^2$$

is rotated about the line $y = -1$.

<u>Solution.</u>   The two curves are parabolas, and the first step is to determine where they intersect. We do this by setting $g(y)$ equal to $h(y)$ and solving for y:

$$3y - 2y^2 = x = y^2 \quad ,$$
$$\text{so} \qquad 0 = 3y - 3y^2 = 3y(1 - y)$$

This gives $y = 0$ and $y = 1$, so, using $x = y^2$, the points of intersection are $(0,0)$

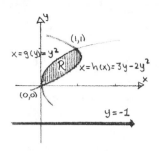

and (1, 1) . The region R is pictured to

the right.

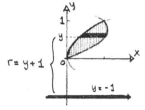

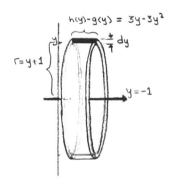

Rotating the infinitesimal slice of R

shown to the left about the line $y = -1$ will

yield an infinitesimal cylindrical shell with

approximate volume

$$dV = 2\pi r \,[\, h(y) - g(y) \,]\, dy$$

$$= 2\pi (y + 1) \,[\, 3y - 3y^2 \,]\, dy$$

where $r = y - (-1) = y + 1$ is the (positive)

distance from the slice at y to the axis of

rotation $y = -1$. Thus the total volume obtained

by rotating R about the line $y = -1$ will be the "sum" of all the infinitesimal volumes dV

as y takes on all values in the interval $0 \le y \le 1$, i.e.,

$$V = \int dV = \int_0^1 2\pi (y + 1)\,[\, 3y - 3y^2 \,]\, dy = \ldots = 3\pi/2 \qquad \square$$

$\uparrow$

┌─────────────────────┐
│ The details are left │
│   for you to do.   │
└─────────────────────┘

   Notice how the infinitesimal approach allows us to switch easily from rotating about

vertical lines to rotating about horizontal lines, and to handle regions between curves. Again

we suggest that you use this approach with all cylindrical shells computations.

6.4.1

Exercises. Use the infinitesimal cylindrical shells approach to find the volume of the solid
of revolution obtained by rotating the given region about the indicated axis of revolution.

1. The region under $y = (x - 1)^3$ for $3 \leq x \leq 4$ about $x = 1$

2. The region under $x = -y^2 + 2y + 3$ for $-1 \leq y \leq 3$ about $y = -1$

3. The region between the curves $y = x^2 - 2$ and $y = 2x - 2$ about $y = -2$

4. The region between the curves $x = \sqrt{y} + 4$ and $x = y^3 + 4$ about $x = 4$

Answers. 1) $422\pi/5$ 2) $128\pi/3$ 3) $64\pi/15$ 4) $5\pi/14$

Section 6.4.   Length of a Plane Curve

1. Finding arclength: another application of that abstract Riemann Sum   definition of the definite
integral. The definition of the arclength L of a smooth curve $y = f(x)$ from $(a, f(a))$ to
$(b, f(b))$

Definition of arc length	$$L = \int_a^b \sqrt{1 + [f'(x)]^2}\, dx$$

is another use of the "abstract" Riemann Sum definition of the definite integral (Anton's Definition
5.7.2). As Anton's discussion makes clear, that definition turns what appears to be a non-
integration problem into an integration problem. Another "abstract" result, the Mean Value
Theorem (Theorem 4.10.2), also plays a critical role in the development of the arclength
formula. (It is used following formula (1).) All that "abstract" mathematics is proving to be
very "applied."

Note that the assumption that f is a smooth function, i.e., that f' exists and is continuous, is essential in making the definition of arclength. Since f' is continuous, so is

the function $\sqrt{1 + [f'(x)]^2}$; hence (by Theorem 5.7.3(a)) the integral $\int_a^b \sqrt{1 + [f'(x)]^2}$

exists. Without the assumption of smoothness, the definition of L might not make sense!

2. <u>Remembering the arclength formula: think infinitesimals.</u> As we have seen several times

already (e. g. , in §§6.2.1 and 6.3.2 of <u>The Companion</u>) , formulas are sometimes easy to

remember, and understand, if we "think infinitesimally. " This is true of the arclength formula

as well:

The length of an infinitesimal piece of the

curve at x is given (approximately)

by

$$ds = \sqrt{(dx)^2 + (dy)^2}$$

where dx and dy are the (infinitesimal)

changes in x and y over that piece.

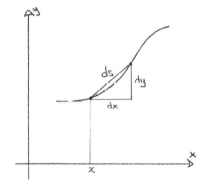

Since

$$ds = \sqrt{(dx)^2 + (dy)^2} = \sqrt{\left(1 + \left(\frac{dy}{dx}\right)^2\right)(dx)^2} = \sqrt{1 + [f'(x)]^2}\, dx$$

the arclength L from (a, f(a)) to (b, f(b)) may be obtained by "adding up" all these

infinitesimal pieces, i. e. , by integrating over all x-values, $a \leq x \leq b$:

$$L = \int_a^b \sqrt{1 + [f'(x)]^2}\, dx$$

Here is an application of this approach.

6.4.3

Example A. Find the arclength of the curve $y = 2x^{3/2}$ between $x = 1/3$ and $x = 5/3$.

Solution. The length of an infinitesimal

piece of the curve at x is approximately

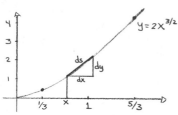

$$ds = \sqrt{(dx)^2 + (dy)^2} = \sqrt{1 + \left(\frac{dy}{dx}\right)^2}\ dx = \sqrt{1 + [f'(x)]^2}\ dx = \sqrt{1 + (3\sqrt{x})^2}\ dx$$

$$= \sqrt{1 + 9x}\ dx \quad .$$

Thus the arclength L from $(1/3, 2\sqrt{3}/9)$ to $(5/3, 10\sqrt{5/3}\sqrt{3})$ is

$$L = \int_{1/3}^{5/3} \sqrt{1 + 9x}\ dx = \frac{1}{9}\int_{1/3}^{5/3} (1 + 9x)^{\frac{1}{2}} 9\, dx = \frac{2}{27}(1 + 9x)^{3/2}\Big|_{1/3}^{5/3}$$

$$= \frac{2}{27}(64 - 8) = 112/27 \qquad \square$$

3. The truth of the matter: evaluating arclength integrals isn't easy. Unfortunately most arc-length integrals encountered outside of textbooks are not as easy to evaluate as the one in Example A above. In fact, most arclength integrals out in the "real world" cannot be evaluated in "closed form," i.e., by integral formulas involving only elementary functions. When this happens, the best you can do is approximate the value of the integral, using methods such as those to be developed in §9.9. For now, here is an example to illustrate our point:

Example B. Find the arclength of the curve $y = \sin x$ between $x = 0$ and $x = \pi$.

<u>Solution.</u>   The arclength formula says that the

arclength   L   of   f(x) = sin x   from   (0,0)

to   (π,0)   is

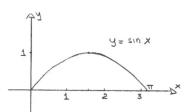

$$ L = \int_a^b \sqrt{1 + [f'(x)]^2}\, dx = \int_0^\pi \sqrt{1 + \cos^2 x}\, dx $$

But we do not know how to evaluate this integral! Moreover, there are no formulas on the

inside covers of the textbook which tell us how. The best we can hope for is an approximation

of L....   HELP!!! (Never fear! Help will come later in §9.9.)   □

4.   <u>The geometric meaning of "smoothness."</u> <u>(Optional)</u>   Consider the curve $f(x) = |x|$,

- 1 ≦ x ≦ 1 ,   the graph of which is

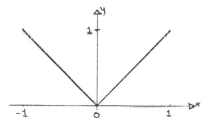

The most obvious feature of this graph is its sharp corner at the origin. The graph is not

"smooth" at that point.

What causes this lack of smoothness? We can see by calculating the derivative f' :

$$ f(x) = |x| = \begin{cases} x & \text{for } 0 \leq x \leq 1 \\[2mm] -x & \text{for } -1 \leq x < 0 \end{cases} \qquad \text{so} \qquad f'(x) = \begin{cases} 1 & \text{for } 0 < x < 1 \\[2mm] \text{undefined for } x = 0 \\[2mm] -1 & \text{for } -1 < x < 0 \end{cases} . $$

Observe that the derivative is not defined at x = 0 and changes abruptly (from - 1 to 1)

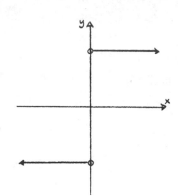

as x passes from negative to positive values. The graph of y = f'(x) is shown to the right. Thus f' is <u>not</u> <u>continuous</u> at x = 0 since it is not even defined at x = 0! That's what causes f to be "not smooth" at the origin.

A more bizarre example is as follows: consider the function

$$g(x) = \begin{cases} x^2 \sin(\frac{1}{x}) & \text{for} \quad x \neq 0 \\ \\ 0 & \text{for} \quad x = 0 \end{cases}$$

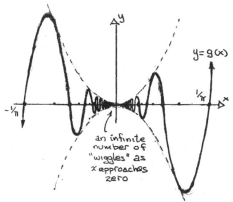

Although bounded between the curves $y = x^2$ and $y = -x^2$, the graph of g fluctuates wildly as x approaches zero, and it would be reasonable to claim that g is not "smooth" in any geometric sense at x = 0. (Howard Ant would get very dizzy riding a roller coaster along this curve into x = 0!) It can be shown, however, that g <u>is</u> differentiable at every value of x, including x = 0. The derivative is

$$g'(x) = \begin{cases} 2x \sin(\frac{1}{x}) - \cos(\frac{1}{x}) & \text{for} \quad x \neq 0 \\ \\ 0 & \text{for} \quad x = 0 \end{cases}$$

But notice that $\lim_{x \to 0} g'(x)$ is <u>undefined</u>! Thus, although the derivative g' is defined everywhere, it is <u>not continuous</u> at x = 0. That's what causes g to be "not smooth" at the origin.

Based on these examples, it is very natural to <u>define</u> a smooth function to be a function whose derivative is continuous, that is

Definition of a smooth function	$y = f(x)$ is <u>smooth</u> on $[a, b]$ if $f'(x)$ is continuous on $[a, b]$.

The name "smooth" is absolutely perfect: our examples should indicate that you won't find any

sharp corners or points where the graph behaves weirdly on the graph of a smooth function.

Section 6.5.   Area of a Surface of Revolution.

1.   Finding surface area: another application of that abstract Riemann Sum definition of the definite

integral.   The definition of the surface area   S   of the solid of revolution obtained by rotating

the curve $y = f(x)$ over $[a, b]$ about the   x-axis

Definition of Surface Area of a Solid of Revolution	$$S = \int_a^b 2\pi f(x) \sqrt{1 + [f'(x)]^2}\ dx$$

is once again an application of the "abstract" Riemann Sum definition of the definite integral

(Definition 5.7.2).   That definition makes an integration problem out of what is apparently a

non-integration problem.   And, once again, the "abstract" Mean Value Theorem

(Theorem 4.10.2) and Intermediate Value Theorem (Theorem 3.7.8) are used in a prominent

way. Note their uses following Anton's formulas (2) and (3).

2. <u>Remembering the surface area formula: think infinitesimals.</u> Once again, an infinitesimal

approach can help to understand and remember the surface area formula given above. Here

is the procedure:

(1)   The length   ds   of an infinitesimal

piece of the curve   y = f(x)   at   x

is approximately

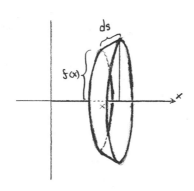

$$ds = \sqrt{dx^2 + dy^2} = \sqrt{\left[1 + \left(\frac{dy}{dx}\right)^2\right]dx^2} = \sqrt{1 + [f'(x)]^2} \; dx$$

(2)   Thus the area   dS   of an infinitesimal band

of the surface is approximately

$$dS = \left\{\begin{array}{c} \text{circumference} \\ \text{of circle} \\ \text{of radius} \quad f(x) \end{array}\right\} \times \left\{\begin{array}{c} \text{width} \\ ds \\ \text{of band} \end{array}\right\}$$

$$= 2\pi f(x) \; ds$$

$$= 2\pi f(x) \sqrt{1 + [f'(x)]^2} \; dx$$

(3)   Then "add up" all the areas of the infinitesimal bands, i.e., integrate over all possible

x   values,   a ≤ x ≤ b,   to get the surface area   S   of the solid

$$S = \int_a^b 2\pi f(x) \sqrt{1 + [f'(x)]^2} \; dx$$

We hope this derivation will make the formula less mysterious. Here is this approach

applied in an example.

Example A.   The arc of the curve $y = x^3$ above $[0,2]$ is rotated about the x-axis. Find

the area of the surface generated.

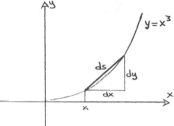

Solution.   The length of an infinitesimal

piece of the curve $y = x^3$ at x is

approximately

$$ds = \sqrt{dx^2 + dy^2} = \sqrt{1 + [f'(x)]^2}\ dx = \sqrt{1 + (3x^2)^2}\ dx = \sqrt{1 + 9x^4}\ dx$$

Thus the area of an infinitesimal band of

the surface is approximately

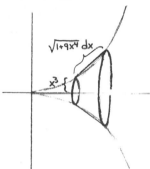

$$dS = 2\pi f(x)\ ds = 2\pi x^3 \sqrt{1 + 9x^4}\ dx .$$

Hence, "adding up" all the areas of infinitesimal

bands over $0 \leq x \leq 2$, we obtain

$$S = \int_a^b 2\pi f(x) \sqrt{1 + [f'(x)]^2}\ dx = \int_0^2 2\pi x^3 \sqrt{1 + 9x^4}\ dx$$

$$= \frac{2\pi}{36} \int_{x=0}^{x=2} u^{1/2}\ du \qquad (\text{with}\quad u = 1 + 9x^4,\quad du = 36x^3\ dx)$$

$$= \frac{2\pi}{36} \frac{2}{3} (1 + 9x^4)^{3/2} \Big|_0^2 = \frac{\pi}{27} [(145)^{3/2} - 1]$$

One can, of course, solve Example A by plugging the appropriate functions and numbers

into the surface area formula. However, that requires that you memorize the formula! This

infinitesimal approach is more flexible (i. e., it can apply to various situations) and its

geometric nature makes it easy to remember.   □

3. <u>The lateral area of the frustum of a cone</u> (Optional).   The critical approximation step in applying the definition of the definite integral to define the surface area of a solid of revolution uses the formula for the lateral area   S   of the frustum of a cone

<table>
<tr><td>Surface
Area of a
Cone Frustum</td><td>$$S = \pi(r_1 + r_2)\,\ell$$</td></tr>
</table>

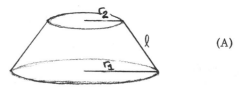

(A)

where r_1 and r_2 are the radii of the bases and ℓ is the <u>slant</u> height.   This formula is derived as follows:

We start by determining the surface area   A   of a cone of slant height   L   and base radius   r ,   as shown to the right.   Cut along a lateral edge and laid flat, the cone becomes a disk of circumference $2\pi r$ with one sector removed, as shown to the right.   Let θ be the central angle of the sector that remains.

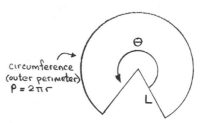

Clearly we have the following proportionality:

$$\frac{\text{area of sector}}{\text{total area of disk}} = \frac{\text{central angle of sector}}{\text{total central angle of disk}}$$

$$\frac{A}{\pi L^2} = \frac{\theta}{2\pi}$$

Hence $$A = L^2\,\theta/2$$ (B)

However, we would prefer to express   A   in terms of   L   and   r ,   not   L   and θ .   To do this, observe that the outer perimeter P

of our sector must satisfy the proportionality

$$\frac{P}{2\pi L} = \frac{\theta}{2\pi}$$

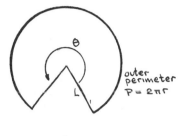

outer
perimeter
$P = 2\pi r$

But P is also the circumference of the base

of the original cone, so $P = 2\pi r$. Thus

$$\frac{2\pi r}{2\pi L} = \frac{\theta}{2\pi}$$

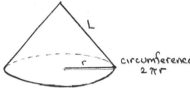

circumference
$2\pi r$

which gives $\theta = 2\pi r/L$. Putting this into (B) yields

$$A = \frac{L^2}{2}\left(\frac{2\pi r}{L}\right)$$

or $\boxed{A = 2rL}$ (C)

We now compute the lateral area S of the frustum of a cone with base radii r_1

and r_2 and slant height ℓ . This area is the difference

of the surface areas of the (big) cone with radius r_1

and slant height L and the (small) cone with

radius r_2 and slant height $L - \ell$. From

formula (C) this means

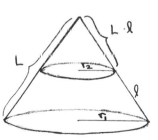

$$S = \pi r_1 L - \pi r_2 (L - \ell)$$

But by similar triangles

$$\frac{r_1}{L} = \frac{r_2}{L - \ell}$$

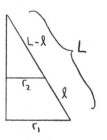

or $\;\; L = r_1 \ell / (r_1 - r_2)$. Thus

$$S = \pi r_1 L - \pi r_2 (L - \ell) = \pi L (r_1 - r_2) + \pi r_2 \ell$$

$$= \pi \left(\frac{r_1 \ell}{r_1 - r_2} \right) (r_1 - r_2) + \pi r_2 \ell = \boxed{\pi (r_1 + r_2) \ell}$$

This proves formula (A).

Section 6.6. Work

The importance of the concept of "work" to physics and engineering cannot be over-emphasized. The solution of work problems is a major application of the definite integral.

1. <u>The basic law: work equals force times distance.</u> The solution to every work problem depends on the following basic definition:

> if an object moves a distance   d   along a line subject to a
> <u>constant</u> force   F ,   the work   W   done is the product of
> the force and the distance. That is,

Basic
Work
Formula

$$\boxed{W \;=\; F \cdot d} \tag{A}$$

For instance, if Howard Anton exerts a constant force of 200 lbs to push his automobile a distance of 100 ft, the work he does is

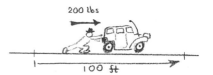

$$W \;=\; 200 \cdot 100 \;=\; 20,000 \;\; \text{ft/lbs.}$$

If he leaves the car's emergency brake on, so that it requires a constant force of 10,000 lbs. to push the car, the work he does is

$$W = 10,000 \cdot 100 = 1,000,000 \quad \text{ft/lbs.}$$

Because of his stupidity, Howard has worked harder! He would also work harder if he pushed the car farther; if he pushed the car (with the emergency brake <u>off</u>) a distance of 1200 ft, the work he does would be

$$W = 200 \cdot 1200 = 240,000 \quad \text{ft/lbs.}$$

Anyone who has ever pushed a car understands the essential truth of "more distance means more work" and "more force means more work." The basic law $W = F \cdot d$ merely expresses this mathematically.

The more general work problems which are encountered in applications tend to be of two types: <u>variable force</u> and <u>decomposable object</u> work problems. Their solutions require the definite integral, and are derived from the basic law $W = F \cdot d$. In this way the law $W = F \cdot d$ plays the same role in general work problems that the formula $V = A \cdot h$ (for the volume of a right cylinder) plays in volume-of-revolution problems: the more general laws are derived from the basic law.

We will treat each of our two types of work problems in detail.

2. <u>Variable force work problems.</u>

An object moves from $x = a$ to $x = b$ under the action of a <u>variable force</u> in the direction of motion, i.e., a force $F(x)$ which depends on the position x of the object.

Variable
Force

For instance, suppose Howard Anton pushes his automobile a distance of 100 ft, as before, but now it is winter and there are some patches of ice on the road. At those points, the force

required to push the car is substantially less. Thus the force required <u>varies</u>. It is expressed as a function F(x) of position, i.e., F(x) is the force required to push the car when it is at position x.

The variable force work problem is solved as follows:

if an object moves along a line from x = a to x = b subject to a <u>variable</u> force F(x) in the direction of motion, the work W done is the definite integral of F(x) from x = a to x = b , i.e.,

<table>
<tr><td>
Definition of
work from
a variable
force
</td><td>

$$W = \int_a^b f(x)\,dx$$

</td><td>(B)</td></tr>
</table>

This formula and its derivation are easy to remember and to understand if we "think infinitesimally." The work done by the force in moving the object over the infinitesimal distance dx from x to x + dx is (approximately) equal to

$$dW = F(x)\,dx$$

Here we have considered the force to be a <u>constant</u> value F(x) over the infinitesimal interval [x , x + dx] and used our basic formula (A). The total work W may then be obtained by "adding up" all the infinitesimal pieces of work, i.e., by integrating over all x values,

$a \le x \le b$:

$$W = \int_a^b dW = \int_a^b F(x)\, dx$$

[Anton's motivation for formula (B) is similar to the argument above except that he does not use infinitesimals. Instead, he uses the Riemann Sum definition of the definite integral (Definition 5.7.2). The key step is the approximation of a variable force by a constant force over small distances.]

Spring Problems.   Problems involving Hooke's law for springs, such as Anton's Example 2, are examples of variable force work problems. Here is another spring problem:

Example A.   A spring has a natural length of 20 inches, and a 40 lb force is required to compress it to a length of 18 inches. How much work is done on the spring in compressing it from 18 inches to 16 inches?

Solution.   Let x represent the distance in inches that the compressed spring differs from its natural equilibrium position. Then we have an "object" (the end of the spring) moving along a line from $x = 20 - 18 = 2$ to $x = 20 - 16 = 4$, subject to a variable force F(x) given by Hooke's Law. (Hooke's Law applies equally well to compression as to extension of springs.)

(1) Compute the force F(x) at any given x-value. By Hooke's law we have

$$F(x) = kx$$

where k is the spring constant for our spring. We must determine k from the given information. This is easy, since we are given

$$F = 40 \text{ lbs} \quad \text{when} \quad x = 20 - 18 = 2 \text{ inches}$$

Thus $40 = F(2) = 2k$,   or $k = 20$,   proving

$$F(x) = 20x$$

(2) <u>Compute the work</u> done in moving the spring from $x = 2$ to $x = 4$ by using the formula $W = \int_a^b F(x)\, dx$:

$$W = \int_2^4 20x\, dx = 10x^2 \Big|_2^4 = 10(16 - 4) = 120 \text{ in - ft} \qquad \square$$

<u>Lifting Problems.</u>   Situations in which an object undergoes a weight change during a lift are also examples of variable force work problems.   Here is an illustration:

<u>Example B.</u>   A bag of sand originally weighing 144 lb. is lifted at a constant rate.   The sand leaks out uniformly at such a rate that half of the sand is lost when the bag has been lifted 18 ft.   Find the work done in lifting the bag this distance.

<u>Solution.</u>   Let x represent the distance in feet of the bag from the ground.   Then we have an object (the bag of sand) moving along a vertical line from $x = 0$ to $x = 18$,   subject to a variable force $F(x)$,   the weight of the bag when it is at height x .

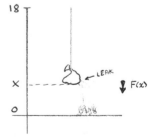

(1) <u>Compute the force</u> (weight) $F(x)$ at the height x .   Since 72 lbs. of sand are lost over the 18 foot lift (and the rate of lift and the rate of sand loss are both constant), then $F(x)$ drops by $72/18 = 4$ lbs per foot.   Thus

$$F(x) = 144 - 4x$$

(2) Compute the work done in moving the bag from $x = 0$ to $x = 18$ by using the

formula $W = \int_a^b F(x)\, dx$:

$$W = \int_0^{18} (144 - 4x)\, dx = (144x - 2x^2)\Big|_0^{18}$$

$$= 2592 - 648 = 1944 \quad \text{ft.-lbs.} \qquad \qquad \sqcap$$

3. Decomposable Object Work Problems.

Decomposable Object Work Problems

An object is composed of (infinitesimal) pieces, each of which moves

its own distance under the action of its own (constant) force in the

direction of motion. Thus both the distance traveled and the force

vary from point to point in the object.

The basic technique for solving this type of work problem is to decompose the object into

(infinitesimal) pieces each of which moves a fixed distance, compute the (infinitesimal) work

dW done on each piece and "add them up" (i. e. , integrate) to get the total work W :

$$W = \int dW \qquad \qquad (C)$$

Pumping Problems.   The following example illustrates the solution technique we

recommend for decomposable object work problems:

<u>Example C.</u>   A cone-shaped water reservoir is 20 ft. in diameter across the top and 15 ft. deep.

If the reservoir is filled to a depth of 10 ft. , how much work is required to pump the top 5 feet

of the water to the top of the reservoir? (The density of water is $\rho = 62.4 \ \mathrm{lb/ft}^3$.)

<u>Solution.</u>   This is a decomposable object work

problem because each horizontal circular "sheet"

of water must be pumped a different distance (from

where it is to the top of the reservoir).

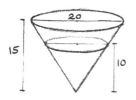

    Our approach will be to divide the water

into "infinitesimal disks" of thickness   dx   [so

thin that all portions of a given disk can be considered

to move the same distance].   Then consider the

infinitesimal disk at height   x :

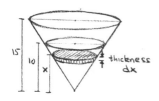

    (1) <u>Compute the surface area</u> of its base.   The

surface area is πr^2 where   r   is the radius at height   x .

To write πr^2 in terms of   x ,   we use similar triangles:

$$\frac{r}{10} \ = \ \frac{x}{15} \qquad \text{so} \qquad r \ = \ \frac{2}{3} \, x$$

Hence the surface area is $\frac{4}{9} \pi x^2$.

    (2) <u>Compute the infinitesimal volume</u>   d V   of the infinitesimal disk of thickness   dx :

$$d V \ \approx \ (4/9) \pi x^2 \, dx \qquad \text{(i. e. , base area times thickness)}$$

    (3) <u>Compute the infinitesimal force</u> (weight)   d F   of the infinitesimal disk.   Recalling

the basic relationship

$$\boxed{\text{weight} \;=\; \text{density times volume}}$$

we obtain

$$dF \;=\; \rho\, dV$$

$$=\; (62.4)\,(4/9)\,\pi x^2\, dx$$

$$=\; 27.73\,\pi x^2\, dx$$

(4) <u>Compute the infinitesimal work</u> dW done in pumping the infinitesimal disk to the top of the reservoir by using the basic work formula $W = F \cdot d$. The infinitesimal disk at height x moves a distance of $15 - x$ to reach the top of the reservoir. Thus

$$dW \;=\; (15 - x)\,dF \;=\; (15 - x)\,(27.73)\,\pi x^2\, dx$$

(5) <u>Compute the total work</u> W done in pumping the top five feet of water to the top of the reservoir by using $W = \displaystyle\int dW$. We are moving all the infinitesimal disks that correspond to x values <u>between 5 and 10</u>! "Adding up" all the corresponding infinitesimal amounts of work will yield

$$W \;=\; \int dW \;=\; \int_5^{10} (15 - x)\,(27.73)\,\pi x^2\, dx$$

$$=\; 27.73\,\pi \int_5^{10} (15 x^2 - x^3)\, dx \;=\; 27.73\,\pi\,(5 x^3 - x^4/4)\Big|_5^{10}$$

$$=\; 27.73\,\pi\,(5000 - 2500 - 625 + 156.25) \;\cong\; 56{,}326.6\,\pi \;\cong\; 176{,}955 \text{ ft.-lbs.} \qquad \square$$

<u>Lifting Problems in which different parts move different distances</u> are also examples of

decomposable object work problems. For instance,

Example D.   A 20 foot chain weighs 2 lbs/ft. How much work is done in taking it by one end and

lifting it until the other end is 30 feet off the ground? (Disregard the horizontal motion of the

chain along the ground.)

Decomposable Object Solution.   Once the chain has cleared the ground (i. e. , when the top is´

20 feet up), the weight of the chain is constant   (20 x 2 = 40 lbs) and the work done in lifting

it the final 30 feet is   W = F · d = (40) (30) = 1200 ft - lbs. However, to get the total work,

we must add to this the work done in lifting the (top of the) chain the first 20 feet. This can be

considered as a decomposable object work problem because each piece of the chain must move

a different vertical distance (remember: we are

disregarding horizontal motion).

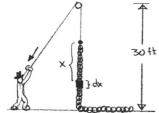

    We can solve the problem as follows:

Consider the infinitesimal piece of the chain of

length   dx   located   x   feet from the lifted end of the chain:

    (1) Compute the infinitesimal force (weight) dF   of the infinitesimal piece at   x :

$$dF = 2\,dx \qquad (2 \text{ lbs/ft times } dx \text{ feet})$$

    (2) Compute the infinitesimal work dW   done in lifting the infinitesimal piece of chain

by using the basic work formula   W = F · d .

When the lifted end of the chain reaches a height

of 20 feet, then the infinitesimal piece   dx   at   x

will have been lifted a distance   20 - x   (see

figure to the right).   Thus

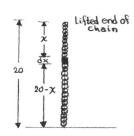

$$dW = (20 - x)\, dF = 2(20 - x)\, dx$$

(3) <u>Compute the total work</u>   W   done in lifting the chain to the point where the top is

20 feet off the ground by using $W = \int dW$. We are moving all the infinitesimal chain lengths

that correspond to   x   values between   0   and   20 . "Adding up" all the corresponding

infinitesimal amounts of work will yield

$$W = \int dW = \int_0^{20} 2(20 - x)\, dx$$

$$= (40x - x^2)\Big|_0^{20} = 400 \ \text{ft-lbs}$$

Hence the total work done in lifting the chain until the bottom end is 30 feet off the ground is

$$1200 + 400 = 1600 \ \text{ft-lbs} \qquad\qquad \square$$

Some problems can be done using either variable force or decomposable object methods.
Example D is in fact such a problem. The decomposable object solution has been given above.
Here is a variable force solution.

<u>Variable force solution for Example D.</u>   Once again, 1200 ft-lbs. of work are required to lift

the chain once it is off the ground. (It is a simple $W = F \cdot d$ type problem.) To find the

work done in lifting the (top of the) chain the first 20 feet, the top of the chain may be regarded

as an object which <u>gains weight as it rises</u> (i. e. , as it rises, there is more of the chain off the

ground, causing it to weight more).

(1) <u>Compute the force</u> (weight) $F(x)$ of the chain when (the top of) the chain is x feet off the ground. Since the chain weighs 2 lbs. per foot, then

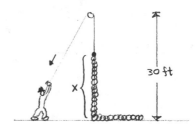

$$F(x) = 2x$$

(2 lbs. per foot times x feet).

(2) <u>Compute the work</u> done in lifting the chain from $x = 0$ to $x = 20$ by using the formula $W = \int_a^b F(x)\,dx$:

$$W = \int_0^{20} 2x\,dx = x^2 \Big|_0^{20} = (20)^2 = 400 \text{ ft-lbs.}$$

So, once again, the total work done is $1200 + 400 = 1600$ ft-lbs. □

4. <u>Overview.</u> Here is a tabular summary of the various types of work problems. In each case we calculate the work done in moving an object over a distance subject to a force:

Type of Work Problem	Equation	Characteristics
Basic	(A)	Constant force is applied over a fixed distance
Variable force	(B)	Variable force is applied over a fixed distance
Decomposable object	(C)	The force applied and distance travelled vary from point to point in the object

<u>Exercises.</u>   In solving these additional work problems, first decide which type of work problem

it is and proceed accordingly.

1.   If the spring constant is 100 lb/in , find the work done in compressing a spring of

natural length 20 in. from a length of 19 in. to a length of 15 in.

2.   Find the work required to empty a conical reservoir of radius 6 ft at the top and height

8 ft if it is full of a liquid of weight density δ lb/ft^3 and the liquid must be lifted 4 ft above

the top of the reservoir.

3.   Solve Problem 2 if the reservoir is filled only to a depth of 4 ft.

4.   The pull needed to bring a water skier out of the water decreases remarkably as the

skier rises. Suppose the pull is $120 - x^2/4$ lb   (for $0 \leq x \leq 20$)   where   x   is the distance

in feet that the skier has moved from his starting point and that he is completely out of the water

after being pulled 20 ft.   What is the force on the tow rope (a) just after the boat starts and

(b) after the boat and the skier have gone 20 ft ? (c) How much work is done on the skier during

the first 10 ft. of travel and (d) during the first 20 ft. of travel?

5. a) Find the work done in lifting a 10,000 -lb. elevator 50 ft. if the weight of the cables is

ignored.

    b) Find the work done in lifting a 10,000 - lb. elevator 50 ft. if the cables weight 20 lb/ft.

6.   An anchor for a large ship weighs 5,000 lb and the anchor chain weighs 200 lb/ft.   Find

the work required to lift the anchor from a point 30 ft below the surface of the water to a point

20 ft above the surface of the water.

<u>Answers.</u>   1.   100 ft - lbs.   2.   576 $\pi \delta$ ft - lbs.   3.   108 $\pi \delta$ ft - lbs.   4.   (a) 120 lbs.

(b) 20 lbs.   (c) 1,116. 67 ft - lbs.   (d) 1,733. 33 ft - lbs.   5.   (a) 500,000 ft - lbs.

(b) 525,000 ft - lbs   6.   500,000 ft -lbs.

Section 6.7.   Liquid Pressure and Force

1.   The basic principles.   If you are building boats or dams, it is obviously important to know how to compute the fluid force (or liquid pressure) on a surface.   As is usual in calculus, we start with a basic equation

Basic
Equation
for fluid
pressure

$$p \;=\; \rho h \tag{A}$$

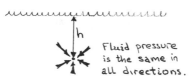

where   p   is the pressure (force per unit area) exerted by a liquid of density ρ (weight per unit volume) at a depth   h .   Fluid pressure is the same in all directions (Pascal's Principle), so no matter how a submerged surface   S   is oriented (horizontal, vertical, slanted), any point   a   at a depth   h   will feel the pressure $p = \rho h$.

As we use (A) to compute the total fluid force on various types of submerged surfaces, we will keep the development simple by considering only flat surfaces.   However, the basic principles which we will develop apply to non-flat surfaces as well.

2.   Horizontal surfaces.   The fluid force on a flat surface of area   A   submerged horizontally at a depth   h   in a fluid is easy to compute, since, by equation (A), the pressure on the surface is constant.   Thus

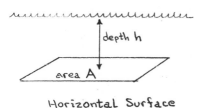

Horizontal Surface

$$\text{fluid force} = \begin{bmatrix} \text{force per unit area} \\ \text{(i. e., the pressure)} \end{bmatrix} \times [\text{area}]$$

which gives, by equation (A),

┌─────────────┐
│ Fluid   │
│ Pressure on │ $$\boxed{F = \rho h A}$$ (B)
│ horizontal │
│ flat surface│
└─────────────┘

where A is the area of the surface. Anton's Example 1 is a straightforward application

of equation (B).

3. <u>Vertical surfaces.</u>   Now the plot thickens,

for fluid pressure is NOT constant over

a flat surface submerged <u>vertically</u>: it

is a function of the (varying) depth h .

We will need calculus to solve this problem!

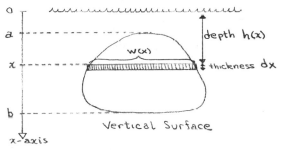

First we set up a vertical x-axis as shown in the diagram above*, and then, for each

x value, consider the "infinitesimal strip" of width w(x) and "infinitesimal thickness" dx .

This is at a depth h(x) in the fluid, and hence the pressure, given by (A), is a constant on

the (horizontal) strip:

$$\text{pressure} = \rho h(x)$$

─────────────

*
 Actually you can label your axis with any variable you choose, so long as that variable is
 carried along through the rest of the solution. In Example A below, we use a y-axis
 instead of an x-axis because it seems more natural in that problem.

Thus

$$\begin{pmatrix} \text{infinitesimal} \\ \text{fluid force} \\ \text{on strip} \end{pmatrix} = \begin{bmatrix} \text{force per unit area} \\ \text{(i. e. , the pressure)} \end{bmatrix} \times \begin{bmatrix} \text{infinitesimal} \\ \text{area of} \\ \text{strip} \end{bmatrix}$$

which gives, by equation (A),

$$dF = (\rho h(x)) (w(x) \, dx)$$

or

$$dF = \rho h(x) w(x) \, dx$$

Finally, we compute the total fluid force F on the full surface by "adding up" the infinitesimal forces for the strips at each x value between a and b, i. e., we integrate dF from $x = a$ to $x = b$:

Fluid Pressure on vertical flat surface	

$$F = \int dF = \int_a^b \rho h(x) w(x) \, dx \qquad (C)$$

This is Anton's equation (3), derived by the "infinitesimal approach" rather than his "Riemann sums" method. You have the choice in specific problems of merely applying (C) or of using the "infinitesimal approach" to get your solution. Here is an example of the latter method:

Example A.   A dam across a small stream has a vertical face which has roughly the shape of a parabola. It is 36 feet across the top and 9 feet deep at the center. Find the maximum force that the water can exert on the face of this dam.

<u>Solution.</u>   Obviously the maximum force occurs during flood time when the water is at the top

of the dam and is just about to overflow.   The density

of water is ρ = 62.4 lb. /ft^3 .   We place the

water and the dam on a coordinate system as

shown to the right.   The equation of the

parabolic border of the dam is $y = cx^2$

and, since the curve passes through   (18,9),

we have $9 = c(18)^2$ or $c = 1/36.$ Thus the equation of the border of the dam is $y = x^2/36$

or, in the first quadrant, $x = 6\sqrt{y}$.

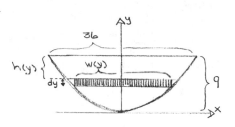

We divide the dam into infinitesimal horizontal strips of thickness dy .   Then

(1)   The <u>width</u> of the infinitesimal strip at y is $w(y) = 2x = 12\sqrt{y}$ and its depth

below the surface of the water is $h(y) = 9 - y$.

(2)   The <u>infinitesimal surface area</u> dA of the strip is

$$dA = w(y)\, dy = 12\sqrt{y}\ dy \qquad \text{(width times infinitesimal thickness)}$$

(3)   The <u>infinitesimal force</u> dF on the strip is

$$dF = p\, dA = \rho h(y)\, dA = \rho(9 - y)\, 12\sqrt{y}\ dy$$

(4)   The <u>total force</u> on the dam is

$$F = \int_0^9 12\, \rho\, (9 - y)\sqrt{y}\ dy = 12\, \rho \int_0^9 (9y^{1/2} - y^{3/2})\ dy$$

$$= 12\, \rho \left[6(9)^{3/2} - \frac{2}{5}(9)^{5/2} \right] = 3888\, \rho/5 = 777.77\, \rho$$

With $\rho = 62.4$,   this gives   F = 48,532. 85 lbs. or 24. 3 tons.   [Clearly, this problem could

also be solved by plugging the values of a, b, ρ, $h(x)$ and $w(x)$ into equation (C) .]   □

4. <u>Tilted surfaces</u> (optional).   If you choose to solve fluid force problems on vertical surfaces merely by applying formula (C), you are rather at a loss when it comes to determining the fluid force on a <u>tilted</u> surface: formula (C) does not apply in that situation.   However, the infinitesimal approach, if handled with care, easily generalizes to tilted surfaces as the next example shows:

<u>Example B.</u>   Suppose that in Example A the face of the dam on the water side is built out to make a flat surface that is slanted at a 45° angle.   Find the maximum force that the water can now exert on the face of this dam.

<u>Solution.</u>   A cross section of the dam (i. e. , a view from the side) is shown to the right.   The straight-on view of the face of the dam still appears as in Example A.

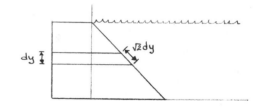

As in Example A , we divide the dam into infinitesimal horizontal strips by dividing the y-axis into segments of infinitesmial thickness dy .   However, now the infinitesimal width of our strip is $\sqrt{2}$ dy , NOT dy . (See the 45° - 45° - 90° triangle in the diagram to the right.)

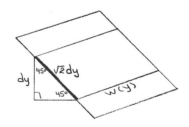

The rest of the derivation parallels that of Example A :

(1) The <u>width</u> of the infinitesimal strip at   y   is

$$w(y) \;=\; 12\sqrt{y}$$

and its depth below the surface of the water is

$$h(y) \;=\; 9 - y$$

(2) The <u>infinitesimal surface area</u>   dA   of the strip is

$$dA \;=\; w(y)(\sqrt{2}\;dy) \;=\; 12\sqrt{2}\;dy$$

(3) The <u>infinitesimal force</u>   dF   on the strip is

$$dF \;=\; p\,dA \;=\; \rho h(y)\,dA$$

$$=\; 12\sqrt{2}\;\rho\,(9 - y)\,\sqrt{y}\;dy$$

(4) The <u>total force</u> on the dam is

$$F \;=\; \int_{0}^{9} 12\sqrt{2}\;\rho\,(9 - y)\,\sqrt{y}\;dy \;=\; \ldots \;\cong\; 34.4\ \text{tons}\qquad\qquad \square$$

we'll leave the
details to you

Section 6. 8.   An Overview of Chapter 6.   (Optional)

In Sections 6. 1 - 6. 7 you should have observed the general pattern that runs through all these applications of definite integration: <u>definite integrals appear because the desired quantities are limits of Riemann Sums:</u>

Riemann Sum Definition of the Definite Integral

$$\int_a^b f(x)\,dx = \lim_{\max \Delta x_k \to 0} \sum_{k=1}^{n} f(x_k^*)\,\Delta x_k$$

(Definition 5. 7. 2).   In more detail, the pattern can be described as follows:

1.   <u>The General Riemann Sums Pattern.</u>   We need to measure a quantity Q which is "evenly distributed" over an interval [a, b] of the x-axis. That is, when Q is broken up into small pieces, each piece can be determined using a basic law:

The Basic Law

Q has the property that for each subinterval $[x_{k-1}, x_k]$ of $[a, b]$ there exists a constant ρ_k such that

$$\Delta Q_k = \rho_k \, \Delta x_k$$

where ΔQ_k = the amount of quantity Q   over $[x_{k-1}, x_k]$

Δx_k = the length of subinterval $[x_{k-1}, x_k]$

Step 1.   The interval [a, b] is partitioned into n   subintervals by the points $a < x_1 < x_2 < \ldots < x_{n-1} < b$. The subintervals have widths $\Delta x_1, \Delta x_2, \ldots, \Delta x_n$, respectively.

Step 2.   The quantity Q is broken up into n pieces $\Delta Q_1, \Delta Q_2, \ldots, \Delta Q_n$,   one for

each of the n subintervals,   in such a way that

Q is the sum of these n pieces, i.e.,

$$Q = \sum_{k=1}^{n} \Delta Q_k$$

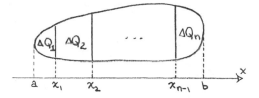

Step 3.   Each subpiece ΔQ_k is approximated using The Basic Law, i.e., we find a

constant ρ_k such that

$$\Delta Q_k \approx \rho_k \Delta x_k$$

In all of the applications discussed in   Chapter 6   (and in most applications of definite

integration) the constants ρ_k arise as follows:

<table>
<tr><td>

The
Continuity
Assumption

</td><td>

There exists a continuous function ρ on $[a,b]$

such that each ρ_k can be defined by

$$\rho_k = \rho(x_k^*)$$

for some point x_k^* in $[x_{k-1}, x_k]$.

</td></tr>
</table>

Thus the approximation for ΔQ_k will be of the form

$$\Delta Q_k \approx \rho(x_k^*) \Delta x_k$$

Step 4.   The quantity Q is now approximated by the sum of the approximations of its

pieces.   This gives

$$Q \approx \sum_{k=1}^{n} \rho(x_k^*) \Delta x_k$$

which is a Riemann Sum for the continuous function ρ from a to b.

<u>Step 5.</u> Assuming this approximation improves as the widths Δx_k of the subintervals

approach zero, then it is reasonable to believe Q is equal to the limit of this process, i. e.,

$$Q = \lim_{\max \Delta x_k \to 0} \sum_{k=1}^{n} \rho(x_k^*) \Delta x_k = \int_a^b \rho(x)\, dx$$

For example, here is how this five-step pattern applies to the <u>Method of Disks</u>,

used in §6.2 to compute volumes for solids of revolution:

Let R be the region under the continuous curve $y = f(x)$

and over the interval $[a, b]$. Let $Q = V$, where

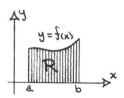

$$V = \left\{ \begin{array}{c} \text{volume generated} \\ \text{by rotating } R \\ \text{about the } x\text{-axis} \end{array} \right\}$$

<table>
<tr><td>The
Basic
Law</td></tr>
</table>

V is such that (thin) slices of it are volumes of

right circular cylinders, i. e., $f(x)$ is a

constant r_k on the (small) subinterval

$[x_{k-1}, x_k]$. Thus

$$\Delta V_k = \pi r_k^2 \Delta x_k$$

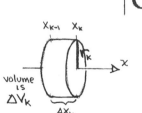

(Here the Basic Law is the volume formula for a right circular cylinder, so that $\rho_k = \pi r_k^2$.)

<u>Step 1.</u> Partition $[a, b]$ by points $a < x_1 < x_2 < \ldots < x_{n-1} < b$ into n sub-

intervals with widths $\Delta x_1, \Delta x_2, \ldots, \Delta x_n$, respectively.

Step 2. V is broken up into n corresponding pieces $\Delta V_1, \ldots, \Delta V_n$ (volumes of right circular cylinders).

Step 3. Each subvolume ΔV_k is approximated by the Basic Law, where the radius r_k is taken to be

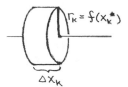

$r_k = f(x_k^*)$, for x_k^* some point in $[x_{k-1}, x_k]$. Thus

$$\Delta V_k \approx \pi r_k^2 \Delta x_k = \pi f(x_k^*)^2 \Delta x_k$$

(The Continuity Assumption is satisfied by $\rho(x) = \pi f(x)^2$, a nice, continuous function.)

Step 4. Thus $V \approx \sum_{k=1}^{n} \pi f(x_k^*)^2 \Delta x_k$, a Riemann Sum for $\rho(x) = \pi f(x)^2$ from a to b.

Step 5. Hence, taking the limit as $\max \Delta x_k$ goes to zero, we obtain

$$V = \int_a^b \pi f(x)^2 \, dx$$

This is the Method of Disks formula for volume.

As you can see, the most important part of the general pattern is "The Basic Law" used in Step 3. In the applications studied in this chapter the Basic Laws are determined either by simple geometry (for area, volume, arclength and surface area) or by elementary principles of physics (for work and fluid force). Indeed, the whole procedure can be viewed as a way to take a simple "Basic Law" and, via Riemann Sums and the definite integral, transform it into a formula which applies to a much wider range of problems.

Here's a summary of the "Basic Laws" and their generalizations which we have considered in this chapter:

6.8.5

Section	The Basic Law	The generalized formula
6.1	Area of a rectangle $$A = \ell \cdot w$$	Area between two curves $$A = \int_a^b [f(x) - g(x)]\, dx$$
6.2	Volume of a right cylinder $$V = A \cdot h$$	Volume of a solid by cross-sections $$V = \int_a^b A(x)\, dx$$
	Volume of a right circular cylinder $$V = \pi r^2 h$$	Volume of a solid of revolution by Method of Disks $$V = \pi \int_a^b f(x)^2\, dx$$
6.3	Volume of a cylindrical shell $$V = 2\pi \left(\frac{r_2 + r_1}{2} \right) h(r_2 - r_1)$$	Volume of a solid of revolution by Method of Cylindrical Shells $$V = 2\pi \int_a^b x\, f(x)\, dx$$
6.4	Length of a line segment $$L = \sqrt{(x_1 - x_0)^2 + (y_1 - y_0)^2}$$	Arclength of a curve $$L = \int_a^b \sqrt{1 + [f'(x)]^2}\, dx$$
6.5	Surface area of the frustum of a cone $$S = \pi(r_1 + r_2)\ell$$	Surface area of a surface of revolution $$S = 2\pi \int_a^b f(x) \sqrt{1 + [f'(x)]^2}\, dx$$
6.6	Work done by a constant force $$W = F \cdot d$$	Work done by a variable force $$W = \int_a^b F(x)\, dx$$

6.7   Fluid force on a horizontal plate   Fluid force on a vertical plate

$$F = \rho h A \qquad\qquad F = \int_a^b \rho h(x)\, w(x)\, dx$$

2.   The Infinitesimal Approach.   Whenever the general pattern applies in calculating a quantity Q, there is a corresponding "infinitesimal approach" which can be useful in remembering and understanding the formula for Q:

Step A.   Use the Basic Law to find a formula for dQ,

                        the infinitesimal amount of Q distributed

                        over an infinitesimal interval of length dx:

$$dQ = \rho(x)\, dx$$

Step B.   Q is thus the "sum" (i.e., the definite integral)

                        of all the infinitesimal dQ from x = a to x = b:

$$Q = \int_a^b dQ = \int_a^b \rho(x)\, dx$$

Keep in mind that this "infinitesimal approach" is not rigorous mathematics... but it is a simple and intuitive way to summarize and remember the methods of this chapter!

        In each section of The Companion for Chapter 6 we have explicitly written out the appropriate infinitesimal method and have applied it in specific problems. The kind of reasoning seen in these problems is used over and over again in the physical and social sciences.

Chapter 7:   Logarithm and Exponential Functions

Section 7.2.   The Natural Logarithm

1.   <u>Logarithms ... again!</u>   Somewhere in your mathematical past, you learned about logarithms:

> a logarithm is an exponent.

Specifically, if   x   and   b   are positive numbers, then the <u>logarithm to the base   b   of   x</u>

is defined to be

$$\log_b x = \left\{ \begin{array}{l} \text{that power to which} \quad b \quad \text{must} \\ \text{be raised to produce} \quad x \end{array} \right\}$$

In symbols we have

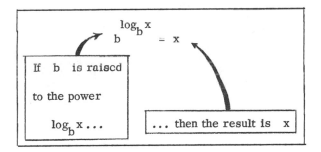

The algebraic properties of the logarithm are summarized in Theorem 7.2.1. <u>We</u>

<u>cannot overemphasize how important it is for you to memorize these properties and become</u>

<u>proficient in their use!!</u> You will be seriously handicapped in subsequent work if you do not

do so. Here are some computations with these rules that are typical of what you will need

to do in coming sections:

Example A. Without using a calculator or the tables, find the exact value of $\log_8(1/2)$.

Solution. $\log_8(1/2) = -\log_8 2$ by Theorem 7.2.1e. By definition, $\log_8 2$ is the power to which 8 must be raised to produce 2. Since $8^{1/3} = 2$, then $\log_8 2 = 1/3$, which gives

$$\log_8(1/2) = \boxed{-1/3}$$

Example B. Solve for x without using a calculator or the tables:

$$\log_3(x+1)^2 - \log_3 3\sqrt{x+1} = 2$$

Solution.

$$\log_3(x+1)^2 - \log_3 3\sqrt{x+1}$$

$$= \log_3 \frac{(x+1)^2}{3\sqrt{x+1}} \qquad \text{by Theorem 7.2.1c}$$

$$= \log_3 \frac{(x+1)^{3/2}}{3}$$

Thus

$$\log_3 \frac{(x+1)^{3/2}}{3} = 2$$

which, by the definition of $\log_3$, means

$$\frac{(x+1)^{3/2}}{3} = 3^2$$

$$(x+1)^{3/2} = 3^3$$

$$x + 1 = \left(3^3\right)^{2/3} = 3^2 = 9$$

$$\boxed{x = 8}$$

Warning: Examine Theorem 7.2.1 very carefully. Notice that there are no rules listed for

$\log_b (a + c)$ or $\log_b (a - c)$. Do <u>not</u> "make up" rules for $\log_b (a + c)$ or $\log_b (a - c)$.

2. <u>The number e.</u> As Anton points out, the two most widely used bases b for logarithm

functions are

$$b = 10, \text{ producing the } \underline{\text{common logarithm}} \quad \log_{10} x$$
$$b = e, \text{ producing the } \underline{\text{natural logarithm}} \quad \log_e x$$

The number <u>e is a constant</u>; in that respect it is similar to the number π which you

are accustomed to using. The value of e is a bit less than 3 , specifically

$$e \approx 2.718281828\ldots$$

to nine decimal places. The reason for the importance of such an unusual number will be

discussed in the subsequent sections.

3. <u>The integral definition of ln x.</u> The integral

$$\int_1^2 t^{-1}\, dt = \int_1^2 \frac{1}{t}\, dt$$

looks so simple that it is hard to believe we can't evaluate it ! We know this integral <u>exists</u> by

Theorem 5.7.3 , but at the moment we can't <u>find</u> what it is. If the power of t were <u>any</u> rational

number other than - 1 , we could evaluate the integral with ease using $\int t^r\, dt = t^{r+1}/(r+1)+C$,

but that darn - 1 is the <u>one</u> case where that formula doesn't work. In view of the fact that it is the single exception, it is reasonable to believe that this case will be important and that a way should be found to deal with it.

The solution? We simply call $\int_1^2 \frac{1}{t}\, dt$ the <u>natural logarithm of 2</u> and denote it $\ln 2$; that is, we identify this one-case exception as being important and give it a name! Then we follow through by studying its properties.

In general, we define the <u>natural logarithm $\ln x$ of x</u> to be

Definition of the Natural Logarithm

$$\ln x = \int_1^x \frac{1}{t}\, dt \qquad \text{for} \qquad x > 0$$

Remember that

(i) t is a "dummy variable" and x is the variable that matters, so $\ln x$ is a function of x [See §5.10.1 of <u>The Companion</u>.]

(ii) $\ln x$ is defined only for $x > 0$

(iii) we cannot evaluate the integral, so at the moment we have difficulty finding specific values for $\ln x$. (We can only <u>approximate</u> them and now our only technique for doing that is to use Riemann sums; later, in §9.9 and 11.11, we will develop important techniques which will enable us to approximate $\ln x$ very closely.)

It is amazing but true that the natural logarithm function $\ln x$ just defined via the integral of $1/t$ is the same as the logarithm function $\log_e x$. Anton will show this in §7.4.

The definition of $\ln x$ is extremely important and you must learn it well. Moreover, <u>you must study the properties of $\ln x$ until they become second nature to you;</u> failure to do so will cause you to have a lot of trouble later on! (This advice - or more accurately, warning - cannot be overstated.)

In this section we study the basic derivative and integral formulas of the natural logarithm.

4. <u>Differentiation and ln x.</u>   The 2nd FTC (Theorem 5.10.1) tells us that $y = \ln x$ is

differentiable for $x > 0$ and that its derivative is

$$\frac{d}{dx} [\ln x] = \frac{1}{x} \quad \text{if} \quad x > 0$$

Moreover, in Example 3, Anton extends this rule to $\ln |x|$:

| The Derivative of ln |x| |

$$\frac{d}{dx} [\ln |x|] = \frac{1}{x} \quad \text{if} \quad x \neq 0$$

If this formula is applied in conjunction with the Chain Rule to a composite function

$\ln g(x)$ where $g(x)$ is a function of x, we obtain

$$\frac{d}{dx} [\ln |g(x)|] = \frac{g'(x)}{g(x)} \quad \text{if} \quad g(x) \neq 0$$

This is simply an alternate (and useful) way to write Anton's Equation (5):

$$\frac{d}{dx} [\ln u] = \frac{d}{dx} [\ln |u|] = \frac{1}{u} \cdot \frac{du}{dx}$$

where $u = g(x) > 0$.

Anton's Examples 2, 3 and 4 illustrate these formulas. Here are three other examples:

<u>Example C.</u>   Find $\frac{d}{dx} [\ln (x^2 - 1)]$.

<u>Solution.</u>   This is $\frac{d}{dx} [\ln g(x)]$ where $g(x) = x^2 - 1$, so

$$\frac{d}{dx} [\ln (x^2 - 1)] = \frac{2x}{x^2 - 1}$$

Note that $\ln(x^2 - 1)$ is defined only for $x^2 - 1 > 0$, i.e., $x > 1$ or $x < -1$. Hence

its derivative is also defined only for these values.   □

Example D. Find $\dfrac{d}{dx}\,[\,x^2\ln 3x\,]$.

Solution. By the Product Rule

$$\frac{d}{dx}\,[\,x^2\ln 3x\,] = x^2\,\frac{d}{dx}\,[\,\ln 3x\,] + (\ln 3x)\,\frac{d}{dx}\,[\,x^2\,]$$

$$= x^2\,\frac{d}{dx}\,[\,\ln 3x\,] + (\ln 3x)\,(2x)$$

The remaining derivative is $\dfrac{d}{dx}\,[\,\ln g(x)\,]$ with $g(x) = 3x$, so

$$\frac{d}{dx}\,[\,x^2\ln 3x\,] = x^2\,(\frac{3}{3x}) + 2x\ln 3x = x + 2x\ln 3x$$

Note that $x^2\ln 3x$ is defined only when $x > 0$, so its derivative is also defined only for

$x > 0$. ☐

Example E. Find $\dfrac{dy}{dx}$ by implicit differentiation if $\ln x + 2\ln y = 4$.

Solution. Differentiating both sides with respect to x, we obtain (regarding y as $g(x)$)

$$\frac{1}{x} + 2\,\frac{1}{y}\,\frac{dy}{dx} = 0$$

or

$$\frac{dy}{dx} = \frac{-y}{2x}$$

Note that $\ln x + 2\ln y = 4$ is defined only when $x > 0$ or $y > 0$. Hence $\dfrac{dy}{dx}$ is also

defined only for those values. ☐

5. **Integration and $\ln x$.** The formulas of the last subsection need only be read backwards to

obtain the integration formulas

$$\int \frac{dx}{x} = \ln |x| + C$$

| ln |
| Integration |
| Formulas |

and

$$\int \frac{g'(x)}{g(x)} \, dx = \ln |g(x)| + C$$

Note that the second of these formulas can be obtained from the first by writing the first formula

with the variable   u   (instead of   x)   and then substituting $u = g(x)$.

These formulas are very important. Many integrals need   ln   in order to be evaluated

(that's one <u>very important</u> use of   ln x). Anton's Examples 5 and 6 are good illustrations.

Here are two more:

<u>Example F.</u>   Evaluate $\displaystyle\int_{-3}^{-1} \frac{x^2}{3x - 1} \, dx$.

<u>Solution.</u>   If we let $u = 3x - 1$,   then

$$\int_{-3}^{-1} \frac{x^2}{3x - 1} \, dx = \int_{-10}^{-4} \frac{\left(\frac{u + 1}{3}\right)^2}{u} \, \frac{du}{3}$$

| $u = 3x - 1$ |
| $du = 3\,dx$ |
| $\dfrac{du}{3} = dx$ |
| $\dfrac{u + 1}{3} = x$ |

$$= \frac{1}{27} \int_{-10}^{-4} \left(u + 2 + \frac{1}{u} \right) du$$

$$= \frac{1}{27} \left(\frac{u^2}{2} + 2u + \ln |u| \right) \Bigg|_{-10}^{-4}$$

$$= \frac{1}{27} \left(\ln 4 - 30 - \ln 10 \right)$$

In §7.3 we will learn that $\ln 4 - \ln 10 = \ln \frac{4}{10} = \ln \frac{2}{5}$,   but for now the answer above

suffices.   □

Example G.   Evaluate $\displaystyle\int \sec x \, dx$.

Solution.   This problem is easy if we use the following trick: Multiply the integrand by

$$1 \; = \; \frac{\sec x \,+\, \tan x}{\sec x \,+\, \tan x}$$

(How's that for pulling a rabbit out of a hat?) When we do, we get

$$\int \sec x \, dx \; = \; \int \frac{(\sec x \,+\, \tan x) \sec x}{\sec x \,+\, \tan x} \; dx$$

$$= \; \int \frac{(\sec^2 x \,+\, \sec x \tan x) \, dx}{\sec x \,+\, \tan x}$$

$$\boxed{\begin{array}{l} u = \sec x + \tan x \\[4pt] \dfrac{du}{dx} = \sec x \tan x + \sec^2 x \end{array}} \longrightarrow \quad = \; \int \frac{du}{u}$$

$$= \; \ln \left| u \right| \,+\, C$$

$$= \; \ln \left| \sec x \,+\, \tan x \right| \,+\, C \qquad\qquad \square$$

Note that Anton's Example 6 and our Example G show that ln is needed to evaluate the integrals of even very simple trigonometric functions. We told you it was important as an integration tool!

Section 7.3.   Properties of the Natural Logarithm

1.   The algebraic properties and the graph of ln x: know them cold! The ln properties

summarized in Theorem 7.3.1 are very important; you must know them cold! If you do not

learn them, they will come back to haunt you again and again. Here they are in English:

Theorem 7.3.1
in English

a) ln of 1 is zero:

$$\ln 1 = 0$$

b) ln of a product is the sum of the ln's:

$$\ln (ac) = \ln a + \ln c$$

c) ln of a quotient is the ln of the numerator minus the ln of
the denominator:

$$\ln \left(\frac{a}{c}\right) = \ln a - \ln c$$

d) ln of a power is the power times the ln:

$$\ln (a^r) = r \ln a$$

e) ln of a reciprocal is the negative of the ln:

$$\ln \left(\frac{1}{c}\right) = - \ln c$$

These properties probably look familiar to you. That is because they are properties

of any logarithm $\log_b x$, the very properties stated in Theorem 7.2.1 of the last section!

This should not surprise you since we know (see §7.2.2 above) that

$$\ln x = \log_e x$$

Anton will prove this following Theorem 7.4.4 in the next section.

There are some other properties of ln which you must know:

$$\boxed{\begin{array}{l} \ln x \quad \text{is undefined for} \quad x \leq 0 \\[4pt] \ln x < 0 \quad \text{if} \quad 0 < x < 1 \\[4pt] \ln x > 0 \quad \text{if} \quad x > 1 \end{array}}$$

These properties are probably best remembered by memorizing the <u>basic shape of the graph</u>

<u>of y = ln x</u>. (See Anton's Figure 7.3.3 for a more careful graph):

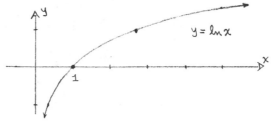

Once you have memorized the basic shape of this graph, you should also have no trouble

remembering the following facts:

$$\boxed{\begin{array}{l} y = \ln x \quad \text{is an increasing function} \\[6pt] \lim_{x \to 0^+} \ln x = -\infty \\[6pt] \lim_{x \to +\infty} \ln x = +\infty \end{array}}$$

2. <u>Using the properties of ln x.</u>   The properties of ln x can be very helpful. For instance, they can be used to obtain new information from old (or given) information, as in these two examples:

<u>Example A.</u>   Given that ln 2 ≈ .6931 and ln 3 ≈ 1.0986, find $\ln \sqrt{72}$.

<u>Solution.</u>   (The letter references are to appropriate parts of Theorem 7.3.1.)

$$\ln \sqrt{72} \; = \; \ln (72)^{1/2} \underset{(d)}{=} \frac{1}{2} \ln 72 = \frac{1}{2} \ln (2^3 \cdot 3^2)$$

$$\underset{(b)}{=} \frac{1}{2} [\ln 2^3 + \ln 3^2] \underset{(d)}{=} \frac{1}{2} [3 \ln 2 + 2 \ln 3]$$

$$\approx \frac{1}{2} [3(.6931) + 2(1.0986)] = \boxed{2.1383} \qquad \square$$

<u>Example B.</u>   Suppose ln a = -2. Solve the equation $\ln x^2 - 8$ for x in terms of a.

<u>Solution.</u>   (The letter references are to appropriate parts of Theorem 7.3.1.)

$$\ln x^2 = 8 = (-4)(-2) = (-4) \ln a \underset{(d)}{=} \ln (a^{-4})$$

Since the graph of ln x is increasing, the only way to have $\ln x^2 = \ln \frac{1}{a^4}$ is to have $x^2 = \frac{1}{a^4}$, or $x = \pm \frac{1}{a^2}$. $\qquad \square$

The properties of the logarithm can also be used to simplify an expression like ln f(x) before differentiating or integrating it. Anton illustrates this in Example 2. Here are two more examples:

Example C.   Find $\dfrac{d}{dx}\left[\ln\sqrt{\dfrac{x^2+1}{x^2-1}}\right]$

Solution.   Before differentiating, we simplify the function:

$$\ln\sqrt{\frac{x^2+1}{x^2-1}} \;=\; \ln\left(\frac{x^2+1}{x^2-1}\right)^{1/2}$$

$$=\; \frac{1}{2}\ln\left(\frac{x^2+1}{x^2-1}\right) \qquad\qquad \text{(Theorem 7.3.1 d)}$$

$$=\; \frac{1}{2}\left[\ln(x^2+1)-\ln(x^2-1)\right] \qquad \text{(Theorem 7.3.1 c)}$$

Therefore

$$\frac{d}{dx}\ln\sqrt{\frac{x^2+1}{x^2-1}} \;=\; \frac{1}{2}\frac{d}{dx}\ln(x^2+1)-\frac{1}{2}\ln(x^2-1)$$

$$=\; \frac{1}{2}\left[\frac{2x}{x^2+1}-\frac{2x}{x^2-1}\right] = \frac{-2x}{x^4-1}$$

<u>Example D.</u>   Find $\int \dfrac{\ln x^4}{x}\, dx$.

<u>Solution.</u>   First we simplify $\ln x^4$ to $4 \ln x$ using Theorem 7.3.1 d. Then we integrate:

$$\int \frac{\ln x^4}{x} = 4 \int \frac{\ln x}{x}\, dx$$

$$\boxed{\begin{array}{l} u = \ln x \\[4pt] du = \dfrac{1}{x}\, dx \end{array}} \nearrow \quad = 4 \int u\, du$$

$$= 2 u^2 + C$$

$$= 2 (\ln x)^2 + C \qquad\qquad \square$$

Finally, the properties can be used in graphing:

<u>Example E.</u>   Draw the graph of the function $y = \ln \sqrt{x+2}$.

<u>Solution.</u>   Of course we <u>could</u> use the full range of graphing techniques developed in Chapter 4 to graph this function. However, the following approach avoids a lot of tedious computations. By 7.3.1 d , $y = \ln \sqrt{x+2} = \ln (x+2)^{1/2} = \frac{1}{2} \ln (x+2)$. The translation principles (see Appendix E) tell us that the graph of $y = \ln (x+2)$ is the graph of $y = \ln x$ shifted two units to the left:

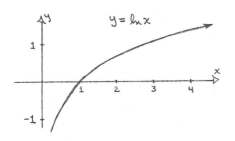
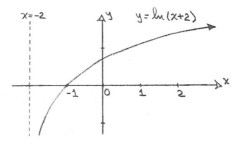

7.3.6

Then multiplication by $\frac{1}{2}$ "flattens" the graph of $y = \ln (x + 2)$ by halving each of its y values:

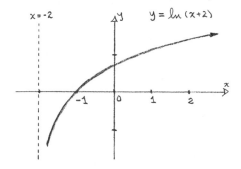

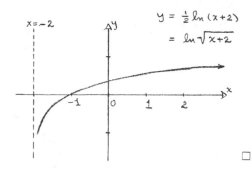

3. <u>Logarithmic Differentiation.</u> Anton's Examples 3 and 4 illustrate a technique known as <u>logarithmic differentiation.</u> It works like this:

Logarithmic
Differentiation

> Suppose we want to find $\dfrac{dy}{dx}$ when $y = f(x)$ in a messy quotient and/or product of terms with exponents.
>
> <u>First take $\ln$ of both sides:</u>
>
> $$\ln y = \ln f(x) \longleftarrow \boxed{\text{an expression which can frequently be simplified using the properties of logarithms}}$$
>
> <u>Then differentiate both sides with respect to x:</u>
>
> $$\frac{1}{y} \frac{dy}{dx} = \frac{d}{dx} [\ln f(x)]$$
>
> so $$\frac{dy}{dx} = f(x) \frac{d}{dx} [\ln f(x)]$$

Here are two additional examples:

Example F.   Find $\dfrac{dy}{dx}$ if $y = \dfrac{(x^2 + 1)^{1/2} (6x + 5)^{1/3}}{(x^2 - 1)^{1/2}}$.

Solution.   Using logarithmic differentiation, we first take $\ln$ of both sides and simplify:

$$\ln y = \ln \frac{(x^2 + 1)^{1/2} (6x + 5)^{1/3}}{(x^2 - 1)^{1/2}}$$

$$= \frac{1}{2} \ln (x^2 + 1) + \frac{1}{3} \ln (6x + 5) - \frac{1}{2} \ln (x^2 - 1)$$

using parts b), c) and d) of Theorem 7.3.1.   Then we differentiate both sides with respect to x :

$$\frac{1}{y} \frac{dy}{dx} = \frac{x}{x^2 + 1} + \frac{2}{6x + 5} - \frac{x}{x^2 - 1}$$

Finally, multiplying through by y and using the initial expression for y in terms of x, we obtain

$$\frac{dy}{dx} = \frac{(x^2 + 1)^{1/2} (6x + 5)^{1/3}}{(x^2 - 1)^{1/2}} \cdot \left(\frac{x}{x^2 + 1} + \frac{2}{6x + 5} - \frac{x}{x^2 - 1} \right) \qquad \square$$

Example G.   Find $\dfrac{dy}{dx}$ if $y^3 = x(x - 1)$.

Solution.   We use logarithmic differentiation.   First we take $\ln$ of both sides:

$$\ln y^3 = \ln [x(x - 1)]$$

or

$$3 \ln y = \ln x + \ln (x - 1)$$

Then differentiating both sides with respect to x , we obtain

$$\frac{3}{y} \frac{dy}{dx} = \frac{1}{x} + \frac{1}{x - 1}$$

or

$$\frac{3}{y}\frac{dy}{dx} = \frac{2x-1}{x(x-1)}$$

Then, since $y^3 = x(x-1)$, multiplying through by $\frac{y}{3}$ yields

$$\frac{dy}{dx} = \frac{y}{3} \cdot \frac{2x-1}{y^3} = \frac{2x-1}{3y^2}$$

[If an answer in terms of x alone is required, it can be obtained using $y^2 = (y^3)^{2/3} = [x(x-1)]^{2/3}$]. $\qquad\qquad\qquad$ □

Section 7.4. The number e; the Functions a^x and e^x.

As in the previous section, there is a lot which must be memorized in this section. Failure to learn the various facts and formulas well at this stage will come back to haunt you later on. We recommend that you keep a list of key results and that you study it carefully. (Such a list is given in Subsection 7 below.) You may even want to make a set of flash cards. But how you memorize is for you to decide; our message is that it must be done.

1. The definition of e. This section begins with the formal definition of the special number e:

Definition of e	e is that number such that $\ln e = 1$

Thus there are two particular values of the natural logarithm $\ln$ which you should know

$\ln 1 = 0$	(Theorem 7.3.1a)
$\ln e = 1$	(The definition of e)

As we pointed out in §7.2.2, the value of e is a little less than 3, specifically

$$e \approx 2.718281828\ldots$$

to 9 decimal places.

     You should be curious about how such an approximation of e is obtained. Here are two methods for approximating e which will be derived in subsequent chapters:

1.   In Chapter 10 it will be shown that

$$e = \lim_{x \to +\infty} \left(1 + \frac{1}{x}\right)^x$$

(This is Equation (4) in §10.3 and its derivation requires L'Hôpital's rule.) This equation means that the quantity $\left(1 + \frac{1}{x}\right)^x$ is a good approximation of e for large values of x ; Table 7.2.1 gives values of this quantity for various values of x .

2.   In Chapter 11 it will be shown that

$$e = 1 + 1 + \frac{1}{2!} + \frac{1}{3!} + \frac{1}{4!} + \cdots + \frac{1}{n!} + \cdots$$

(This is Equation (1) of §11.11 and its derivation requires Maclaurin Series.) This equation means that the quantity

$$1 + 1 + \frac{1}{2!} + \frac{1}{3!} + \cdots + \frac{1}{n!}$$

is a good approximation of e for large values of n, and even for not-so-large values of n. For instance, Example 1 of §11.11 shows that with just n = 8 , this quantity gives an approximation of e which is accurate to four decimal places.

7.4.3

2. **The definition of a^x.** From Anton's Definition 7.4.2 we have

Definition
of a^x

Suppose a is any <u>positive</u> real number

and x is <u>any</u> real number. Then a^x

is defined to be that real number such that

$$\ln a^x = x \ln a$$

As Anton shows, this rule holds when x is a rational number. Hence it is reasonable to want the rule to be true as well when x is an irrational number.

<u>Example A.</u>   Approximate $2^{\sqrt{2}}$.

<u>Solution.</u>   Here we have $a = 2$ and $x = \sqrt{2}$, so the definition says $2^{\sqrt{2}}$ is that real number such that $\ln 2^{\sqrt{2}} = \sqrt{2} \ln 2$. Thus, since $\ln 2 \approx .6931$ from Table 3 of Anton's Appendix 3 , we find

$$\ln 2^{\sqrt{2}} = \sqrt{2} \ln 2 \approx (1.414)(.6931) \approx .9800$$

Furthermore, the value of n in Table 3 whose logarithm is closest to .9800 is n = 2.7. Thus $2^{\sqrt{2}} \approx 2.7$. (For another method of approximating $2^{\sqrt{2}}$, see Example C of Subsection 3 below.)   □

Anton's Theorem 7.4.3 may be summarized as follows:

Laws of Exponents
for real numbers

All the usual laws of exponents hold

when the exponents are arbitrary

<u>real</u> numbers.

If the laws of exponents listed in Theorem 7.4.3 are not already second nature to you, you should memorize them once and for all. <u>Few things cause as much trouble in calculus as the faulty use of laws of exponents!!!</u>

3.   <u>The definition of e^x</u>. The definition of a^x given above holds when a is <u>any</u> positive number, but its most important special case occurs when a is the number e :

Suppose x is any real number. Then e^x is

defined to be that number such that

$$\ln e^x = x$$

Definition

of e^x

The function $f(x) = e^x$, called the <u>exponential function</u> (or the <u>natural exponential function</u>), is simply the constant e "raised to the power x."

The two most basic equations for the exponential function are:

Relations

between

$\ln x$ and e^x

$$\ln e^x = x \text{ for } -\infty < x < +\infty$$

$$e^{\ln x} = x \text{ for } x > 0$$

The first is simply the defining equation for e^x while the second is proved in Anton's Theorem 7.4.4. The equations show that <u>the natural logarithm and the exponential function undo the effect of one another.</u> You have already seen this type of relationship with the functions

$$f(x) = x^n \qquad \text{and} \qquad g(x) = x^{1/n}$$

for when x is positive, then

$$f(g(x)) = (x^{1/n})^n = x$$

$$g(f(x)) = (x^n)^{1/n} = x$$

Hence   f   and   g   undo the effect of one another (in terminology to be introduced in §8.1, they are <u>inverses of one another</u>). Thus, using this terminology, the relations between   ln x   and e^x can be summarized by saying

> The exponential function and the natural logarithm
>
> function are inverses of each other.

Here is an example showing how this critical relationship can be used:

<u>Example B.</u>   Solve $e^{2x-3} = .5$ for   x .

<u>Solution.</u>   First we want to "undo" e^{2x-3} to recover $2x - 3$,   so we take   ln   of both sides of the equation and get

$$2x - 3 = \ln e^{2x-3} = \ln .5$$

Since $\ln (.5) \approx -.6931$ from   Table 3   of Anton's   Appendix 3 , we solve for   x   to find

$$x \approx (-.6931 + 3)/2 = 1.15345 \qquad \square$$

*   *   *   *   *

Let's return to the <u>general</u> exponential a^x . As noted earlier, a^x is defined to be that number such that

$$\ln a^x = x \ln a$$

Using the formula $e^{\ln c} = c$ we can now obtain a direct equation for a^x:

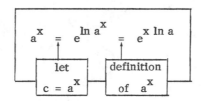

This means that the value of a^x for any a > 0 and any x can be approximated using only the tables for e^x and ln x (Tables 2 and 3 of Anton's Appendix 3)!

Example C.   Approximate $8^{\sqrt{3}}$.

Solution. $8^{\sqrt{3}} = e^{\sqrt{3}\, \ln 8} \approx e^{(1.73)(2.0794)} \approx e^{3.6} \approx 36.598$

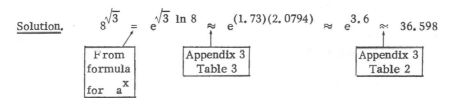

4.   The derivatives of e^x and a^x .   There are four basic derivative formulas discussed in this section:

Derivative Rules

Suppose a is a positive constant, and

   u is a differentiable function of x.

i. $\dfrac{d}{dx}[e^x] = e^x$ iii. $\dfrac{d}{dx}[e^u] = e^u \cdot \dfrac{du}{dx}$

ii. $\dfrac{d}{dx}[a^x] = a^x \ln a$ iv. $\dfrac{d}{dx}[a^u] = a^u \ln a \cdot \dfrac{du}{dx}$

Formulas i and ii are the most basic of the four:

i. $\frac{d}{dx}[e^x] = e^x$ means

the derivative of e^x is itself

ii. $\frac{d}{dx}[a^x] = a^x \ln a$ means

the derivative of a^x is itself <u>multiplied by</u> $\ln a$

Formulas iii and iv are merely the versions of i and ii, respectively, which incorporate the Chain Rule.

    Anton's Examples 3 and 4 illustrate all four of these formulas. Here are some additional examples involving the formulas, in some cases combining them with other rules of differentiation:

<u>Example D.</u>   Find $\frac{d}{dx}[x^2 e^{-3x}]$.

<u>Solution.</u>

$$\frac{d}{dx}[x^2 e^{-3x}] = x^2 \frac{d}{dx}[e^{-3x}] + e^{-3x}\frac{d}{dx}[x^2] \qquad \text{(product rule)}$$

$$= x^2(-3e^{-3x}) + e^{-3x}(2x) \qquad \text{(formula iii for } \frac{d}{dx}[e^u]$$
$$\text{with } u = -3x)$$

$$= x e^{-3x}(2 - 3x) \qquad\qquad\qquad \square$$

<u>Example E.</u>   Find $\frac{d}{dx}[e^{\sin x^2}]$.

<u>Solution.</u>

$$\frac{d}{dx}[e^{\sin x^2}] = e^{\sin x^2}\frac{d}{dx}[\sin x^2] \qquad \text{(formula iii for } \frac{d}{dx}[e^u]$$
$$\text{with } u = \sin x^2)$$

$$= 2x(\cos x^2)e^{\sin x^2} \qquad\qquad\qquad \square$$

Example F.   Find $\frac{d}{dx}[(\sqrt{3})^x]$.

Solution.

$$\frac{d}{dx}[(\sqrt{3})^x] = (\sqrt{3})^x \ln\sqrt{3} \qquad\qquad \text{(formula ii   with   } a = \sqrt{3})$$

$$= \frac{1}{2}(\ln 3)(\sqrt{3})^x \qquad\qquad\qquad\qquad \square$$

Example G.   Find $\frac{d}{dx}[3^{x^2-1}]$.

Solution.

$$\frac{d}{dx}[3^{x^2-1}] = (\ln 3)\,3^{x^2-1}\,\frac{d}{dx}[x^2-1] \qquad \text{(formula iv   with   } a = 3,$$
$$u = x^2 - 1)$$

$$= (\ln 3)(2x)\,3^{x^2-1} \qquad\qquad\qquad\qquad \square$$

5.   **The power rule (again).**   From Theorem 7.4.6 we have

The Power Rule	$\dfrac{d}{dx}\left[x^r\right] = r\,x^{r-1}$ where r is a real number.

This formula was originally established for r an integer (Theorem 3.2.2) and then for r a rational number (Theorem 3.2.7). Now the formula is true when r is any real number. This important extension of the formula enables us to differentiate functions like x^π (see Anton's Example 5) and $x^{\sqrt{2}}$:

Example H.   Find $\frac{d}{dx}[x^{\sqrt{2}}]$.

Solution.

$$\frac{d}{dx}[x^{\sqrt{2}}] = \sqrt{2}\,x^{\sqrt{2}-1} \qquad\qquad\qquad\qquad \square$$

WARNING: You must be careful to distinguish between the functions _____

$$x^a \qquad\qquad \text{and} \qquad\qquad a^x$$

(the constant as   (the variable as
the exponent)   the exponent)

Their derivatives are

$$\frac{d}{dx}[x^a] = a\,x^{a-1} \qquad\qquad \text{(the power rule)}$$

and

$$\frac{d}{dx}[a^x] = a^x \ln a \qquad\qquad \text{(formula iii)}$$

and they are always different !!

Example I.   Find the derivatives of $x^{\sqrt{3}}$ and $(\sqrt{3})^x$.

Solution.

$$\frac{d}{dx}[x^{\sqrt{3}}] = \sqrt{3}\; x^{\sqrt{3}-1}$$

and

$$\frac{d}{dx}[(\sqrt{3})^x] = (\sqrt{3})^x \ln \sqrt{3}$$

What about the third possibility

$$x^x ? \qquad \text{(both the exponent and the base are variables)}$$

Well, we can differentiate the function $y = x^x$ using logarithmic differentiation as follows:

$$\ln y = \ln x^x = x \ln x \qquad\qquad \text{(take } \ln \text{ of both sides)}$$

$$\frac{1}{y}\frac{dy}{dx} = x\frac{d}{dx}[\ln x] + (\ln x)\frac{d}{dx}[x] \qquad\qquad \text{(differentiate)}$$

$$\frac{dy}{dx} = y\left[x(\frac{1}{x}) + \ln x\right] \qquad\qquad \text{(solve for } \frac{dy}{dx})$$

or $$\frac{d}{dx}[x^x] = x^x[1 + \ln x] \qquad\qquad \text{(since } y = x^x)$$

In general, you may be asked to differentiate functions of the following three types:

the simple one $\dfrac{d}{dx}[f(x)^a] = a\,f(x)^{a-1}\,f'(x)$

the new one $\dfrac{d}{dx}[a^{g(x)}] = (\ln a)\,a^{g(x)}\,g'(x)$

the combination $\dfrac{d}{dx}[f(x)^{g(x)}]$ ...   use logarithmic differentiation!

Here is an example of the third type:

Example J.   Find $\dfrac{d}{dx}[(x^2-1)^{x^3}]$.

Solution.   Using logarithmic differentiation, let $y = (x^2-1)^{x^3}$, and take $\ln$ of both sides of this equation:

$$\ln y = \ln (x^2-1)^{x^3}$$
$$= x^3 \ln (x^2-1) \qquad \text{(Theorem 7.3.1 d)}$$

Then differentiate both sides with respect to x :

$$\frac{1}{y}\frac{dy}{dx} = x^3 \cdot \frac{d}{dx}[\ln (x^2-1)] + \frac{d}{dx}[x^3]\ln (x^2-1)$$

$$= x^3 \cdot \frac{2x}{x^2-1} + 3x^2 \ln (x^2-1) \qquad \left(\text{using } \frac{d}{dx}[\ln u] = \frac{1}{u}\frac{du}{dx}\right.$$
$$\left.\text{with } u = x^2-1\right)$$

$$= \frac{2x^4}{x^2-1} + 3x^2 \ln (x^2-1)$$

Finally, multiplying through by $y = (x^2-1)^{x^3}$, we obtain

$$\frac{dy}{dx} = (x^2-1)^{x^3}\left(\frac{2x^4}{x^2-1} + 3x^2 \ln (x^2-1)\right) \qquad \square$$

7.4.11

6. <u>The integrals of e^x and a^x.</u> Since integration is the reverse process of differentiation, we can read the derivative formulas of Subsection 4 backwards to get the following integration formulas:

Integral Rules

Suppose   a   is a positive constant, and

  u   is a differentiable function of   x .

i. $\int e^x\,dx = e^x + C$

ii. $\int a^x\,dx = \dfrac{a^x}{\ln a} + C$

iii. $\int e^u \cdot \dfrac{du}{dx}\,dx = e^u + C$

iv. $\int a^u \cdot \dfrac{du}{dx}\,dx = \dfrac{a^u}{\ln a} + C$

Formulas i and ii are the most basic of the four:

i. $\int e^x\,dx = e^x + C$ means   the integral of e^x is itself

ii. $\int a^x\,dx = \dfrac{a^x}{\ln a} + C$ means   the integral of a^x is itself divided by $\ln a$

Formulas iii and iv are merely the versions of i and ii which incorporate a u-substitution; for this reason there is little reason to memorize them.

Anton's Examples 6 through 10 illustrate these formulas. Here are some additional examples:

Example K.   Find $\displaystyle\int x\, 3^{x^2}\, dx$.

Solution.   The function 3^{x^2} is of the form a^u which suggests formula iv (i.e., a

u-substitution and formula ii):

$$\int x\, 3^{x^2}\, dx \;=\; \frac{1}{2}\int 3^u\, du$$

$$\boxed{\begin{array}{l} u = x^2 \\ du = 2\,x\,dx \\ \dfrac{1}{2}\,du = x\,dx \end{array}} \qquad = \frac{3^u}{2\ln 3} + C \qquad\qquad \text{(formula ii)}$$

$$= \frac{3^{x^2}}{2\ln 3} + C \qquad\qquad\qquad \square$$

Example L.   Find $\displaystyle\int_0^1 (x^2 - 2)\, e^{x^3 - 6x}\, dx$.

Solution.   The function e^{x^3-6x} is of the form e^u which suggests formula iii (i.e., a

u-substitution and formula i):

$$\int_0^1 (x^2 - 2)\, e^{x^3 - 6x}\, dx \;=\; \frac{1}{3}\int_0^{-5} e^u\, du$$

$$\boxed{\begin{array}{l} u = x^3 - 6x \\ du = (3x^2 - 6)\,dx \\ \dfrac{1}{3}\,du = (x^2 - 2)\,dx \end{array}} \qquad = \frac{1}{3}\, e^u \Big|_0^{-5} \qquad \text{(formula i)}$$

$$= \frac{1}{3}\, e^{-5} - \frac{1}{3} \qquad\qquad \square$$

7.   The facts and formulas you must memorize.   Here is a summary of the facts and formulas

from §7.4 which you must memorize.   Note that, in light of the techniques suggested above,

the integral formulas for $e^u \dfrac{du}{dx}$ and $a^u \dfrac{du}{dx}$ are not listed:

Definition of e   e is that number such that $\ln e = 1$

Definition of a^x a^x is that number such that $\ln a^x = x \ln a$ or $a^x = e^{x \ln a}$

(The special case when $a = e$: e^x is that number such that $\ln e^x = x$)

Inverse Function Relationship of e^x and $\ln x$ $\ln e^x = x$ and $e^{\ln x} = x$

Derivative of e^x $\dfrac{d}{dx}[e^x] = e^x$

(when combined with the Chain Rule: $\dfrac{d}{dx}[e^u] = e^u \cdot \dfrac{du}{dx}$)

Derivative of a^x $\dfrac{d}{dx}[a^x] = a^x \ln a$

(when combined with the Chain Rule: $\dfrac{d}{dx}[a^u] = (\ln a)\, a^u \dfrac{du}{dx}$)

Integral of e^x $\displaystyle\int e^x\, dx = e^x + C$

Integral of a^x $\displaystyle\int a^x\, dx = \dfrac{a^x}{\ln a} + C$

Section 7.5.   Additional Properties of e^x.

1.   Graphing exponential functions.   As Anton's Figure 7.5.2 makes clear, if you know the basic shape of the graph of the natural logarithm function $y = \ln x$

The graph
of $y = \ln x$

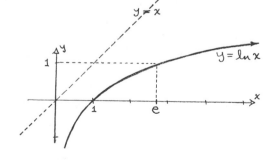

then the basic shape of the graph of the exponential function $y = e^x$ is easily obtained by

reflecting it over the line $y = x$

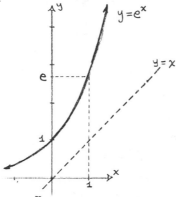

The graph
of $y = e^x$

This is because $y = \ln x$ and $y = e^x$ are inverse functions (as explained in §7.4.3 of The

Companion). The graphs of any pair of inverse functions are related in this way as we will

discover in §8.1.

Knowing the basic shape of the graph of $y - e^x$ is important; for one thing, it provides

an easy way to remember the following facts:

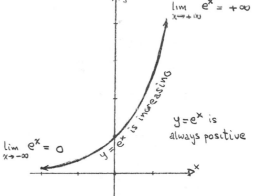

$y = e^x$ is an increasing function

$e^x > 0$ for all x

$\lim\limits_{x \to +\infty} e^x = +\infty$

$\lim\limits_{x \to -\infty} e^x = 0$

The basic shape of the graph of $y = e^x$ can also be useful in graphing other exponential functions,

at least simple ones of the form $y = c e^{kx} + d$ where c, d and k are constants. Anton's

Example 1 is an illustration;* here is another:

Example A. Sketch the graph of $y = 2e^{\frac{1}{3}x}$.

Solution.   The graph of $y = e^{\frac{1}{3}x}$ can be obtained by <u>expanding</u> the graph of $y = e^x$ by a

factor of 3 along the x-axis. (See the "expansion/contraction principles" in Appendix G.4):

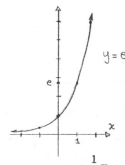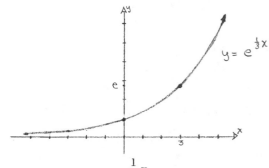

Then the graph of $y = 2e^{\frac{1}{3}x}$ is obtained from the graph of $y = e^{\frac{1}{3}x}$ by doubling the height

at every point:

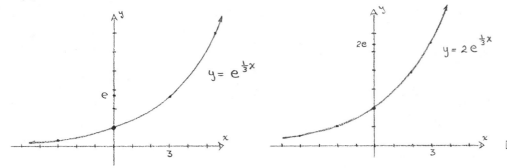

---

* While you are learning basic shapes, we recommend that you also memorize the basic

shape of the graph of $y = e^{-x}$ discussed in Example 1:

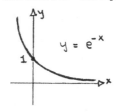

Of course more complicated exponential graphs require the full range of the graphing techniques developed in Chapter 4. Anton's Example 2 illustrates how they are used for the function $y = e^{-x^2/2}$. Here is another example; in solving it, we will use the graphing procedure summarized in §4.4.3 of The Companion (it is just a modified version of the polynomial graphing procedure Anton gives in §4.4):

Example B.   Sketch the graph of $y = x e^{-x}$.

Solution.   The function is defined for every value of x and hence its domain is the whole x-axis.

Step 1.   The derivatives. $f'(x) = x e^{-x}(-1) + e^{-x} = (1 - x) e^{-x}$

$f''(x) = (1 - x) e^{-x}(-1) - e^{-x} = (x - 2) e^{-x}$

Both of these derivatives are defined everywhere.

Step 2.   Increasing and decreasing.   Since $e^x > 0$ for all x, we know $e^{-x} > 0$ for all x. Hence $f'(x)$ equals zero only when $x = 1$, so $x = 1$ is the only stationary point for $f(x)$. Therefore we check the sign of f' on the intervals $(-\infty, 1)$ and $(1, +\infty)$:

	$(1 - x)$	e^{-x}	$= f'(x).$	Thus $f(x)$ is ...
$(-\infty, 1)$	+	+	+	... increasing if $x < 1$
$(1, +\infty)$	-	+	-	... decreasing if $x > 1$

This proves that $x = 1$ gives an <u>absolute maximum</u> value for f.

<u>Step 3.</u> <u>Concavity.</u> Since $e^{-x} > 0$ for all x, then $f''(x)$ equals zero only when $x = 2$. Therefore we check the sign of f'' on the intervals $(-\infty, 2)$ and $(2, +\infty)$:

	$(x - 2)$	e^{-x}	$= f''(x).$	Thus $f(x)$ is ...
$(-\infty, 2)$	-	+	-	... concave down if $x < 2$
$(2, +\infty)$	+	+	+	... concave up if $x > 2$

This proves that $x = 2$ gives an <u>inflection point</u> for f.

<u>Step 4.</u> <u>Endpoint limits and graph.</u> We have only the endpoint limits at $+\infty$ and $-\infty$ to consider:

$$\lim_{x \to -\infty} x\,e^{-x} = \left(\lim_{x \to -\infty} x \right) \left(\lim_{x \to -\infty} e^{-x} \right)$$

tends toward $+\infty$

tends toward $-\infty$ Hence this limit is $-\infty$.

$$\lim_{x \to +\infty} x\,e^{-x} = \left(\lim_{x \to +\infty} x \right) \left(\lim_{x \to +\infty} e^{-x} \right)$$

tends toward 0

tends toward $+\infty$ OOPS! This method of analysis fails!

For the second limit we observe that when $x > 2$, then

f is positive, decreasing and concave up; intuitively the

graph must therefore approach some finite, non-negative

limit L. Using techniques to be developed in Chapter 10

(L'Hôpital's Rule) it can be shown that this limit is $L = 0$

(also see Exercise 44 in Exercise Set 7.5).

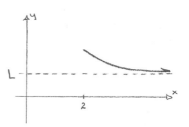

We now compute some points on the graph:

x	0	1	2	- 1
f(x)	0	$\frac{1}{e} \approx .37$	$\frac{2}{e^2} \approx .27$	$- e \approx - 2.72$

Plotting all these points, adding in the auxiliary f' and

f'' axes, and "following the instructions" produces the

graph as shown.

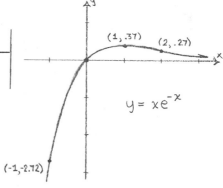

$$y = xe^{-x}$$

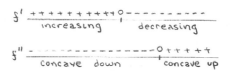

2. <u>Limits as $x \to \pm\infty$</u>. As was the case in Example B, in many applications you will be required

to find the limit of an exponential function as $x \to +\infty$ or $x \to -\infty$. When the exponential

function is simple, (i.e., of the form $y = ce^{kx}$ where c and k are constants) it is

usually easiest to sketch a rough graph of the function and "read off" the limit:

<u>Example C.</u> Find $\lim_{x \to -\infty} 2e^{-3x}$.

<u>Solution.</u> First we draw a rough sketch of the graph:

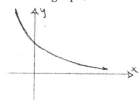

From this it is apparent that $\lim\limits_{x \to -\infty} 2e^{-3x} = +\infty$.   □

Example C illustrates two important general principles:

$$\lim_{x \to -\infty} e^x = 0 \quad \text{means that} \quad \underline{\text{an exponential tends to } 0 \text{ if}}$$
$$\underline{\text{its exponent tends to negative infinity,}}$$

and

$$\lim_{x \to +\infty} e^x = +\infty \quad \text{means that} \quad \underline{\text{an exponential tends to } +\infty \text{ if}}$$
$$\underline{\text{its exponent tends to positive infinity.}}$$

These general rules, when combined with the techniques developed in earlier chapters to compute limits of rational functions, will allow us to find the limits of more complicated expressions involving exponentials. Here are some examples:

<u>Example D.</u>   Find $\lim\limits_{x \to +\infty} \dfrac{2e^x - 3}{4 - e^x}$.

<u>Solution.</u>   The obvious first approach is to use "the limit of a quotient is the quotient of the limits." However, that fails because (using the second principle given above) the limit of the numerator is $+\infty$ while the limit of the denominator is $-\infty$. We must transform our expression into a quotient of two functions with <u>finite</u> limits as x tends toward $+\infty$. We do this by dividing the numerator and denominator by e^x :

$$\lim_{x \to +\infty} \frac{2e^x - 3}{4 - e^x} = \lim_{x \to +\infty} \frac{(2e^x - 3)/e^x}{(4 - e^x)/e^x} = \lim_{x \to +\infty} \frac{2 - 3e^{-x}}{4e^{-x} - 1}$$

$$= \frac{2 - 3 \lim_{x \to +\infty} e^{-x}}{4 \lim_{x \to +\infty} e^{-x} - 1}$$

$$= \frac{2 - 3(0)}{4(0) - 1} = \boxed{-2}$$

          □

Example E.   Find $\displaystyle \lim_{x \to -\infty} \frac{e^x - e^{-x}}{e^x + e^{-x}}$.

Solution.   Since the limits of the numerator and denominator are both infinite, we must algebraically manipulate them into forms with finite limits. Multiplying both the numerator and the denominator by e^x , we obtain

$$\lim_{x \to -\infty} \frac{e^x - e^{-x}}{e^x + e^{-x}} = \lim_{x \to -\infty} \frac{e^x - e^{-x}}{e^x + e^{-x}} \cdot \frac{e^x}{e^x}$$

$$= \lim_{x \to -\infty} \frac{e^{2x} - 1}{e^{2x} + 1}$$

$$= \frac{\lim_{x \to -\infty} e^{2x} - 1}{\lim_{x \to -\infty} e^{2x} + 1}$$

From the first principle given above we see that $\displaystyle \lim_{x \to -\infty} e^{2x} = 0$. Thus

$$\lim_{x \to -\infty} \frac{e^x - e^{-x}}{e^x + e^{-x}} = \frac{0 - 1}{0 + 1} = -1$$

          □

Example F.   Find $\lim\limits_{x \to -\infty} e^{e^{2x}}$.

Solution.   As   x   tends toward negative infinity, e^{2x} tends toward   0   from the first principle given above.   Hence

$$\lim_{x \to -\infty} e^{e^{2x}} = e^{\lim\limits_{x \to -\infty} e^{2x}} = e^{0} = 1$$

The first step in this computation is merely   Theorem 3.7.6   applied to the <u>continuous</u> function $f(x) = e^{x}$:   if $\lim g(x) = L$ is a finite number, then

$$\lim e^{g(x)} = e^{\lim g(x)} = e^{L} \qquad\qquad \Box$$

Section 7.6.   The Hyperbolic Functions

1.   <u>What do I need to memorize?</u>   As Anton indicates, it is convenient to give special names, collectively known as "the hyperbolic functions," to certain combinations of e^{x} and e^{-x} which occur frequently in applications.   Although this section contains a lot of new terms and definitions, it is not necessary to memorize more than a few basic facts.   There are two reasons for this:

    1)   In Anton's <u>Calculus,</u> the hyperbolic functions are used to illustrate concepts, but they are not a fundamental part of the text.   Thus you will not <u>need</u> to have complete command of them to read Anton.   (You can always look up what you don't remember.)

    2)   All of the hyperbolic function results can be easily derived from a few basic facts.

What you <u>do</u> need to know are the definitions of the two basic hyperbolic functions,

sinh x   and   cosh x :

Definitions
of   sinh x
and   cosh x

$$\sinh x \ = \ (e^x \ - \ e^{-x})/2$$

$$\cosh x \ = \ (e^x \ + \ e^{-x})/2$$

It is also useful to know the basic shapes of their graphs:

Graphs of
y = sinh x
and
y = cosh x

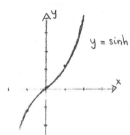

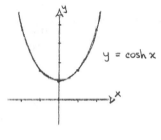

(As Anton shows, these can be derived quite easily from the graphs of $y = e^x$ and $y = e^{-x}$.)

From these graphs, the following information can be "read off" :

y = sinh x   is an increasing function

$$\lim_{x \to +\infty} \sinh x \ = \ |\infty$$

$$\lim_{x \to -\infty} \sinh x \ = \ -\infty$$

y = cosh x $\geq$ 1   for all   x

$$\lim_{x \to +\infty} \cosh x \ = \ \lim_{x \to -\infty} \cosh x \ = \ +\infty$$

You should also remember that

> except for a few differences in signs,
>
> the hyperbolic functions behave like
>
> the trigonometric functions.

For instance, the definitions of the other four hyperbolic functions in terms of sinh x and cosh x are exactly the same as those of the "last" four trigonometric functions; the defining formulas are simply the trigonometric formulas with an "h" added to each trigonometric function (see the first blue box of the section).

The various hyperbolic identities (Anton's Equations (1) through (9)) are now almost exact replicas of the corresponding trigonometric identities; the only differences are in an occasional plus or minus sign. In fact, it is convenient to put together a short table which separates the formulas with sign changes from those without sign changes:

no change in sign from the corresponding trigonometric formula	changes in sign from the corresponding trigonometric formula
	$\cosh^2 - \sinh^2 x = 1$ hence also: $1 - \tanh^2 x = \text{sech}^2 x$ $\coth^2 x - 1 = \text{csch}^2 x$
$\sinh(x+y) = \ldots$ hence also: $\sinh 2x = \ldots$	$\cosh(x+y) = \cosh x \cosh y + \sinh x \sinh y$ hence also: $\cosh 2x = \cosh^2 x + \sinh^2 x$
$\cosh(-x) = \cosh x$ $\sinh(-x) = -\sinh x$	
derivatives of $\sinh x$ $\tanh x$ $\coth x$ $\text{csch } x$	$\dfrac{d}{dx}[\cosh x] = +\sinh x$ $\dfrac{d}{dx}[\text{sech } x] = -\text{sech } x \tanh x$
integrals of $\cosh x$ $\text{sech}^2 x$ $\text{csch}^2 x$ $\text{csch } x \coth x$	$\displaystyle\int \sinh x \, dx = \cosh x + C$ $\displaystyle\int \text{sech } x \tanh x \, dx = -\text{sech } x + C$

This table makes it easier to remember the basic hyperbolic identities; however, equations (1)-(9) need not be memorized. Any given identity can be verified, usually quite easily, from the basic definitions in terms of $\sinh x$ and $\cosh x$. (See Subsection 2 below.) Even the derivative and integral formulas can be derived quickly from the basic definitions of $\sinh x$ and $\cosh x$. However, you will probably find it useful to memorize the derivative and integral formulas for $\sinh x$ and $\cosh x$:

7.6.5

$$\frac{d}{dx}\,[\,\sinh x\,] \;=\; \cosh x \qquad\qquad \int \sinh x\;dx \;=\; \cosh x \;+\; C$$

and

$$\frac{d}{dx}\,[\,\cosh x\,] \;=\; \sinh x \qquad\qquad \int \cosh x\;dx \;=\; \sinh x \;+\; C$$

2. <u>Verifying hyperbolic identities.</u>   Verifying a given hyperbolic identity is not usually difficult, although it is sometimes tedious, as the following examples illustrate:

<u>Example A.</u>   Verify the identity

$$\cosh (x - y) \;=\; \cosh x \cosh y \;-\; \sinh x \sinh y$$

<u>Solution.</u>   By definition

$$\cosh (x - y) \;=\; \frac{e^{x-y} + e^{-(x-y)}}{2} \;=\; \frac{e^{x}e^{-y} + e^{-x}e^{y}}{2}$$

and

$$\cosh x \cosh y - \sinh x \sinh y \;=\; \left(\frac{e^{x}+e^{-x}}{2}\right)\!\left(\frac{e^{y}+e^{-y}}{2}\right) \;-\; \left(\frac{e^{x}-e^{-x}}{2}\right)\!\left(\frac{e^{y}-e^{-y}}{2}\right)$$

$$= \frac{1}{4}\left(e^{x}e^{y} + e^{-x}e^{y} + e^{x}e^{-y} + e^{-x}e^{-y}\right)$$

$$-\; \frac{1}{4}\left(e^{x}e^{y} - e^{-x}e^{y} - e^{x}e^{-y} + e^{-x}e^{-y}\right)$$

$$= \frac{1}{4}\left(2\,e^{-x}e^{y} + 2\,e^{x}e^{-y}\right)$$

$$= \frac{e^{-x}e^{y} + e^{x}e^{-y}}{2}$$

Thus the two sides of the given identity are equal.   ☐

Example B.   Verify the identity

$$\tanh (x + y) = \frac{\tanh x + \tanh y}{1 + \tanh x \tanh y}$$

Solution.   Direct verification of this identity by plugging in

$$\tanh x = \frac{\sinh x}{\cosh x} = \frac{e^x - e^{-x}}{e^x + e^{-x}}$$

is possible, but extremely tedious and unenlightening.   Instead we can use the formulas for

$\sinh (x + y)$ and $\cosh (x + y)$:

$$\frac{\tanh x + \tanh y}{1 + \tanh x \tanh y} = \frac{\dfrac{\sinh x}{\cosh x} + \dfrac{\sinh y}{\cosh y}}{1 + \dfrac{\sinh x}{\cosh x} \cdot \dfrac{\sinh y}{\cosh y}}$$

$$= \frac{\dfrac{\sinh x \cosh y + \cosh x \sinh y}{\cosh x \cosh y}}{\dfrac{\cosh x \cosh y + \sinh x \sinh y}{\cosh x \cosh y}}$$

$$= \frac{\sinh x \cosh y + \cosh x \sinh y}{\cosh x \cosh y + \sinh x \sinh y}$$

$$= \frac{\sinh (x + y)}{\cosh (x + y)}$$

$$= \tanh (x + y) \qquad \qquad \square$$

3.   Differentiating and integrating hyperbolic functions.   Anton's Examples 1 and 2 illustrate that differentiating and integrating hyperbolic functions is simply a matter of applying known techniques to these new functions.   Here are two more examples:

Example C.   Find $\dfrac{dy}{dx}$ if $y = \operatorname{sech}^3 x$.

7. 6. 7

Solution. By definition $\operatorname{sech} x = \dfrac{1}{\cosh x}$ so

$$y = \cosh^{-3} x$$

Thus

$$\frac{dy}{dx} = -3 \cosh^{-4} x \, \frac{d}{dx} [\cosh x]$$

$$= -3 \cosh^{-4} x \, \sinh x$$

$$= -3 \, \frac{\sinh x}{\cosh x} \, \frac{1}{\cosh^3 x}$$

$$= -3 \tanh x \, \operatorname{sech}^3 x$$

Alternate Solution.

$$\frac{dy}{dx} = 3 \operatorname{sech}^2 x \, \frac{d}{dx} [\operatorname{sech} x]$$

$$= 3 \operatorname{sech}^2 x \, (-\operatorname{sech} x \tanh x)$$

$$= -3 \operatorname{sech}^3 x \tanh x \qquad \square$$

Example D. Find $\displaystyle\int \cosh (3x - 2) \, dx$.

Solution. If we let $u = 3x - 2$, then $\dfrac{du}{dx} = 3$ so $dx = \dfrac{1}{3} du$. Then

$$\int \cosh (3x - 2) \, dx = \frac{1}{3} \int \cosh u \, du$$

$$= \frac{1}{3} \sinh u + C$$

$$= \frac{1}{3} \sinh (3x - 2) + C \qquad \square$$

Section 7.7.   First-Order Differential Equations and Applications

1.   The terminology of differential equations.   The topic of differential equations probably has

more applications than any other studied in mathematics.   In calculus, we barely scratch the

surface - there are whole courses on differential equations as you probably are aware - but

it is important to learn the elementary techniques of the subject.   First we'll give a short

glossary of the terms used in differential equations, using Anton's Example 3 as our model:

- a differential equation is an equation involving one or more derivatives

$$\frac{dy}{dx} = -4xy^2 \quad \longleftarrow \boxed{\text{A DIFFERENTIAL EQUATION}}$$

- the order of a differential equation is the order of the highest derivative appearing in

it. * (As the title of this section indicates, we consider only first-order differential

equations.)

$$\frac{dy}{dx} = -4xy^2 \quad \boxed{\text{The differential equation is of FIRST-ORDER}}$$

- a solution of a differential equation is a function which solves the differential equation

(i. e. , when it is "plugged into" the differential equation, the equation is satisfied).

$$y = \frac{1}{2x^2} \quad \longleftarrow \boxed{\text{A SOLUTION of the given differential equation}}$$

Reason:

$$\frac{dy}{dx} = \frac{d}{dx}\left[\frac{1}{2}x^{-2}\right] = \frac{1}{2}(-2)x^{-3} = -\frac{1}{x^3} \longleftarrow$$

$$-4xy^2 = -4x\left[\frac{1}{2x^2}\right]^2 = -4x\left[\frac{1}{4x^4}\right] = -\frac{1}{x^3} \longleftarrow$$

$$\Big]\text{equal}$$

---
* Recall that $\frac{dy}{dx}$ has order 1, $\frac{d^2y}{dx^2}$ has order 2, etc.

Note that <u>a solution to a differential equation is a function</u> (or a whole <u>family</u> of functions), not just a number as you are accustomed to with "regular" equations. We speak of two types of solutions:

- <u>the general solution</u> of a differential equation is a solution (containing arbitrary constants) from which <u>all</u> solutions of the differential equation can be obtained (by assigning values to the arbitrary constants). Usually,

the number of arbitrary constants in the general solution equals the order of the differential equation.

Thus

> general solutions of first-order differential equations will have <u>one</u> arbitrary constant.

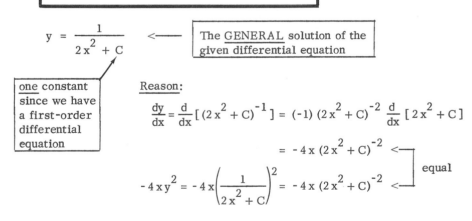

$$y = \frac{1}{2x^2 + C}$$ <——— The GENERAL solution of the given differential equation

one constant since we have a first-order differential equation

Reason:

$$\frac{dy}{dx} = \frac{d}{dx}[(2x^2 + C)^{-1}] = (-1)(2x^2 + C)^{-2}\frac{d}{dx}[2x^2 + C]$$

$$= -4x(2x^2 + C)^{-2} \;<\!\!\!-\!\!\!-\!\!\!\!\rceil$$

$$-4xy^2 = -4x\left(\frac{1}{2x^2 + C}\right)^2 = -4x(2x^2 + C)^{-2} \;<\!\!\!-\!\!\!\!\rceil \;\text{equal}$$

- a particular solution of a differential equation is a solution obtained by assigning a

value to each arbitrary constant in the general solution.

$$y = \frac{1}{2x^2}, \quad y = \frac{1}{2x^2 - 4}, \quad y = \frac{1}{2x^2 + 1} \quad \longleftarrow \boxed{\begin{array}{l} \text{PARTICULAR solutions} \\ \text{of the given differential equation} \end{array}}$$

The arbitrary constants of the general solution are closely related to initial conditions:

- an initial condition (or initial value) of a differential equation is a specification of

the value of the general solution at some point. Each initial condition determines

the value of one arbitrary constant in the general solution. (Hence in this section

on first-order differential equations whose general solutions have only one arbitrary

constant, we will need only one initial condition to determine a particular solution

from the general solution.)

$$y(0) = 1 \quad \longleftarrow \boxed{\begin{array}{l} \text{an INITIAL CONDITION for the} \\ \text{given differential equation} \end{array}}$$

Using this initial condition, we determine the constant C in the

general solution as follows:

$$y = \frac{1}{2x^2 + C}$$

$$\text{so} \quad 1 = y(0) = \frac{1}{2(0)^2 + C} = \frac{1}{C}$$

and hence $C = 1$

Thus the particular solution corresponding to this initial condition is

$$y = \frac{1}{2x^2 + 1}$$

We are thus confronted with two general problems:

---

1. Given a differential equation, how do we
   determine the general solution?

2. Given an initial condition for a differential equation,
   how do we use this initial condition to determine
   the correct particular solution from the
   general solution?

Since the second question is the easier of the two, we will consider it first...

2. Using initial conditions to evaluate constants in a general solution.   In Anton's Example 3 ,
   the initial condition $y(0) = 1$ is used to determine the value of the one arbitrary constant   C
   in the general solution. Here are three more examples:

Example A.   The general solution of the first-order differential equation $x \dfrac{dy}{dx} - 2y = 2$ is
$y = Cx^2 - 1$. (How this general solution was obtained will be discussed later.) Use the initial
condition $y(3) = 4$ to determine the value of the arbitrary constant C .

Solution.   The initial condition says

$$4 = y(3) = C(3)^2 - 1 = 9C - 1$$

so $C = \dfrac{5}{9}$. Thus

$$y = \frac{5}{9} x^2 - 1$$

is the particular solution corresponding to the initial condition $y(3) = 4$.   □

Example B.   Repeat Example A for the initial condition $y(1) = 3$.

Solution.   The initial condition says

$$3 = y(1) = C(1)^2 - 1 = C - 1$$

so   C = 4 . Thus

$$y = 4x^2 - 1$$

is the particular solution corresponding to the initial condition   y(1) = 3 .   □

Example C.   Repeat Example A for the initial condition   y(-1) = 3 .

Solution.   The initial condition says

$$3 = y(-1) = C(-1)^2 - 1 = C - 1$$

so   C = 4   as in Example B.   □

Note that Examples A , B and C illustrate that

usually, but not always, different initial conditions lead to different values of the arbitrary constant(s) and hence to different particular solutions.

*   *   *   *   *

As you can see, using initial conditions to determine a particular solution from a general solution is really pretty easy. However, our other major question,

"How do we find the general solution?"

is much more difficult!! In fact, for many differential equations there is no way to express the general solution in "closed form," i.e., in terms of elementary functions.

Fortunately there are several types of differential equations for which solution methods do exist. Two important such types considered by Anton are <u>separable differential equations and first-order linear differential equations.</u> We will examine each type individually.

3. <u>Separable differential equations.</u> The easiest kind of first-order differential equation to solve is the <u>separable</u> differential equation:

Separable Differential Equation

> A first-order differential equation is <u>separable</u>
>
> if it can be written as
> $$h(y)\, dy \ = \ g(x)\, dx$$
> i. e. , we can put all the y's on one side of the
>
> equation and all the x's on the other.

(Anton's Equation (11))

The solution technique for first-order separable differential equations is straightforward:

Solution Technique for Separable Differential Equations

> Step 1: <u>Separate the variables,</u> i. e. , put the equation into the form
> $$h(y)\, dy \ = \ g(x)\, dx$$
>
> Step 2: <u>Integrate both sides</u> of the equation, i. e. ,
> $$\int h(y)\, dy \ = \ \int g(x)\, dx \ + \ C \qquad \boxed{\text{A CRITICAL CONSTANT}}$$
> This will give an equation relating x and y ,
>
> Step 3: <u>Solve for y</u> as a function of x (if possible).
>
> <u>Determine the constant</u> if given an initial value.

Anton illustrates this technique in Examples 1-3. Here are several other examples, the first presented with a very detailed solution:

Example D.   Solve the initial-value problem

$$x^2 \frac{dy}{dx} = x^4 y^2 + y^2 , \qquad y(1) = 3$$

Step 1: Separate the variables.   When given a differential equation, the only way to tell if it is separable is to attempt to put it into the separable form

$$h(y) \, dy = g(x) \, dx$$

If the equation cannot be put into this form, then it is not separable and some other solution technique must be found (see Example F below). However, the equation at hand can be put into the desired form:

$$x^2 \frac{dy}{dx} = (x^4 + 1) y^2$$

$$\frac{dy}{dx} = (x^2 + x^{-2}) y^2$$

$$\boxed{y^{-2} \, dy = (x^2 + x^{-2}) \, dx}$$

Step 2: Integrate both sides.

$$\int y^{-2} \, dy = \int (x^2 + x^{-2}) \, dx + C$$

DON'T FORGET
THE CONSTANT!!

$$- y^{-1} = \frac{1}{3} x^3 - x^{-1} + C$$

Actually, we were very lucky in that we could perform both of the necessary integrations! If you cannot perform one of the integrations ... well, then you're in big trouble (see Example E below)!

<u>Step 3</u>: <u>Solve for y</u>.

$$-\frac{1}{y} = \frac{1}{3}x^3 - \frac{1}{x} + C$$

$$y = \frac{1}{-\frac{1}{3}x^3 + \frac{1}{x} - C}$$

$$y = \frac{3x}{3 - 3Cx - x^4} \qquad \text{(by multiplying by } \frac{3x}{3x}\text{)}$$

Thus

$$\boxed{y = \frac{3x}{3 + Kx - x^4}}$$

is our general solution, where K (= -3C) is an arbitrary constant.

Again we were very lucky in that we could solve for y as an (explicit) function of x .
In many instances the integration in Step 2 results in an equation in x and y which cannot
be solved for y as a function of x (see Anton's Example 2). In such a case we say that
the final equation "implicitly defines" y as a function of x . Implicitly defined functions were
discussed at length in §3.6.

<u>Determine the constant.</u>   The initial condition says

$$3 = y(1) = \frac{3}{3 + K - 1} = \frac{3}{2 + K}$$

Thus 6 + 3K = 3 , so that K = -1. The desired solution is therefore

$$\boxed{y = \frac{3x}{3 - x - x^4}}$$

□

Example E.   Solve the differential equation

$$\ln (y^5 + 1) \frac{dy}{dx} - e^{-x^2} = 1$$

Step 1. Separate the variables.   This is a separable equation because we can separate the variables:

$$\ln (y^5 + 1) \frac{dy}{dx} = 1 + e^{-x^2}$$

$$\boxed{\ln (y^5 + 1)\, dy = \left(1 + e^{-x^2}\right) dx}$$

? Step 2.   Integrate both sides?

$$\int \ln (y^5 + 1)\, dy = \int \left(1 + e^{-x^2}\right) dx + C$$

$$\qquad\quad ??!? \qquad\qquad\qquad ?!!?!$$

We cannot evaluate either of these two integrals, and so we are stuck. This differential equation cannot be solved "in closed form. "   □

Example F.   Solve the initial-value problem

$$e^{-t} \frac{dy}{dt} + 2y e^{-t} = 1, \qquad y(0) = 2$$

? Step 1.   Separate the variables?

$$\frac{dy}{dt} + 2y = e^{t} \qquad\qquad \text{(multiplying by } e^{t})$$

$$\frac{dy}{dt} = e^{t} - 2y$$

$$dy = (e^{t} - 2y)\, dt$$

$$\boxed{\text{OOPS! NO } y \text{ SHOULD BE HERE} \ldots}$$

Try as we may, there is no way to separate the variables in this equation, and hence it cannot be solved by the techniques just discussed. However it can be transformed into a <u>first-order linear differential equation,</u> and can thus be solved by techniques which we'll now discuss (see Example G)... ☐

4.   <u>Linear Differential Equations.</u>   The second kind of (first-order) differential equation studied in this section is the first-order linear differential equation:

Linear
Differential
Equation

A first-order differential equation is <u>linear</u>

if it can be put in the form

$$\frac{dy}{dx} + p(x)y = q(x)$$

(Anton's Equation (13))

where $p(x)$ and $q(x)$ are functions

solely of x (they may be constants).

Anton summarizes the solution procedure for first-order linear differential equations in a 3-step method just prior to   Example 4 .   Actually, the procedure can be reduced to a 2-step method:

Solution
Technique
for Linear
Differential
Equations

After writing the differential equation in the form

$$\frac{dy}{dx} + p(x)y = q(x)$$

Step 1.   Find the <u>integrating factor</u> $\rho = e^{\int p(x)\, dx}$
(using 0 as the constant of integration).

Steps 2/3.   The <u>general solution</u> is $y = \dfrac{1}{\rho}\, [\int \rho\, q(x)\, dx + C]$

<u>Determine the constant</u> if given an initial value.

We have simply combined Anton's last two steps into one!

Anton's Examples 4 and 5 illustrate this procedure; note that in Example 5, it is necessary to "work" the given equation into the desired first-order linear form. Here is one more example:

Example G.   Solve the initial-value problem

$$e^{-t} \frac{dy}{dt} + 2y e^{-t} = 1, \qquad y(0) = 2$$

Solution.   This is the equation from Example F which was not separable. Multiplication by e^t transforms the equation into

$$\frac{dy}{dt} + 2y = e^t$$

which is a first-order linear differential equation with $p(t) = 2$ and $q(t) = e^t$. Thus we can solve it as follows:

Step 1.   The integrating factor is

$$\rho = e^{\int p(t)\, dt} = e^{\int 2\, dt} = e^{2t}$$

Steps 2/3.   The general solution is

$$y = \frac{1}{\rho} \left[\int \rho\, q(t)\, dt + C \right]$$

$$= \frac{1}{e^{2t}} \left[\int e^{2t}\, e^t\, dt + C \right]$$

$$= e^{-2t} \int e^{3t}\, dt + C e^{-2t}$$

$$= e^{-2t} \left(\frac{1}{3} e^{3t} \right) + C e^{-2t}$$

Thus

$$y = \frac{1}{3} e^t + C e^{-2t}$$

is the general solution.   The initial condition says

$$2 = y(0) = \frac{1}{3} e^0 + C e^{-2(0)} = \frac{1}{3} + C$$

so   C = 5/3 .   Hence

$$y = \frac{1}{3} e^t + \frac{5}{3} e^{-2t}$$

is the <u>particular solution</u> corresponding to the initial condition   y(0) = 2 .   □

$$* \qquad * \qquad * \qquad * \qquad *$$

<u>Sometimes a first-order differential equation will be both separable and linear.</u>   This occurs, for example, when a first-order equation can be put in the linear form $\frac{dy}{dx} + p(x)y = q(x)$ and one of the following occurs: (1)   p(x) = 0 ,   (2)   q(x) = 0   or (3)   p(x)   and   q(x)   are proportional.   In any of these cases you have a choice of solution techniques; either the separable procedure or the first-order linear procedure may be used. Generally the separable procedure will be easier to use (it is a bit more elementary than the linear procedure), but which technique to use is really a matter of taste.   Here is an example of a differential equation which can be solved in either manner:

<u>Example H.</u>   Solve the differential equation

$$xy' - 2y = 2$$

<u>Solution as a separable equation.</u>

Step 1. Separate the variables.

$$x\,y' = 2y + 2$$

$$x\,\frac{dy}{dx} = 2y + 2$$

$$\boxed{\frac{dy}{2y + 2} = \frac{dx}{x}}$$ Thus we do have a separable
differential equation.

Step 2. Integrate both sides.

$$\frac{1}{2}\int \frac{dy}{y + 1} = \int \frac{dx}{x} + C$$

$$\frac{1}{2}\ln\left|y + 1\right| = \ln\left|x\right| + C$$

$$\ln\left|y + 1\right| - 2\ln\left|x\right| + 2C$$

Step 3. Solve for y .   To extract the y from the logarithm we "exponentiate" both sides of
the equation:

$$e^{\ln\left|y + 1\right|} = e^{2\ln\left|x\right| + 2C}$$

$$\left|y + 1\right| = e^{2\ln\left|x\right|}\, e^{2C} = e^{2C}\left|x\right|^{2} = e^{2C}\,x^{2}$$

$$y + 1 = \pm\, e^{2C}\,x^{2}$$

Thus

$$\boxed{y = -1 + K x^{2}}$$

is the general solution, where K ($= \pm\, e^{2C}$) is an arbitrary constant.

Solution as a linear equation.   Divide through by x so that the coefficient of y' becomes 1 :

$$y' - \frac{2}{x}\,y = \frac{2}{x}$$

Thus we obtain a first-order linear differential equation with $p(x) = -\frac{2}{x}$ and $q(x) = \frac{2}{x}$

which can be solved as follows:

Step 1. The <u>integrating factor</u> is

$$\rho = e^{\int p(x)\,dx} = e^{\int \frac{-2}{x}\,dx} = e^{-2\ln x} = e^{\ln x^{-2}} = x^{-2}$$

Steps 2/3. The <u>general solution</u> is

$$
\begin{aligned}
y &= \frac{1}{\rho}\left[\int \int \rho\, q(x)\,dx + C\right] \\[2mm]
&= \frac{1}{x^{-2}}\left[\int \int x^{-2}\,\frac{2}{x}\,dx + C\right] \\[2mm]
&= 2x^{2}\int x^{-3}\,dx + C\,x^{2} \\[2mm]
&= 2x^{2}\left(\frac{x^{-2}}{-2}\right) + C\,x^{2}
\end{aligned}
$$

Thus

$$\boxed{\,y = -1 + C\,x^{2}\,}$$

is the general solution where   C   is an arbitrary constant.   (Of course this

agrees with the separable equation solution.)   □

5.   <u>Word problems involving first-order differential equations.</u>   As we have mentioned, differential

equations occur in many practical applications of calculus. Most of the time they will be

encountered as <u>word problems</u> (Oh dear!). The general procedure for attacking such a word

problem is:

<table>
<tr><td>

Solving
Differential
Equations
Word
Problems

</td></tr>
</table>

Step 1. Translate the problem into a differential equation involving the unknowns. Generally there will be an initial condition given.

Step 2. Find the general solution and then, if there is an initial condition, the particular solution corresponding to it.

Step 3. Use the solution obtained in Step 2 to answer the question(s) asked. Usually this will simply be a matter of "plugging in" certain values into the solution.

As you probably suspect, Step 1 is the major (and usually the most difficult) step. However, don't despair; you will get better at it with practice! *

Derivatives find their way into word problems in several ways. Anton's examples illustrate two of them:

- the slope of a line (e. g. , a tangent line) is a derivative (Example 6)

- the rate of change (e. g. , the rate of growth or the rate of decay) of a quantity is a derivative (Examples 7 - 10). Most often, but not always, the rate of change of the quantity is with respect to time.

We will consider two important types of word problems in detail...

6. Mixing problems. A common application of first-order differential equations occurs in Mixing Problems. In Example 7, Anton solves a mixing problem involving salt dissolved in water; for the sake of simplicity we will continue using salt and water. Keep in mind, however, that these techniques also apply to more general situations.

---

* If you are uncomfortable with word problems we suggest that you read at least part of Appendix H ("Word Problems") in The Companion.

Suppose at time $t = 0$ we have a
tank containing V_0 gallons of water in
which y_0 pounds of salt are dissolved (i. e.,
V_0 gallons of salt water containing y_0
pounds of salt). Further suppose that another
salt water (or perhaps clear water) solution is
flowing into the tank, and that we are draining
salt water out of the tank. The question is:

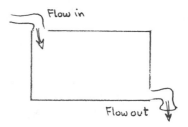

V_0 = volume (gallons)

y_0 = amount of salt
(pounds)

Flow in

Flow out

> how much salt is contained in the
>
> tank at any time $t \geq 0$?

The key idea in answering this question is that the rate at which the amount of salt in
the tank is changing is the rate at which it is coming in minus the rate at which it is going out.
That is, if we let $y = y(t)$ be the amount of salt (in pounds) dissolved in the tank at time t ,

The
Mixing
Equation

$$\begin{bmatrix} \text{rate of} \\ \text{change} \\ \text{in } y \end{bmatrix} = \begin{bmatrix} \text{rate at which} \\ \text{salt enters tank} \end{bmatrix} \text{ minus } \begin{bmatrix} \text{rate at which} \\ \text{salt leaves tank} \end{bmatrix}$$

(lbs/min)   (lbs/min)   (lbs/min)

This can be abbreviated to read

$$\frac{dy}{dt} = \text{rate in} - \text{rate out}$$

This will yield a differential equation in y and t with an initial condition $y = y_0$ when
$t = 0$. Solving this initial-value problem will give us the desired answer.

CAUTION! "Rate in" and "rate out" are in pounds per minute, NOT gallons per minute, i.e.,

they are measuring a change in the amount of salt (i.e., mass or weight), NOT a change in the

amount of water, (i.e., volume). However, as you will see in the example below, the total

volume $V(t)$ of salt water in the tank at time t does play a role in the solution.

Example 1. A tank with a 2000-gallon capacity initially contains 500 gal of brine containing 100 lbs

of salt. Starting at time $t = 0$, brine containing 0.1 lb of salt per gallon is added at a rate

of 60 gal/min and the mixed solution is drained off at a rate of 40 gal/min. How much salt is

in the tank when it reaches the point of overflowing?

Solution.   TRANSLATE. As described above, we let

$$y = y(t) = \text{amount (pounds) of salt dissolved in}$$
$$\text{the tank at time } t$$

$$V = V(t) = \text{volume (gallons) of salt water in the}$$
$$\text{tank at time } t$$

The problem asks us to find the value of $y(t)$ at some particular time ("the point of over-

flowing"), so we will first determine what $y = y(t)$ is at any time t . To do so we will use

the mixing equation

$$\frac{dy}{dt} = \text{rate in } - \text{ rate out}$$

Given this general set-up, let's organize our data into three categories:

initial conditions	inflow data	outflow data
$y_0 = 100$ lb	0.1 lb/gal	
$V_0 = 500$ gal	60 gal/min	40 gal/min

Let's consider the <u>inflow</u> data: 60 gallons of salt solution flow into the tank each minute, and each gallon contains 0. 1 lb of salt. That's simple enough: $60 \times 0.1 = 6$ pounds of salt flow into the tank each minute. Hence

$$\text{rate in} = \left(60 \ \frac{\text{gal}}{\text{min}}\right)\left(0.1 \ \frac{\text{lb}}{\text{gal}}\right) = 6 \ \frac{\text{lb}}{\text{min}}$$

Now look at the <u>outflow</u> data: 40 gallons of salt solution are withdrawn from the tank each minute, and each gallon contains ... ? how much salt? Well, at time t there are $y = y(t)$ pounds of salt dissolved in $V = V(t)$ gallons of solution. Thus we have y/V pounds of salt in each gallon. Hence

$$\text{rate out} = \left(40 \ \frac{\text{gal}}{\text{min}}\right)\left(\frac{y}{V} \ \frac{\text{lb}}{\text{gal}}\right) = \frac{40y}{V}\left(\frac{\text{lb}}{\text{min}}\right)$$

Thus our mixing equation $\frac{dy}{dt} = (\text{rate in}) - (\text{rate out})$ becomes

$$\frac{dy}{dt} = 6 - \frac{40y}{V}, \qquad y(0) = 100$$

But wait a minute .... We have some more information about V. We can express the volume V as a function of time (and we need to do this in order to be able to solve the initial-value problem). Every minute we add 60 gal and remove 40 gal, and hence every minute there is a net increase of 20 gal in the volume. Remembering that 500 gal were present initially (i. e., $V_0 = 500$), this means

$$V = V(t)$$

$$V = V_0 + 20t$$

$$\boxed{V = 500 + 20t}$$

The final version of our initial-value problem is thus

$$\boxed{\frac{dy}{dt} = 6 - \frac{40y}{500 + 20t}, \qquad y(0) = 100}$$

* * * * *

<u>FIND THE SOLUTION.</u> Now we solve this equation. When placed in the form

$$\frac{dy}{dt} + \left(\frac{40}{500 + 20t}\right)y = 6$$

we see that it is a first-order linear differential equation with $p(t) = 40/(500 + 20t)$ and $q(t) = 6$.

<u>Step 1.</u> The integrating factor is

$$\rho = e^{\int p(t)\,dt} = e^{\int \frac{40}{500 + 20t}\,dt} = e^{\frac{40}{20}\ln(500 + 20t)}$$

$$= e^{2\ln(500 + 20t)} = (500 + 20t)^2$$

<u>Steps 2/3.</u> The general solution is

$$y = \frac{1}{\rho}\left[\int \rho\, q(t)\,dt + C\right]$$

$$= \frac{1}{(500 + 20t)^2}\left[\int (500 + 20t)^2\, 6\,dt + C\right]$$

$$= \frac{1}{(500 + 20t)^2}\left[(6)\left(\tfrac{1}{3}\right)(500 + 20t)^3\left(\tfrac{1}{20}\right) + C\right]$$

$$= \frac{1}{(500 + 20t)^2}\left[\tfrac{1}{10}(500 + 20t)^3 + C\right]$$

7.7.20

Thus

$$y = 50 + 2t + C(500 + 20t)^{-2}$$

is the general solution.

The initial condition says

$$100 = y(0) = 50 + C(500)^{-2}$$

so $C = 12,500,000$. Hence

$$y = 50 + 2t + 12,500,000(500 + 20t)^{-2}$$

is the particular solution corresponding to the initial condition $y(0) = 100$.

* * * * *

ANSWER THE QUESTION. "How much salt is in the tank when it reaches the point of over-flowing?" Well, we have the amount of salt expressed as a function of time, $y = y(t)$. Hence we need to know at what time t_0 the tank reaches the point of overflow, i.e., the time t_0 when $V = V(t_0)$ equals the full volume of the tank (2000 gal). We previously computed

$$V = 500 + 20t$$

Hence we need to find t_0 such that

$$2000 = 500 + 20t_0$$

i.e., $t_0 = 75$ minutes. Thus the amount of salt at the point of overflow is

$$y = y(75) = 50 + 2(75) + \frac{12,500,000}{(500 + 1500)^2} = \boxed{203 \tfrac{1}{8} \text{ lb}} \qquad \square$$

7.   <u>Exponential growth or decay problems.</u>   The second type of word problem involving first-order

differential equations considered by Anton is the <u>Exponential Growth or Decay Problem.</u>   These

are very prevalent in applications because they occur whenever (Definition 7.7.1)

> a quantity increases (or decreases)
>
> in proportion to the amount present.

That is, if $y(t)$ is the amount of a quantity at time t ,   then

the rate of change of y is proportional to y

or

| Exponential Growth or Decay Equation |

$$\frac{dy}{dt} = ky \qquad \text{where} \quad y(t) = \text{amount of quantity at time } t$$
$$k = \text{proportionality constant}$$

This is an easy equation to solve, either as a separable or as a first-order linear

differential equation.   Assuming an initial condition of

$$y(0) = y_0$$

(i.e., at time $t = 0$ there is y_0 of the quantity present), Anton obtains the following solution

for the Exponential Growth or Decay Equation:

| Exponential Growth or Decay Solution |

$$y(t) = y_0 e^{kt}$$ (Anton's Equation (24))

<u>This is a result worth memorizing!</u>   It will make solving exponential growth or decay

problems quite a bit easier than having to work from scratch.   Anton illustrates this procedure in

Example 8. Here is another illustration:

Example J.   A culture of bacteria grows so that the rate of change of the population is proportional to the population. Suppose the population was originally 100 and two days later it is 100,000. How many days does it take for the population to reach 10^{11}?

Solution. TRANSLATE.   Let $y(t)$ equal the number of bacteria at time t days. Then the various phrases in the problem translate into equations as follows:

"... the rate of change of the population is proportional to the population. "...	$\dfrac{dy}{dt} = ky$ for some positive constant k
"... the population was originally 100... "...	$y(0) = 100$
"... and two days later it is 100,000. "...	$y(2) = 100,000$
"How many days does it take for the population to reach 10^{11}?"...	$t = ?$ when $y(t) = 10^{11}$

We thus have an Exponential Growth initial-value problem

$$\frac{dy}{dt} = ky, \qquad y(0) = 100$$

FIND THE SOLUTION.   This is the easy part since we have memorized the solution (have you?):

$$y = y_0 e^{kt} = 100 e^{kt}$$

Now we need to evaluate the constant k. To do this, we use the fact that

$$y(2) = 100,000$$

Putting this into the solution above we can determine k:

$$100,000 = y(2) = 100\, e^{2k}$$

$$1000 = e^{2k}$$

$$\ln 1000 = \ln(e^{2k}) = 2k$$

$$k = \frac{1}{2}\ln 1000 - \frac{1}{2}\ln 10^3 = \ln 10^{3/2}$$

Thus

$$\boxed{k = \ln 10^{3/2}}$$

(Notice how many of the logarithm and exponential rules were needed in this derivation!). Our solution to the initial-value problem then becomes

$$y(t) = 100\, e^{tk} = 100\, e^{t(\ln 10^{3/2})}$$

$$= 100\,(10^{3/2})^t \quad \text{since} \quad e^{t\ln a} = (e^{\ln a})^t = a^t$$

$$= 100\,(10^{3t/2})$$

$$= 10^2\,(10^{3t/2})$$

$$\boxed{y(t) = 10^{2+3t/2}}$$

ANSWER THE QUESTION.   The question is:

$$\text{"}\ t = ? \quad \text{when} \quad y(t) = 10^{11}\ \text{"}$$

We have only to use our equation for $y(t)$:

$$10^{11} = y(t) = 10^{2 + 3t/2}$$

$$11 \ln 10 = (2 + 3t/2) \ln 10 \qquad \text{(taking the logarithm of both sides of the equation)}$$

$$11 = 2 + 3t/2$$

$$3t/2 = 9$$

$$\boxed{t = 6 \text{ days}} \qquad \qquad \Box$$

* * * * *

Doubling Time and Halving Time. Anton's Examples 9 and 10 also illustrate the procedure for solving exponential growth or decay problems. In each case we are asked to find the doubling time (the time it takes a growing quantity to double) or the halving time or half-life (the time it takes a decaying quantity to halve). Anton derives formulas for doubling time and halving time (Equations (27) and (28)) :

Doubling Time $\qquad\qquad T = \dfrac{1}{k} \ln 2$
(when $k > 0$)

Halving Time (Half-Life) $\qquad T = -\dfrac{1}{k} \ln 2$
(when $k < 0$)

However, it is not necessary (or desirable!) to memorize these formulas. Instead, all questions which involve doubling time or halving time (or tripling time, etc.) can be treated as we did in the previous Example J. In that example there was an "extra" initial condition $(y(2) = 100,000)$ which we used to find the proportionality constant k; then we used k to find a value of t . Usually, doubling time and halving time can be treated in the same way:

solve $y(T) = 2y_0$ for the doubling time,

$y(T) = \frac{1}{2} y_0$ for the halving time, etc.

after finding the proportionality constant k.

Here is an example:

Example K.   The rate of decay of a radioactive material is proportional to the amount present.   Find

the half-life of the material if it takes 5 years for a third of the material to decay.

Solution.   TRANSLATE.   Let $y(t)$ equal the amount of material present at t years.   Then

the various phrases in the problem translate into equations as follows:

"The rate of decay ... is proportional to the amount present ... "	$\frac{dy}{dt} = ky$ for some negative constant k
"Find the half-life ... "	$T = ?$ when $y(T) = \frac{1}{2} y_0$
"... if it takes 5 years for a third of the material to decay. "	$y(5) = \frac{2}{3} y_0$

FIND THE SOLUTION.   We recall that the solution to the differential equation $\frac{dy}{dt} = ky$ is

simply

$$y = y_0 e^{kt}$$

To use it, we need to know the value of the constant k.   To find k we use the "extra" condition

$y(5) = (2/3)y_0$:

$$\frac{2}{3} y_0 = y(5) = y_0 e^{5k}$$

Hence

$$\frac{2}{3} = e^{5k}$$

$$\ln\left(\frac{2}{3}\right) = \ln e^{5k} = 5\,k$$

$$k = \frac{1}{5}\ln\left(\frac{2}{3}\right) = \ln\left(\frac{2}{3}\right)^{1/5}$$

Thus

$$\boxed{k = \ln\left(\frac{2}{3}\right)^{1/5}}$$

Our solution to the differential equation then becomes

$$y(t) = y_0\, e^{tk} = y_0\, e^{t\,\ln\left(\frac{2}{3}\right)^{1/5}}$$

$$= y_0\left(\left(\frac{2}{3}\right)^{1/5}\right)^{t} \quad \text{since} \quad e^{t\,\ln a} = \left(e^{\ln a}\right)^{t} = a^{t}$$

$$\boxed{y(t) = y_0\left(\frac{2}{3}\right)^{t/5}}$$

ANSWER THE QUESTION.   The question is:

$$"\,T = ? \quad \text{when} \quad y(T) = \frac{1}{2}\, y_0\,"$$

i. e. , "Find the half-life T." We can, at this point, simply use Anton's formula

$$T = -\frac{1}{k}\ln 2$$

since we have computed k . However, if you do not wish to memorize this formula, it is easy

enough just to solve the equation $y(T) = \frac{1}{2}\, y_0$ for T , as follows:

$$\frac{1}{2} y_0 = y(T) = y_0 \left(\frac{2}{3}\right)^{T/5}$$

$$\frac{1}{2} = \left(\frac{2}{3}\right)^{T/5}$$

$$- \ln 2 = \frac{T}{5} (\ln 2 - \ln 3) \qquad \text{(Theorem 7.3.1)}$$

Thus

$$T = \frac{-5 \ln 2}{\ln 2 - \ln 3} = \frac{-5 (.6931)}{.6931 - 1.0986} \approx \boxed{8.55 \text{ years}} \qquad \Box$$

using Appendix 3 , Table 3
or a hand calculator

Chapter 8:   Inverse Trigonometric and Hyperbolic Functions

Introduction.   At this point we have been introduced to the two major operations of calculus -

differentiation and integration - and we have applied them to a growing collection of "elementary

functions":

      1.   polynomial and rational functions

      2.   trigonometric functions

      3.   exponential functions

      4.   logarithm functions

      5.   hyperbolic functions .

Before going on to discuss integration further in Chapter 9 , we take time now to add some

important new functions to our list: the "inverse functions" of the trigonometric and hyperbolic

functions.

Section 8. 1.   Inverse Functions

    This section forms the basis for the entire chapter.   Subsequent sections will give you

less trouble if you understand this section well.

1.   Definition and existence of inverse functions.   The definition (Definition 8. 1. 1) of inverse

functions is not difficult:

8. 1. 2

In §7.4.3 of <u>The Companion</u> we showed that

$$f(x) = x^n \qquad \text{and} \qquad g(x) = x^{1/n} \qquad (\text{for} x > 0)$$

are inverse functions and that

$$f(x) = e^x \qquad \text{and} \qquad g(x) = \ln x$$

are inverse functions. Here are two additional examples:

<u>Example A</u>.   Determine if $f(x) = e^{2x-1}$ and $g(x) = \ln \sqrt{x} + \frac{1}{2}$ are inverse functions.

<u>Solution</u>.   Simple calculations show

$$f(g(x)) = f(\ln \sqrt{x} + \frac{1}{2}) = e^{2(\ln \sqrt{x} + \frac{1}{2}) - 1} = e^{2 \ln \sqrt{x}} = e^{\ln (\sqrt{x})^2} = e^{\ln x} = x$$

and

$$g(f(x)) = g(e^{2x-1}) = \ln \sqrt{e^{2x-1}} + \frac{1}{2} = \frac{1}{2} \ln e^{2x-1} + \frac{1}{2} = \frac{1}{2}(2x - 1) + \frac{1}{2} = x$$

Thus we see that f and g "undo" each other and hence are inverse functions. □

<u>Example B</u>.   Determine if $f(x) = 2x + 1$ and $g(x) = \frac{1}{2}x + 1$ are inverse functions.

<u>Solution</u>.   A simple calculation shows

$$f(g(x)) = f(\frac{1}{2}x + 1) = 2(\frac{1}{2}x + 1) + 1 = x + 3 \neq x$$

so the function f does not "undo" the function g. For that matter, g does not "undo"

f either since

$$g(f(x)) = g(2x + 1) = \frac{1}{2}(2x + 1) + 1 = x + \frac{3}{2} \neq x$$

The failure of either one of these conditions (i.e., $f(g(x)) = x$ or $g(f(x)) = x$) is enough

to show that f and g are NOT inverse functions of each other.   □

*   *   *   *   *

As we said in the Introduction, our purpose in studying inverse functions is to add to

our list of elementary functions by adding the inverses of other functions already on the list.

But not all functions have inverse functions! When given a function f, how can we

determine if an inverse function $g = f^{-1}$ exists for f? The answer to this question is given

in Anton's Theorem 8.1.3 :

<table>
<tr><td>

Inverse
exists
if and only if
one-to-one

</td><td>

A function f has an inverse function $g = f^{-1}$

if and only if f is one-to-one, i.e.,

no two distinct x-values are

sent by f to the same y-value

</td></tr>
</table>

Geometrically this can be expressed as follows:

Take any horizontal line $y = y_0$. If f has an

inverse function, then this line can intersect the

graph of $y = f(x)$ in at most one point.

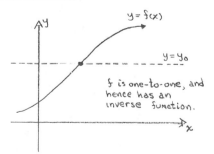

f is one-to-one, and hence has an inverse function.

8.1.4

(If there were <u>two</u> (or more) intersection points,
then there would be two (or more) x-values
sent by f to the single y-value y_0 , and
f would not be one-to-one.) Thus Theorem 8.1.3
can be restated geometrically as the <u>Horizontal Line Test</u>   (See Definition 8.1.2):

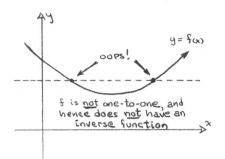

f is <u>not</u> one-to-one, and
hence does <u>not</u> have an
inverse function

<table>
<tr><td>Horizontal
Line
Test</td><td>A function f has an inverse function $g = f^{-1}$

if and only if every horizontal line intersects

the graph of f in <u>at most</u> one point.</td></tr>
</table>

As Anton points out, any function f which is always increasing (or
always decreasing) on an interval must be
one-to-one (or, equivalently, must pass
the Horizontal Line Test). This proves:

Theorem 8.1.4:

<table>
<tr><td>Increasing
and
Decreasing
functions</td><td>Any function f which is always increasing

or always decreasing on an interval must

have an inverse function on the interval.</td></tr>
</table>

This is a very useful result because we know how to determine where a function is increasing
and decreasing by using the first derivative!

Thus we have three ways to check if a function f has an inverse function $g = f^{-1}$
<u>without actually computing</u> f^{-1} !! Anton illustrates the One-To-One Test in Example 3 , the
Horizontal Line Test in Examples 2, 3 and 4 are the Increasing/Decreasing Test in Example
5 . We'll illustrate all three methods in the next example:

<u>Example C.</u>   Determine whether $f(x) = \sqrt{x-1}$ has an inverse function.

First Solution: One-to-one. Suppose $f(x_1) = f(x_2)$, i. e., suppose x_1 and x_2 are points in $[1, +\infty)$, the domain of f, which take on the same f-value. Then

$$\sqrt{x_1 - 1} = f(x_1) = f(x_2) = \sqrt{x_2 - 1}$$

Squaring both sides gives

$$x_1 - 1 = x_2 - 1$$

so that

$$x_1 = x_2$$

This shows that there can NOT be two <u>DISTINCT</u> x-values, x_1 and x_2, which are sent by f to the same y-value. Hence f is one-to-one and therefore it has an inverse.

Second Solution: Horizontal Line Test.   Recalling that the square root symbol $\sqrt{}$ (with no minus sign in front of it) always means the non-negative square root, the graph of $y = \sqrt{x - 1}$ is easily drawn as shown to the right. Since any horizontal line intersects the graph

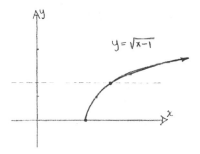

at most once, $f(x) = \sqrt{x - 1}$ passes the Horizontal Line Test, proving that it does have an inverse.

Third Solution: Increasing Function. $f(x) = \sqrt{x - 1}$ can be "seen" to be an increasing function from its graph above. However, we do not need the graph because we can prove f is increasing on $[1, +\infty)$ simply by considering the derivative

$$f'(x) = \frac{1}{2} (x - 1)^{-1/2}$$

Since f' is always positive when $x > 1$, f is always increasing on the open interval $(1, +\infty)$ [Theorem 4. 2. 2]. Moreover, the value of f at 1 equals zero (i. e., $f(1) = 0$)

8.1.6

which is less than the value of f for $x > 1$. Hence f is increasing on its full domain

$[1, +\infty)$, proving that f does have an inverse. □

Example D. Determine whether $f(x) = \sqrt{x^2 - 1}$ has an inverse function.

Solution. The domain of this function is

$$D = \{x : x^2 - 1 \geq 1\} = (-\infty, -1] \cup [1, +\infty)$$

and its graph is

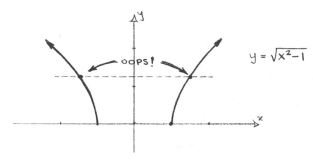

From the graph, it is clear that f does not pass the Horizontal Line Test and hence does not

have an inverse. Even without the graph, we can see that f is not one-to-one: since x

appears only as a square (i.e., as x^2), x and -x will always be sent to the same

y-value! Thus we know that f cannot have an inverse. □

2. Computation of inverse functions.   Even if you have established that a function   f   has an

inverse function f^{-1} ,   there remains the question,

how do you find the inverse

function f^{-1} of a function   f ?

Actually, in the cases of greatest interest to us, there will be no way to "find" the inverse

function f^{-1} if we interpret this to mean "express f^{-1} in terms of functions we already

know. " Our major purpose in this chapter is to obtain a whole collection of inverse functions

which cannot be expressed in terms of "functions we already know!"

However, in many situations inverse functions can be found. Anton suggest a procedure

for finding f^{-1} prior to Example 6 :

$$\boxed{\text{Solve the equation } f(f^{-1}(x)) = x \text{ for } f^{-1}(x)}$$

Anton's Examples 6 and 7 illustrate the use of this technique and Example 8 illustrates the

situation in which f^{-1} is known to exist but in which it cannot be "found " by this method.

Here are two more examples:

Example E.   Find f^{1} if $f(x) = \sqrt{x - 1}$.

Solution.   We determined that   f   has an inverse function f^{-1} in Example C. To compute

a formula for $f^{-1}(x)$ we replace   x   by $f^{-1}(x)$ in the formula for $f(x)$. Thus

$$x = f(f^{-1}(x)) = \sqrt{f^{-1}(x) - 1}$$

| by the definition | by the formula |
| of inverse function | for   f(x) |

Squaring both sides of the equation will allow us to solve for $f^{-1}(x)$:

$$x^2 = f^{-1}(x) - 1$$

$$\boxed{f^{-1}(x) = x^2 + 1}$$

☐

If, given a function f, you <u>compute</u> the inverse function f^{-1}, then you have also verified the <u>existence</u> of an inverse function! Thus the calculation of f^{-1} in Example E makes the existence proof in Example C unnecessary. Conversely, if by using this inverse calculation procedure you come up with two or more $f^{-1}(x)$ -values for a given x-value, then no inverse can exist. Here is an example:

<u>Example F.</u>   Determine if $f(x) = \sqrt{x^2 - 1}$ has an inverse. If so, find it.

<u>Solution.</u>   In Example D we determined that f does not have an inverse because it fails the Horizontal Line Test. However, in this case attempting to calculate an inverse will also show that such an inverse cannot exist:

$$x = f(f^{-1}(x)) = \sqrt{(f^{-1}(x))^2 - 1}$$
$$x^2 = (f^{-1}(x))^2 - 1$$
$$(f^{-1}(x))^2 = x^2 + 1$$
$$?!? \quad f^{-1}(x) = \pm\sqrt{x^2 + 1} \quad ?!?$$

Oops! We have obtained <u>two</u> $f^{-1}(x)$-values for every x-value. Hence (since functions are <u>single</u> valued) $f^{-1}(x)$ does not exist.

☐

3. <u>Graphs of inverse functions.</u>   As Anton proves early in this section, the graph of $y = f^{-1}(x)$

can be obtained from that of $y = f(x)$ by "reflection about the line $y = x$." Specifically,

Anton shows

$$
\left\{ \begin{array}{c} (a\,,\,b)\ \ \text{is a point} \\ \text{on the graph of} \\ y\ =\ f(x) \end{array} \right\} \quad \begin{array}{c} \text{if and} \\ \text{only if} \end{array} \quad \left\{ \begin{array}{c} (b\,,\,a)\ \ \text{is a point} \\ \text{on the graph of} \\ y\ =\ f^{-1}(x) \end{array} \right\}
$$

To see how this works geometrically, perform the following three-step procedure on the graph

of $y = f(x)$:

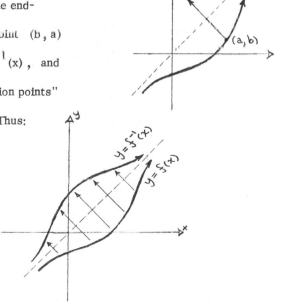

Step 1. take any point $(a\,,\,b)$ on the graph of $y = f(x)$,

Step 2. draw the line segment which starts at

   $(a\,,\,b)$ and ends with a perpendicular inter-

   section on the line $y = x$, and

Step 3. extend this line segment the same distance on the

   other side of the line $y = x$. The end-

   point of this segment will be the point $(b\,,\,a)$

   which lies on the graph of $y = f^{-1}(x)$, and

   the collection of all these "reflection points"

   gives the graph of $y = f^{-1}(x)$. Thus:

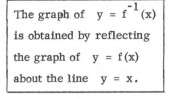

The graph of $y = f^{-1}(x)$ is obtained by reflecting the graph of $y = f(x)$ about the line $y = x$.

Example G.   Suppose $f(x) = 2x - 4$.   Find the graph of $y = f^{-1}(x)$.

First Solution: Compute f^{-1} .   If we were not familiar with the "reflection principle" for graphing $y = f^{-1}(x)$,   our only approach to graphing $y = f^{-1}(x)$ would be to compute $f^{-1}(x)$ first:

$$x = f(f^{-1}(x)) = 2f^{-1}(x) - 4$$

so that
$$f^{-1}(x) = \frac{1}{2}x + 2$$

The graph of this function is a line with slope $\frac{1}{2}$ and   y-intercept   2 ,   as shown to the right.

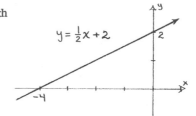

Second solution: Reflect f in y = x .   In this approach we take the graph of $y = f(x)$ and simply reflect it in the line $y = x$.   The graph of $y = f(x) = 2x - 4$ is a line with slope 2 and   y-intercept -4 , as shown below. Reflecting it in the line $y = x$ produces the graph of $y = f^{-1}(x)$. As in the first solution, we obtain the line with slope $\frac{1}{2}$ and   y-intercept 2 .

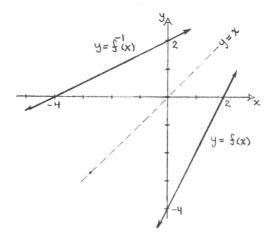

4. <u>Derivatives of inverse functions</u>. Suppose f has an inverse function f^{-1}. A natural and

important question is

> If f is differentiable, then is f^{-1} differentiable?

Anton's answer is a qualified "yes," and is contained in Theorem 8. 1. 5 :

<table>
<tr>
<td>

When is an
inverse function
differentiable?

</td>
<td>

Suppose f has an inverse function f^{-1} and is

differentiable on an open interval I.

Then f^{-1} is differentiable at x if $f^{-1}(x)$ is

a point in I at which f' is non-zero.

</td>
</tr>
</table>

This result looks a lot more formidable than it really is: roughly what it says is

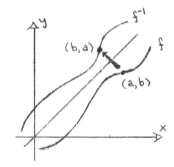

> if y = f(x) is differentiable at (a , b) ,
>
> then $y = f^{-1}(x)$ is differentiable at the
>
> corresponding reflected point (b, a)
>
> ... as long as f'(a) ≠ 0

The reason for the qualification that f'(a) ≠ 0 is simple geometrically: if f'(a) = 0 ,

then y = f(x) has a horizontal tangent line at (a , b). Thus, because of the reflection,

$y = f^{-1}(x)$ will have a <u>vertical</u> tangent line at (b, a). Hence f^{-1} would not be differentiable

at (b, a) !!

Having answered "When will f^{-1} be differentiable?", the next question is

> How do we compute the derivative of f^{-1} ?

Theorem 8. 1. 5 again provides an answer, in the form of a formula for the derivative of $f^{-1}(x)$:

$$\boxed{(f^{-1})'\,(x) \;=\; \frac{1}{f'\,(f^{-1}(x))}}$$

This rather strange-looking formula can be put in a more memorable form as follows. Let $y = f^{-1}(x)$, so that $x = f(y)$. Then

$$\frac{dy}{dx} \;=\; (f^{-1})'\,(x) \;=\; \frac{1}{f'\,(f^{-1}(x))} \;=\; \frac{1}{dx/dy}$$

so that

$$\boxed{\frac{dy}{dx} \;=\; \frac{1}{dx/dy}}$$

These formulas can be used to compute the derivative of an inverse function, but many times it is preferable to employ implicit differentiation:

Computation of the derivative of an inverse function	Let $y = f^{-1}(x)$, so that $f(y) = x$. Then compute dy/dx by implicit differentiation of $f(y) = x$.

Example H. Find the derivative of the inverse function of

$$f(x) \;=\; e^{2x} + \ln x \qquad (x > 0)$$

Solution. Since $f'(x) = 2e^{2x} + 1/x$ is always positive for $x > 0$, then f is increasing on $(0, +\infty)$. Hence f has an inverse on $(0, +\infty)$ by Theorem 8. 1. 4. Moreover, since

$f'(x)$ is never zero, then Theorem 8. 1. 5 guarantees that the inverse f^{-1} is differentiable wherever it is defined. We will compute $(f^{-1})'$ in two ways:

Use of dy/dx formula: Let $y = f^{-1}(x)$, so that

$$x = f(f^{-1}(x)) = f(y) = e^{2y} + \ln y$$

Then
$$\frac{dx}{dy} = \frac{d}{dy} [e^{2y} + \ln y] = 2e^{2y} + 1/y$$

so using the dy/dx formula yields

$$\frac{dy}{dx} = \frac{1}{dx/dy} = \frac{1}{2e^{2y} + 1/y}$$

Use of implicit differentiation: Let $y = f^{-1}(x)$, so that

$$x = f(f^{-1}(x)) = f(y) = e^{2y} + \ln y$$

Then
$$\frac{d}{dx}(x) = \frac{d}{dx} [e^{2y} + \ln y]$$

$$1 = 2e^{2y} \frac{dy}{dx} + \frac{1}{y} \frac{dy}{dx}$$

$$1 = [2e^{2y} + 1/y] \frac{dy}{dx}$$

$$\frac{dy}{dx} = \frac{1}{2e^{2y} + 1/y}$$

This answer agrees with our previous solution.   □

Notice that our answer for dy/dx in Example H is in terms of y. We would prefer that our answer be in terms of x, but this is not possible since we cannot solve the equation $x = e^{2y} + \ln y$ for y in terms of x. This is a common situation when computing derivatives of inverse functions.

Section 8.2.   Inverse Trigonometric Functions

1.   <u>The definitions of the inverse trigonometric functions.</u>   As we saw in the last section, a
function has an inverse if and only if it passes the Horizontal Line Test (i. e. , any horizontal
line intersects its graph at most once).   It is apparent from the graphs of the six basic trigonometric functions

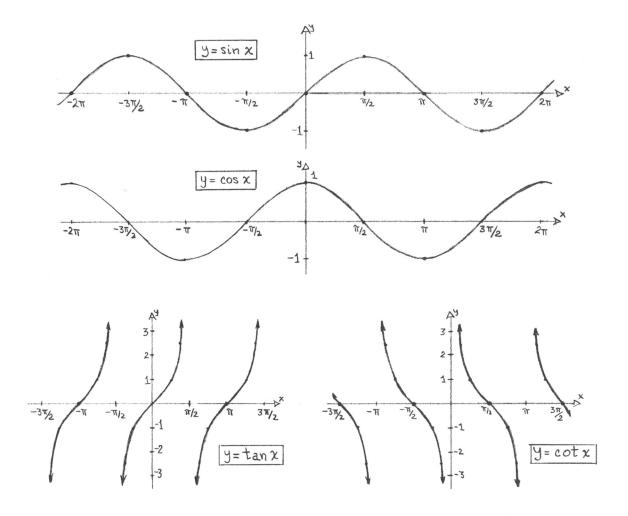

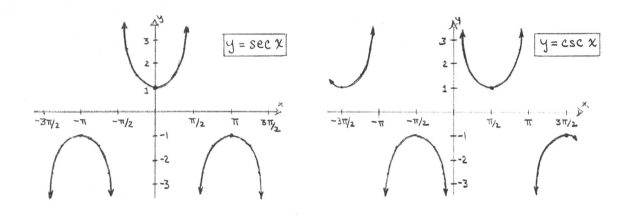

that <u>none of these graphs passes the Horizontal Line Test,</u> so in order to obtain a function

which has an inverse function, we have to <u>restrict the domains.</u> Anton restricts the domains

of the trigonometric functions as follows:

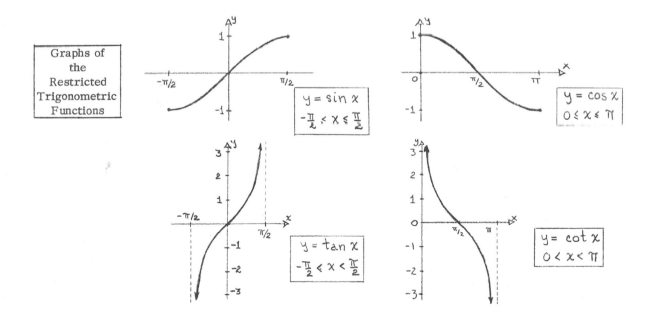

8. 2. 3

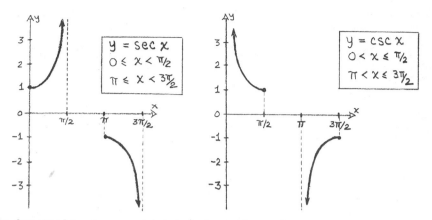

Once the domains have been restricted as shown above, <u>the six restricted trigonometric functions pass the Horizontal Line Test and hence have inverse functions.</u> Those inverse functions are denoted $\sin^{-1} x$, $\cos^{-1} x$, $\tan^{-1} x$, etc. Because it is important, we will repeat a warning that Anton makes:

<u>WARNING</u>: $\sin^{-1} x$ is the inverse function of $y = \sin x$, $-\dfrac{\pi}{2} \leq x \leq \dfrac{\pi}{2}$; it is <u>NOT</u> $1/\sin x$. The same goes for the other five inverse trigonometric functions!!

2. <u>General Properties.</u>   Now we will apply the general inverse function results of §8.1 to the inverse trigonometric functions. For the sake of simplicity we will deal exclusively with $\tan^{-1} x$. However, keep in mind that <u>these results can be modified to apply equally well to the other five inverse trigonometric functions!</u>

As Anton shows, we have

$$\tan(\tan^{-1} x) = x \quad \text{for all} \quad -\infty < x < +\infty \tag{A}$$
$$\text{(range of tangent)}$$

$$\tan^{-1}(\tan x) = x \quad \text{for all} \quad -\pi/2 < x < \pi/2 \tag{B}$$
$$\text{(restricted domain of tangent)}$$

This leads to Anton's Theorem 8. 2. 2 :

$$\boxed{\begin{array}{l} \text{If} \quad -\infty < x < +\infty \quad \text{and} \quad -\pi/2 < y < \pi/2, \quad \text{then} \\[2mm] y = \tan^{-1} x \quad \text{is equivalent to} \quad \tan y = x. \end{array}}$$

(C)

Thus $y = \tan^{-1} x$ is an $\underline{\text{ANGLE}}$; specifically it is $\underline{\text{the}}$ angle y which lies in $-\pi/2 < y < \pi/2$ and whose tangent is x.

Example A.   Evaluate (a) $\tan^{-1} 1$, and (b) $\sin^{-1}(.643)$.

Solution.   (a) Let $y = \tan^{-1} 1$. From Theorem 8.2.2 (our equation C) this equation is equivalent to

$$\tan y = 1, \quad -\pi/2 < y < \pi/2$$

Recalling the basic values of the tangent function, we see that $y - \pi/4$ is the only such angle. Thus

$$\tan^{-1} 1 = y = \pi/4$$

(b)   Let $y = \sin^{-1}(.643)$. From Theorem 8.2.1 (the analogue of equation C for $\sin^{-1} x$) this equation is equivalent to

$$\sin y = .643, \quad -\pi/2 \le y \le \pi/2$$

From Table 1 in Anton's Appendix 3 we see that $y = 40^{\circ} = 2\pi/9$ is the angle in the specified domain whose sine is approximately .643. Thus

$$\sin^{-1}(.643) \approx y = 2\pi/9$$

Example B.   Evaluate

8.2.5

    a. $\tan(\tan^{-1} 2)$

    b. $\sin(\sin^{-1}(1/2))$

    c. $\sin(\sin^{-1} 3)$

    d. $\sin^{-1}(\sin \pi/7)$

    e. $\tan^{-1}(\tan 3\pi/5)$

Solution.

a. $\tan(\tan^{-1} 2) = 2$ from Equation A, since $x = 2$ is in the range of tangent.

b. $\sin(\sin^{-1}(1/2)) = 1/2$ from the analogue of Equation A for the sine, since

        $x = 1/2$ is in the range of sine (i.e., $-1 \le x \le 1$).

c. $\sin(\sin^{-1} 3) = 3$??? ... NOPE! Since $x = 3$ is NOT in the range of sine

        (i.e., $-1 \le x \le 1$), $\sin^{-1} 3$ is not defined, and hence

              $\sin(\sin^{-1} 3)$ is UNDEFINED!

        As you can see, the restrictions on the variables in Equations A and B are
important!

d. $\sin^{-1}(\sin \pi/7) = \pi/7$ from the analogue of Equation B for the sine, since $x = \pi/7$

        is in the restricted domain of the sine (i.e., $-\pi/2 \le x \le \pi/2$).

e. $\tan^{-1}(\tan 3\pi/5) = 3\pi/5$??? ... NOPE! Since $x = 3\pi/5$ is NOT in the restricted

        domain of $\tan x$ (i.e., $-\pi/2 < x < \pi/2$), then Equation B does not directly apply.

        However, recall that the tangent function is <u>periodic with period</u> π (Equations 24 ab

        in Anton's Appendix 1); thus

             $\tan 3\pi/5 = \tan(3\pi/5 - \pi) = \tan(-2\pi/5)$.

        Since $x = -2\pi/5$ <u>is</u> in the restricted domain of $\tan x$ (i.e., $-\pi/2 < x < \pi/2$),

        then $\tan^{-1}(\tan 3\pi/5) = \tan^{-1}(\tan(-2\pi/5))$

                        $= -2\pi/5$ by Equation B. $\square$

Equations A , B and C must become second nature to you! The equations themselves are easy ("tan and $\tan^{-1}$ undo each other"). However as parts c and e of Example B show, you must be careful to observe the restrictions on the variables!

*   *   *   *   *

The graph of $y = \tan^{-1}x$ is easily obtained from the graph of (the restricted version of) $y = \tan x$ by reflection over the line $y = x$:

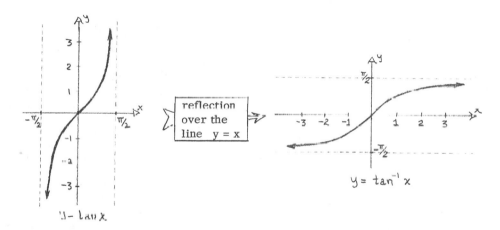

$y = \tan x$

$y = \tan^{-1} x$

Knowing equations A , B and C , and the graph of $\tan^{-1}x$, is crucial in the remainder of the chapter! The analogues of these results for the other trigonometric functions are obtained by using the appropriate restricted domains. For example,

$$\sec(\sec^{-1}x) = x \quad \text{for all} \quad |x| \geq 1 \tag{A'}$$
(range of secant)

$$\sec^{-1}(\sec x) = x \quad \text{for all} \quad 0 \leq x < \pi/2 \quad \text{or} \quad \pi \leq x < 3\pi/2 \tag{B'}$$
(restricted domain of secant)

$$\text{If} \quad |x| \geq 1 \quad \text{and} \quad 0 \leq y < \pi/2 \quad \text{or} \quad \pi \leq y < 3\pi/2, \quad \text{then} \tag{C'}$$
$$y = \sec^{-1}x \quad \text{is equivalent to} \quad \sec y = x$$

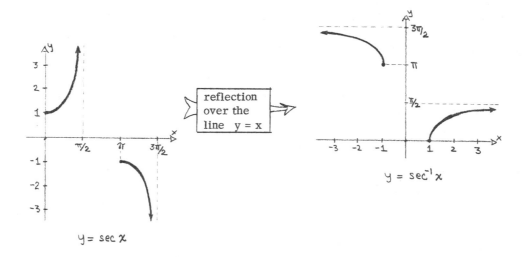

$y = \sec x$

$y = \sec^{-1} x$

3. <u>Computational techniques; The Triangle Method.</u> We will now evaluate expressions involving inverse trigonometric functions which are more complicated than those in Examples A and B.

<u>Example C.</u>   Find the exact value of $\tan[\sec^{-1}(3/2)]$.

<u>Solution.</u>   As in Anton's Examples 2 and 3, we could use a trigonometric identity to evaluate this expression (in this case $\tan^2\theta = \sec^2\theta - 1$). However, it is generally easier to construct a triangle containing the angle $\theta = \sec^{-1}(3/2)$, and then "read off" the desired number $\tan\theta$ with the help of the Pythagorean Theorem. Here's how:

By Equation C' , $\theta = \sec^{-1}(3/2)$ is

equivalent to $\sec\theta = 3/2$, where θ

is in the restricted domain of θ. Since

$$\sec\theta = \frac{1}{\cos\theta} = \frac{\text{hypotenuse}}{\text{adjacent}}$$

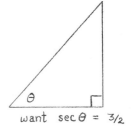

want $\sec\theta = 3/2$

if we set

$$\text{hypotenuse} = 3$$

$$\text{adjacent} = 2$$

then   sec θ will equal   3/2 ,   as desired.

The third (opposite) side   s   of the triangle is easily calculated by the Pythagorean

Theorem:

$$2^2 + s^2 = 3^2$$

$$s^2 = 9 - 4 = 5$$

$$s = \sqrt{5}$$

Once we have both the opposite and the adjacent sides of the triangle, it is easy to find the

tangent of θ :

$$\tan \theta = \frac{\text{opposite}}{\text{adjacent}} = \boxed{\frac{\sqrt{5}}{2}} \qquad \square$$

Remark: As you can see, the angle θ in Example C lies in the interval $0 \leq \theta < \pi/2$.

Fortunately, this is part of the restricted domain of the secant, as required by $\theta = \sec^{-1}(3/2)$.

Example C illustrates the use of <u>The Triangle Method</u> for evaluating expressions of

the form

$$\boxed{\text{trig}_1 \ (\text{inverse trig}_2 \ (x))}$$

i. e. , a trigonometric function applied to an inverse trigonometric function. Written out, the

method is:

The Triangle Method	Step 1.	Construct a right triangle with an angle θ = inverse $\text{trig}_2(x)$ by specifying lengths for two of the three sides of the triangle.
	Step 2.	Use the Pythagorean Theorem to compute the length of the third side of the triangle.
	Step 3.	"Read off" $\text{trig}_1\,\theta$ from the triangle. This is the desired answer.

In using The Triangle Method there is one point about which you must be very careful: the triangle must be constructed so that θ = inverse $\text{trig}_2(x)$ is in the restricted domain of $\text{trig}_2(x)$. This can require the use of one of the following four triangles:

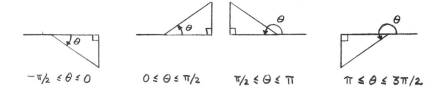

$-\pi/2 \leq \theta \leq 0 \qquad 0 \leq \theta \leq \pi/2 \qquad \pi/2 \leq \theta \leq \pi \qquad \pi \leq \theta \leq 3\pi/2$

Anton illustrates The Triangle Method following Examples 2 and 3. Here is one additional illustration:

Example D.   Simplify the expression $\sin(2\tan^{-1}x)$.

Solution.   At first glance the presence of the 2 might appear troublesome. But it causes no problems! Merely use the double angle formula for the sine (Appendix G, Equation G9):

$$\sin 2\theta = 2\sin\theta\cos\theta$$

so that
$$\sin(2 \tan^{-1}x) = 2 \sin(\tan^{-1}x) \cos(\tan^{-1}x)$$

We'll evaluate both $\sin(\tan^{-1}x)$ and $\cos(\tan^{-1}x)$ by The Triangle Method:

Step 1.   Letting $\theta = \tan^{-1}x$,

we need a right triangle with angle θ so that

$$x = \tan\theta = \frac{\text{opposite}}{\text{adjacent}}$$

Thus we will set

opposite = x

adjacent = 1

as shown in the triangle to the right.

Step 2.   By the Pythagorean Theorem we compute

the hypotenuse:

$$h = \sqrt{1 + x^2}$$

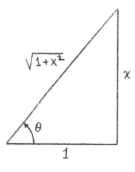

Step 3. $\sin\theta = \dfrac{\text{opposite}}{\text{hypotenuse}} = \dfrac{x}{\sqrt{1 + x^2}}$

$\cos\theta = \dfrac{\text{adjacent}}{\text{hypotenuse}} = \dfrac{1}{\sqrt{1 + x^2}}$

Thus

$$\sin(2\tan^{-1}x) = 2\sin\theta\cos\theta = \boxed{\dfrac{2x}{1 + x^2}} \qquad \square$$

Suppose we had to construct a right triangle with $\theta = \sin^{-1}(-.3)$. Then

$\sin \theta = -.3$ and $-\pi/2 \le \theta \le \pi/2$. However, since

the sine is negative only on the negative half of

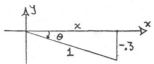

$[-\pi/2, \pi/2]$, then our triangle would look as shown

to the right. Using Pythagoras yields $x = \sqrt{1 - .09} = \sqrt{.91} \cong .954$

$$* \quad * \quad * \quad * \quad *$$

We have been considering expressions of the form

$$\text{trig}_1 \, (\text{inverse trig}_2 \, (x))$$

What about expressions of the form

$$\text{inverse trig}_1 \, (\text{trig}_2 \, (x))$$

e. g., $\tan^{-1}[\sin(-\pi/2)]$ or $\sin^{-1}[\cot(\pi/3)]$? The method of evaluation is generally to

compute the inner trigonometric function first, and then to compute the outer inverse

trigonometric function.

Example E.   Compute $\sin^{-1}[\cot(\pi/3)]$.

Solution.

$$\cot(\pi/3) = \frac{\cos(\pi/3)}{\sin(\pi/3)} = \frac{1/2}{\sqrt{3}/2} \approx .577$$

Thus

$$\sin^{-1}[\cot(\pi/3)] \approx \sin^{-1}(.577) \approx 35^\circ = 7\pi/36 \qquad \square$$

$$\uparrow$$

$$\text{(Table 1 in Appendix 3)}$$

Occasionally an inverse trigonometric identity helps in solving this type of problem. For an illustration, see Example G below.

4. Graphs of inverse trigonometric functions.   From §8. 1 we know

$$\left\| \begin{array}{l} \text{the graph of } y = f^{-1}(x) \text{ is obtained} \\ \text{by reflecting the graph of } y = f(x) \\ \text{about the line } y = x \, . \end{array} \right\|$$

Thus the graphs of the inverse trigonometric functions are all easily obtained from the graphs of the (restricted) trigonometric functions by reflection about the line y = x. (See Anton's Figures 8. 2. 2 , 8. 2. 4 and 8. 2. 6.)

Knowing the graphs of the inverse trigonometric functions makes it relatively easy to obtain graphs for simple variants, as the next example shows.

Example F.   Sketch the graph of $y = \sin^{-1} 2x$.

Solution.   The graph of

$$y = \sin^{-1} 2x \qquad *$$

should certainly be related to the graph of

$$y = \sin^{-1} x \qquad **$$

In fact, to get the same y-value from both * and **, we have only to take an x-value in * which is 1/2 as large as the x-value needed in **. Thus the graph of * is just a contraction by 1/2 along the x-axis of the graph of **. This yields the graphs

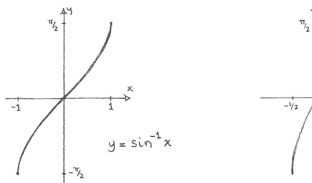

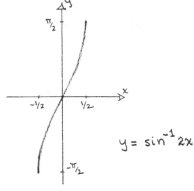

[This general phenomenon is discussed in Appendix G §4 of <u>The Companion</u> (the expansion/ contraction principles).]   □

5.   <u>Identities involving inverse trigonometric functions.</u>   Just as the trigonometric functions can

be combined to form identities, the inverse trigonometric functions can also be combined to

form identities. * Anton gives six of them at the end of the section.   The first three   (Equations

(5) - (7)) can be summarized by saying

> the inverse of a trigonometric function
>
> and the inverse of its co-function
>
> add to $\pi/2$.

For this reason $\cos^{-1}$, $\cot^{-1}$ and $\csc^{-1}$ are all of minor importance since they are all

easily derived by subtracting $\sin^{-1}$, $\tan^{-1}$ and $\sec^{-1}$, respectively, from $\pi/2$.

---

*
  It is important to remember that <u>the identities involving inverse trigonometric functions</u>
  <u>depend on the domains chosen for the restricted trigonometric functions.</u>   A change in those
  domains may make the identities no longer valid.

<u>Example G</u>.   Compute $\cot^{-1}[\tan(\pi/3)]$.

<u>Solution</u>.   From our observation above we know that

$$\cot^{-1}x = \pi/2 - \tan^{-1}x$$

Thus, letting $x = \tan(\pi/3)$,   we obtain

$$\cot^{-1}[\tan(\pi/3)] = \pi/2 - \tan^{-1}[\tan(\pi/3)]$$

$$= \pi/2 - \pi/3 = \boxed{\pi/6}$$

by Equation B since

$\theta = \pi/3$ is in $-\pi/2 < \theta < \pi/2$,

the restricted domain of $\tan\theta$.   □

<u>Section 8.3</u>.   Derivatives and Integrals Involving Inverse Trigonometric Functions

1.   <u>Derivatives and Integrals of inverse trigonometric functions</u>.   In §8.1, we developed a

formula (Theorem 8.1.5) which enables us to find the derivative of an inverse function f^{-1}

of a function f in terms of the derivative of f:

$$(f^{-1})'(x) = \frac{1}{f'(f^{-1}(x))}$$

or

$$\frac{dy}{dx} = \frac{1}{dx/dy} \qquad \text{where} \qquad y = f^{-1}(x)$$

This formula is exactly what is needed to find the derivatives of the inverse trigonometric

functions. In the proof of Theorem 8.3.1, Anton applies this formula to compute the derivatives

of

$$\tan^{-1}x \qquad \text{and} \qquad \sec^{-1}x$$

8. 3. 2

In these proofs, notice that in each of the three cases the denominator term dx/dy is transformed into an expression of the form

$$\text{trig}_1 \,[\text{ inverse trig}_2 (x)\,]$$

That is why we considered such expressions in §8.2.

The derivatives of the remaining three inverse trigonometric functions

$$\cos^{-1} x, \quad \cot^{-1} x \quad \text{and} \quad \csc^{-1} x$$

can be obtained in a similar way. However, as Anton shows in Theorem 8.3.1, the derivatives of these three functions are more easily obtained by using inverse trigonometric identities.

Anton summarizes the derivatives of the six inverse trigonometric functions in equations (1) through (6). <u>We recommend memorizing three of those six derivatives</u>:

$$\frac{d}{dx}\,[\sin^{-1} x\,] = \frac{1}{\sqrt{1 - x^2}}$$

$$\frac{d}{dx}\,[\tan^{-1} x\,] = \frac{1}{1 + x^2}$$

$$\frac{d}{dx}\,[\sec^{-1} x\,] = \frac{1}{x\sqrt{x^2 - 1}}$$

and then remembering that

the derivatives of the respective inverse <u>co</u>-functions are just the negatives of those derivatives.

For example,

$$\frac{d}{dx}\,[\cos^{-1} x] = \text{the negative of } \frac{d}{dx}\,[\sin^{-1} x]$$

$$= -\,\frac{1}{\sqrt{1 - x^2}}$$

Anton's Examples 1 and 2 illustrate the use of these formulas; in each case the key step involves The Chain Rule. Here is one more example:

<u>Example A.</u>   Find $\frac{dy}{dx}$ if $y = x - \frac{1}{2}(x^2 + 4)\tan^{-1}(x/2)$.

<u>Solution.</u>   Applying the product rule to the second term, we obtain

$$\frac{dy}{dx} = 1 - \frac{1}{2}(x^2 + 4)\frac{d}{dx}[\tan^{-1}(x/2)] - x\tan^{-1}(x/2)$$

Now we must apply The Chain Rule to the composite function $\tan^{-1}(x/2)$. Using the format discussed in §3.5.2 of <u>The Companion</u> we have

$$\frac{d}{dx}[\tan^{-1}(x/2)] = \ldots$$ "The derivative of a composition is ...

$$= \frac{1}{1 + (\ldots)^2}\ldots$$ "... the derivative of the outer function ...

$$= \frac{1}{1 + (x/2)^2}\ldots$$ "... (evaluated at the inner function) ...

$$= \frac{1}{1 + (x/2)^2}\cdot(\frac{1}{2})$$ "... times the derivative of the inner function. "

$$= \frac{1}{1 + x^2/4}\cdot(\frac{1}{2})$$

$$= \frac{1}{2 + x^2/2}$$

Thus we have

$$\frac{dy}{dx} = 1 - \frac{1}{2}(x^2 + 4)\frac{1}{2 + x^2/2} - x\tan^{-1}(x/2)$$

$$= 1 - \frac{x^2 + 4}{x^2 + 4} - x\tan^{-1}(x/2)$$

$$= -x\tan^{-1}\frac{x}{2}$$

8.3.4

*   *   *   *   *

As usual, each of the derivative formulas can be "read backwards" to produce a corresponding integral formula. For instance, the three integral formulas corresponding to the three derivative formulas for $\sin^{-1}x$, $\tan^{-1}x$ and $\sec^{-1}x$ given above are Anton's Equations (7)-(9):

$$\int \frac{dx}{\sqrt{1-x^2}} = \sin^{-1}x + C$$

$$\int \frac{dx}{1+x^2} = \tan^{-1}x + C$$

$$\int \frac{dx}{x\sqrt{x^2-1}} = \sec^{-1}x + C$$

Anton's formulas (10)-(12), in which a^2 appears in the integrals instead of 1, can be obtained from the above formulas using the substitution

$$x = au$$

For instance, $\int \frac{dx}{\sqrt{a^2-x^2}}$ may be evaluated by making this substitution as follows:

$$\int \frac{dx}{\sqrt{a^2-x^2}} = \int \frac{a\,du}{\sqrt{a^2-(au)^2}}$$

$$\boxed{\begin{array}{l} x = au \\ dx = a\,du \end{array}} \qquad = \int \frac{du}{\sqrt{1-u^2}}$$

$$= \sin^{-1}u + C \qquad \text{[the first integral formula above]}$$

$$= \sin^{-1}\frac{x}{a} + C$$

This is Anton's Formula (10). In Example 5, Anton derives Formula (11) in a similar fashion.

Anton's Examples 3, 4 and 6 illustrate the use of these integral formulas. Here are two more examples:

<u>Example B.</u>   Evaluate $\displaystyle\int \frac{dx}{\sqrt{1-4x^2}}$.

<u>Solution.</u>   This integral will be in a recognizable form if we can "turn" $4x^2$ into the square of a single variable u. To this end, we let $u = 2x$ (i.e., $x = u/2$) to obtain

$$\int \frac{dx}{\sqrt{1-4x^2}} = \frac{1}{2} \int \frac{du}{\sqrt{1-u^2}}$$

$$\boxed{\begin{array}{l} x = u/2 \\ dx = du/2 \end{array}} \qquad = \frac{1}{2}\,\sin^{-1}u + C$$

$$= \frac{1}{2}\,\sin^{-1}2x + C \qquad\qquad \square$$

<u>Example C.</u>   Evaluate $\displaystyle\int \frac{\cos x\,dx}{4 + \sin^2 x}$.

<u>Solution.</u>   This integral will be in a recognizable form if we can "turn" $4 + \sin^2 x$ into $4 + 4u^2 = 4(1+u^2)$ in such a way that the numerator $\cos x\,dx$ becomes a multiple of du. To this end, we let $u = \frac{1}{2}\sin x$ (i.e., $\sin x = 2u$) to obtain

$$\int \frac{\cos x\,dx}{4+\sin^2 x} = \frac{2}{4} \int \frac{du}{1+u^2}$$

$$\boxed{\begin{array}{l} \sin x = 2u \\ \cos x\,dx = 2\,du \end{array}} \qquad = \frac{1}{2}\,\tan^{-1}u + C$$

$$= \frac{1}{2}\,\tan^{-1}(\tfrac{1}{2}\sin x) + C \qquad\qquad \square$$

2. <u>Word problems involving inverse trigonometric functions.</u>   Inverse trigonometric functions frequently occur in word problems. For example, here is a <u>related rate problem</u> (solved using the four-step procedure discussed in §4. 1):

<u>Example D.</u>   A balloon is released at a point on the ground 400 ft from Howard Anton. If the balloon rises at a constant rate of 20 ft/sec, find the rate at which the angle of Howard's line of sight to the balloon is changing 10 seconds after the balloon is released.

<u>Solution.</u>

<u>Step 1: Diagram and labels.</u>   We draw the picture to the right, where θ is the angle of Howard's line of sight to the balloon and x is the height of the balloon; both are functions of t (time). We are given

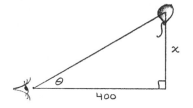

$$\boxed{\frac{dx}{dt} = 20 \text{ ft/sec}}$$

and we want to find

$$\boxed{\frac{d\theta}{dt} = ? \quad \text{when} \quad t = 10}$$

<u>Step 2: A relating equation.</u>   We need an equation relating θ and x. From our diagram we see that $\tan \theta = x/400$, or

$$\boxed{\theta = \tan^{-1} \frac{x}{400}}$$

Step 3: Differentiate.   We differentiate both sides of our equation with respect to t :

$$\frac{d\theta}{dt} = \frac{1}{1 + (\frac{x}{400})^2} \ \frac{d}{dt}\left[\frac{x}{400}\right]$$ [using The Chain Rule
as in Example A]

$$= \frac{(400)^2}{(400)^2 + x^2} \cdot (\frac{1}{400}) \frac{dx}{dt}$$ [using The Chain Rule again ...
since x is a function of t]

Thus

$$\boxed{\frac{d\theta}{dt} = \frac{400}{(400)^2 + x^2} \cdot \frac{dx}{dt}}$$

Step 4: Evaluate.   When $t = 10$, $x = (10)(20) = 200$ ft   [using the familiar formula

distance = (rate)(time)].   Thus when $t = 10$ (so $x = 200$)   and $\frac{dx}{dt} = 20$,   we have

$$\frac{d\theta}{dt} = \frac{400}{(400)^2 + (200)^2} (20) = \boxed{\frac{1}{25} \ \text{radians/sec}}$$ □

Here is a max-min problem involving an inverse trigonometric function (solved using the

five-step procedure discussed in §§4. 7 and 4. 8):

Example E.   A sign 20 ft high stands on a slight rise so that the bottom of the sign is 20 ft

above the horizontal plane of the highway.   If the eye of a driver is 4 ft above the road, at what

horizontal distance from the sign will the sign appear to be largest to the driver (i. e. , at what

distance will it subtend the largest angle)?

8.3.8

Solution.

Step 1: Diagram and labels.   We draw the
picture to the right, letting   x   be the
horizontal distance from the sign to the
driver and θ the angle subtended by the
sign at the driver's eye.   This picture leads
to the triangles shown below it, where we have
also labeled the angle ϕ as shown.   We are asked to

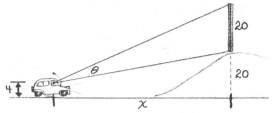

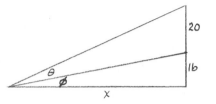

$$\boxed{\text{find the }\ x\text{-value}\ \text{that maximizes}\quad \theta}$$

Step 2: Formula.   From our diagram we see that

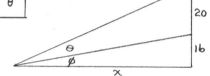

$$\tan (\theta + \phi) = \frac{16 + 20}{x} = \frac{36}{x}$$

so

$$\theta + \phi = \tan^{-1}\left(\frac{36}{x}\right)$$

Thus
$$\boxed{\theta = \tan^{-1}\left(\frac{36}{x}\right) - \phi}$$

Step 3: Formula in one variable.   We need to eliminate ϕ from our formula for θ.   This is
easy using inverse tangents since from our diagram we see that

$$\tan \phi = \frac{16}{x}$$

so

$$\phi = \tan^{-1}\left(\frac{16}{x}\right)$$

Thus
$$\boxed{\theta = \tan^{-1}\left(\frac{36}{x}\right) - \tan^{-1}\left(\frac{16}{x}\right)}$$

Step 4: Interval. The range of allowable x-values is clearly

$$0 < x < \infty$$

Step 5: Max-min techniques.   We differentiate the function of Step 3 with respect to   x :

$$\frac{d\theta}{dx} = \frac{1}{1 + (\frac{36}{x})^2} \frac{d}{dx} (\frac{36}{x}) - \frac{1}{1 + (\frac{16}{x})^2} \frac{d}{dx} (\frac{16}{x})$$

(using The Chain
Rule as in
Example A)

$$= \frac{x^2}{x^2 + (36)^2} \left(\frac{-36}{x^2}\right) - \frac{x^2}{x^2 + (16)^2} \left(\frac{-16}{x^2}\right)$$

$$= \frac{16}{x^2 + (16)^2} - \frac{36}{x^2 + (36)^2}$$

To find the critical points we set this equal to zero and solve for   x :

$$0 = \frac{16}{x^2 + (16)^2} - \frac{36}{x^2 + (36)^2}$$

$$36(x^2 + (16)^2) = 16(x^2 + (36)^2)$$

$$20x^2 = 16(36)^2 - 36(16)^2 = (16)(36)(20)$$

$$x = \pm 24$$

Ruling out   x = -24   because   x   is always positive, we see that the sign appears largest to
the driver when he is 24 ft away from it. *   □

─────────────

*   Since the physical set-up of the problem guarantees that a maximum (but not a minimum)
value exists,   x = 24   <u>must</u> give this value. You can also check this using the First or
Second Derivative Tests.

8. 3. 10

3. <u>What do I need to memorize?</u>   We conclude this section with a summary of the formulas which must be memorized.   There are <u>only three</u>:

$$(1) \quad \frac{d}{dx}\left[\sin^{-1}x\right] = \frac{1}{\sqrt{1-x^2}}$$

$$(2) \quad \frac{d}{dx}\left[\tan^{-1}x\right] = \frac{1}{1+x^2}$$

$$(3) \quad \frac{d}{dx}\left[\sec^{-1}x\right] = \frac{1}{x\sqrt{x^2-1}}$$

Once you have memorized these three, the other formulas of this section can be remembered as follows:

- the derivatives of their respective inverse co-functions are the <u>negatives</u> of those derivatives

- the integral formulas involving 1 are obtained by "reading the derivative formulas backwards"

- the integral formulas in which a^2 appears instead of 1 are easily obtained using the substitution $x = au$.

That's a nice short summary list!

Section 8.4.   Inverse Hyperbolic Functions

1.   Definitions and graphs of the inverse hyperbolic functions.   The procedure used for defining

the inverse hyperbolic functions is the same as was used for the inverse trigonometric functions:

> - if the hyperbolic function is one-to-one (i.e. , if it
>
>   passes the Horizontal Line Test), then it has an
>
>   inverse function (Theorem 8.1.3)

> - if the hyperbolic function is not one-to-one (i.e. , if it
>
>   fails the Horizontal Line Test), then we restrict its
>
>   domain to obtain a closely-related function which is
>
>   one-to-one.   That function has an inverse function
>
>   (Theorem 8.1.3).

As Anton shows, $\sinh x$ needs no restriction on its domain to be one-to-one, while
$\cosh x$ does need to be restricted.   The restriction chosen is $x \geq 0$ as shown in Figure
8.4.2(a).   The function $\operatorname{sech} x$ is the only other hyperbolic function which needs a domain
restriction (it is also chosen to be $x \geq 0$).

The graphs of the inverse hyperbolic functions are obtained by the same procedure used
for obtaining the graphs of the inverse trigonometric functions (§8.1) :

> - the graph of $y = f^{-1}(x)$ is obtained by reflecting
>
>   the graph of $y = f(x)$ over the line $y = x$.

8.4.2

For example, the graphs of $y = \sinh^{-1} x$ and $y = \cosh^{-1} x$ are

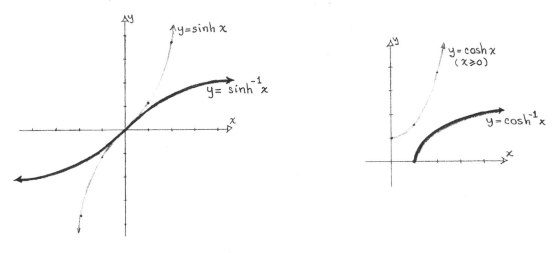

These same procedures are used to define and draw the graphs of the other four inverse hyperbolic functions.

2. <u>The inverse hyperbolic functions as natural logarithms.</u> Unlike the inverse trigonometric functions, the inverse hyperbolic functions can all be expressed in terms of other functions, principally <u>the natural logarithm</u> function. Since (see §7.6)

> the hyperbolic functions are combinations

> of the exponential function e^x

it should not be surprising that

> the <u>inverse</u> hyperbolic functions are

> combinations of the natural logarithm

> function $\ln x$ (which is the <u>inverse</u> function of e^x).

For example, Anton shows (equation (1)) that

$$\sinh^{-1} x = \ln\left(x + \sqrt{x^2 + 1}\right), \quad -\infty < x < +\infty$$

Here is another example:

Example A.   Show that

$$\cosh^{-1} x = \ln\left(x + \sqrt{x^2 - 1}\right), \quad x \geqq 1$$

Solution.   By definition (see §7.6)

$$y = \cosh x = \frac{e^x + e^{-x}}{2}, \quad x \geqq 0$$

(As we mentioned in the previous subsection, the domain of $y = \cosh x$ must be restricted to $x \geqq 0$ for it to be one-to-one and hence have an inverse function.) To find $y = \cosh^{-1} x$, we use the procedures of §8.1 (in particular, see §8.1.2 of The Companion):

First we replace x by $y - \cosh^{-1} x$ in the formula for $\cosh x$:

$$x = \cosh(\cosh^{-1} x) = \cosh y = (e^y + e^{-y})/2$$

| by the definition | | by the formula |
| of inverse function | | for $f(x)$ |

Thus

$$2x = e^y + e^{-y}$$

$$e^y - 2x + e^{-y} = 0$$

$$e^{2y} - 2x e^y + 1 = 0 \qquad \text{(multiply by } e^y\text{)}$$

This is a quadratic equation in e^y (i.e., it is of the form $a(e^y)^2 + b e^y + c = 0$ with $a = c = 1$ and $b = -2x$) so it may be solved by the quadratic formula (see Appendix D):

8.4.4

$$e^y = \frac{2x \pm \sqrt{4x^2 - 4}}{2} = x \pm \sqrt{x^2 - 1}$$

We will show in a moment that $e^y = x - \sqrt{x^2 - 1}$ is impossible when $y \geq 0$ and $x > 1$.

Thus

$$e^y = x + \sqrt{x^2 - 1}$$

Now take $\ln$ of both sides to get

$$\boxed{y = \ln\left(x + \sqrt{x^2 - 1}\right), \qquad x \geq 1}$$

To finish the argument we must show that $e^y = x - \sqrt{x^2 - 1}$ is impossible when

$y \geq 0$ and $x > 1$. Since $y \geq 0$, then $e^y \geq 1$. However, with $x > 1$, then

$x - \sqrt{x^2 - 1} = \dfrac{1}{x + \sqrt{x^2 - 1}} < \dfrac{1}{1 + 0} = 1.$ Thus $x - \sqrt{x^2 - 1} < 1.$ Hence

$e^y = x - \sqrt{x^2 - 1}$ is impossible since $e^y \geq 1$ while $x - \sqrt{x^2 - 1} < 1.$ □

   Anton's Theorem 8.4.2 lists all six inverse hyperbolic functions written as natural

logarithms.

3. <u>Derivatives and integrals of inverse hyperbolic functions.</u>   Inverse hyperbolic functions are

studied primarily because <u>they are useful in integration,</u> i. e. , certain integrals can be evaluated

most easily in terms of inverse hyperbolic functions. As usual, the integral formulas are

derived by first developing the corresponding derivative formulas and then "reading them

backwards. "

   When we want to differentiate an inverse hyperbolic function, we have <u>a choice:</u>

- we can write the inverse hyperbolic function as a natural

  logarithm (as done in the previous subsection) and differentiate

  it directly, or

- we can find the derivative of the inverse hyperbolic function

  by first finding the derivative of the corresponding hyperbolic

  function and using $\dfrac{dy}{dx} = \dfrac{1}{dx/dy}$

- we can use implicit differentiation as we did in §8.1

Anton uses the first of these approaches in deriving his equation (2)

$$\frac{d}{dx} [\sinh^{-1} x] = \frac{1}{\sqrt{x^2 + 1}}$$

The same formula can be derived using the $\dfrac{1}{dx/dy}$ approach as follows (See The Companion

§8.1.4):

Example B.   Compute $\dfrac{d}{dx} [\sinh^{-1} x]$ using the formula $\dfrac{dy}{dx} = \dfrac{1}{dx/dy}$.

Solution.   We let $y = \sinh^{-1} x$ so that $x = \sinh y$.

Then

$$\frac{dx}{dy} = \cosh y \qquad\qquad\qquad (\frac{d}{dx} [\sinh x] \text{ formula})$$

so

$$\frac{dy}{dx} = \frac{1}{dx/dy}$$ (Theorem 8.1.5)

$$= \frac{1}{\cosh y}$$ (the answer in terms of y)

$$= \frac{1}{\cosh (\sinh^{-1} x)}$$ [converting to x's using
$$y = \sinh^{-1} x]$$

$$= \frac{1}{\sqrt{1 + \sinh^2 (\sinh^{-1} x)}}$$ [using $\cosh^2 u - \sinh^2 u = 1$
(equation (1) of §7.6)]

$$= \frac{1}{\sqrt{1 + x^2}}$$ [$\sinh (\sinh^{-1} x) = x$
by Definition 8.1.1 of
inverse function] $\square$

Anton's Theorem 8.4.3 gives the derivatives of all six inverse hyperbolic functions.
Each may be derived via any of our three methods. As Anton remarks, be sure to note the
slight but important distinction in the derivatives of $\tanh^{-1} x$ and $\coth^{-1} x$:

$$\frac{d}{dx} [\tanh^{-1} x] = \frac{1}{1 - x^2}$$ where $|x| < 1$

$$\frac{d}{dx} [\coth^{-1} x] = \frac{1}{1 - x^2}$$ where $|x| > 1$

That is,

$\tanh^{-1} x$ and $\coth^{-1} x$ have the same
derivative $\frac{1}{1 - x^2}$, but they have
different domains, $|x| < 1$ and $|x| > 1$,
respectively.

Here are two examples illustrating the use of these derivative formulas:

<u>Example C.</u>   Find $\dfrac{dy}{dx}$ if $y = \cosh^{-1}\sqrt{x^2 + 1}$.

<u>Solution.</u>   The domain of $y = \cosh^{-1}x$ is $x \geq 1$ so we require $\sqrt{x^2 + 1} \geq 1$ or $x \geq 0$
for the given function to be defined.   Then

$$\frac{dy}{dx} = \frac{1}{\sqrt{\left(\sqrt{x^2 + 1}\right)^2 - 1}} \ \frac{d}{dx}\left[\sqrt{x^2 + 1}\right] \qquad\qquad (\frac{d}{dx}\,[\cosh^{-1}x]\ \text{formula}$$

$$\text{and The Chain Rule)}$$

$$= \frac{1}{\sqrt{x^2 + 1 - 1}}\ \left(\frac{1}{2}\ \frac{2x}{\sqrt{x^2 + 1}}\right)$$

$$= \frac{1}{\sqrt{x^2 + 1}}$$

This derivative is valid for $x > 0$.   ⌐

<u>Example D.</u>   Find $\dfrac{dy}{dx}$ if $y = \tanh^{-1}(x^2 - 1)$.

<u>Solution.</u>   The domain of $y = \tanh^{-1}u$ is $|u| < 1$ so we require $|x^2 - 1| < 1$ or
$0 < |x| < \sqrt{2}$ for the given function to be defined.   Then

$$\frac{dy}{dx} = \frac{1}{1 - (x^2 - 1)^2}\ \frac{d}{dx}\,[x^2 - 1] \qquad\qquad (\frac{d}{dx}\,[\tanh^{-1}x]\ \text{formula}$$

$$\text{and The Chain Rule)}$$

$$= \frac{2x}{2x^2 - x^4}$$

$$= \frac{2}{2x - x^3}$$

This derivative is valid for $0 < |x| < \sqrt{2}$.   ☐

Anton's Theorem 8.4.4 lists the 5 integral formulas obtained by "reading backwards" the derivative formulas for the inverse hyperbolic functions (Theorem 8.4.3). For example,

$$\int \frac{dx}{1-x^2} = \begin{cases} \tanh^{-1} x + C & \text{if} \quad |x| < 1 \\ \coth^{-1} x + C & \text{if} \quad |x| > 1 \quad * \end{cases}$$

It is important to realize that these integrals can also be evaluated by other techniques which we will develop in Chapter 9. Anticipating the Chapter 9 results, here is a list:

(1) $\displaystyle\int \frac{dx}{\sqrt{1+x^2}}$ Use the trigonometric substitution $x = \tan\theta$ (§9.5)

(2) $\displaystyle\int \frac{dx}{\sqrt{x^2-1}}$ Use the trigonometric substitution $x = \sec\theta$ (§9.5)

(3) $\displaystyle\int \frac{dx}{1-x^2}$ Use partial fractions (§9.7)

(4) $\displaystyle\int \frac{dx}{x\sqrt{1-x^2}}$ Use the trigonometric substitution $x = \sin\theta$ (§9.5)

(5) $\displaystyle\int \frac{dx}{x\sqrt{1+x^2}}$ Use the trigonometric substitution $x = \tan\theta$ (§9.5)

Thus, while they are convenient, the formulas for the integrals of the inverse hyperbolic functions are not crucial to integration theory; we will soon develop the other ways of evaluating these integrals.

Here are two examples illustrating the use of these formulas:

Example E. Evaluate $\displaystyle\int \frac{dx}{\sqrt{4+9x^2}}$.

----

* Note the split-level form of $\displaystyle\int \frac{dx}{1-x^2}$ stemming from the distinction between $\frac{d}{dx}[\tanh^{-1} x]$ and $\frac{d}{dx}[\cosh^{-1} x]$ discussed above.

<u>Solution.</u>   This integral will be in a recognizable form if we can "turn $4 + 9x^2$ into

$4(1 + u^2)$." To that end, we write $4 + 9x^2 = 4(1 + \frac{9}{4}x^2)$, which suggests the use of

$u = \frac{3}{2}x$. Then we have on substituting

$$\int \frac{dx}{\sqrt{4 + 9x^2}} = \frac{2}{3} \int \frac{du}{\sqrt{4 + 9(\frac{2}{3}u)^2}}$$

$$\boxed{\begin{array}{l} x = \frac{2}{3}u \\[2mm] dx = \frac{2}{3}du \end{array}} \qquad = \frac{2}{3} \int \frac{du}{\sqrt{4(1 + u^2)}}$$

$$= \frac{1}{3} \int \frac{du}{\sqrt{1 + u^2}}$$

$$= \frac{1}{3} \sinh^{-1} u + C$$

$$= \frac{1}{3} \sinh^{-1}\left(\frac{3}{2}x\right) + C \qquad \square$$

<u>Example F.</u>   Derive the formula

$$\int \frac{dx}{a^2 - x^2} = \frac{1}{a} \tanh^{-1} \frac{x}{a} + C, \qquad |x| < a$$

where   a   is a positive constant.

<u>Solution.</u>   We "turn $a^2 - x^2$ into $a^2(1 - u^2)$" by the substitution $u = \frac{x}{a}$. Then we have

$$\int \frac{dx}{a^2 - x^2} = a \int \frac{du}{a^2 - (au)^2}$$

$$\boxed{\begin{array}{l} x = au \\ dx = adu \end{array}} \qquad = \frac{1}{a} \int \frac{du}{1 - u^2}$$

$$= \frac{1}{a} \tanh^{-1} u + C \qquad\qquad \text{[since } |x| < a \text{ implies}$$

$$= \frac{1}{a} \tanh^{-1} \frac{x}{a} + C \qquad\qquad\qquad |u| = \frac{|x|}{|a|} < 1\text{]}$$

$$\square$$

As with the inverse trigonometric functions, the substitution used in Example F
$(x = a u)$ can also be used to write the other integral formulas of Theorem 8.4.4 in a form
in which a^2 appears instead of 1.

Chapter 9:   Techniques of Integration

Section 9.1.   A Brief Review

In this section Anton pauses and takes stock of what we have learned so far about the two major operations of Calculus, differentiation and integration.

For the most part, the functions we have considered thus far are known as the elementary functions. Specifically, the elementary functions are defined to be those functions which can be formed by algebraic operations (i.e., sums, differences, products, quotients and n-th roots) and compositions from

> polynomials
>
> rational functions
>
> trigonometric functions
>
> $\ln x$ and e^x
>
> inverse trigonometric functions

Thus, for example,

$$\frac{x^2 + 1}{\sin x} \, , \quad \sqrt[3]{\ln (\cos x)} \, , \quad e^{\frac{x-1}{x+1}} \quad \text{and} \quad \tan^{-1} (\ln 6x)$$

are all elementary functions.

Our knowledge about differentiation of the elementary functions is complete. We can systematically differentiate any elementary function and the derivative will be another elementary function. (Such a solution function is said to be "in closed form" since it can be written down explicitly as an elementary function.) We can do this because

(1)   we know the derivatives of all the elementary functions, and

(2)   we have derivative rules (sum, difference, product, quotient and Chain rules) which we can use to differentiate combinations of these functions.

When we come to <u>integration of the elementary functions</u>, however, we are not in nearly as good shape:

(1)   <u>The integral of an elementary function is not necessarily another elementary function</u>! (Thus all integrals of elementary functions cannot be expressed "in closed form" - we might even have to "make up" a new function to express a given integral.)

(2)   Even when there is a "closed form" answer for an integration, <u>we cannot give a systematic procedure which is guaranteed to produce the answer in every case.</u> *

Because of this lack of a systematic procedure, integration is much more of an <u>art form</u> than differentiation ... you need a bag of "tricks" and some imagination. Each integration must be approached as a puzzle!

The "basic tricks" fall into four general categories:

(1)   All specific derivative formulas reverse to give <u>specific integration formulas</u> (these are given in Anton's review list in §9. 1).

(2)   Most of the basic derivative rules reverse in some way to give <u>integration rules</u>:

- the sum and difference rules reverse to give similar sum and difference rules for integration (Theorem 5. 2. 3)

---

* The lack of a systematic procedure for integrating is the reason for tables of integrals such as that found on the inside covers of Anton's book. When a complicated integral is evaluated and the answer is verified, it is recorded so that others do not have to search for it. Because of the systematic procedures for differentiation, there is no need for a lengthy "Table of Derivatives. "

- as we will see, the product rule reverses to give "integration

  by parts" (§9. 2)

- The Chain Rule reverses to give   "u-substitution"   (§5. 3)

(3)   Specific techniques can be developed to convert certain types of

difficult and complicated integrals into simpler ones (e. g. , in

§9. 7 the method of "partial fractions" is developed for simplifying

integrals of rational functions).

(4)   Numerical techniques can be developed to <u>approximate</u> the values

of definite integrals (the most basic are considered in §9. 9).

These techniques are especially valuable in dealing with integrals

whose solutions cannot be expressed in closed form.

This chapter will develop further these four aspects of "systematic integration" and will

equip you with a bag of "tricks" to enable you to handle as many integrals as possible.

<u>Section 9. 2.</u>   <u>Integration By Parts</u>

1.   <u>Integration by parts: the reverse of the product rule.</u>   The extremely useful technique known

as <u>integration by parts</u> will seem less mysterious if you keep firmly in mind that

> integration by parts is the reverse of the
>
> product rule for differentiation.

In fact, if you happen to forget the integration by parts formula (which, frankly, is not likely since it is so simple), you can quickly derive it as follows:

Let u and v be functions of x .

Then by the Product Rule (Theorem 3. 2. 5), written in

differential notation as in Anton's table in §3. 4 ,

$$d(uv) = u\, dv + v\, du$$

Integrating both sides yields

$$\int d(uv) = \int u\, dv + \int v\, du$$

Since integration "undoes" differentiation, the integral

on the left is simply uv , i. e. ,

$$uv = \int u\, dv + \int v\, du$$

Hence

The Integration By Parts Formula

$$\int u\, dv = uv - \int v\, du$$

This simple formula is all the "theory" there is to integration by parts ... learn it well!! (In particular, remember the minus sign!)

2. "How should I choose u and dv ?" Although integration by parts is simple to describe, it can be hard to use, simply because there are so many choices! For instance, in his Example 1 , Anton considers the integral

$$\int x\, e^{x}\, dx$$

He (cleverly) lets

$$u = x \qquad \text{and} \qquad dv = e^x \, dx$$

and it works! But he could have let

$$u = e^x \qquad \text{and} \qquad dv = x \, dx$$

or perhaps

$$u = x \, e^x \qquad \text{and} \qquad dv = dx$$

or even

$$u = 1 \qquad \text{and} \qquad dv = x \, e^x \, dx \quad .$$

Why did Anton make the choice he did? The answer is that he followed some simple
guidelines for the choices of u and dv . These guidelines are based on a simple principle:

> generally, differentiation simplifies a
>
> function while integration complicates it.

The whole idea of applying integration by parts to an integral $\int u \, dv$ is to end up with a
simpler integral $\int v \, du$!! Hence, in view of our Simple Principle, we ideally look for a choice
of u and dv in which

Guidelines for choice of u and dv

1. differentiating u produces a simpler function du , and

2. integrating dv produces a no-more-complicated function v

Let's examine Anton's Example 1 using these guidelines. We have indicated four
possible choices for splitting-up the integrand $x \, e^x \, dx$:

9.2.4

<u>Choice 1</u> (Anton's choice): $\quad u = x \quad$ and $\quad dv = e^x\,dx$

Then $\qquad\qquad\qquad du = dx \quad$ and $\quad v = \displaystyle\int e^x\,dx$

$$v = e^x \qquad *$$

Ahh ... $du = dx$ is somewhat simpler than $u = x$ (we have converted the x into 1), so $u = x$ is a FAIR choice for u. Moreover, the function $v = e^x$ is certainly no more complicated than $dv = e^x\,dx$ (in fact, the e^x remains unchanged!) Hence $dv = e^x\,dx$ is a GOOD choice for dv.

So as you can see, Anton made a FAIR/GOOD pair of choices for u/dv. (It is rare that you can make unqualified GOOD choices for both u and dv.) As we will now show, the other three possibilities for u and dv don't fare so well:

<u>Choice 2</u>: $\qquad\qquad\qquad u = e^x \quad$ and $\quad dv = x\,dx$

Then $\qquad\qquad\qquad du = d(e^x) \quad$ and $\quad v = \displaystyle\int x\,dx$

$$du = e^x\,dx \qquad\qquad v = x^2/2 \quad .$$

Hmm ... since $du = e^x\,dx$ represents no simplification over $u = e^x$, then $u = e^x$ is a POOR choice for u. Moreover, $v = x^2/2$ gives a somewhat-more-complicated function than $dv = x\,dx$, and thus $dv = x\,dx$ is at best a FAIR choice for u. Thus Choice 2 is a POOR/FAIR combination for u/dv, clearly not to be preferred over Anton's Choice 1!

<u>Choice 3</u>: $\qquad\qquad\qquad u = x\,e^x \qquad\qquad$ and $\quad dv = dx$

Then $\qquad\qquad\qquad du = d(x\,e^x) \qquad$ and $\quad v = \displaystyle\int dx$

$$du = (e^x + x\,e^x)\,dx \qquad\qquad v = x$$

---

[*] As Anton points out, it is not necessary to use an integration constant C when obtaining v from dv.

Blaaa! Another POOR/FAIR combination for u/dv , clearly not to be preferred
over Anton's Choice 1!

<u>Choice 4 (stupid choice)</u>: $u = 1$ and $dv = x\,e^x\,dx$

Then $du = d(1)$ and $v = \int x\,e^x\,dx$

                              $du = 0\,dx = 0$ $v = ?$

Oh, for crying out loud! This is not helping at all! If we could evaluate $v = \int x\,e^x\,dx$,
we would have evaluated it in the original problem and be done by now! This combination
for u/dv is clearly POOR/POOR. [<u>Note</u>: The choice of $u = 1$ will <u>always</u> be a
stupid choice because then $du = 0$ so the integration by parts formula
$\int u\,dv = uv - \int v\,du$ becomes $\int dv = v$. Thus you are left with the original
integral to evaluate. No progress has been made!]

Now Anton's choice of u and dv in Example 1 should not appear so mysterious: he made
the best possible choice for u and dv under the guidelines stated above.

                    *   *   *   *   *

We can use our guidelines to make a table of GOOD, FAIR and POOR choices for u and
dv. Since differentiation and integration reverse each other, it should not come as a surprise
that the GOOD choices for u will be the POOR choices for dv , while the POOR choices for
u will be the GOOD choices for dv.

	choices for u		
	GOOD	FAIR	POOR
Table of choices for u and dv	logarithms inverse trigonometric functions	powers	exponentials trigonometric functions
	POOR	FAIR	GOOD
		choices for dv	

Example A.   Evaluate $\displaystyle\int x \cos x \, dx$.

Solution.   The integrand $x \cos x$ is a product of two functions, which should make you suspect

that an integration by parts might be helpful.   Since we have a power   ("x")   and a trigonometric

function   ("cos x") , our table above suggests letting

$$u = x \quad \text{(FAIR)} \quad \text{and} \quad dv = \cos x \, dx \quad \text{(GOOD)}$$

Then $\qquad\qquad du = dx \qquad\qquad$ and $\quad v = \sin x$,

so the differentiation   (u   going to   du)   produces some simplification, while the integration

(dv   going to   v)   produces no additional complexity.   The actual integration then proceeds as

follows:

$$\boxed{\text{MINUS SIGN!!}}$$

$$\int x \cos x \, dx \;=\; u v \;\overset{\downarrow}{-}\; \int v \, du$$

$$\boxed{\begin{array}{ll} u = x & dv = \cos x \\ du = dx & v = \sin x \end{array}} \;\longrightarrow\; = \;\; x \sin x \;-\; \int \sin x \, dx$$

$$\boxed{\text{Ahh} \ldots \text{a "doable" integral!}}$$

$$= \;\; x \sin x \;-\; (- \cos x) \;+\; C$$

$$= \;\; x \sin x \;+\; \cos x \;+\; C \qquad\qquad \boxed{\text{BINGO!}}$$

We now check our answer (of course!):

$$\frac{d}{dx} [\, x \sin x \;+\; \cos x \;+\; C \,]$$

$$= \;\; \underbrace{\sin x \;+\; x \cos x}_{} \;-\; \sin x \;+\; 0$$

$$\text{using the Product Rule on} \quad x \sin x$$

$$= \;\; x \cos x , \quad \text{as desired!} \qquad\qquad \square$$

It is instructive to see what happens when an "ill-advised" choice is made for the pair u and dv. We'll examine two such choices for Example A:

Non-solution using $u = \cos x$ and $dv = x\,dx$. Since you know the guidelines, you would not make this choice since

$$u = \cos x, \quad \text{a trigonometric function, is a}$$
$$\text{POOR choice for } u, \text{ while}$$

$$dv = x\,dx, \quad \text{a power (i.e., } x = x^1), \text{ is only a}$$
$$\text{FAIR choice for } dv.$$

Since it is possible to make a FAIR/GOOD choice for u/dv as done in the original solution, why would we want to use a POOR/FAIR choice? Well, we wouldn't!!... because look at what would happen:

$$\int x \cos x \, dx \qquad = uv - \int v\,du$$

$u = \cos x$	$dv = x\,dx$
$du = -\sin x\,dx$	$v = x^2/2$

$$= \cos x \left(\frac{x^2}{2}\right) - \int \left(\frac{x^2}{2}\right)(-\sin x)\,dx$$

$$= \frac{1}{2} x^2 \cos x + \frac{1}{2} \int x^2 \sin x \, dx$$

$$= ? \ldots ?? \ldots \# \% \,!! \, \$ \, ?! \ldots \text{YUK!}$$

We have produced a more complicated integral (x^2 instead of x times a trigonometric function) so our POOR/FAIR choice for u/dv got us nowhere, as we should have expected!

Non-solution using $u = x \cos x$ and $dv = dx$. If $\cos x$ was a POOR choice for u, as we just demonstrated, then $x \cos x$ should be even worse! [According to our table, $dv = dx$, which is a power (i.e., $1 = x^0$), is still only a FAIR choice for dv.] If we try it, we can see that it is worse:

9.2.8

$$\int x \cos x \, dx \qquad = uv - \int v \, du$$

$u = x \cos x$	$dv = dx$
$du = (-x \sin x + \cos x) \, dx$	$v = x$

$$= (x \cos x) x - \int x (-x \sin x + \cos x) \, dx$$

$$= x^2 \cos x + \int x^2 \sin x \, dx - \int x \cos x \, dx$$

$$= ? ? ?$$

We have produced the same more complicated integral $\int x^2 \sin x \, dx$ that we obtained with the

$u = \cos x$, $dv = x \, dx$ non-solution above, and we still have the original integral $\int x \cos dx$

as well. We have clearly moved in the wrong direction!

3. "If one dose helps, then a second may cure." Sometimes two applications of integration by

parts can be just what the doctor ordered. Anton's Example 2 illustrates this situation. Here

is another illustration:

Example B. Evaluate $\int x^2 \cos x \, dx$.

Solution. The integrand $x^2 \cos x$ is a product of two functions, so an integration by parts

might be useful. Our table suggests letting

$$u = x^2 \quad \text{(FAIR)} \quad \text{and} \quad dv = \cos x \, dx \quad \text{(GOOD)} .$$

Then $\qquad\qquad du = 2x \, dx \qquad\qquad \text{and} \qquad v = \sin x$

so the differentiation (u going to du) produces some simplification, while the integration

(dv going to v) produces no additional complexity. The actual integration by parts then

proceeds as follows:

$$\int x^2 \cos x \, dx \qquad = uv - \int v \, du$$

$$\boxed{\begin{array}{ll} u = x^2 & dv = \cos x \, dx \\ du = 2x \, dx & v = \sin x \end{array}} \longrightarrow = x^2 (\sin x) - \int (\sin x)(2x \, dx)$$

$$= x^2 \sin x - 2 \int x \sin x \, dx$$

$$= \, ? \ldots \# \& \, !! \, \$ \ldots \text{ now wait a minute!}$$

Although we cannot instantly evaluate the new integral, we have made progress: instead of having the square x^2 times a trigonometric function, we have the simpler first power $x = x^1$ times a trigonometric function. Perhaps using a second integration by parts on the new integral $\int x \sin x \, dx$ will solve our problem. Our table suggests letting

$$u = x \quad \text{(FAIR)} \qquad \text{and} \qquad dv = \sin x \, dx \quad \text{(GOOD)} \, .$$

Then

$$\int x \sin x \, dx \qquad = uv - \int v \, du$$

$$\boxed{\begin{array}{ll} u = x & dv = \sin x \, dx \\ du = dx & v = -\cos x \end{array}} \longrightarrow = x(-\cos x) - \int (-\cos x) \, dx$$

$$= -x \cos x + \int \cos x \, dx$$

$$\boxed{\text{Ahh} \ldots \text{ a simple integral!}}$$

$$= -x \cos x + \sin x + C$$

Hence, plugging this result into our earlier equation for $\int x^2 \cos x \, dx$ yields

$$\int x^2 \cos x \, dx = x^2 \sin x - 2 \, [\, -x \cos x + \sin x + C \,]$$

$$= x^2 \sin x + 2x \cos x - 2 \sin x + C_1 \qquad \boxed{\text{BINGO!}} \qquad \qquad \Box$$

Example B illustrates the following principle:

> if after using integration by parts, the remaining integral is simpler than the original one, it is frequently helpful to use integration by parts again to evaluate the new integral.

4. Logarithms and inverse trigonometric functions.   Integrals that contain logarithms and inverse trigonometric functions are often good candidates for integration by parts. The reason is that, as our guidelines indicate, both types of functions are GOOD choices for u since their derivatives are so much simpler than the original functions. Anton's Examples 3 and 4 illustrate the use of integration by parts with two such integrals:

Example 3: $\qquad \displaystyle\int \ln x \, dx \, \ldots$ let $\qquad u = \ln x \qquad\qquad dv = dx$

$\qquad\qquad\qquad\qquad\qquad\qquad\qquad\qquad du = \dfrac{1}{x} \, dx \qquad v = x$

Example 4: $\qquad \displaystyle\int \tan^{-1} x \, dx \, \ldots$ let $\qquad u = \tan^{-1} x \qquad dv = dx$

$\qquad\qquad\qquad\qquad\qquad\qquad\qquad\qquad du = \dfrac{1}{x^2 + 1} \, dx \qquad v = x$

Here is one more example:

Example C.   Evaluate $\displaystyle\int x \ln x \, dx$.

Solution.   We have a product of two functions, one of which is a logarithm ... two good indicators of integration by parts! The choice of $u = \ln x$ and $dv = x \, dx$ then rates as GOOD/FAIR   according to our guidelines. Here's the integration:

$$\int x \ln x \, dx \qquad = uv - \int v \, du$$

$$\boxed{\begin{array}{ll} u = \ln x & dv = x \, dx \\ du = \dfrac{1}{x} \, dx & v = \dfrac{x^2}{2} \end{array}} \longrightarrow = (\ln x)(x^2/2) - \int \left(\frac{x^2}{2}\right)\left(\frac{1}{x} \, dx\right)$$

$$= \frac{1}{2} x^2 \ln x - \frac{1}{2} \int x \, dx \qquad \boxed{\text{a simple integral!}}$$

$$= \frac{1}{2} x^2 \ln x - \frac{1}{2} \left(\frac{x^2}{2} \right) + C$$

$$= \frac{1}{2} x^2 \ln x - \frac{1}{4} x^2 + C \qquad \boxed{\text{BINGO!}} \qquad \square$$

5. **The Reappearing Integral Technique.** In Anton's Example 5, the integral to be evaluated

reappears after two integrations by parts, and then it can be solve for. This situation is

actually quite common. Here is another example:

Example D.   Evaluate $\displaystyle\int \frac{1}{x} \ln x \, dx$.

Solution.   As in Example C we have a <u>product</u> of two functions, one of which is a <u>logarithm</u>...

two good indications of integration by parts! The choice of $u = \ln x$ and $dv = \frac{1}{x} dx$ then

rates as a GOOD/... ?... err, $dv = \frac{1}{x} dx$ does not look like such a promising choice! After

all, integrating $dv = \frac{1}{x} dx$ yields $v = \ln x$, which is a lot more complicated than the

original $1/x$. Hence the choice $u = \ln x$ and $dv = \frac{1}{x} dx$ can only be given a GOOD/POOR

rating.

But don't discard this approach yet! Oftentimes GOOD/POOR choices for u/dv will

work out... although sometimes in bizarre ways! So we persevere ...

$$\int \frac{1}{x} \ln x \, dx \qquad\qquad = uv - \int v \, du$$

$$\boxed{\begin{array}{ll} u = \ln x & dv = \frac{1}{x} dx \\ du = \frac{1}{x} dx & v = \ln x \end{array}} \;\longrightarrow\; = (\ln x)(\ln x) - \int \ln x \left(\frac{1}{x} dx \right)$$

$$= (\ln x)^2 - \int \frac{1}{x} \ln x \, dx$$

$$\boxed{\text{the integral we started with!!}}$$

Hmm... . Have we simply ended up back where we started? NO... because the two appearances

of $\displaystyle\int \frac{1}{x} \ln x \, dx$ fortunately do not cancel each other out!   Hence, by treating the integral as an

"unknown," we can solve our equation for it:

$$2 \int \frac{1}{x} \ln x \, dx = (\ln x)^2$$

$$\int \frac{1}{x} \ln x \, dx = \frac{1}{2} (\ln x)^2 + C \qquad \boxed{\text{BINGO!}}$$

> Don't forget to add the integration constant.

There is also another lesson to be learned from Example D : <u>the most efficient way to evaluate an integral might not be the method which first comes to mind!</u> Example D illustrates this point because the fastest way to evaluate the integral $\int \frac{1}{x} \ln x \, dx$ is <u>not</u> to use integration by parts, but rather u-substitution :

$$\int \frac{1}{x} \ln x \, dx \quad = \int u \, du = \frac{1}{2} u^2 + C$$

$$\boxed{\begin{array}{l} u = \ln x \\ du = \frac{1}{x} dx \end{array}} \qquad = \frac{1}{2} (\ln x)^2 + C \quad .$$

This is one reason why integration is almost more of an art form than a science!

6. <u>When all else fails, try integration by parts.</u> Sometimes integration by parts works when nothing else will. Because it can work in unpredictable, almost mysterious ways, we recommend

> when an integral does not seem susceptible to any other techniques, try integration by parts. It may surprise you and work!

For example, suppose we need to evaluate an integral $\int f(x) \, dx$ where $f(x)$ is a function which we know how to differentiate but not how to integrate. Then we might attack it by choosing $u = f(x)$ and $dv = dx$. This is exactly what is done in Anton's Examples 3 and 4, as well as in the following example:

Example E.   Evaluate $\displaystyle\int \sin^{-1}(1 - x^2)\, dx$.

Solution.   Noting that $f(x) = \sin^{-1}(1 - x^2)$ is a function which we cannot integrate (immediately) but can differentiate (immediately), we try integration by parts with $u = \sin^{-1}(1 - x^2)$ and $dv = dx$ (this choice has a GOOD/FAIR rating by the table in §9.2.2 of The Companion). Using the Chain Rule, $\dfrac{du}{dx} = \dfrac{d}{dx}[\sin^{-1}(1 - x^2)] = \dfrac{-2x}{\sqrt{1 - (1 - x^2)^2}} = \dfrac{-2}{\sqrt{2 - x^2}}$, so

$$\int \sin^{-1}(1 - x^2)\, dx \qquad = uv - \int v\, du$$

$$\boxed{\begin{array}{ll} u = \sin^{-1}(1-x)^2 & dv = dx \\[2mm] du = \dfrac{-2}{\sqrt{2 - x^2}}\, dx & v = x \end{array}} \;\longrightarrow\; = \sin^{-1}(1 - x^2)\, x - \int x\, \frac{-2}{\sqrt{2 - x^2}}\, dx$$

$$= x \sin^{-1}(1 - x^2) - \int \frac{-2x}{\sqrt{2 - x^2}}\, dx$$

$$\boxed{\begin{array}{l} w = 2 - x^2 \\ dw = -2x\, dx \end{array}} \;\longrightarrow\; = x \sin^{-1}(1 - x^2) - \int \frac{dw}{\sqrt{w}}$$

$$\boxed{\text{a simple integral}}$$

$$= x \sin^{-1}(1 - x^2) - 2\sqrt{w} + C$$

$$\boxed{\begin{array}{l}\text{don't forget the} \\ \text{integration constant}\end{array}}$$

$$= x \sin^{-1}(1 - x^2) - 2\sqrt{2 - x^2} + C \qquad\qquad \square$$

Here is one more example in which integration by parts with the right choice of u and dv yields a solution when nothing else seems to work:

Example F.   Evaluate $\displaystyle\int \frac{x e^x}{(x + 1)^2}\, dx$.

Solution.   A very odd choice of u and dv solves this integration:

9.2.14

$$\int \frac{x\,e^x}{(x+1)^2}\,dx \qquad = uv - \int v\,du$$

$u = x\,e^x$	$dv = \dfrac{1}{(x+1)^2}\,dx$
$du = (x+1)e^x\,dx$	$v = -\dfrac{1}{x+1}$

$$= \frac{-x\,e^x}{x+1} + \int e^x\,dx$$

$$= \frac{-x\,e^x}{x+1} + e^x + C$$

don't forget ...

$$= \frac{e^x}{x+1} + C \qquad \square$$

7. The reduction formulas: learn the technique, but not the formulas. Integration by parts can be used to develop reduction formulas like Anton's Equations (5) and (6):

$$\int \cos^n x\,dx = \frac{1}{n}\cos^{n-1}x\,\sin x + \frac{n-1}{n}\int \cos^{n-2}x\,dx$$

$$\int \sin^n x\,dx = -\frac{1}{n}\sin^{n-1}x\,\cos x + \frac{n-1}{n}\int \sin^{n-2}x\,dx$$

[The first of these "reduces" the problem of integrating $\int \cos^n x\,dx$ to the problem of integrating $\int \cos^{n-2}x\,dx$. If we use the formula repeatedly, then eventually $\int \cos^n x\,dx$ cand be "reduced" to $\int \cos x\,dx = \sin x + C$ (if n is odd) or $\int dx = x + C$ (if n is even)].

We do not recommend that you memorize these reduction formulas; instead, you should learn the technique by which they are developed:

The Reduction Procedure	To obtain a reduction formula for $\int [f(x)]^n\,dx$ where $f(x) = \sin x$ or $\cos x$,
	Step 1. Let $u = [f(x)]^{n-1}$ and $dv = f(x)\,dx$. Then integrate by parts.
	Step 2. Use the identity $\sin^2 x + \cos^2 x = 1$ to obtain the original integral on the right side of the equation.

Step 3.   Solve for the integral to obtain the reduction formula.

To illustrate, here is the derivation of Anton's formula (6). Although the computations are messy and a bit tedious, notice that they follow The Reduction Procedure perfectly:

Example G.   Derive the reduction formula

$$\int \sin^n x \, dx = -\frac{1}{n} \sin^{n-1} \cos x + \frac{n-1}{n} \int \sin^{n-2} x \, dx \quad .$$

Solution.   Using The Reduction Procedure,

Step 1.   We integrate by parts with $u = \sin^{n-1} x$:

$$\int \sin^n x \, dx = -\sin^{n-1} x \cos x + (n-1) \int \sin^{n-2} x \cos^2 x \, dx$$

$u = \sin^{n-1} x$	$dv = \sin x \, dx$
$du = (n-1) \sin^{n-2} x \cos x \, dx$	$v = -\cos x$

Step 2.   Then we use the trigonometric identity $\sin^2 x + \cos^2 x = 1$ to obtain the integral $\int \sin^n x \, dx$ on both sides of the equation:

$$\int \sin^n x \, dx = -\sin^{n-1} x \cos x + (n-1) \int \sin^{n-2} x \cos^2 x \, dx$$

$$= -\sin^{n-1} x \cos x + (n-1) \int \sin^{n-2} x \, (1 - \sin^2 x) \, dx$$

$$= -\sin^{n-1} x \cos x + (n-1) \int \sin^{n-2} x \, dx - (n-1) \int \sin^n x \, dx$$

Step 3.   Then we solve for the integral $\int \sin^n x \, dx$:

9.3.1

$$\int \sin^n x \, dx + (n-1) \overbrace{\int \sin^n x \, dx}^{\text{moved from the right side of the equation}} = -\sin^{n-1} x \cos x + (n-1) \int \sin^{n-2} x \, dx$$

$$n \int \sin^n x \, dx = -\sin^{n-1} x \cos x + (n-1) \int \sin^{n-2} x \, dx$$

so

$$\int \sin^n x \, dx = -\frac{1}{n} \sin^{n-1} x \cos x + \frac{n-1}{n} \int \sin^{n-2} x \, dx$$

as desired. □

Example A of the next section (§9.3) illustrates the particular case of Example G when $n = 5$. Anton's derivation of his formula (5) also illustrates The Reduction Procedure, although it is not broken down and labeled in our three steps.

There is one obvious but important point to be made about The Reduction Procedure:

> Reduction formulas are useful only if you know how to evaluate the "reduced" integral. You may need to use The Reduction Procedure again, or some other technique, to evaluate it.

Example A of §9.3 illustrates using The Reduction Procedure more than once.

Section 9.3.   Integrating Powers of Sine and Cosine

1.   The Table for $\int \sin^m x \cos^n x \, dx$.   Integrals of this form occur quite often in science and engineering, and hence learning how to evaluate them is important.   Unfortunately, there is not one integration technique which "works" on every such integral; instead, the best technique to use varies from case to case.   We can summarize them in the following table (which is an expanded version of Anton's Table 9.3.1):

How to evaluate $\int \sin^m x \cos^n x \, dx$

Case A: $\boxed{n \text{ odd}}$

$$\int \sin^m x \, \cos^n x \, dx = \int \sin^m x \, \underline{\cos^{n-1} x} \, (\cos x \, dx)$$

This is now an <u>even</u> power of $\cos x$; convert to $\sin^2 x$ terms by using

$$\cos^2 x = 1 - \sin^2 x$$

The result will be ⸺

$$= \int P(\sin x)(\cos x \, dx) \quad \longleftarrow$$

where $P(\sin x)$ is a polynomial in $\sin x$. *
This is easily evaluated by the u-substitution

$$u = \sin x$$
$$du = \cos x \, dx$$

Case B: $\boxed{m \text{ odd}}$

$$\int \sin^m x \, \cos^n x \, dx = \int \underbrace{\sin^{m-1} x} \cos^n x \, (\sin x \, dx)$$

This is now an <u>even</u> power of $\sin x$; convert to $\cos^2 x$ terms by using

$$\sin^2 x = 1 - \cos^2 x$$

The result will be ⸺

$$= \int P(\cos x)(\sin x \, dx) \quad \longleftarrow$$

where $P(\cos x)$ is a polynomial in $\cos x$. This is easily evaluated by the u-substitution

$$u = \cos x$$
$$du = -\sin x \, dx$$

---

* Examples of polynomials in $\sin x$ are:

$$\sin^2 x + 1, \quad 3\sin^3 x - \sin x, \quad \sin^2 x (1 - \sin^2 x)^3$$

Case C: ┌m even┐ ┌n even┐

$$\int \sin^m x \; \cos^n x \; dx \; = \; \int (\sin^2 x)^{m/2} \; (\cos^2 x)^{n/2} \; dx$$

Use the identities

i. $\sin^2 x = \dfrac{1}{2} (1 - \cos 2x)$

ii. $\cos^2 x = \dfrac{1}{2} (1 + \cos 2x)$

iii. $(1 - \cos 2x)(1 + \cos 2x) = 1 - \cos^2 2x$
$$= \sin^2 2x$$

to convert the integral into a sum of integrals of the form

$\int \sin^r 2x \cos^s 2x \, dx$. Continue in this way until all

resulting integrals fall into Cases A or B, or are powers

of $\sin x$ or $\cos x$.

2. The odd powers are easy! When you see at least one odd power in $\int \sin^m x \cos^n x \, dx$ you

should rejoice, for then you are in either Case A or Case B, and these are really very easy

(once you've had a little practice!). * Anton illustrates Case A in Example 4 and Case B

in Example 5. When both powers are odd, then you have a choice of treating the integral as

either Case A or Case B. Here is an example of such a situation:

Example A. Evaluate $\int \sin^3 x \cos^3 x \, dx$.

Case A Solution. This is $\int \sin^m x \cos^n x \, dx$ with $m = n = 3$. In particular, $n = 3$ is

odd so we are in Case A. Hence

---

* In fact, when one power is odd, the other power can be any number, i.e., it need NOT be
an integer! This will be illustrated in Example G.

$$\int \sin^3 x \cos^3 x \, dx = \int \sin^3 x \underbrace{\cos^2 x}_{\text{use } \cos^2 x \, = \, 1 - \sin^2 x} (\cos x \, dx)$$

$$= \int \sin^3 x \underbrace{(1 - \sin^2 x)}_{} \cos x \, dx$$

a polynomial in $\sin x$.
Now use the u-substitution $\quad u = \sin x$
$\qquad\qquad\qquad\qquad\qquad\qquad\qquad du = \cos x \, dx$

$$= \int u^3 (1 - u^2) \, du$$

$$= \int (u^3 - u^5) \, du$$

$$= \frac{1}{4} u^4 - \frac{1}{6} u^6 + C$$

$$= \boxed{\frac{1}{4} \sin^4 x - \frac{1}{6} \sin^6 x + C}$$

Case B Solution. $m = 3$ is also odd, so we are also in Case B. Hence

$$\int \sin^3 x \cos^3 x \, dx = \int \cos^3 x \underbrace{\sin^2 x}_{\text{use } \sin^2 x \, = \, 1 - \cos^2 x} (\sin x \, dx)$$

$$= \int \cos^3 x \underbrace{(1 - \cos^2 x)}_{} \sin x \, dx$$

a polynomial in $\cos x$.
Now use the u-substitution $\quad u = \cos x$
$\qquad\qquad\qquad\qquad\qquad\qquad\qquad du = - \sin x \, dx$

$$= - \int u^3 (1 - u^2) \, du$$

$$= \int (- u^3 + u^5) \, du$$

$$= - \frac{1}{4} u^4 + \frac{1}{6} u^6 + C$$

$$= \boxed{- \frac{1}{4} \cos^4 x + \frac{1}{6} \cos^6 x + C}$$

You might now be ready to protest "The Companion has obtained two different answers for the same integral; doesn't one have to be wrong?!?". NO!... because the answers are really the same, but just have a different form! To prove this, we'll use the identity $\sin^2 x = 1 - \cos^2 x$:

$$\text{first answer} = \frac{1}{4} \sin^4 x - \frac{1}{6} \sin^6 x + C$$

$$= \frac{1}{4} (\sin^2 x)^2 - \frac{1}{6} (\sin^2 x)^3 + C$$

$$= \frac{1}{4} (1 - \cos^2 x)^2 - \frac{1}{6} (1 - \cos^2 x)^3 + C$$

$$= \frac{1}{4} (1 - 2 \cos^2 x + \cos^4 x)$$

$$\qquad - \frac{1}{6} (1 - 3 \cos^2 x + 3 \cos^4 x - \cos^6 x) + C$$

$$= (\frac{1}{4} - \frac{1}{6}) + (-\frac{2}{4} + \frac{3}{6}) \cos^2 x + (\frac{1}{4} - \frac{3}{6}) \cos^4 x + \frac{1}{6} \cos^6 x + C$$

$$= -\frac{1}{4} \cos^4 x + \frac{1}{6} \cos^6 x + (C + \frac{1}{12})$$

$$= \text{second answer}   (\text{since}   C + \frac{1}{12}   \text{is still just an arbitrary constant!})$$

Hence our two answers do agree! (Whew!)   ☐

Example A illustrates an important point:

> When evaluating an integral, the form of the answer
> may depend on the method used. This is especially
> true of integrals involving trigonometric functions.

Thus, for example, in doing an exercise you might obtain a solution which looks quite different from the answer given in the back of Anton's text. Although your answer could be incorrect (or Anton's could be incorrect!!), the difference might have occurred because you and Anton used different solution methods! If this is the case, appropriate trigonometric identities can be used to show that the answers are equivalent. This can be a complicated and time-consuming task, however. A better approach is to check your answer (by differentiating) and <u>prove</u> that it is correct.

3. <u>The even powers are a pain-in-the neck!</u> Things get harder when <u>both</u> powers in $\int \sin^m x \cos^n x \, dx$ are even. This is Case C, which Anton illustrates in Example 6. The essential idea is to use appropriate trigonometric identities (see the table in §9.3.1 above) to obtain integrals which fall into Cases A or B, or which are powers of $\sin x$ or $\cos x$. The computations can be long and tedious. Here is another example:

<u>Example B.</u>   Evaluate $\displaystyle\int \sin^6 x \cos^4 x \, dx$.

<u>Solution.</u>   This is $\int \sin^m x \cos^n x \, dx$ with $m = 6$ and $n = 4$ so we are in Case C. Hence

$$\int \sin^6 x \cos^4 dx = \int \underbrace{(\sin^2 x)}^3 \underbrace{(\cos^2 x)}^2 dx \quad [\text{write as powers of } \sin^2 x \text{ and } \cos^2 x]$$

use $\cos^2 x = \dfrac{1}{2}(1 + \cos 2x)$

use $\sin^2 x = \dfrac{1}{2}(1 - \cos 2x)$

$$= \int \left(\frac{1}{2}(1 - \cos 2x)\right)^3 \left(\frac{1}{2}(1 + \cos 2x)\right)^2 dx$$

Combine as many of these terms as possible. In this case, we can combine two factors of each term:

$$= \frac{1}{32} \int (1 - \cos 2x)\left[(1 - \cos 2x)(1 + \cos 2x)\right]^2 dx$$

Now use the identity

$$(1 - \cos 2x)(1 + \cos 2x) = 1 - \cos^2 2x = \sin^2 2x$$

$$= \frac{1}{32} \int (1 - \cos 2x)(\sin^2 2x)^2 \, dx$$

$$= \frac{1}{32} \int \sin^4 2x \, dx - \frac{1}{32} \int \sin^4 2x \cos 2x \, dx$$

get rid of the $2x$ by the   u-substitution $u = 2x$

$du = 2 \, dx$

$$= \frac{1}{64} \int \sin^4 u \, du - \frac{1}{64} \int \sin^4 u \cos u \, du$$

The second of these two integrals is a straightforward Case A integral: $m = 4$ (even) and $n = 1$ (<u>odd</u> !). Its evaluation is thus very easy:

$\boxed{\begin{array}{l}\text{second}\\ \text{integral}\end{array}}$
$$\int \sin^4 u \cos u \, du = \int v^4 \, dv = \frac{1}{5} v^5 + C_2$$

$\boxed{\begin{array}{l} v = \sin u \\ dv = \cos u \, du \end{array}}$
$$= \frac{1}{5} \sin^5 u + C_2$$

$$= \frac{1}{5} \sin^5 2x + C_2 \qquad \text{since} \quad u = 2x \quad .$$

The first integral is, however, the fourth power of $\sin x$. It can be evaluated in two ways:

1.   Use the reduction formula (6) of §9.2 for $\int \sin^m x \, dx$, or

2.   Consider $\int \sin^4 x \, dx$ as another Case C integral (!) where

$m = 4$ and $n = 0$.

As we will show in Example D below, method (2) will yield

$\boxed{\begin{array}{l}\text{first}\\ \text{integral}\end{array}}$
$$\int \sin^4 x \, dx = \frac{3}{8} x - \frac{1}{4} \sin 2x + \frac{1}{32} \sin 4x + C_1$$

Thus, from our equation above for $\int \sin^6 x \cos^4 x \, dx$, we obtain

$$\int \sin^6 x \cos^4 x \, dx - \frac{1}{64} \left[\frac{3}{8} x - \frac{1}{4} \sin 2x + \frac{1}{32} \sin 4x + C_1 \right]$$

$$- \frac{1}{64} \left[\frac{1}{5} \sin^5 2x + C_2 \right]$$

$$= \boxed{\frac{3}{512} x - \frac{1}{256} \sin 2x + \frac{1}{2048} \sin 4x - \frac{1}{320} \sin^5 2x + C}$$

As we said, evaluating Case C integrals can be long and tedious! □

4. <u>Integrating $\int \sin^m x \, dx$ and $\int \cos^n x \, dx$ without reduction formulas.</u> From §9.2 we have the reduction formulas:

$$\int \sin^m x \, dx = -\frac{1}{m} \sin^{m-1} x \cos x + \frac{m-1}{m} \int \sin^{m-2} x \, dx$$

$$\int \cos^n x \, dx = \frac{1}{n} \cos^{n-1} x \sin x + \frac{n-1}{n} \int \cos^{n-2} x \, dx$$

These formulas enable you (in theory) to evaluate any power of $\sin x$ and $\cos x$. There are, however, some drawbacks. First of all, you have to look these formulas up (it does not make sense to memorize them!). Second, the formulas merely drop the power by 2; you must continue using the formula until the power is 1 or 0. This would clearly be an insane procedure to use for a <u>large</u> power, say $\int \sin^{87} x \, dx$!

Fortunately there is an alternative to using the reduction formulas. You can simply treat $\int \sin^m x \, dx$ and $\int \cos^n x \, dx$ as Case A, B or C integrals, depending on whether m or n is even or odd. In particular, for <u>odd</u> powers of $\sin x$ or $\cos x$, this alternate method is <u>greatly</u> to be preferred over the reduction formulas!

Anton illustrates this alternate method in three examples:

Example 1: $\displaystyle\int \sin^2 x\, dx$ as a Case C integral $(m = 2 , n = 0)$

$\displaystyle\int \cos^2 x\, dx$ as a Case C integral $(m = 0 , n = 2)$

Example 2: $\displaystyle\int \cos^4 x\, dx$ as a Case C integral $(m = 0 , n = 4)$

Example 3: $\displaystyle\int \sin^3 x\, dx$ as a Case B integral $(m = 3 , n = 0)$

$\displaystyle\int \cos^3 x\, dx$ as a Case A integral $(m = 0 , n = 3)$

In particular, notice how easy the Example 3 solutions are; these odd powers of $\sin x$ and $\cos x$ are very easy to evaluate as Case A or Case B integrals.

Here are two additional examples of this alternate method:

Example C.   Evaluate $\displaystyle\int \sin^5 x\, dx$.

Solution.   Using Case B $(m = 5 , n = 0)$ this evaluation is straightforward:

$$\int \sin^5 x\, dx = \int \sin^4 x\, (\sin x\, dx)$$

$$= \int (\sin^2 x)^2\, (\sin x\, dx)$$

$$\text{use}\quad \sin^2 x = 1 - \cos^2 x$$

$$= \int (1 - \cos^2 x)^2\, (\sin x\, dx)$$

A polynomial in $\cos x$. Use the
u-substitution $u = \cos x$
$du = - \sin x\, dx$

$$= - \int (1 - u^2)^2\, du$$

$$= -\int (1 - 2u^2 + u^4)\, du$$

$$= -(u - \frac{2}{3} u^3 + \frac{1}{5} u^5) + C$$

$$= \boxed{- \cos x + \frac{2}{3} \cos^3 x - \frac{1}{5} \cos^5 x + C}$$ $\square$

__Example D.__   Evaluate $\displaystyle\int \sin^4 x\, dx$.

__Solution.__   We can use Case C $(m = 4, n = 0)$ on this integral, but it still takes some work;
the even powers are always more troublesome:

$$\int \sin^4 x\, dx = \int (\underbrace{\sin^2 x})^2\, dx \qquad [\,\text{write as powers of } \sin^2 x\,]$$

$$\qquad\qquad\qquad\quad \text{use } \sin^2 x = \frac{1}{2}(1 - \cos 2x)$$

$$= \int (\frac{1}{2}(1 - \cos 2x))^2\, dx$$

$$= \frac{1}{4} \int (1 - 2 \cos 2x + \cos^2 2x)\, dx$$

$$= \frac{1}{4} \int dx - \frac{1}{2} \int \cos 2x\, dx + \frac{1}{4} \int \cos^2 2x\, dx$$

$$[\,\text{easy}\,] \qquad\qquad [\,\text{easy}\,] \qquad\qquad [\,\text{use } \cos^2 2x = \frac{1}{2}(1 + \cos 4x)\,]$$

$$= \frac{1}{4} x - \frac{1}{4} \sin 2x + \frac{1}{8} \int (1 + \cos 4x)\, dx$$

$$= \frac{1}{4} x - \frac{1}{4} \sin 2x + \frac{1}{8} x + \frac{1}{32} \sin 4x + C$$

$$= \boxed{\frac{3}{8} x \quad \frac{1}{4} \sin 2x + \frac{1}{32} \sin 4x + C}$$ $\square$

5.   "What must I memorize?"

    1.   You must remember the three evaluation techniques for integrals

        of the form $\int \sin^m x \cos^n x \, dx$ given in Anton's Table 9.3.1 and in the

        table in §9.3.1 of The Companion.

    2.   You must remember the two evaluation methods for integrals of the

        form $\int \sin^m x \, dx$ or $\int \cos^n x \, dx$:

           i.   reduction formulas, or

           ii.   treatment as integrals of the form $\int \sin^m x \cos^n x \, dx$

              where $n = 0$ or $m = 0$.

On the other hand, we recommend that you NOT memorize the reduction formulas for $\int \sin^m x \, dx$ and $\int \cos^n x \, dx$ (you can look them up when needed ... or rederive them your-self!). Moreover, we certainly recommend that you NOT memorize the six formulas that Anton places in the blue boxes (Equations (1),(2) and (4)-(7)). You must be able to derive them when needed.

6.   Integrals of the form $\int \sin mx \cos nx \, dx$, $\int \sin mx \sin nx \, dx$ and $\int \cos mx \cos nx \, dx$.

Integrals of this form can be handled via the following simple procedure:

Integrating $\int \sin mx \cos nx \, dx$, etc.

Use the appropriate trigonometric product formula

$$\sin \alpha \cos \beta = \frac{1}{2} [\, \sin (\alpha - \beta) + \sin (\alpha + \beta)\,],$$

$$\sin \alpha \sin \beta = \frac{1}{2} [\, \cos (\alpha - \beta) - \cos (\alpha + \beta)\,],$$

$$\cos \alpha \cos \beta = \frac{1}{2} [\, \cos (\alpha - \beta) + \cos (\alpha + \beta)\,],$$

to reduce the integral to sums of integrals of the form

$$\int \sin (m \pm n)x \, dx \quad \text{or} \quad \int \cos (m \pm n)x \, dx$$

Anton's Example 7 illustrates this procedure. Here is another example:

Example E.   Evaluate $\int \cos 2x \cos 3x \, dx$.

Solution.   By the third trigonometric product formula,

$$\cos 2x \cos 3x = \frac{1}{2} [\cos (2x - 3x) + \cos (2x + 3x)]$$

$$= \frac{1}{2} [\cos (-x) + \cos 5x]$$

$$= \frac{1}{2} \cos x + \frac{1}{2} \cos 5x$$

since $\cos (-x) = \cos x$
[Equation 10 b , Appendix 1 , Unit I]

Thus

$$\int \cos 2x \cos 3x \, dx = \frac{1}{2} \int \cos x \, dx + \frac{1}{2} \int \cos 5x \, dx$$

use $u = 5x$, $du = 5 \, dx$

$$= \frac{1}{2} \sin x + \frac{1}{10} \sin 5x + C$$ □

Obviously your success with this procedure depends on your ability to recall the trigonometric product formulas. These formulas are not so difficult to memorize if you keep the following facts in mind:

i)   each has a factor of $\frac{1}{2}$

ii)   each is a <u>sine or cosine of $\alpha - \beta$</u> plus or minus a <u>sine or cosine of $\alpha + \beta$</u>

iii) $\sin \alpha \cos \beta$ is $\frac{1}{2}$ the <u>sum</u> of two <u>sines</u>

$\sin \alpha \sin \beta$ is $\frac{1}{2}$ the <u>difference</u> of two <u>cosines</u>

$\cos \alpha \cos \beta$ is $\frac{1}{2}$ the <u>sum</u> of two <u>cosines</u>

iv) $\sin \alpha \sin \beta$ and $\cos \alpha \cos \beta$ are identical except for the middle sign

(-   for $\sin \alpha \sin \beta$ and   +   for $\cos \alpha \cos \beta$)

7.   <u>Variations on the theme</u>...   We've spent a great deal of time on the "basic" integral forms

$\int \sin^m x \cos^n x \, dx$, $\int \sin mx \cos nx \, dx$,   etc.   However, in many applications (and in many of

Anton's exercises), these integrals arise in a "disguised" form, e.g., a   u-substitution   or

an integration by parts is needed to put the integral into the "basic" form.   Here are two

examples:

<u>Example F.</u>   Evaluate $\displaystyle\int_0^{\pi/6} \sin^3 2\theta \cos^2 2\theta \, d\theta$.

<u>Solution.</u>   You should recognize the basic form of this integral to be $\int \sin^m x \cos^n x \, dx$ ...

except that we have a 2θ in place of the   x.   This calls for the obvious substitution $x = 2\theta$:

$$\int_0^{\pi/6} \sin^3 2\theta \cos^2 2\theta \, d\theta \;=\; \frac{1}{2} \int_0^{\pi/3} \sin^3 x \cos^2 x \, dx$$

This integral is evaluated in
Anton's Example 5 as a
Case B integral.   This yields:

$$\begin{array}{c} x = 2\theta \\ dx = 2\,d\theta \\ dx/2 = d\theta \\ x = \pi/3 \text{ when } \theta = \pi/6 \\ x = 0 \quad \text{when } \theta = 0 \end{array}$$

$$= \frac{1}{2} \left[\frac{1}{5} \cos^5 x - \frac{1}{3} \cos^3 x + C \right]_0^{\pi/3}$$

$$= \frac{1}{2} \left[\frac{1}{5} \left(\frac{1}{2}\right)^5 - \frac{1}{3} \left(\frac{1}{2}\right)^3 + C \right]$$

$$- \frac{1}{2} \left[\frac{1}{5} (1)^5 - \frac{1}{3} (1)^3 + C \right]$$

$$= \frac{47}{960} \;\approx\; \boxed{.0490} \qquad \qquad \Box$$

Example G.   Evaluate $\displaystyle\int \sin^{1/3} x \cos^3 x \, dx$.

Solution.   This resembles a Case A integral of the form $\displaystyle\int \sin^m x \cos^n x \, dx$ with $n = 3$, an odd positive integer. The problem is, the other power, $m = 1/3$, is <u>not a positive integer.</u> This poses no difficulty, however, since the same Case A solution technique will work <u>no matter</u> what value the exponent m takes on!

$$\int \sin^{1/3} x \cos^3 x \, dx = \int \sin^{1/3} x \underbrace{\cos^2 x} (\cos x \, dx)$$

$$\uparrow\!\!\!\!\!-\text{use}\quad \cos^2 x = 1 - \sin^2 x$$

$$= \int \sin^{1/3} x \, (1 - \sin^2 x)(\cos x \, dx)$$

use the u-substitution $u = \sin x$
$$du = \cos x \, dx$$

$$= \int u^{1/3} (1 - u^2) \, du$$

$$= \int (u^{1/3} - u^{7/3}) \, du$$

$$= \frac{3}{4} u^{4/3} - \frac{3}{10} u^{10/3} + C$$

$$= \boxed{\frac{3}{4} \sin^{4/3} x - \frac{3}{10} \sin^{10/3} x + C}$$ $\square$

Section 9.4.   <u>Integrating Powers of Secant and Tangent</u>

1.   The integrals $\displaystyle\int \tan x \, dx$ and $\displaystyle\int \sec x \, dx$: learn the methods, not the formulas.

The elementary integrals $\displaystyle\int \tan x \, dx$ and $\displaystyle\int \sec x \, dx$ can both be evaluated via clever u-substitutions:

	Integral	Method
Integrating $\int \tan x \, dx$ and $\int \sec x \, dx$	$\int \tan x \, dx = \ln \left\| \sec x \right\| + C$	Write $\tan x = \dfrac{\sin x}{\cos x}$ and use * the u-substitution $u = \cos x$ $du = -\sin x \, dx$
	$\int \sec x \, dx = \ln \left\| \sec x + \tan x \right\| + C$	Write $\sec x = \sec x \left(\dfrac{\sec x + \tan x}{\sec x + \tan x} \right)$ and use the u-substitution $u = \sec x + \tan x$ $du = (\sec x \tan x + \sec^2 x) \, dx$

Anton works out the details at the beginning of this section. <u>We recommend that you learn the methods rather than memorize the formulas.</u> After you get over the initial shock of that weird substitution $u = \sec x + \tan x$ (or maybe <u>because</u> it is so weird!), you'll be surprised at how easy it is to remember.

2.   <u>The reduction formulas for powers of $\tan x$ and $\sec x$.</u> There are reduction formulas (Anton's formulas (3) and (4)) for integrating powers of $\tan x$ and $\sec x$ which correspond to the reduction formulas for powers of $\sin x$ and $\cos x$ developed in §9.2. The reduction formula for $\int \tan^m x \, dx$ is easy to obtain:

$$\int \tan^m x \, dx = \frac{1}{m-1} \tan^{m-1} x \, - \int \tan^{m-2} x \, dx$$

---

* Notice that $\int \tan x \, dx = \int \sin^1 x \cos^{-1} x \, dx$ so that this integral is of the form $\int \sin^m x \cos^n x \, dx$ with $m = +1$, $n = -1$. The method used to evaluate the integral is then precisely the second case (Case B) in Anton's Table 9.3.1 (or the table in §9.3.1 of <u>The Companion</u>).

The method used to obtain this formula is to split off one $\tan^2 x$ term, express it as $\tan^2 x = \sec^2 x - 1$, and then use the u-substitution $u = \tan x$. Anton works out the details just before Example 1.

The reduction formula for $\int \sec^n x \, dx$,

$$\int \sec^n x \, dx = \frac{1}{n-1} \sec^{n-2} x \tan x + \frac{n-2}{n-1} \int \sec^{n-2} x \, dx$$

is not as easy to obtain. We derive it via an integration by parts which is similar in many ways to the methods used in §9.2 to derive the reduction formulas for $\int \sin^n x \, dx$ and $\int \cos^n x \, dx$. Here are the details:

Example A.   Derive the reduction formula for $\int \sec^n x \, dx$ given above.

Solution.   (Compare this solution with Example G in §9.2 of The Companion.)

Step 1.   We integrate by parts with $u = \sec^{n-2} x$:

$$\int \sec^n x \, dx = \sec^{n-2} x \tan x - (n-2) \int \sec^{n-2} x \tan^2 x \, dx$$

$$\boxed{\begin{array}{ll} u = \sec^{n-2} x & dv = \sec^2 x \, dx \\ du = (n-2) \sec^{n-2} x \tan x \, dx & v = \tan x \end{array}}$$

(We use $dv = \sec^2 x \, dx$ rather than $dv = \sec x \, dx$ because $\int \sec^2 x \, dx$ is SIMPLE while $\int \sec x \, dx$ is COMPLICATED to evaluate!)

Step 2.   Then we use the trigonometric identity $\tan^2 x + 1 = \sec^2 x$ to obtain the integral $\int \sec^n x \, dx$ on both sides of the equation:

$$\int \sec^n x \, dx \;=\; \sec^{n-2} x \tan x \;-\; (n-2) \int \sec^{n-2} x \tan^2 x \, dx$$

$$=\; \sec^{n-2} x \tan x \;-\; (n-2) \int \sec^{n-2} x \, (\sec^2 x - 1) \, dx$$

$$=\; \sec^{n-2} x \tan x \;-\; (n-2) \int \sec^n x \, dx \;+\; (n-2) \int \sec^{n-2} x \, dx$$

<u>Step 3.</u> Then we solve for the integral $\int \sec^n x \, dx$:

$$[\, 1 + (n-2) \,] \int \sec^n x \, dx \;=\; \sec^{n-2} x \tan x \;+\; (n-2) \int \sec^{n-2} x \, dx$$

or

$$\int \sec^n x \, dx \;=\; \frac{1}{n-1} \sec^{n-2} x \tan x \;+\; \frac{n-2}{n-1} \int \sec^{n-2} x \, dx$$

$\Box$

As we shall see below in §9.4.6,

> only the reduction formula for <u>odd powers</u> of
>
> sec x   is absolutely necessary.

Integrals of the <u>even powers of   sec x</u> and of <u>all powers of tan x</u> can be handled reasonably well by other techniques, but integrals of odd powers of   sec x   can't be handled conveniently except by the reduction formula.

We do not recommend that you memorize the reduction formulas for $\int \tan^m x \, dx$ and $\int \sec^n x \, dx$. Instead,

- remember how to derive the reduction formula for $\displaystyle\int \sec^n x \, dx$

(you will need it only when n is odd)

- rely on the methods of Subsection 4 below to handle $\displaystyle\int \sec^n x \, dx$

when n is even and $\displaystyle\int \tan^m x \, dx$ for any $m > 0$.

3. <u>The Table for $\displaystyle\int \tan^m x \, \sec^n x \, dx$.</u>   In evaluating integrals of the form

$$\int \tan^m x \, \sec^n x \, dx$$

the situation is much the same as it was for integrals of the form $\displaystyle\int \sin^m x \, \cos^n x \, dx$ in §9.3. Once again, it is important to know how to evaluate such integrals because they occur frequently in "real world" applications. And, once again, there is not one integration technique which "works" on all such integrals; instead, the best technique depends on the values of m and n.

We can summarize the $\displaystyle\int \tan^m x \, \sec^n x \, dx$ solution techniques in the following table (which is an expanded version of Anton's Table 9.4.1):

How to evaluate $\int \tan^m x \sec^n x \, dx$

Case A: $\boxed{n \text{ even}}$

$$\int \tan^m x \ \sec^n x \ dx \ = \ \int \tan^m x \ \underbrace{\sec^{n-2} x} \ (\sec^2 x \ dx)$$

This is still an <u>even</u> power of $\sec x$; convert to $\tan^2 x$ terms by using
$$\sec^2 x = \tan^2 x + 1 \ .$$

The result will be ⌐

$$= \int P(\tan x)\,(\sec^2 x \ dx) \qquad \longleftarrow\lrcorner$$

where $P(\tan x)$ is a polynomial in $\tan x$. *
This is easily evaluated by the u-substitution
$$u = \tan x$$
$$du = \sec^2 x \ dx$$

Case B: $\boxed{m \text{ odd}}$

$$\int \tan^m x \ \sec^n x \ dx \ = \ \int \underbrace{\tan^{m-1} x} \ \sec^{n-1} x \ (\sec x \tan x \ dx)$$

This is now an <u>even</u> power of $\tan x$; convert to $\sec^2 x$ terms by using
$$\tan^2 x = \sec^2 x - 1 \ .$$

The result will be ⌐

$$= \int P(\sec x)\,(\sec x \tan x \ dx) \ \longleftarrow\lrcorner$$

where $P(\sec x)$ is a polynomial in $\sec x$.
This is easily evaluated by the u-substitution
$$u = \sec x$$
$$du = \sec x \tan x \ dx$$

---

* Examples of polynomials in $\tan x$ are:
$$\tan^2 x + 1 , \quad 3\tan^4 x + \tan x , \quad \tan^3 x \,(\tan^2 x + 1)^2$$

Case C: m even n odd

$$\int \tan^m x \; \sec^n x \; dx = \int \underbrace{(\tan^2 x)}^{m/2} \sec^n x \; dx$$

This $\tan^2 x$ can be converted to $\sec^2 x$ terms by using

$$\tan^2 x = \sec^2 x - 1$$

The resulting integrals will all be integrals of powers of $\sec x$.

4. <u>The cases "m odd" or "n even" are easy!</u> In these situations the integral $\int \tan^m x \; \sec^n x \; dx$ falls into either Case A or Case B, and these are really easy (once you've had a little practice!)* Anton illustrates Case A in Example 3 and Case B in Example 4. These cases are straight-forward enough that we will not give additional examples. You should, however, compare Anton's two solutions with the Case A and Case B procedures as we have given them above.

5. <u>The case "m even and n odd"</u> can be tedious. This is Case C, which Anton illustrates in Example 5. Anton's example is the simplest possible for Case C; we'll give a nastier one to illustrate the problems which can be encountered here:

<u>Example B.</u> Evaluate $\int \tan^4 x \; \sec^3 x \; dx$.

<u>Solution.</u> This is $\int \tan^m x \; \sec^n x \; dx$ with $m = 4$ and $n = 3$ so we are in Case C. Hence

---

* In fact, when n is even, then m can be <u>any number</u>, i.e., it need not be an integer. Conversely, when m is odd, then n can be <u>any number</u>. These are the observations needed to solve Anton's Exercises 25 and 26!

9.4.8

$$\int \tan^4 x \sec^3 x \, dx = \int (\tan^2 x)^2 \sec^3 x \, dx \qquad [\text{write as powers of } \tan^2 x]$$

$$\downarrow \boxed{\text{use } \tan^2 x = \sec^2 x - 1}$$

$$= \int (\sec^2 x - 1)^2 \sec^3 x \, dx$$

$$= \int (\sec^4 x - 2 \sec^2 x + 1) \sec^3 x \, dx$$

Thus

(*) $$\int \tan^4 x \sec^3 x \, dx = \int \sec^7 x \, dx - 2 \int \sec^5 x \, dx + \int \sec^3 x \, dx$$

Hence our original integral has been rewritten as a sum of integrals of <u>powers of sec x</u>; we may evaluate these integrals by the reduction formula

$$\int \sec^n x \, dx = \frac{1}{n-1} \sec^{n-2} x \tan x + \frac{n-2}{n-1} \int \sec^{n-2} x \, dx$$

which was derived in Example A above. Thus

$$\int \sec^7 x \, dx = \frac{1}{6} \sec^5 x \tan x + \frac{5}{6} \int \sec^5 x \, dx$$

$$\int \sec^5 x \, dx = \frac{1}{4} \sec^3 x \tan x + \frac{3}{4} \int \sec^3 x \, dx$$

$$\int \sec^3 x \, dx = \frac{1}{2} \sec x \tan x + \frac{1}{2} \int \sec x \, dx$$

Finally, we evaluate $\int \sec x \, dx$ as in the beginning of this section to obtain

$$\int \sec x \, dx = \ln |\sec x + \tan x|$$

Plugging these results into Equation (*) above, we obtain the final answer

$$\int \tan^4 x \sec^3 x \, dx = \frac{1}{6} \sec^5 x \tan x - \frac{7}{24} \sec^3 x \tan x$$

$$+ \frac{3}{48} \sec x \tan x + \frac{3}{48} \ln \left| \sec x + \tan x \right| + C$$

(We'll leave the arithmetic details to you.)   □

6. Integrating $\int \sec^n x \, dx$ (n even) and $\int \tan^m x \, dx$ without reduction formulas. As we

mentioned in §9.3.4 of The Companion, there are some drawbacks associated with using

reduction formulas to evaluate integrals. Hence, if possible, convenient alternatives to the

use of reduction formulas are always desirable. Fortunately such alternatives exist for the

even powers of sec x and for all powers of tan x .

For even powers of sec x we have only to observe that

$$\int \overset{\boxed{\text{n even}}}{\sec^n x} \, dx = \int \tan^m x \, \overset{\boxed{\text{n even}}}{\sec^n x} \, dx \qquad \text{for} \qquad m = 0$$

Hence we have Case A , and the even powers of sec x can be integrated using the Case A

procedure. Anton illustrates this in Example 6 .

A similar trick can be used for any power of tan x !! To see this simply observe that

$$\int \tan^m x \, dx = \int \tan^m x \, \overset{\boxed{\text{n = 0 is even!}}}{\sec^0 x} \, dx$$

which shows that we again have Case A . The solution method therefore will proceed as follows:

$$\tan^m x \, dx = \int \tan^m x \, \sec^{-2} x \, (\sec^2 x \, dx)$$

$$= \int \frac{\tan^m x}{\tan^2 x + 1} \, (\sec^2 x \, dx)$$

At this point we would continue with the u-substitution u = tan x . Here's an example:

Example C.   Evaluate $\displaystyle\int \tan^6 x \, dx$.

Solution.   We will treat this as $\displaystyle\int \tan^6 x \sec^0 x \, dx$, a Case A type of the integral $\displaystyle\int \tan^m x \sec^n x \, dx$ since $m = 6$ and $n = 0$ (even!). Thus

$$\int \tan^6 x \, dx = \int \tan^6 x \sec^{-2} x \, (\sec^2 x \, dx)$$

$$\text{use} \quad \sec^2 x = \tan^2 x + 1$$

$$= \int \frac{\tan^6 x}{\tan^2 x + 1} \, (\sec^2 x \, dx)$$

use the   u-substitution $u = \tan x$
$$du = \sec^2 x \, dx$$

$$= \int \frac{u^6}{u^2 + 1} \, du$$

We now divide $u^2 + 1$ into $u^6 \ldots$
(See Appendix D. 2 if this gives you trouble.)

$$= \int \left(u^4 - u^2 + 1 - \frac{1}{u^2 + 1} \right) du$$

$$= \frac{1}{5} u^5 - \frac{1}{3} u^3 + u - \tan^{-1} u + C$$

Since $u = \tan x$, we have $\tan^{-1} u = x$, so ...

$$= \boxed{\; \frac{1}{5} \tan^5 x - \frac{1}{3} \tan^3 x + \tan x - x + C \;} \qquad \square$$

7.   Integrating integrals of the form $\displaystyle\int \cot^m x \csc^n x \, dx$.   As Anton remarks at the end of the section, integrals of the form $\displaystyle\int \cot^m x \csc^n x \, dx$ can be treated by the same techniques we have used on $\displaystyle\int \tan^m x \sec^n x \, dx$. Here is an example:

Example D.   Evaluate $\displaystyle\int \cot^3 x \csc x \, dx$.

Solution.   We can treat this like a Case B integral: $m = 3$ (odd) and $n = 1$. Thus

$$\int \cot^3 x \csc x \, dx = \int \cot^2 x \, (\csc x \cot x \, dx)$$

$$\text{use} \quad \cot^2 x = \csc^2 x - 1$$

$$= \int (\csc^2 x - 1)(\csc x \cot x \, dx)$$

use the u-substitution $u = \csc x$

$$du = - \csc x \cot x \, dx$$

$$= - \int (u^2 - 1) \, du$$

$$= - \frac{1}{3} u^3 + u + C$$

$$= - \frac{1}{3} \csc^3 x + \csc x + C \qquad\qquad \square$$

Section 9.5.   Trigonometric Substitutions

1. Simplifying the expressions $\sqrt{a^2 - x^2}$, $\sqrt{x^2 + a^2}$ and $\sqrt{x^2 - a^2}$. By using clever but simple

substitutions involving trigonometric functions, we can eliminate the radicals in the terms

$\sqrt{a^2 - x^2}$, $\sqrt{x^2 + a^2}$ and $\sqrt{x^2 - a^2}$. The idea is simple:

find appropriate trigonometric substitutions

$$\text{"} x = a \text{ trig } \theta \text{ "} \quad *$$

so that $a^2 - x^2$, $x^2 + a^2$ and $x^2 - a^2$ will

become <u>perfect squares</u>!

The trigonometry needed is elementary. Begin with the basic identity

$$1 = \sin^2 \theta + \cos^2 \theta$$

---

* "a trig θ" is read "a times a trigonometric function of θ."

(everybody knows that, right?) Solving this for $\cos^2\theta$ and multiplying by a^2 yields

$$\boxed{a^2 - a^2 \sin^2\theta = a^2 \cos^2\theta} \tag{A}$$

That's the first equation we'll need. Returning to the basic identity, if we divide through by $\cos^2\theta$ we obtain

$$\frac{1}{\cos^2\theta} = \frac{\sin^2\theta}{\cos^2\theta} + \frac{\cos^2\theta}{\cos^2\theta}$$

or $\qquad \sec^2\theta = \tan^2\theta + 1 \quad .$

Multiplying this result by a^2 yields the two other necessary equations

$$\boxed{\begin{aligned} a^2 \tan^2\theta + a^2 &= a^2 \sec^2\theta \\ a^2 \sec^2\theta - a^2 &= a^2 \tan^2\theta \end{aligned}} \qquad \begin{aligned} &\text{(B)} \\ &\text{(C)} \end{aligned}$$

We now can pick the appropriate substitution to simplify each of the radicals $\sqrt{a^2 - x^2}$, $\sqrt{x^2 + a^2}$ and $\sqrt{x^2 - a^2}$:

$\boxed{\sqrt{a^2 - x^2}}$ Compare this radical with equation (A) :

$$\sqrt{a^2 - x^2}$$
$$\updownarrow$$
$$a^2 - a^2 \sin^2\theta = a^2 \cos^2\theta \tag{A}$$

Hmm .... the substitution $\boxed{x = a \sin\theta}$ looks promising, since then $a^2 - x^2$ becomes a perfect square:

$$\sqrt{a^2 - x^2} = \sqrt{a^2 - a^2 \sin^2\theta}$$

$$= \sqrt{a^2 \cos^2\theta} \qquad \text{from equation (A)}$$

$$= a \cos\theta \qquad \text{if we restrict the } \theta\text{-values to}$$

$$\boxed{-\frac{\pi}{2} \le \theta \le \frac{\pi}{2}}^{*}$$

$\boxed{\sqrt{x^2 + a^2}}$ Equation (B) looks tailor-made for this radical:

$$\sqrt{x^2 + a^2}$$
$$\updownarrow$$
$$a^2 \tan^2\theta + a^2 = a^2 \sec^2\theta \qquad\qquad\qquad\qquad (B)$$

Our substitution should certainly be $\boxed{x = a \tan\theta}$,   for then

$$\sqrt{x^2 + a^2} = \sqrt{a^2 \tan^2\theta + a^2}$$

$$= \sqrt{a^2 \sec^2\theta} \qquad \text{from equation (B)}$$

$$= a \sec\theta \qquad \text{if we restrict the } \theta\text{-values to}$$

$$\boxed{-\frac{\pi}{2} < \theta < \frac{\pi}{2}}$$

---

*The restrictions on θ values in all three of our cases are needed to guarantee that the resulting trigonometric functions are non-negative, e.g., $\cos\theta \ge 0$ when $-\frac{\pi}{2} \le \theta \le \frac{\pi}{2}$.

$\boxed{\sqrt{x^2 - a^2}}$ Equation (C) is our choice now:

$$\sqrt{x^2 - a^2}$$

$$\uparrow$$

$$a^2 \sec^2\theta - a^2 = a^2 \tan^2\theta \tag{C}$$

Our substitution will then be $\boxed{x = a \sec \theta}$,   which yields

$$\sqrt{x^2 - a^2} = \sqrt{a^2 \sec^2\theta - a^2}$$

$$= \sqrt{a^2 \tan^2\theta} \qquad\qquad \text{from equation (B)}$$

$$= a \tan \theta \qquad\qquad \text{if we restrict the } \theta\text{-values to}$$

$$\boxed{0 \leq \theta < \frac{\pi}{2} \ \text{ or } \ \pi \leq \theta < \frac{3\pi}{2}}$$

*   *   *   *   *

You can, of course, memorize the appropriate substitution for each of the three radicals given above. However, you will be less likely to make mistakes if you learn the <u>method for determining the substitutions</u> rather than the substitutions themselves! That's why we spent so much time on the method.

2.   <u>Integrals involving $\sqrt{a^2 - x^2}$, $\sqrt{x^2 + a^2}$ or $\sqrt{x^2 - a^2}$.</u>   The general procedure for evaluating integrals of this form is as follows:

Step 1.   Determine the appropriate trigonometric substitution

$$"x \; = \; a \; \text{trig} \; \theta \; "$$

(as described in the previous subsection) to eliminate

the radical.

Step 2.   Use the substitution of Step 1 to obtain an integral of

trigonometric functions in θ. This should be evaluated

using the techniques of §§9.3 and 9.4.

Step 3.   The result of Step 2 will generally be a trigonometric

function of θ. To convert back to x use

$$"\theta \; = \; \text{trig}^{-1} \left(\frac{x}{a}\right) " \; *$$

along with the Triangle Method of §8.2.3 of The Companion.

Anton illustrates this procedure in Examples 1, 2 and 3. Here is another example, with the three steps carefully spelled out:

Example A.   Evaluate $\displaystyle\int \frac{x^2}{\sqrt{4 - x^2}} \; dx$.

Step 1.   We wish to turn the expression $4 - x^2$ into a perfect square. Recalling the three important identities,

$$a^2 - a^2 \sin^2\theta = a^2 \cos^2\theta \; ,$$

$$a^2 \tan^2\theta + a^2 = a^2 \sec^2\theta \; ,$$

$$a^2 \sec^2\theta - a^2 = a^2 \tan^2\theta \; ,$$

_____

* "trig^{-1} $\left(\frac{x}{a}\right)$" is read "the inverse trigonometric function of $\frac{x}{a}$."

the form of $4 - x^2$ suggests that we use the <u>first</u> of these identities and set

$$4 - x^2 = a^2 - a^2 \sin^2\theta = a^2 \cos^2\theta \quad .$$

Hence $a = 2$ and

$$\boxed{\begin{array}{c} x = 2 \sin\theta \\[2mm] \sqrt{4 - x^2} = 2 \cos\theta \end{array}} \quad \text{for} \quad -\frac{\pi}{2} \le \theta \le \frac{\pi}{2} . \quad *$$

<u>Step 2.</u> Making this substitution into the original integral yields

$$\int \frac{x^2}{\sqrt{4 - x^2}} \, dx \quad = \int \frac{4 \sin^2\theta}{2 \cos\theta} (2 \cos\theta \, d\theta)$$

$$\boxed{\begin{array}{l} x = 2 \sin\theta \\[2mm] dx = 2 \cos\theta \, d\theta \\[2mm] \sqrt{4 - x^2} = 2 \cos\theta \end{array}} \quad = 4 \int \sin^2\theta \, d\theta$$

Ahh ... an integral of a trigonometric function in θ. We can evaluate this integral by using
$$\sin^2\theta = \frac{1}{2}(1 - \cos 2\theta)$$

$$= 2 \int (1 - \cos 2\theta) \, d\theta$$

$$= 2\theta - \sin 2\theta + C$$

use the identity $\sin 2\theta = 2 \sin\theta \cos\theta$

$$= 2\theta - 2 \sin\theta \cos\theta + C$$

<u>Step 3.</u> We need to convert our answer back into the variable x. Since $x = 2 \sin\theta$, then

$$\boxed{\sin\theta = \frac{x}{2} \quad \text{and} \quad \theta = \sin^{-1}\left(\frac{x}{2}\right)}$$

---

* The θ-restriction ensures that $\cos\theta \ge 0$ and $\theta = \sin^{-1}\left(\frac{x}{2}\right)$.

Hence the answer from Step 2 becomes

$$\int \frac{x^2}{\sqrt{4-x^2}}\, dx = 2\sin^{-1}\left(\frac{x}{2}\right) - 2\left(\frac{x}{2}\right)\cos\left[\sin^{-1}\left(\frac{x}{2}\right)\right] + C$$

We have only to evaluate $\cos\left[\sin^{-1}\left(\frac{x}{2}\right)\right]$, and this is conveniently done using the Triangle

Method discussed in §8. 2. 3 of The Companion :

1. We draw a right triangle with an

    angle $\theta = \sin^{-1}(x/2)$, i.e.,

    $$\sin\theta = x/2$$

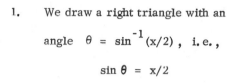

2. We fill in the third side of the triangle

    using the Pythagorean Theorem. In this

    case the third (adjacent) side will be

    $$\sqrt{2^2 - x^2} = \sqrt{4 - x^2}$$

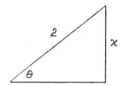

3. We "read off" the desired trigonometric

    value. In this case,

    $$\cos\theta = \frac{\text{adjacent}}{\text{hypotenuse}} = \frac{\sqrt{4 - x^2}}{2}$$

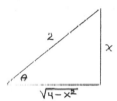

Thus our final answer is

$$\int \frac{x^2}{\sqrt{4-x^2}}\, dx = \boxed{\; 2\sin^{-1}\left(\frac{x}{2}\right) - \frac{x}{2}\sqrt{4 - x^2} + C \;}$$

$\square$

3. <u>Using trigonometric substitutions when the coefficient of x^2 is not 1.</u> With very little added difficulty the procedure of using trigonometric substitutions to turn expressions into perfect squares can be used on integrals containing expressions of the form

$$\sqrt{a^2 - b^2 x^2} \, , \quad \sqrt{b^2 x^2 + a^2} \quad \text{and} \quad \sqrt{b^2 x^2 - a^2}$$

where $a > 0$ and $b > 0$. (In the cases we have considered so far $b = 1$, but now we allow it to be <u>any positive number.</u>) The following example will illustrate this procedure:

<u>Example B.</u> Evaluate $\displaystyle\int \frac{x^2}{\sqrt{2x^2 - 3}} \, dx$.

<u>Solution.</u> The easiest way to handle the unwanted "2" in front of the x^2 is to "u-substitute it away," i.e., to let

$$u^2 = 2x^2 \, , \quad \text{i.e.,} \quad u = \sqrt{2} \, x$$

Here's what will happen:

$$\int \frac{x^2}{\sqrt{2x^2 - 3}} \, dx = \int \frac{u^2/2}{\sqrt{u^2 - 3}} \left(\frac{du}{\sqrt{2}} \right)$$

$$\boxed{\begin{array}{c} u^2 = 2x^2 \\ u = \sqrt{2} \, x \\ du = \sqrt{2} \, dx \end{array}} = \frac{1}{2\sqrt{2}} \int \frac{u^2}{\sqrt{u^2 - 3}} \, du \qquad\qquad (D)$$

We have now reduced the problem to the evaluation of the integral $\displaystyle\int \frac{u^2}{\sqrt{u^2 - 3}} \, du$ which is of the type previously considered, so we proceed as in Example A:

<u>Step 1.</u> We wish to turn the expression $u^2 - 3$ into a perfect square. Recalling the three

identities

$$a^2 - a^2 \sin^2 \theta = a^2 \cos^2 \theta \ ,$$

$$a^2 \tan^2 \theta + a^2 - a^2 \sec^2 \theta \ ,$$

$$a^2 \sec^2 \theta - a^2 = a^2 \tan^2 \theta \ ,$$

the form of $u^2 - 3$ suggests that we use the <u>third</u> of these identities and set

$$u^2 - 3 = a^2 \sec^2 \theta - a^2 = a^2 \tan^2 \theta$$

Hence $a = \sqrt{3}$ and $\boxed{\begin{array}{c} u = \sqrt{3} \ \sec \theta \\ \\ \sqrt{u^2 - 3} = \sqrt{3} \ \tan \theta \end{array}}$ for $0 \leq \theta < \dfrac{\pi}{2}$ or $\pi \leq \theta < \dfrac{3\pi}{2}$ *

<u>Step 2.</u>   Making this substitution in the integral yields

$$\int \frac{u^2}{\sqrt{u^2 - 3}} \ du \quad = \quad \int \frac{3 \sec^2 \theta}{\sqrt{3} \ \tan \theta} \ (\sqrt{3} \ \sec \theta \tan \theta \ d\theta)$$

$$\boxed{\begin{array}{l} u = \sqrt{3} \ \sec \theta \\ du = \sqrt{3} \ \sec \theta \tan \theta \ d\theta \\ \sqrt{u^2 - 3} = \sqrt{3} \ \tan \theta \end{array}} \quad = \quad 3 \int \sec^3 \theta \ d\theta$$

This is the integral of a trigonometric
function in θ. We can evaluate it
using the reduction formula for
$\int \sec^n x \ dx \dots$

$$= 3 \left(\frac{1}{2} \sec \theta \tan \theta + \frac{1}{2} \ln \left| \sec \theta + \tan \theta \right| \right) + C$$

$$= \frac{3}{2} \sec \theta \tan \theta + \frac{3}{2} \ln \left| \sec \theta + \tan \theta \right| + C \qquad \text{(E)}$$

---

*   The θ-restrictions ensure that $\tan \theta \geq 0$ and $\theta = \sec^{-1} \left(\dfrac{u}{\sqrt{3}} \right)$.

Step 3.   We need to convert back into the variable   u.   Since $u = \sqrt{3}\,\sec\theta$ then

$$\sec\theta \;=\; u/\sqrt{3}$$

However, in Equation (E) we also have the term $\tan\theta$; to evaluate this we will use the Triangle Method of §8. 2. 3 :

$$\sec\theta \;=\; u/\sqrt{3} \quad\text{gives this triangle} \ldots$$

... from which we obtain

$$\tan\theta \;=\; \frac{\text{opposite}}{\text{adjacent}} \;=\; \frac{\sqrt{u^2 - 3}}{\sqrt{3}}$$

Thus Equation (E) becomes

$$\int \frac{u^2}{\sqrt{u^2 - 3}}\; du \;=\; \frac{3}{2}\left(\frac{u}{\sqrt{3}}\right)\left(\frac{\sqrt{u^2 - 3}}{\sqrt{3}}\right) + \frac{3}{2}\ln\left|\frac{u}{\sqrt{3}} + \frac{\sqrt{u^2 - 3}}{\sqrt{3}}\right| + C$$

$$=\; \frac{1}{2}\,u\sqrt{u^2 - 3} + \underbrace{\frac{3}{2}\ln\left|u + \sqrt{u^2 - 3}\right| - \frac{3}{2}\ln\sqrt{3}}\; + C$$

$$\text{using}\quad \ln\left(\frac{a}{b}\right) = \ln a - \ln b$$

$$=\; \frac{1}{2}\,u\sqrt{u^2 - 3} + \frac{3}{2}\ln\left|u + \sqrt{u^2 - 3}\right| + C_1$$

Finally we need to convert this answer back into the original variable   x   by substituting $u = \sqrt{2}\,x$

$$\int \frac{u^2}{\sqrt{u^2 - 3}} \, du = \frac{1}{2} (\sqrt{2}\, x) \sqrt{2x^2 - 3} + \frac{3}{2} \ln \left| \sqrt{2}\, x + \sqrt{2x^2 - 3} \right| + C_1$$

$$= \frac{\sqrt{2}}{2} x \sqrt{2x^2 - 3} + \frac{3}{2} \ln \left| \sqrt{2}\, x + \sqrt{2x^2 - 3} \right| + C_1$$

Hence the answer to the <u>original</u> integral is (from Equation (D))

$$\int \frac{x^2}{\sqrt{2x^2 - 3}} \, dx = \frac{1}{2\sqrt{2}} \int \frac{u^2}{\sqrt{u^2 - 3}} \, du$$

$$= \frac{1}{2\sqrt{2}} \left(\frac{\sqrt{2}}{2} x \sqrt{2x^2 - 3} + \frac{3}{2} \ln \left| \sqrt{2}\, x + \sqrt{2x^2 - 3} \right| + C_1 \right)$$

$$\boxed{ = \quad \frac{1}{4} x \sqrt{2x^2 - 3} + \frac{3}{4\sqrt{2}} \ln \left| \sqrt{2}\, x + \sqrt{2x^2 - 3} \right| + C_2 } \qquad \square$$

4. <u>Look before you leap!</u> The methods developed in this section for handling integrals involving $\sqrt{a^2 - x^2}$, $\sqrt{x^2 + a^2}$ and $\sqrt{x^2 - a^2}$ can lead to some fairly tedious computations (as you have seen!). For that reason you should always look for simpler methods before slugging away with a trigonometric substitution. Anton illustrates this point with a definite integral in Example 4. Here is an example with an indefinite integral:

<u>Example C.</u>   Evaluate $\displaystyle\int \frac{x}{\sqrt{3x^2 - 2}} \, dx$.

<u>Solution.</u>   Yes, you could first eliminate the 3 by using $u = \sqrt{3}\, x$ as in Example B. And then you could go through the trigonometric substitution $u^2 - 2 = a^2 \sec^2 \theta - a^2$, i.e.,

$$u = \sqrt{2} \; \sec \theta$$

$$du = \sqrt{2} \; \sec \theta \tan \theta \; d\theta$$

$$\sqrt{u^2 - 2} = \sqrt{2} \; \tan \theta \quad,$$

and so on .... But all this is unnecessary if you observe a very simple u-substitution :

$$\int \frac{x}{\sqrt{3x^2 - 2}} \qquad = \int \frac{1}{\sqrt{u}} \left(\frac{du}{6} \right) \; = \; \frac{1}{6} \int u^{-1/2} \, du$$

$$\boxed{\begin{aligned} u &= 3x^2 - 2 \\ du &= 6x \, dx \\ \frac{du}{6} &= x \, dx \end{aligned}} \qquad = \; \frac{1}{6} \left(\frac{u^{1/2}}{1/2} \right) + C$$

$$= \; \frac{1}{3} u^{1/2} + C$$

$$= \; \boxed{\frac{1}{3} \sqrt{3x^2 - 2} + C} \qquad \qquad \square$$

5.   Variations on the theme ....   In many applications (and in some of Anton's exercises), the integrals we've discussed in this section arise in a "disguised" form, e. g., a u-substitution or an integration by parts is needed to put the integral into a recognizable form. Here is an example:

Example D.   Evaluate $\displaystyle \int \sqrt{e^{2t} - 25} \; dt$.

Solution.   The basic shape of this integral should make you suspect an   "x-substitution" of the form

$$x^2 - 25 = e^{2t} - 25 \, , \quad \text{i. e.,} \quad x = e^t \quad .$$

Using this substitution we obtain:

$$\int \sqrt{e^{2t} - 25} \; dt \;=\; \int \sqrt{x^2 - 25} \left(\frac{dx}{x}\right)$$

$$\boxed{\begin{array}{l} x = e^t \\[4pt] dx = e^t \, dt \\[4pt] \dfrac{dx}{x} = dt \end{array}} \qquad = \int \frac{\sqrt{x^2 - 25}}{x} \; dx$$

Ahh... now we have an integral in a form which we "recognize." In fact, this is the integral from Anton's Example 3 (what a strange coincidence!). From that example we have

$$= \; \sqrt{x^2 - 25} \; - \; 5\, \sec^{-1}\!\left(\frac{x}{5}\right) + C$$

$$= \; \boxed{\sqrt{e^{2t} - 25} \; - \; 5\, \sec^{-1}\!\left(\frac{e^t}{5}\right) + C} \qquad \square$$

Section 9.6.   Integrals Involving $ax^2 + bx + c$

1. <u>Completing the square.</u>   The integrals discussed in this section involve

$$ax^2 + bx + c$$

when this expression <u>cannot</u> be factored as a product

of real linear factors (or, equivalently, when the roots

of $ax^2 + bx + c = 0$ are <u>complex</u> numbers).

(Handling certain cases in which $ax^2 + bx + c$ <u>can</u> be factored as a product of real linear

factors will be discussed in §9.7.)

The major step in dealing with such integrals is to <u>complete the square</u> in the expression

$ax^2 + bx + c$:

Step 1.   Complete the square to rewrite the expression

$a x^2 + b x + c$ in the form

$$a (x + r)^2 + s$$

where   r   and   s   are constants.

Step 2.   Make the   u-substitution

$$u = x + r$$

$$du = dx$$

Then $a x^2 + b x + c = a(x + r)^2 + s$ will take

on the more familiar form $a u^2 + s$.

Step 3.   The resulting integral will often be of a form

studied previously.   Evaluate the integral by

methods learned earlier (e.g., by trigonometric

substitutions).

Integrals
containing
$a x^2 + b x + c$

The process of completing the square is discussed in Appendix D of The Companion and you should be completely familiar with it.   Two examples (Examples 13 and 14) are given in that Appendix.   Here is one more:

Example A.   Complete the square in $3 x^2 - 12 x + 15$.

Solution.   First we factor the coefficient of x^2 out of the whole expression:

$$3 x^2 - 12 x + 15 = 3 (x^2 - 4 x + 5)$$

Then inside the parentheses we add in and subtract out the square of one-half the   x   coefficient:

$$3(x^2 - 4x + 5) = 3(x^2 - 4x + (-2)^2 + 5 - (-2)^2)$$

| add in | | subtract out |

$$= 3[(x-2)^2 + 1]$$

$$= 3(x-2)^2 + 3 \qquad \square$$

2.   Examples.   Anton's Examples 1-3 all illustrate the completing the square procedure. Here are two additional examples:

Example B.   Evaluate $\displaystyle\int \frac{dx}{3x^2 - 12x + 15}$.

Solution.

Step 1.   The expression $3x^2 - 12x + 15$ has no real linear factors. * Hence we complete the square of this quadratic ... but we just did this in Example A, and found

$$3x^2 - 12x + 15 = 3(x-2)^2 + 3 \qquad .$$

Step 2.   We make the u-substitution $u = x - 2$:

$$\int \frac{dx}{3x^2 - 12x + 15} = \int \frac{dx}{3(x-2)^2 + 3}$$

| $u = x - 2$ |
| $du = dx$ |

$$= \int \frac{du}{3u^2 + 3} = \frac{1}{3}\int \frac{du}{u^2 + 1}$$

---

* A quick check to determine whether a quadratic $ax^2 + bx + c$ has real linear factors is to compute the expression $\sqrt{b^2 - 4ac}$. If this is imaginary, then $ax^2 + bx + c$ has no real linear factors. In Example B we have

$$\sqrt{b^2 - 4ac} = \sqrt{144 - 4(3)(15)} = \sqrt{-36}$$

Since this is imaginary, $3x^2 - 12x + 15$ has no real linear factors.

Step 3. We are lucky in that the resulting integral is quite easy to evaluate:

$$\int \frac{dx}{3x^2 - 12x + 15} = \frac{1}{3}\int \frac{du}{u^2 + 1} = \frac{1}{3}\tan^{-1}u + C$$

$$= \boxed{\frac{1}{3}\tan^{-1}(x - 2) + C} \qquad \square$$

Example C. Evaluate $\displaystyle\int \frac{e^t}{3e^{2t} - 12e^t + 15}\, dt$.

Solution. At first this does not appear to be an integral of the form we are studying in this section ... but, wait a minute! Examining the integral reveals a lot of e^t terms; in fact, t never appears <u>except</u> in the form e^t! This makes the substitution $x = e^t$ a good bet:

$$\int \frac{e^t}{3e^{2t} - 12e^t + 15}\, dt = \int \frac{1}{3x^2 - 12x + 15}\, dx$$

$$\boxed{\begin{array}{l} x = e^t \\ dx = e^t\, dt \\ x^2 = e^{2t} \end{array}}$$

Well, how about that? This <u>is</u> an integral of the form we are

considering in this section. So we forge ahead and complete the square in $3x^2 - 12x + 15$, to obtain ... etc, etc. The details may be found in Example B, for (just by coincidence) this integral is the one considered there!

$$\int \frac{e^t}{3e^{2t} - 12e^t + 15}\, dt = \int \frac{1}{3x^2 - 12x + 15}\, dx$$

$$= \frac{1}{3}\tan^{-1}(x - 2) + C$$

$$\boxed{\text{From Example B}}$$

$$= \boxed{\frac{1}{3}\tan^{-1}(e^t - 2) + C} \qquad \square$$

3. What if $ax^2 + bx + c$ CAN be factored into real linear factors? In that case other methods

will often prove more fruitful than completing the square. * However, one notable exception

occurs when the quadratic term $ax^2 + bx + c$ appears under a radical, i.e., as

$$\sqrt{ax^2 + bx + c} \quad .$$

In that situation it is often advantageous to complete the square even if $ax^2 + bx + c$ does

factor! The resulting integral will then have a term of one of the following types:

$$\sqrt{a^2 - u^2} \quad , \quad \sqrt{u^2 + a^2} \quad \text{or} \quad \sqrt{u^2 - a^2} \quad .$$

Evaluating such integrals was discussed in the previous section.

Although Anton doesn't mention this situation in the text, he does include it in an example:

the quadratic term $5 - 4x - 2x^2$ in Example 2 DOES factor into real linear terms since it has

REAL roots (found using the quadratic formula $x = \left(-b \pm \sqrt{b^2 - 4ac} \right) / 2a$) :

$$-1 + \frac{1}{2}\sqrt{14} \quad \text{and} \quad -1 - \frac{1}{2}\sqrt{14}$$

However, this quadratic appears under a radical sign, and, as you can see in the solution to

Example 2, starting off by completing the square does lead to a solution.

Anton's Exercises 2, 3 and 7 all involve quadratic terms which do factor into real

linear terms, but for which completion of the square is the best way to start. Here is another

such example:

Example D.   Evaluate $\displaystyle\int \frac{dx}{\sqrt{x^2 + 2x - 3}}$.

---

*   Some of these methods will be discussed in §9.7.

Solution. The quadratic term $x^2 + 2x - 3$ DOES factor,

$$x^2 + 2x - 3 = (x - 1)(x + 3)$$

but since it appears under a radical (and no better method appears available) we will start by completing the square.

Step 1. Complete the square:

$$x^2 + 2x - 3 = (x^2 + 2x + 1) - 3 - 1$$
$$= (x + 1)^2 - 4$$

Step 2. We make the u-substitution $u = x + 1$:

$$\int \frac{dx}{\sqrt{x^2 + 2x - 3}} = \int \frac{dx}{\sqrt{(x + 1)^2 - 4}}$$

$$\boxed{\begin{array}{l} u = x + 1 \\ du = dx \end{array}} = \int \frac{du}{\sqrt{u^2 - 4}}$$

Step 3. Evaluate the integral. If you have studied §8.4 on the inverse hyperbolic functions then you can make quick work of this integral:

$$\int \frac{du}{\sqrt{u^2 - 4}} = \cosh^{-1}\left(\frac{u}{2}\right) + C \qquad \text{from Theorem 8.4.4}$$

$$= \boxed{\cosh^{-1}\left(\frac{x + 1}{2}\right) + C \;.}$$

However, even without the hyperbolic functions, this integral can be evaluated by the techniques of §9.5: We wish to turn the expression $u^2 - 4$ into a perfect square. Recalling the three important identities of §9.5:

$$a^2 - a^2 \sin^2 \theta = a^2 \cos^2 \theta \ ,$$

$$a^2 \tan^2 \theta + a^2 = a^2 \sec^2 \theta \ ,$$

$$a^2 \sec^2 \theta - a^2 = a^2 \tan^2 \theta \ ,$$

the form of $u^2 - 4$ suggests that we use the <u>third</u> of these identities and set

$$u^2 - 4 = a^2 \sec^2 \theta - a^2 = a^2 \tan^2 \theta$$

Hence $a = 2$ and

$$\boxed{\begin{array}{l} u = 2 \sec \theta \\[6pt] \sqrt{u^2 - 4} = 2 \tan \theta \end{array}}$$

for $0 \le \theta < \pi/2$ or $\pi \le \theta < 3\pi/2$ *

Making this substitution in the integral yields

$$\int \frac{du}{\sqrt{u^2 - 4}} = \int \frac{2 \sec \theta \tan \theta \, d\theta}{2 \tan \theta}$$

$$\boxed{\begin{array}{l} u = 2 \sec \theta \\ du = 2 \sec \theta \tan \theta \, d\theta \\[6pt] \sqrt{u^2 - 4} = 2 \tan \theta \end{array}} = \int \sec \theta \, d\theta$$

This is the integral of a trigonometric function in θ, and we learned it in §9.4...

$$= \ln \left| \sec \theta + \tan \theta \right| + C$$

To convert back to the variable u, we note that since $u = 2 \sec \theta$, then

$$\sec \theta = \frac{u}{2} \quad \text{and} \quad \theta = \sec^{-1}\left(\frac{u}{2}\right)$$

Hence the answer above becomes

$$\int \frac{du}{\sqrt{u^2 - 4}} = \ln \left| \frac{u}{2} + \tan\left(\sec^{-1}\left(\frac{u}{2}\right)\right) \right| + C$$

---
* The θ-restrictions ensure that $\tan \theta \ge 0$ and $\theta = \sec^{-1}\left(\frac{u}{2}\right)$.

It only remains to evaluate $\tan(\sec^{-1}(\frac{u}{2}))$ and this is done via the Triangle Method (see §8.2.3 of The Companion):

$$\tan\left(\sec^{-1}\left(\frac{u}{2}\right)\right) = \tan\theta = \frac{\sqrt{u^2 - 4}}{2}$$

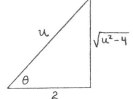

Thus our answer above becomes

$$\int \frac{dx}{\sqrt{x^2 + 2x - 3}} = \int \frac{du}{\sqrt{u^2 - 4}} \quad \ln\left|\frac{u}{2} + \frac{\sqrt{u^2 - 4}}{2}\right| + C$$

$$= \ln\left|u + \sqrt{u^2 - 4}\right| - \ln 2 + C$$

$$\underbrace{\qquad\qquad\qquad\qquad\qquad}$$

using $\ln\left(\frac{a}{b}\right) = \ln a - \ln b$

$$= \ln\left|u + \sqrt{u^2 - 4}\right| + C_1$$

Finally, we convert this answer back into the original variable x by substituting $u = x + 1$:

$$\int \frac{dx}{\sqrt{x^2 + 2x - 3}} = \boxed{\ln\left|x + 1 + \sqrt{x^2 + 2x - 3}\right| + C_1} \qquad \square$$

4. Look before you leap...

Example E.   Evaluate $\displaystyle\int \frac{x + 1}{\sqrt{x^2 + 2x - 3}}\, dx$.

Solution.   This looks very much like the integral in Example D... so we should start by completing the square, and then head off for a trigonometric substitution, right? WRONG!

Look (... for a simpler method...) before you leap (... into a complicated method)! The

x + 1   term in the numerator makes possible a successful   u-substitution:

$$\int \frac{x+1}{\sqrt{x^2+2x-3}} = \int \frac{du/2}{\sqrt{u}}$$

Whoopee! The rest is
a cinch!

$$\boxed{\begin{array}{l} u = x^2 + 2x - 3 \\ du = (2x+2)\,dx \\ du/2 = (x+1)\,dx \end{array}} = \frac{1}{2} \int u^{-1/2}\,du$$

$$= \frac{1}{2}\left(2u^{1/2}\right) + C$$

$$= \boxed{\sqrt{x^2 + 2x - 3} + C} \qquad \square$$

Section 9. 7.   Integrating Rational Functions; Partial Fractions

1.   Factoring and long division of polynomials.   To be successful with the techniques of this section
you must be proficient at

       - factoring polynomials, and

       - dividing one polynomial by another.

Both of these topics are covered in Appendix D of The Companion and you should look there
if you need a review:

      Appendix D §2   Division of polynomials

              §5   Factoring quadratic (i. e. , second degree) polynomials

              §6   Factoring general polynomials

Factoring polynomials is made easier by the two theorems (Theorems 9.7.1 and 9.7.2) Anton includes in the optional section at the end of §9.7. Slightly reworded, they are

The Factor Theorem

> ### The Factor Theorem
>
> The linear term $x - r$ is a factor of a polynomial $p(x)$
>
> if and only if r is a root of $p(x)$, i.e., $p(r) = 0$

and

The Rational Root Test

> ### The Rational Root Test *
>
> Suppose $p(x) = a_n x^n + a_{n-1} x^{n-1} + \cdots + a_1 x + a_0$ is a
>
> polynomial with <u>integer</u> coefficients and $r = \dfrac{c}{d}$ is a rational
>
> number where c and $d \neq 0$ are integers and $\dfrac{c}{d}$ is expressed
>
> in lowest terms. Then r can be a root of $p(x)$ only
>
> if c divides a_0 and d divides a_n.

---

* This is part (b) of Theorem 9.7.2. Note that part (a) of that theorem is the special case when $c = r$ and $d = 1$.
Also note that in his statement of Theorem 9.7.2, Anton writes the polynomial as

$$p(x) = a_0 x^n + a_1 x^{n-1} + \cdots + a_{n-1} x + a_n \qquad (a_0 \text{ is the leading coefficient and}$$

$$\underset{\uparrow}{\boxed{\text{NOTE}}} \qquad\qquad\qquad \underset{\uparrow}{\boxed{\text{NOTE}}} \qquad a_n \text{ is the constant})$$

instead of as

$$p(x) = a_n x^n + a_{n-1} x^{n-1} + \cdots + a_1 x + a_0 \qquad (a_n \text{ is the leading coefficient and}$$

$$\underset{\uparrow}{\boxed{\text{NOTE}}} \qquad\qquad\qquad \underset{\uparrow}{\boxed{\text{NOTE}}} \qquad a_0 \text{ is the constant})$$

Do not let this difference in notation confuse you. (In either case the theorem says $r = \dfrac{c}{d}$ is a root if and only if <u>c divides the constant term</u> and <u>d divides the leading coefficient.</u>)

Anton's Examples 6 and 7 (and Examples 11 and 12 in Appendix D) illustrate how useful these theorems can be. Here is one more illustration:

Example A.   Factor $p(x) = x^3 - 5x^2 + 8x - 6$.

Solution.   First we use The Rational Root Test to find the roots of our polynomial. Since the leading coefficient 1 has ± 1 as its only divisors and the constant term -6 has ± 1, ± 2, ± 3 and ± 6 as its only divisors, The Rational Root Test says that the <u>only possible</u> (rational) roots are

$$\frac{\pm 1}{\pm 1}, \; \frac{\pm 2}{\pm 1}, \; \frac{\pm 3}{\pm 1}, \; \frac{\pm 6}{\pm 1}$$

i. e.,

$$1, \; -1, \; 2, \; -2, \; 3, \; -3, \; 6, \; -6$$

To check which, if any, are roots, we plug them into the polynomial $p(x)$:

$$
\begin{aligned}
p(1) &= (1)^3 - 5(1)^2 + 8(1) - 6 = -2 \\
p(-1) &= (-1)^3 - 5(-1)^2 + 8(-1) - 6 = -20 \\
p(2) &= (2)^3 - 5(2)^2 + 8(2) - 6 = -2 \\
p(-2) &= (-2)^3 - 5(-2)^2 + 8(-2) - 6 = -50 \\
p(3) &= (3)^3 - 5(3)^2 + 8(3) - 6 = 0 \\
p(-3) &= (-3)^3 - 5(-3)^2 + 8(-3) - 6 = -102 \\
p(6) &= (6)^3 - 5(6)^2 + 8(6) - 6 = 78 \\
p(-6) &= (-6)^3 - 5(-6)^2 + 8(-6) - 6 = -450
\end{aligned}
$$

Thus $x = 3$ is the only (rational) root of $p(x)$. Applying the Factor Theorem, we then know that $(x - 3)$ is a factor of $p(x)$. To obtain the resulting factorization we divide $x^3 - 5x^2 + 8x - 6$ by $x - 3$ (See Appendix D §2):

$$\begin{array}{r} x^2 - 2x + 2 \\ x-3 \overline{\smash{\big)}\ x^3 - 5x^2 + 8x - 6} \\ \underline{x^3 - 3x^2 } \\ -2x^2 + 8x - 6 \\ \underline{-2x^2 + 6x } \\ 2x - 6 \\ \underline{2x - 6} \\ 0 \end{array}$$

Thus $x^3 - 5x^2 + 8x - 6$ factors as

$$x^3 - 5x^2 + 8x - 6 = (x - 3)(x^2 - 2x + 2)$$

Note that the term $x^2 - 2x + 2$ cannot be factored further into <u>real</u> factors [by the quadratic formula, its roots are $x = \left(2 \pm \sqrt{(-2)^2 - 4(1)(2)}\right)/2 = 1 \pm \sqrt{-1}$, and $\sqrt{-1}$ is not a real number]. $\qquad\qquad\square$

In the factorization of Example A

$$x^3 - 5x^2 + 8x - 6 = (x - 3)(x^2 - 2x + 2)$$

the term

$\qquad\qquad (x - 3)$ is a <u>linear factor</u> $\qquad$ (it has degree 1)

and the term

$\qquad\qquad (x^2 - 2x + 2)$ is a <u>quadratic factor</u> $\qquad$ (it has degree 2)

Moreover, the quadratic factor

$\qquad\qquad (x^2 - 2x + 2)$ is <u>irreducible</u> $\qquad\qquad$ (it cannot be factored further into real linear factors)

2. Partial Fraction Decomposition: the reverse of putting fractions over a common denominator.

One of the basic techniques of algebra is combining several fractions into one by putting them

over a common denominator.   For example,

$$\frac{2}{x-3} - \frac{1}{x^2-2x+2} = \frac{2(x^2-2x+2)}{(x-3)(x^2-2x+2)} + \frac{(-1)(x-3)}{(x-3)(x^2-2x+2)}$$

$$= \frac{2x^2-4x+4-x+3}{(x-3)(x^2-2x+2)}$$

$$= \frac{2x^2-5x+7}{(x-3)(x^2-2x+2)}$$

How do we reverse this process, that is, how do we start with

$$\frac{2x^2-5x+7}{x^3-5x^2+8x-6}$$

and determine that it breaks up into the sum (of partial fractions)

$$\frac{2}{x-3} - \frac{1}{x^2-2x+2} \quad ?$$

The procedure for doing this is called partial fraction decomposition:

Partial fraction decomposition is the reverse of putting fractions over a common denominator and combining them.

Here is a detailed description of the method:

Partial Fraction Decomposition

Suppose $P(x)/Q(x)$ is a <u>rational function</u> (i.e., $P(x)$ and $Q(x)$ are polynomials). Break $P(x)/Q(x)$ into partial fractions as follows:

<u>Step 1.</u> If $P(x)/Q(x)$ is NOT PROPER (i.e., if the degree of the numerator $P(x)$ is NOT less than the degree of the denominator $Q(x)$), divide $P(x)$ by $Q(x)$. The result will be a polynomial plus a proper rational function. In the subsequent steps, we assume that $P(x)/Q(x)$ is proper.

<u>Step 2.</u> <u>Factor the denominator $Q(x)$</u> completely (i.e., into linear and irreducible quadratic factors)

<u>Step 3.</u> <u>Write $P(x)/Q(x)$ as a sum of terms</u> according to the following two rules:

<u>The Linear Factor Rule.</u> For each factor of $Q(x)$ of the form $(ax+b)^m$ where $m \geq 1$, include in the sum the m terms

$$\frac{A_1}{ax+b} + \frac{A_2}{(ax+b)^2} + \cdots + \frac{A_m}{(ax+b)^m}$$

where $A_1, A_2, \ldots, A_m$ are unknown coefficients.[1]

<u>The Irreducible Quadratic Factor Rule.</u> For each factor of $Q(x)$ of the form $(ax^2+bx+c)^n$ where ax^2+bx+c is irreducible and $n \geq 1$, include in the sum the n terms

$$\frac{B_1 x + C_1}{ax^2+bx+c} + \frac{B_2 x + C_2}{(ax^2+bx+c)^2} + \cdots + \frac{B_n x + C_n}{(ax^2+bx+c)^n}$$

where $B_1, C_1, B_2, C_2, \ldots, B_n, C_n$ are unknown coefficients.[1]

<u>Step 4.</u> <u>Determine the unknown coefficients</u> A_i, B_i and C_i by putting the partial fractions over a common denominator, combining them and then equating the coefficients of the numerators with those of like terms in the original numerator $P(x)$.[2]

1. Actually it is usually more convenient to use successive letters of the alphabet as the examples below show.
2. This is Anton's "alternate" method given in his Example 1.

As a first illustration, we will consider the example we were discussing above:

Example B.   Write $\dfrac{2x^2 - 5x + 7}{x^3 - 5x^2 + 8x - 6}$ as a sum of partial fractions.

Solution.

Step 1. Is the function proper? The degree of the numerator (2) is less than the

degree of the denominator (3). Hence this rational function is proper

and no division is necessary.

Step 2. Factor denominator. We factor the denominator $Q(x) = x^3 - 5x^2 + 8x - 6$

completely as

$$(x^3 - 5x^2 + 8x - 6) = (x - 3)(x^2 - 2x + 2)$$

(See Example A above for the details.)

Step 3. Write P(x)/Q(x) as a sum of terms. Since (x - 3) is a linear factor,

the linear factor rule says that we must include in our sum the term

$$\frac{A}{x - 3}$$

Since $(x^2 - 2x + 2)$ is an irreducible quadratic factor, the irreducible

quadratic factor rule says that we must include in our sum the term

$$\frac{Bx + C}{x^2 - 2x + 2}$$

Hence we write

$$\frac{2x^2 - 5x + 7}{x^3 - 5x^2 + 8x - 6} = \frac{A}{x - 3} + \frac{Bx + C}{x^2 - 2x + 2}$$

Step 4. Determine the coefficients. To determine A, B and C, first we put the partial fractions over a common denominator and combine them:

$$\frac{A}{x-3} + \frac{Bx+C}{x^2-2x+2} = \frac{A(x^2-2x+2)}{(x-3)(x^2-2x+2)} + \frac{(Bx+C)(x-3)}{(x-3)(x^2-2x+2)}$$

$$= \frac{(A+B)x^2 + (-2A-3B+C)x + (2A-3C)}{(x-3)(x^2-2x+2)}$$

Equating the coefficients of the numerator with those of like terms in the original numerator $P(x) = 2x^2 - 5x + 7$, we get the three equations:

$$A + B \qquad = \quad 2$$

$$-2A - 3B + C = -5$$

$$2A \qquad - 3C = \quad 7$$

This is a system of 3 linear equations in 3 unknowns. Solving it, * we find

$$A = \quad 2$$

$$B = \quad 0$$

$$C = -1$$

Hence the partial fraction decomposition is

$$\frac{2x^2 - 5x + 7}{x^3 - 5x^2 + 8x - 6} = \frac{2}{x-3} - \frac{1}{x^2-2x+2}$$ □

---

* See Appendix F §3 of The Companion if you need a review of solving systems of linear equations.

Here is a slightly more complicated example:

<u>Example C.</u>   Write $\dfrac{x^5 - x^4 - 3x + 5}{x^4 - 2x^3 + 2x^2 - 2x + 1}$ as a sum of partial fractions.

<u>Solution.</u>

<u>Step 1.</u>   <u>Is the function proper?</u> NO ... since the degree of the numerator (5)

is <u>not</u> less than the degree of the denominator (4). So we use long

division of polynomials to obtain

$$\frac{x^5 - x^4 - 3x + 5}{x^4 - 2x^3 + 2x^2 - 2x + 1} = x + 1 + \frac{-2x + 4}{x^4 - 2x^3 + 2x^2 - 2x + 1}$$

The rational function $(-2x + 4)/(x^4 - 2x^3 + 2x^2 - 2x + 1)$ is proper,

so we can break it into partial fractions as follows:

<u>Step 2.</u>   <u>Factor denominator.</u> We factor the denominator $Q(x) = x^4 - 2x^3 + 2x^2 -$

$2x + 1$ completely as $x^4 - 2x^3 + 2x^2 - 2x + 1 = (x - 1)^2 (x^2 + 1)$. (The

Rational Root Test quickly finds the root $x = 1$. We'll leave the remaining

details to you.)

<u>Step 3.</u>   <u>Write $P(x)/Q(x)$ as a sum of terms.</u> Since $(x - 1)^2$ is a <u>linear term</u>

<u>squared,</u> the linear factor rule says that we must include in our sum the

terms

$$\frac{A}{x - 1} \qquad \text{and} \qquad \frac{B}{(x - 1)^2}$$

Since $(x^2 + 1)$ is an <u>irreducible quadratic factor,</u> the irreducible quadratic

factor rule says that we must include in our sum the term

$$\frac{Cx + D}{x^2 + 1}$$

Hence we write

$$\frac{-2x+4}{x^4 - 2x^3 + 2x^2 - 2x + 1} = \frac{A}{x-1} + \frac{B}{(x-1)^2} + \frac{Cx+D}{x^2+1}$$

Step 4. <u>Determine the coefficients.</u> We put our partial fractions over a common denominator:

$$\frac{A}{x-1} + \frac{B}{(x-1)^2} + \frac{Cx+D}{x^2+1} = \frac{A(x-1)(x^2+1) + B(x^2+1) + (Cx+D)(x-1)^2}{(x-1)^2(x^2+1)}$$

$$= \frac{(A+C)x^3 + (-A+B-2C+D)x^2 + (A+C-2D)x + (-A+B+D)}{(x-1)^2(x^2+1)}$$

Equating coefficients of the numerator with those of like terms in the numerator $-2x+4$ of Step 2, we get the equations

$$A \quad\quad + \quad C \quad\quad\quad = \quad 0$$
$$-A + B - 2C + \quad D = \quad 0$$
$$A \quad\quad + \quad C - 2D = -2$$
$$-A + B \quad\quad\quad + \quad D = \quad 4$$

We can solve this system of four equations in four unknowns by subtracting the first equation from the third to get $D = 1$ and subtracting the second equation from the fourth to get $C = 2$. Then the first equation yields $A = -2$ and the fourth yields $B = 1$. Hence the partial fraction decomposition is

$$\frac{-2x+4}{x^4 - 2x^3 + 2x^2 - 2x + 1} = \frac{-2}{x-1} + \frac{1}{(x-1)^2} + \frac{2x+1}{x^2+1}$$

Thus the <u>original</u> rational function may be written as

$$\frac{x^5 - x^4 - 3x + 5}{x^4 - 2x^3 + 2x^2 - 2x + 1} = \boxed{x + 1 - \frac{2}{x - 1} + \frac{1}{(x - 1)^2} + \frac{2x + 1}{x^2 + 1}}$$

$$\boxed{\text{Don't forget the polynomial obtained in Step 1}}$$

□

3. <u>Using The Method of Partial Fractions in Integration.</u> When confronted with an integral $\int \frac{P(x)}{Q(x)} dx$ of a rational function, The Method of Partial Fractions can be used to integrate it as follows:

<u>Using The Method of Partial Fractions in Integration</u>

> To integrate $\int \frac{P(x)}{Q(x)} dx$, first find the partial fraction decomposition of $\frac{P(x)}{Q(x)}$ and then integrate term-by-term.

The nice thing is that once $\frac{P(x)}{Q(x)}$ has been written as a sum of partial fractions, <u>you will always be able to integrate each of the partial fractions!</u> For The Method of Partial Fractions leads to one of four types of partial fractions, <u>each of which can be integrated</u> as indicated below:

<u>Type 1.</u> $\int \frac{A}{ax + b} dx$ $(a \neq 0)$. This integral is <u>easy</u> to evaluate using the u-substitution $u = ax + b$. The result will be a constant multiple of

$$\ln |ax + b|$$

<u>Type 2.</u> $\int \frac{A}{(ax + b)^m} dx$ $(a \neq 0)$. This integral is <u>very easy</u> to evaluate using the u-substitution $u = ax + b$. The where $m \geq 2$ result will be a constant multiple of

$$\frac{1}{(ax + b)^{m-1}}$$

Type 3. $\displaystyle\int \frac{Bx + C}{ax^2 + bx + c}\, dx$ $(a \neq 0)$.   Using the techniques of §9.6, this integral will in general be a sum of constant multiples of

$$\ln\left| ax^2 + bx + c \right| \quad \text{and} \quad \tan^{-1}\left| dx + e \right|$$

for some $d \neq 0$ and e.

Type 4. $\displaystyle\int \frac{Bx + C}{(ax^2 + bx + c)^m}\, dx$ $(a \neq 0)$.   Using the techniques of §9.6, this integral will in general be a sum of constant multiples of

where $m \geq 2$

$$\frac{1}{(ax^2 + bx + c)^k} \, , \quad \frac{x}{(ax^2 + bx + c)^k} \quad (k \leq m - 1)$$

and

$$\tan^{-1}\left| dx + e \right| \quad \text{for some} \quad d \neq 0 \quad \text{and} \quad e\,.$$

The third and fourth types of integrals can be messy to compute, as you have seen in previous sections. Anton's Example 3 in §9.6 is an example of a Type 3 integral and Example 4 in §9.7 is a simple example of a Type 4 integral. (We will give a more complicated example as Example F below.)

Example D.   Evaluate $\displaystyle\int \frac{2x^2 - 5x + 7}{x^3 - 5x^2 + 8x - 6}\, dx\,.$

Solution.   First we write the rational function as a sum of partial fractions

$$\frac{2x^2 - 5x + 7}{x^3 - 5x^2 + 8x - 6} = \frac{2}{x - 3} - \frac{1}{x^2 - 2x + 2}$$

(See Example B above.) Then

$$\int \frac{2x^2 - 5x + 7}{x^3 - 5x^2 + 8x - 6}\, dx = \int \frac{2}{x - 3}\, dx - \int \frac{1}{x^2 - 2x + 2}\, dx \qquad (*)$$

The first integral is easy to evaluate (it is Type 1):

$$\int \frac{2}{x-3}\,dx \;=\; 2\int \frac{dx}{x-3} \;=\; 2\int \frac{du}{u}$$

$$\boxed{\begin{aligned} u &= x-3 \\ du &= dx \end{aligned}} \qquad = \;2\,\ln\left|u\right| + C$$

$$= \;2\,\ln\left|x-3\right| + C$$

The second integral (a Type 3 integral) can be evaluated by completing the square:

$$\int \frac{1}{x^2 - 2x + 2}\,dx \;=\; \int \frac{1}{(x-1)^2 + 1}\,dx$$

$$\boxed{\begin{aligned} u &= x-1 \\ du &= dx \end{aligned}} \longrightarrow \;=\; \int \frac{1}{u^2 + 1}\,du$$

$$= \;\tan^{-1} u + C$$

$$= \;\tan^{-1}(x-1) + C$$

Hence, equation (*) becomes

$$\int \frac{2x^2 - 5x + 7}{x^3 - 5x^2 + 8x - 6}\,dx \;=\; \boxed{2\,\ln\left|x-3\right| \;-\; \tan^{-1}(x-1) + C} \qquad \qquad \square$$

Example E.   Evaluate $\displaystyle\int \frac{x^5 - x^4 - 3x + 5}{x^4 - 2x^3 + 2x^2 + 2x + 1}\,dx$.

Solution.   First we write the rational function as a sum of partial fractions

$$\frac{x^5 - x^4 - 3x + 5}{x^4 - 2x^3 + 2x^2 + 2x + 1} \;=\; x + 1 \;-\; \frac{2}{x-1} \;+\; \frac{1}{(x-1)^2} \;+\; \frac{2x+1}{x^2 + 1}$$

(See Example C above.)   Then we integrate:

$$\int \frac{x^5 - x^4 - 3x + 5}{x^4 - 2x^3 + 2x^2 + 2x + 1} \, dx = \int x \, dx + \int dx - \int \frac{2}{x - 1} \, dx + \int \frac{1}{(x - 1)^2} \, dx + \int \frac{2x + 1}{x^2 + 1} \, dx$$

$$= \frac{x^2}{2} + x - 2 \ln |x - 1| - \frac{1}{x - 1} + \ln (x^2 + 1) + \tan^{-1} x + C$$

As you can see, in this case all our partial fraction integrals are easy! $\square$

Example F.   Evaluate $\int \dfrac{2x + 3}{(x^2 + 2x + 5)^2} \, dx$.   [OPTIONAL]

Solution.   In this case there is no partial fraction decomposition needed, for $\dfrac{2x + 3}{(x^2 + 2x + 5)^2}$

is already written in the appropriate form $\left(\text{i.e.,} \quad \dfrac{Bx + C}{(ax^2 + bx + c)^2} \right)$. This is a Type 4

integral, and we evaluate it as follows:

First we complete the square in the denominator:

$$x^2 + 2x + 5 = (x + 1)^2 + 4$$

Thus $\displaystyle\int \frac{2x + 3}{(x^2 + 2x + 5)^2} \, dx = \int \frac{2x + 3}{[(x + 1)^2 + 4]^2} \, dx$

$$= \int \frac{2u + 1}{(u^2 + 4)^2} \, du$$

$$\boxed{\begin{aligned} u &= x + 1 \\ du &= dx \\ x &= u - 1 \\ 2x + 3 &= 2u + 1 \end{aligned}}$$

$$= \boxed{2 \int \frac{u}{(u^2 + 4)^2} \, du + \int \frac{du}{(u^2 + 4)^2}} \qquad (*)$$

The first integral is easy to evaluate using $v = u^2 + 4$:

$$\int \frac{u}{(u^2+4)^2}\, du \;=\; \frac{1}{2}\int \frac{dv}{v^2} \;-\; \frac{1}{2}\left(-\frac{1}{v}\right) + C$$

$$\boxed{\begin{array}{l} v = u^2 + 4 \\ dv = 2u\, du \end{array}} \qquad =\; -\frac{1}{2}\left(\frac{1}{u^2+4}\right) + C \;=\; -\frac{1}{2}\left(\frac{1}{(x+1)^2+4}\right) + C$$

$$=\; \boxed{\,-\frac{1}{2}\left(\frac{1}{x^2+2x+5}\right) + C\,}$$

The second integral requires a trigonometric substitution:

$$\int \frac{du}{(u^2+4)^2} \;=\; \int \frac{2\sec^2\theta\, d\theta}{16\sec^4\theta} \;=\; \frac{1}{8}\int \cos^2\theta\, d\theta$$

$$\boxed{\begin{array}{l} u = 2\tan\theta \\ du = 2\sec^2\theta\, d\theta \\ u^2 + 4 = 4\sec^2\theta \end{array}} \qquad =\; \frac{1}{16}\int (1 + \cos 2\theta)\, d\theta$$

$$=\; \frac{1}{16}\theta + \frac{1}{32}\sin 2\theta + C$$

$$=\; \frac{1}{16}\theta + \frac{1}{16}\sin\theta \cos\theta + C$$

However, since $\tan\theta = u/2$ and $\theta = \tan^{-1}(\frac{u}{2})$, the Triangle Method (see §8.2.3 of

<u>The Companion</u>) yields

$$\sin\theta \;=\; \frac{u}{\sqrt{u^2+4}}\;, \qquad \cos\theta \;=\; \frac{2}{\sqrt{u^2+4}}$$

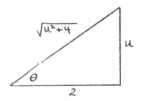

Substituting these expressions into our last formula gives

$$\int \frac{du}{(u^2+4)^2} \;=\; \frac{1}{16}\tan^{-1}\left(\frac{u}{2}\right) + \frac{1}{16}\left(\frac{u}{\sqrt{u^2+4}}\right)\left(\frac{2}{\sqrt{u^2+4}}\right) + C$$

$$=\; \frac{1}{16}\tan^{-1}\left(\frac{u}{2}\right) + \frac{1}{8}\left(\frac{u}{u^2+4}\right) + C$$

$$\boxed{u = x+1} \qquad =\; \boxed{\,\frac{1}{16}\tan^{-1}\left(\frac{x+1}{2}\right) + \frac{1}{8}\left(\frac{x+1}{x^2+2x+5}\right) + C\,}$$

Putting our results back into equation (*) and performing some easy simplifications yields the final answer:

$$\int \frac{2x+3}{(x^2+2x+5)^2} \, dx = \boxed{\frac{1}{16} \tan^{-1}\left(\frac{x+1}{2}\right) + \frac{1}{8}\left(\frac{x}{x^2+2x+5}\right) - \frac{7}{8}\left(\frac{1}{x^2+2x+5}\right) + C}$$

As you can see, Type 4 integrals can be unpleasant to evaluate! □

Section 9. 8.   Miscellaneous Substitutions (Optional)

Anton discusses two additional types of substitutions in this section. We will discuss each one briefly.

1. Rational Exponents.   Functions involving <u>rational powers of x</u>

$$x^{m_1/n_1}, \; x^{m_2/n_2}, \; \ldots, \; x^{m_k/n_k}$$

(where $m_1, n_1, m_2, n_2, \ldots, m_k, n_k$ are integers) can frequently be integrated using the substitution

$$\boxed{\begin{array}{c} u = x^{1/n} \\[4pt] \text{where } n \text{ is the least common multiple of } n_1, n_2, \ldots, n_k \\[4pt] \text{(i.e., } n \text{ is the smallest integer which can be divided by} \\[4pt] \text{each of the integers } n_1, n_2, \ldots, n_k \text{)}. \end{array}}$$

Taking the n-th power of both sides, we see that this is really the substitution

$$
\boxed{\begin{array}{l} x \;=\; u^n \\[2mm] dx \;=\; n\,u^{n-1}\,du \end{array}}
$$

Anton's Examples 1 and 2   illustrate the use of this substitution. As you can see in
those examples, the substitution $x = u^n$ will often convert the original integral into an integral
of a <u>rational function</u> of u   (and such integrals can be evaluated by the method of partial
fractions studied in §9. 7). Here is another example:

<u>Example A.</u>   Evaluate $\displaystyle\int \frac{dx}{\sqrt{x}\,(1 + \sqrt[3]{x})}$.

<u>Solution.</u>   This function involves the rational powers of x

$$
x^{1/2} \qquad \text{and} \qquad x^{1/3}
$$

so we use the substitution $u = x^{1/6}$ (since 6 is the least common multiple of 2 and 3).
Solving for x , we find $x = u^6$. Thus

$$
\int \frac{dx}{\sqrt{x}\,(1 + \sqrt[3]{x})} \;=\; 6 \int \frac{u^5}{u^3\,(1 + u^2)}\,du
$$

$$
\boxed{\begin{array}{l} x = u^6 \\[2mm] dx = 6\,u^5\,du \end{array}}
$$

$$
=\; 6 \int \frac{u^2}{1 + u^2}\,du
$$

$$
=\; 6 \int \left(1 - \frac{1}{1 + u^2}\right) du \qquad \text{(division of polynomials)}
$$

$$
=\; 6u - 6\tan^{-1}u + C
$$

$$
=\; \boxed{6x^{1/6} - 6\tan^{-1}(x^{1/6}) + C} \qquad \square
$$

This method can often be modified to handle the situation in which $\sqrt{f(x)}$ appears in an integral, where $f(x)$ is some function of x. If we let

$$u = \sqrt{f(x)}$$

then we can frequently evaluate the integral <u>if we can also find the inverse function</u>

$$x = f^{-1}(u^2)$$

Anton's Example 3 illustrates this substitution with $u = \sqrt{1 + e^x}$ and $x = \ln(u^2 - 1)$. Here is one more example... one which should make your hair stand on end!

<u>Example B.</u> Evaluate $\sqrt{\tan x}\ dx$.

<u>Solution.</u> This looks simple enough (but it's NOT !). We'll try

$$u = \sqrt{\tan x}$$

which has the inverse function

$$x = \tan^{-1}(u^2)$$

so $\quad dx = \dfrac{1}{1 + (u^2)^2}\ (2u\ du) = \dfrac{2u}{1 + u^4}\ du$

Using this substitution we obtain

$$\int \sqrt{\tan x}\ dx = \int u\left(\frac{2u}{1 + u^4}\ du\right) = \int \frac{2u^2}{1 + u^4}\ du$$

which is the integral of a <u>rational function.</u> Thus we can evaluate the integral by the method of partial fractions. (... although it will be messy!)

__Step 1.__ __Is the function proper?__ Yes, ... since degree $(2u^2) = 2$ is less than

degree $(1+u^4) = 4$. Thus we have no division to perform.

__Step 2.__ __Factor the denominator__ $Q(u) = 1 + u^4$. Hmm. This one is not so obvious! However,

notice that $Q(u)$ is always ≥ 1 . Thus it is __never__ zero and hence has no real roots and thus

no linear factors. Therefore it must be a product of two irreducible quadratic terms:

$$1 + u^4 = (a + bu + u^2)(c + du + u^2)$$

Multiplying out the expression on the right and equating like terms on the two sides of the

equation yields

$$a = c = 1$$
$$b = -d = \sqrt{2} \quad .$$

Thus

$$1 + u^4 = (1 + \sqrt{2}\,u + u^2)(1 - \sqrt{2}\,u + u^2)$$

__Step 3.__ Write $2u^2/(1+u^4)$ as a sum of terms.

$$\frac{2u^2}{1+u^4} = \frac{Au + B}{u^2 + \sqrt{2}\,u + 1} + \frac{Cu + D}{u^2 \quad \sqrt{2}\,u + 1}$$

__Step 4.__ __Determine the coefficients.__   Putting the partial fractions over the common denominator

$1 + u^4$, equating the coefficients of the numerator with those of like terms in $1 + u^4$, and then

solving the resulting system of linear equations yields $A = \dfrac{\sqrt{2}}{2}$, $B = 0$, $C = -\dfrac{\sqrt{2}}{2}$ and

$D = 0$. Thus

$$\frac{2u^2}{1+u^4} = \frac{\sqrt{2}}{2}\left(\frac{u}{u^2 - \sqrt{2}\,u + 1}\right) - \frac{\sqrt{2}}{2}\left(\frac{u}{u^2 + \sqrt{2}\,u + 1}\right)$$

Finally we can integrate:

$$\int \sqrt{\tan x}\ dx = \int \frac{2u^2}{1+u^4}\ du$$

$$\boxed{u = \sqrt{\tan x}}$$

$$= \frac{\sqrt{2}}{2} \int \frac{u}{u^2 - \sqrt{2}\,u + 1}\ du - \frac{\sqrt{2}}{2} \int \frac{u}{u^2 + \sqrt{2}\,u + 1}\ du$$

$$= \frac{\sqrt{2}}{2}\left[\frac{1}{2} \ln (u^2 - \sqrt{2}\,u + 1) + \tan^{-1}(\sqrt{2}\,u - 1) \right]$$

$$\boxed{\begin{array}{l}\text{By the techniques}\\ \text{of } \S 9.6\end{array}}$$

$$- \frac{\sqrt{2}}{2}\left[\frac{1}{2} \ln (u^2 + \sqrt{2}\,u + 1) - \tan^{-1}(\sqrt{2}\,u + 1) \right] + C$$

Hence

$$\boxed{\int \sqrt{\tan x}\ dx = \frac{\sqrt{2}}{4} \ln\left(\frac{u^2 - \sqrt{2}\,u + 1}{u^2 + \sqrt{2}\,u + 1} \right) + \frac{\sqrt{2}}{2}\left[\tan^{-1}(\sqrt{2}\,u - 1) + \tan^{-1}(\sqrt{2}\,u + 1) \right] + C}$$

Technically we should resubstitute $u = \sqrt{\tan x}$ into our answer ... but the result would be so ghastly that we won't attempt it! $\square$

2. <u>Rational expressions in sin x and cos x</u>. Rational functions of sin x and cos x can be turned into (ordinary) rational functions of x by using the substitution

$$\boxed{u = \tan\left(\frac{x}{2}\right)}$$

Then the following steps can be taken:

1. Using <u>The Triangle Method</u> of $\S 8.2.3$, we have

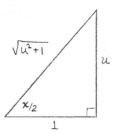

$$\sin \left(\frac{x}{2}\right) = \frac{u}{\sqrt{u^2 + 1}}$$

$$\cos \left(\frac{x}{2}\right) = \frac{1}{\sqrt{u^2 + 1}}$$

2. Using the <u>double angle formulas</u> (see Equations (17 a b) in Anton's Appendix 1.1), we find

$$\sin x = 2 \sin \left(\frac{x}{2}\right) \cos \left(\frac{x}{2}\right) = (2) \frac{u}{\sqrt{u^2 + 1}} \cdot \frac{1}{\sqrt{u^2 + 1}} = \frac{2u}{u^2 + 1}$$

and

$$\cos x = \cos^2 \left(\frac{x}{2}\right) - \sin^2 \left(\frac{x}{2}\right) = \left(\frac{1}{\sqrt{u^2 + 1}}\right)^2 - \left(\frac{u}{\sqrt{u^2 + 1}}\right)^2 = \frac{1 - u^2}{1 + u^2}$$

i.e.,

$$\boxed{\sin x = \frac{2u}{u^2 + 1} \quad \text{and} \quad \cos x = \frac{1 - u^2}{1 + u^2}}$$

3. Using the <u>inverse formula</u>

$$x = 2 \tan^{-1} u$$

we obtain

$$\boxed{dx = \frac{2}{1 + u^2} du}$$

We recommend that you not memorize any of these equations except for the original substitution

$u = \tan \left(\frac{x}{2}\right)$; instead, use the procedures above to derive the expressions for $\sin x$, $\cos x$

and dx in terms of u .

Anton's Example 4 illustrates the use of the substitution $u = \tan\left(\frac{x}{2}\right)$. Here is an additional example:

Example C. Evaluate $\displaystyle\int \frac{dx}{1 + \tan x}$.

Solution.

$$\int \frac{dx}{1 + \tan x} = \int \left(\frac{1}{1 + \tan x}\right)\left(\frac{\cos x}{\cos x}\right) dx$$

$$= \int \frac{\cos x}{\cos x + \sin x}\, dx \qquad\qquad (A)$$

Using the $u = \tan(x/2)$ equations

$$= \int \frac{\left(\dfrac{1 - u^2}{1 + u^2}\right)\left(\dfrac{2}{1 + u^2}\, du\right)}{\dfrac{1 - u^2}{1 + u^2} + \dfrac{2u}{1 + u^2}}$$

$$= \int \frac{2u^2 - 2}{(u^2 - 2u - 1)(u^2 + 1)}\, du$$

Thus we have an integral of a rational function in u to which we apply the method of partial fractions. Leaving the gory details to your verification, here are the major steps:

$$(u^2 - 2u - 1)(u^2 + 1) = (u - 1 + \sqrt{2})(u - 1 - \sqrt{2})(u^2 + 1)$$

Thus

$$\frac{2u^2 - 2}{(u^2 - 2u - 1)(u^2 + 1)} = \frac{A}{u - 1 + \sqrt{2}} + \frac{B}{u - 1 - \sqrt{2}} + \frac{Cu + D}{u^2 + 1}$$

Solving for A, B, C and D (not a pleasant undertaking!) yields

$$A = B = 1/2 , \quad C = -1 \quad \text{and} \quad D = 1$$

Thus

$$\int \frac{2u^2 - 2}{(u^2 - 2u - 1)(u^2 + 1)} \, du \;=\; \frac{1}{2} \int \frac{du}{u - 1 + \sqrt{2}} \;+\; \frac{1}{2} \int \frac{du}{u - 1 - \sqrt{2}} \;-\; \int \frac{u - 1}{u^2 + 1} \, du$$

$$= \frac{1}{2} \ln \left| u - 1 + \sqrt{2} \right| + \frac{1}{2} \ln \left| u - 1 - \sqrt{2} \right|$$

$$- \left(\frac{1}{2} \ln (u^2 + 1) - \tan^{-1} u \right) + C$$

$$= \frac{1}{2} \ln \left| \frac{u^2 - 2u - 1}{u^2 + 1} \right| + \tan^{-1} u + C$$

$$= \frac{1}{2} \ln \left| \frac{2u}{u^2 + 1} + \frac{1 - u^2}{u^2 + 1} \right| + \tan^{-1} u + C$$

$$\boxed{\int \frac{dx}{1 + \tan x} \;=\; \frac{1}{2} \ln \left| \sin x + \cos x \right| + \frac{x}{2} + C}$$

Whew! That's a lot of work to derive such a simple answer.   Could there be an easier way?

Ahh... YES! And here it is: Beginning with integral (A), we have

$$\int \frac{\cos x}{\cos x + \sin x} \, dx \;=\; \int \frac{\cos x - \sin x + \sin x}{\cos x + \sin x} \, dx$$

$$= \int \frac{\cos x - \sin x}{\cos x + \sin x} \, dx + \int \frac{\sin x}{\cos x + \sin x} \, dx$$

$$\boxed{\begin{array}{l} v = \cos x + \sin x \\ dv = (- \sin x + \cos x)\, dx \end{array}} \quad = \int \frac{dv}{v} + \int \frac{\sin x}{\cos x + \sin x} \, dx$$

$$= \ln \left| \sin x + \cos x \right| + \int \frac{\sin x}{\cos x + \sin x} \, dx \qquad\qquad (*)$$

However, evaluating in another way produces

9.8.9

$$\int \frac{\cos x}{\cos x + \sin x}\, dx = \int \frac{\cos x + \sin x - \sin x}{\cos x + \sin x}\, dx$$

$$= \int 1\, dx - \int \frac{\sin x}{\cos x + \sin x}\, dx$$

$$= x - \int \frac{\sin x}{\cos x + \sin x}\, dx \qquad\qquad (**)$$

Hence, solving both (*) and (**) for $\int \dfrac{\sin x}{\cos x + \sin x}\, dx$ and equating the answers, we obtain

$$\int \frac{\cos x}{\cos x + \sin x}\, dx - \ln\left|\sin x + \cos x\right| = x - \int \frac{\cos x}{\cos x + \sin x}\, dx$$

Solving this equation for $\int \dfrac{\cos x}{\cos x + \sin x}\, dx$ yields

$$\int \frac{dx}{1 + \tan x} = \int \frac{\cos x}{\cos x + \sin x}\, dx = \boxed{\ \frac{1}{2} \ln\left|\sin x + \cos x\right| + \frac{x}{2} + C\ }$$

as derived earlier!!   □

This example illustrates how exasperating integration can be, and how elusive the "best" evaluation technique can be. (How would you "dream up" our second solution to Example C without first having seen the answer?) As we said at the beginning of this chapter, integration is more of an <u>art</u> than a <u>science</u>!

Incidentally, there is also a "slicker" solution for Anton's Example 4. Here it is:

<u>Anton's Example 4.</u>   Evaluate $\int \dfrac{dx}{1 + \sin x}$.

Alternate solution.

$$\int \frac{dx}{1 + \sin x} = \int \left(\frac{1}{1 + \sin x}\right) \left(\frac{1 - \sin x}{1 - \sin x}\right) dx$$

$$= \int \frac{1 - \sin x}{1 - \sin^2 x} \, dx$$

$$= \int \frac{1 - \sin x}{\cos^2 x} \, dx$$

$$= \int \sec^2 x \, dx - \int \sec x \tan x \, dx$$

$$= \boxed{\tan x - \sec x + C} \qquad \square$$

This alternate solution illustrates two points

- many times there are better ways to evaluate integrals involving rational functions

  of sines and cosines than by using the substitution $u = \tan\left(\frac{x}{2}\right)$. (The second

  solution to Example C also illustrates this point.)

- as we pointed out in §9.3.1 of The Companion, when evaluating an integral involving

  trigonometric functions, the form of the answer may depend on the method used. As

  you can see, Anton's answer to Example 4 looks very different from the one above!

Section 9.9.   Numerical Integration; Simpson's Rule

1.   Introduction.   We have been waiting for this section! Several times in Chapters 5 - 9 we have

   mentioned the existence of functions whose integrals cannot be evaluated "in closed form," i.e.,

   expressed in terms of our collection of elementary functions.

When this problem arises with a <u>definite integral</u> $\int_a^b f(x)\, dx$, then our approach must be to

$$\boxed{\text{approximate } \int_a^b f(x)\, dx!!!}$$

In this section four methods for approximating definite integrals are discussed; even so, we only scratch the surface of the subject. Integral approximation is studied in much greater depth in the branch of mathematics known as <u>numerical analysis.</u>

$$* \quad * \quad * \quad * \quad *$$

Everything you <u>need to memorize</u> is contained in Anton's four boxed formulas for

> (1) Left-hand Endpoint Approximation
>
> (2) Right-hand Endpoint Approximation
>
> (3) Trapezoidal Approximation
>
> (4) Simpson's Rule

However, these formulas can be messy and unpleasant to use unless you do your algebra and arithmetic carefully, and record your calculations in a legible and logical form. <u>Anton handles this "bookkeeping" by using tables.</u> We recommend that you too set up tables when applying the various approximation techniques of this chapter. However, keep in mind that there is nothing sacred about the form of Anton's tables. If you find another form of bookkeeping more to your liking, then by all means use it !

2. The Trapezoidal Approximation. Although Anton begins this section by discussing the Left-
and Right-hand Endpoint Approximations, the first really important approximation method is
The Trapezoidal Approximation (with n subintervals):

The Trapezoidal Approxi-mation

$$\int_a^b f(x)\,dx \approx \frac{b-a}{2n}\,[y_0 + 2y_1 + \cdots + 2y_{n-1} + y_n]$$

where $\Delta x = \dfrac{b-a}{n}$ (the length of the subintervals)

$x_i = a + i\Delta x$ for $i = 0, 1, \ldots, n$ (the subinterval endpoints)

$y_i = f(x_i)$ for $i = 0, 1, \ldots, n$ (the function values at the subinterval endpoints)

Since the Trapezoidal Approximation is more accurate (and only slightly more complicated) than
the Left-hand and Right-hand Endpoint Approximations, there is little reason to use the latter
two methods.

The Trapezoidal Approximation can be derived easily using the formula for the area of
a trapezoid (see the inside front cover of Anton):

$$A = \frac{1}{2}(\ell_1 + \ell_2)w$$

Then to approximate $\displaystyle\int_a^b f(x)\,dx$ [the area under $y = f(x)$ over $[a, b]$ when $f \geq 0$],

we proceed as follows:

(1)   Divide [a, b] into n equal pieces:

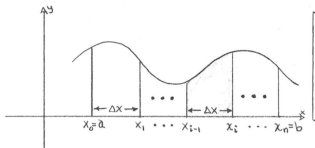

$$\Delta x = \frac{b - a}{n} \quad \text{(length of each subinterval)}$$

$$x_i = a + i\Delta x, \quad i = 0, 1, \ldots, n$$

(the subinterval endpoints)

(2)   Approximate the area over each subinterval $[x_{i-1}, x_i]$ by the area of a trapezoid:

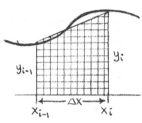

$$\left[\begin{array}{c} \text{area under} \\ y = f(x) \quad \text{over} \\ [x_{i-1}, x_i] \end{array}\right] \approx \frac{1}{2}(y_{i-1} + y_i)\Delta x$$

(3)   Sum the areas of the trapezoids to obtain an approximation of the whole area:

area under
y = f(x) over [a,b]

area of first trapezoid   second trapezoid   · · ·   n-th trapezoid

$$\int_a^b f(x)\,dx \approx \frac{1}{2}(y_0 + y_1)\Delta x \quad + \quad \frac{1}{2}(y_1 + y_2)\Delta x \quad + \cdots + \quad \frac{1}{2}(y_{n-1} + y_n)\Delta x$$

$$= \frac{\Delta x}{2}\left[(y_0 + y_1) + (y_1 + y_2) + \cdots + (y_{n-1} + y_n)\right]$$

$$= \left(\frac{b - a}{2n}\right)\left[y_0 + 2y_1 + \cdots + 2y_{n-1} + y_n\right]$$

Note that all the terms of the sum except the first and last contain the factor 2 .

*   *   *   *   *

As we have pointed out, an essential ingredient for success in using The Trapezoidal Approximation is <u>careful bookkeeping</u> and tables are useful for that purpose. The following numbered diagram illustrates how Anton's Trapezoidal Approximation Table is constructed: <u>Do the operations in the numbered order</u>:

---

(1)   Calculate $\Delta x = \dfrac{b - a}{n}$ (length of subintervals)

(2) the subinterval number i	(3) calculate x_i from $x_i = a + i\,\Delta x$	(4) calculate y_i from $y_i = f(x_i)$	(5) multiplier w_i	(6) multiply $w_i\,y_i$
0	x_0	y_0	1	y_0
1	x_1	y_1	2	$2y_1$
2	x_2	y_2	2	$2y_2$
$\vdots$	$\vdots$	$\vdots$	$\vdots$	
			2	
n	x_n	y_n	1	y_n

(7)   add $y_0 + 2y_1 + \cdots + y_n$

(8)   multiply by $\dfrac{b - a}{2n}$: $\displaystyle\int_a^b f(x)\,dx \approx \dfrac{b - a}{2n}\left[\quad \ldots \quad \right]$

FINAL ANSWER

Anton constructs a Trapezoidal Approximation Table to approximate $\displaystyle\int_1^2 \dfrac{1}{x}\,dx$ when n = 10. Of course, in that case it is possible to evaluate $\displaystyle\int_1^2 \dfrac{1}{x}\,dx$ exactly

$\left[\displaystyle\int_1^2 \dfrac{1}{x}\,dx = \ln 2 - \ln 1 = \ln 2 \approx .6931 \right]$ so the approximation was not really necessary.

(That example was chosen for illustrative reasons because the approximation can be compared with the actual answer.) Here is an example in which the integral <u>cannot</u> be evaluated exactly, and hence we <u>must</u> approximate the answer:

<u>Example A.</u>   Use The Trapezoidal Approximation with   n = 4   to approximate $\int_0^1 \sqrt{1 + x^3}\, dx$.

<u>Solution.</u>   We construct a table in the numbered order using   a = 0 , b = 1 , $f(x) = \sqrt{1 - x^3}$ and   n = 4 :

$$(1) \quad \Delta x = \frac{b - a}{n} = \frac{1 - 0}{4} = \frac{1}{4}$$

(2) i	(3) $x_i = a + i\Delta x = \frac{i}{4}$	(4) $y_i = \sqrt{1 + (\frac{i}{4})^3}$	(5) multiplier	(6) product
0	0	$\sqrt{1 + 0^3} = 1$	1	$1 = 1.0000$
1	$\frac{1}{4}$	$\sqrt{1 + (\frac{1}{4})^3} = \sqrt{65}/8$	2	$\sqrt{65}/4 \approx 2.0156$
2	$\frac{2}{4}$	$\sqrt{1 + (\frac{2}{4})^3} = \sqrt{72}/8$	2	$\sqrt{72}/4 \approx 2.1213$
3	$\frac{3}{4}$	$\sqrt{1 + (\frac{3}{4})^3} = \sqrt{91}/8$	2	$\sqrt{91}/4 \approx 2.3848$
4	$\frac{4}{4}$	$\sqrt{1 + (\frac{4}{4})^3} = \sqrt{2}$	1	$\sqrt{2} \approx 1.4142$

$$(7) \quad \text{sum} = 8.9359$$

(8) Multiply by $\frac{b - a}{2n} = \frac{1}{8}$: $\int_0^1 \sqrt{1 + x^3}\, dx \approx \frac{1}{8}\, [\, 8.9359\,]$

$$\approx \boxed{1.1170}$$

The accuracy of this approximation will be examined later in Example C of Subsection 4.

3. <u>Simpson's Rule.</u>   The second major approximation formula discussed in this section is Simpson's Rule (with n subintervals where n is an <u>even number</u>):

| Simpson's Rule |

$$\int_a^b f(x)\,dx \approx \frac{b-a}{3n}\left[y_0 + 4y_1 + 2y_2 + 4y_3 + 2y_4 + \cdots + 2y_{n-2} + 4y_{n-1} + y_n\right]$$

where $\Delta x = \dfrac{b-a}{n}$ (the length of the subintervals)

$x_i = a + i\Delta x$ for $i = 0, 1, \ldots, n$ (the subinterval endpoints)

$y_i = f(x_i)$ for $i = 0, 1, \ldots, n$ (the function values at the subinterval endpoints)

The comparison between Simpson's Rule and The Trapezoidal Approximation may be summarized as follows:

In general, Simpson's Rule gives a more accurate approximation than the Trapezoidal Rule, although it is slightly more complicated to use.

Simpson's Rule can be derived using the formula for the area under the parabola $y = ax^2 + bx + c$ over an arbitrary interval of width $2h$:

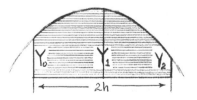

$$A = \frac{h}{3}\left[Y_0 + 4Y_1 + Y_2\right]$$

(Anton gives a careful derivation of this formula which is his Equation (6).) Then to approximate $\int_a^b f(x)\,dx$ [the area under $y = f(x)$ over $[a, b]$ when $f \geq 0$] we proceed as follows:

   (1)   Divide $[a, b]$ into n equal pieces $(n$ <u>even</u>) :

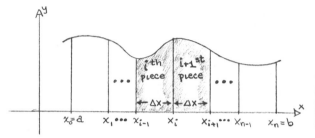

$$\Delta x = \frac{b - a}{n} \quad \text{(length of each subinterval)}$$

$$x_i = a + i\Delta x, \quad i = 0, 1, \ldots, n$$

(the subinterval endpoints)

   (2)   Approximate the area over each <u>pair</u> of pieces by the area under a parabola:

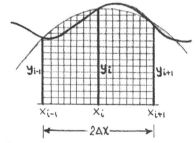

$$\begin{bmatrix} \text{area under} \\ y = f(x) \quad \text{over} \\ [x_{i-1}, x_{i+1}] \end{bmatrix} \approx \frac{\Delta x}{3} [y_{i-1} + 4y_i + y_{i+1}]$$

   (3)   Sum the areas under the parabolas to obtain an approximation of the whole area:

$$\overbrace{\int_a^b f(x)\,dx}^{\substack{\text{area under} \\ y = f(x) \text{ over } [a,b]}} = \overbrace{\frac{\Delta x}{3}\left[y_0 + 4y_1 + y_2\right]}^{\text{area under 1-st parabola}} + \overbrace{\frac{\Delta x}{3}\left[y_2 + 4y_3 + y_4\right]}^{\text{... 2-nd parabola}} + \cdots + \overbrace{\frac{\Delta x}{3}\left[y_{n-2} + 4y_{n-1} + y_n\right]}^{\text{... (n/2)-th parabola}}$$

$$= \frac{\Delta x}{3}\left[(y_0 + 4y_1 + y_2) + (y_2 + 4y_3 + y_4) + \cdots + (y_{n-2} + 4y_{n-1} + y_n)\right]$$

$$= \boxed{\left(\frac{b - a}{3n}\right)\left[y_0 + 4y_1 + 2y_2 + 4y_3 + 2y_4 + \cdots + 2y_{n-2} + 4y_{n-1} + y_n\right]}$$

> Note that all the terms of the sum except the first and
>
> last have factors which alternate between 4 and 2 .

---

*   *   *   *   *

When we compare Anton's table for Simpson's Rule with his table for the Trapezoidal
Approximation, we see that the <u>only differences are</u>

      i)   the multipliers differ, and

     ii)   the sum in the final step is multiplied by

          $(b - a)/3n$ instead of $(b - a)/2n$.

Here is an example using Anton's table:

<u>Example B.</u>   Use Simpson's Rule with $n = 4$ to approximate $\displaystyle\int_0^1 \sqrt{1 + x^3}\ dx$.

<u>Solution.</u>   We construct a table in the numbered order using $a = 0$, $b = 1$, $f(x) = \sqrt{1 + x^3}$,
and $n = 4$:

(1) $\Delta x = \dfrac{b - a}{n} = \dfrac{1 - 0}{4} = \dfrac{1}{4}$

(2) i	(3) $x_i = a + i\Delta x = \dfrac{i}{4}$	(4) $y_i = \sqrt{1 + (\dfrac{i}{4})^3}$	(5) multiplier	(6) product
0	0	$\sqrt{1 + 0^3} = 1$	1	$1 = 1.0000$
1	$\dfrac{1}{4}$	$\sqrt{1 + (\dfrac{1}{4})^3} = \sqrt{65}/8$	4	$\sqrt{65}/2 \approx 4.0311$
2	$\dfrac{2}{4}$	$\sqrt{1 + (\dfrac{2}{4})^3} = \sqrt{72}/8$	2	$\sqrt{72}/4 \approx 2.1213$
3	$\dfrac{3}{4}$	$\sqrt{1 + (\dfrac{3}{4})^3} = \sqrt{91}/8$	4	$\sqrt{91}/2 \approx 4.7697$
4	$\dfrac{4}{4}$	$\sqrt{1 + (\dfrac{4}{4})^3} = \sqrt{2}$	1	$\sqrt{2} \approx 1.4142$

(7) sum $= 13.3363$

(8) Multiply by $\dfrac{b - a}{3n} = \dfrac{1}{12}$: $\displaystyle\int_0^1 \sqrt{1 + x^3}\ dx \approx \dfrac{1}{12}\ [\ 13.3363\ ]$

$$\approx \boxed{1.1114}$$

The accuracy of this approximation will be examined in Example C of Subsection 4. □

4. **Error analysis (optional).** Approximations are of very little use unless you know how accurate they are. Fortunately there are good ways to estimate the error for approximations by the Trapezoidal Approximation and Simpson's Rule using the second and fourth derivatives of f , respectively:

<table>
<tr><td>

Trapezoidal
Error
Bound

</td></tr>
</table>

The absolute value of the error made when $\displaystyle\int_a^b f(x)\,dx$

is approximated by the <u>Trapezoidal Approximation</u> with

n   subdivisions is at most

$$\frac{K(b-a)^3}{12n^2}$$

where   K   is any number such that $\left| f''(x) \right| \le K$

for all   x   in $[a,b]$.

Simpson's
Rule
Error
Bound

The absolute value of the error made when $\displaystyle\int_a^b f(x)\,dx$

is approximated by <u>Simpson's Rule</u> with   n   subdivisions

is at most

$$\frac{M(b-a)^5}{180n^4}$$

where   M   is any number such that $\left| f^{(4)}(x) \right| \le M$

for all   x   in $[a,b]$.

These formulas are contained in Anton's   Exercise 10, but they are not proved until a course

in advanced calculus.   Here is an example showing how they are used:

<u>Example C.</u>   Estimate the absolute value of the maximum error that can occur when approximating

$$\int_0^1 \sqrt{1+x^3}\ dx\quad \text{by}$$

a) the Trapezoidal Approximation with $n = 4$, as was done
in Example A , and

b) Simpson's Rule with $n = 4$, as was done in Example B.

Solution. a) We need only evaluate the expression

$$\frac{K(b - a)^3}{12 n^2} = \frac{K(1 - 0)^3}{12 \cdot 4^2} = \frac{K}{192}$$

Since K is defined to be any number which is greater than $\left| f''(x) \right|$ for x in the interval $[a, b] = [0, 1]$, we must compute $f''(x)$. This computation yields:

$$f''(x) = \frac{3}{4} x \left(1 + x^3 \right)^{-3/2} (x^3 + 4)$$

By choosing the maximum value of each term over the interval $[0, 1]$, we find

$$\left| f''(x) \right| \le \left(\frac{3}{4} \right) (1) (1)^{-3/2} (1 + 4) = \frac{15}{4}$$

Hence we can take $K = 15/4$, so that our maximum error can be at most

$$\frac{15/4}{192} \approx .02$$

b) For Simpson's Rule we must evaluate the expression

$$\frac{M(b - a)^5}{180 n^4} = \frac{M(1 - 0)^5}{180 \cdot 4^4} = \frac{M}{46080}$$

Since M is defined to be any number which is greater than $\left| f^{(4)}(x) \right|$ for all x in the interval $[a, b] = [0, 1]$, we must compute $f^{(4)}(x)$. This (very tedious) computation yields

$$f^{(4)}(x) = \frac{9}{16}\left(1 + x^3\right)^{-7/2}\left(x^8 + 56x^5 - 80x^2\right)$$

Thus on the interval $[0, 1]$ we have

$$\left| f^{(4)}(x) \right| \leq \frac{9}{16}\left(1 + 0\right)^{-7/2}\left(1 + 56\right) = \frac{513}{16}$$

Hence we can take $M = 513/16$,   so that our maximum error will be at most

$$\frac{513/16}{46080} \approx .0007 \qquad\qquad \square$$

Summarizing Examples A , B and C , we find

$$\int_0^1 \sqrt{1 + x^3}\ dx \approx 1.1170 \qquad\qquad \text{with a maximum error of } .02$$

with a maximum error of $.02$
using the Trapezoidal Approximation with $n = 4$,

$$\int_0^1 \sqrt{1 + x^3}\ dx \approx 1.1114$$

with a maximum error of $.0007$
using Simpson's Rule with $n = 4$.

As you can see, we have a much better approximation with Simpson's Rule than with the

Trapezoidal Approximation.

$$*\qquad *\qquad *\qquad *\qquad *$$

In Example C we used the error bounds in a pretty elementary way: given Trapezoidal

and Simpson's Rule approximations using a fixed number of subintervals $(n = 4)$,   we determined

how accurate these approximations are.   However, it is far more important to be able to

answer the following question:

> given a desired accuracy, ϵ, for an
>
> approximation of $\displaystyle\int_a^b f(x)\, dx$, determine
>
> the necessary number of subintervals, n,
>
> which must be used to achieve this accuracy.

This is the main "real world" question which is always asked of any type of approximation

procedure. Fortunately the error bounds given above allow a complete answer to this question

for the Trapezoidal Approximation and Simpson's Rule, as we now illustrate:

Example D. How many subintervals, n, should be taken to ensure that the absolute value of the

error in the approximation of $\displaystyle\int_1^2 \frac{1}{x^2}\, dx$ is at most $\epsilon = .001$ when using

      a)   the Trapezoidal Approximation?

      b)   Simpson's Rule?

Choose the better method and make the approximation.

Solution. We find

$$f(x) = x^{-2}$$

$$f'(x) = -2x^{-3}$$

$$f''(x) = 6x^{-4}$$

$$f^{(3)}(x) = -24x^{-5}$$

$$f^{(4)}(x) = 120x^{-6}$$

$$f^{(5)}(x) = -720x^{-7}$$

Note these two derivatives

Both $f''(x)$ and $f^{(4)}(x)$ are decreasing functions on $[1,2]$ because <u>their</u> derivatives $[f^{(3)}(x) = -24x^{-5}$ and $f^{(5)}(x) = -720x^{-7}]$ are negative on this interval; hence each achieves its maximum on $[1,2]$ at the left endpoint $x = 1$, i.e.,

$$\left| f''(x) \right| \leq \left| f''(1) \right| = 6 \quad \text{for all} \quad 1 \leq x \leq 2$$
$$\left| f^{(4)}(x) \right| \leq \left| f^{(4)}(1) \right| = 120 \quad \text{for all} \quad 1 \leq x \leq 2$$

Thus we will choose $K = 6$ in the Trapezoidal Approximation Error Bound and $M = 120$ in the Simpson's Rule Error Bound.

(a) Using the Trapezoidal Approximation with n subintervals, the absolute value of the error is at most

$$\frac{K(b-a)^3}{12n^2} = \frac{6(2-1)^3}{12n^2} = \frac{1}{2n^2}$$

Thus we want to determine the smallest value of n for which

$$\left| \text{maximum error} \right| \leq .001$$

i.e.,

$$\frac{1}{2n^2} \leq .001$$

We can solve this inequality for n: *

$$\frac{1}{.002} = 500 \leq n^2$$

$$22.36 \leq n \quad \text{(use a hand calculator for } \sqrt{500}\text{)}$$

---

* Often an inequality will be obtained which cannot be "solved" directly for n. In such cases a trial and error procedure is necessary: compute the expression for various values of n until a suitable value is found.

Hence if we take $n = 23$, then we are guaranteed that the absolute value of the error in using the Trapezoidal Approximation for $\displaystyle\int_1^2 \frac{1}{x^2}\,dx$ will be less than or equal to $.001$.

(b) Using Simpson's Rule, the absolute value of the error is at most

$$\frac{M(b-a)^5}{180\,n^4} = \frac{120(2-1)^5}{180\,n^4} = \frac{2}{3\,n^4}$$

Thus we want to determine the smallest value of n for which

$$\frac{2}{3\,n^4} \leq .001$$

Solving this inequality for n we obtain

$$\frac{2}{.003} = 666.67 \leq n^4$$

$$5.08 \leq n \qquad \left[\text{use a hand calculator for}\right.$$

$$\left. \sqrt[4]{666.67} = \sqrt{\sqrt{666.67}}\ \right]$$

Hence if we take $n = 6$, the absolute value of the error in using Simpson's Rule to approximate $\displaystyle\int_1^2 \frac{1}{x^2}\,dx$ will be less than or equal to $.001$.

It is clear that we will save considerable effort in approximating $\displaystyle\int_1^2 \frac{1}{x^2}\,dx$ to the desired degree of accuracy by using Simpson's Rule with $n = 6$ instead of the Trapezoidal Approximation with $n = 23$. We make that approximation in the following table:

(1) $\Delta x = \dfrac{b - a}{n} = \dfrac{2 - 1}{6} = \dfrac{1}{6}$

(2) i	(3) $x_i = a + i\Delta x = 1 + \dfrac{i}{6}$	(4) $y_i = 1/x_i^2$	(5) multiplier	(6) product
0	1	$(1)^2 = 1.0000$	1	1.0000
1	7/6	$\left(\dfrac{6}{7}\right)^2 = .7347$	4	2.9388
2	8/6	$\left(\dfrac{6}{8}\right)^2 = .5625$	2	1.1250
3	9/6	$\left(\dfrac{6}{9}\right)^2 = .4444$	4	1.7776
4	10/6	$\left(\dfrac{6}{10}\right)^2 = .3600$	2	.7200
5	11/6	$\left(\dfrac{6}{11}\right)^2 = .2975$	4	1.1900
6	2	$\left(\dfrac{1}{2}\right)^2 = .2500$	1	.2500

(7)   sum = 9.0014

(8) Multiply by $\dfrac{b - a}{3n} = \dfrac{1}{18}$: $\displaystyle\int_1^2 \dfrac{1}{x^2}\, dx \approx \dfrac{1}{18}\,[\,9.0014\,]$

$$\approx .50008 \qquad \square$$

In this case we can evaluate the integral exactly:

$$\int_1^2 \dfrac{1}{x^2}\, dx = \int_1^2 x^{-2}\, dx = \left.\dfrac{-1}{x}\right|_1^2 = \dfrac{-1}{2} + 1 = .5$$

This confirms that our approximation is made to within the desired accuracy. (Of course, in "real world" applications, approximation methods are useful primarily with integrals which can't be evaluated exactly.)

Exercises.   (These problems cover the error bound material discussed in optional Subsection 4.)

1.   Estimate the absolute value of the maximum error that can occur when approximating

$$\ln \frac{3}{2} = \int_1^{3/2} \frac{dt}{t} \quad \text{by using}$$

   a)   the Trapezoidal Approximation with $n = 4$

   b)   Simpson's Rule with $n = 4$.

2.   Repeat Problem 1 for the integral $\int_1^2 e^{x^2} dx$ and $n = 6$.

3.   How many subintervals, n , should be taken to ensure that the absolute value of the error in the approximation of $\int_1^2 e^{x^2} dx$ is at most $\epsilon = .01$ when using

   a)   the Trapezoidal Approximation?

   b)   Simpson's Rule?

4.   Repeat Problem 3 for the integral $\ln 4 = \int_1^4 \frac{dt}{t}$ and $\epsilon = .0001$.

Answers

1.   a)   .001302   b)   .000017

2.   a)   2.5278   b)   .1161

3.   a)   91   b)   12

4.   a)   213   b)   24

Chapter 10:   Improper Integrals; L'Hôpital's Rule

Section 10.1.   Improper Integrals

1.   "Replace - Integrate - Take the Limit. "   The integral $\int_a^b f(x)\,dx$ is termed improper
     if either

       (i) one or both of the limits of integration (a or b) is infinite, or

      (ii) the function f "becomes infinite" on the interval [a, b], most
            commonly (although not always!) at one of the endpoints a or b .

Anton gives several examples of each type of improper integral and all of these employ the
same basic "Replace - Integrate - Take the Limit" technique:

> To evaluate an improper integral,
>
> 1.   Replace the troublesome limit of integration
>   by ℓ .
>
> 2.   Integrate the resulting definite integral.
>
> 3.   Take the limit of the result as ℓ approaches
>   the number it replaced.

If a finite limit is obtained in (3), then the improper integral is said to converge to this limit.
If no finite limit is obtained in (3), then the improper integral is said to diverge.

     Let's see how the "Replace - Integrate - Take the Limit" technique applies to the various
kinds of improper integrals:

2. <u>Improper integrals with infinite limits of integration</u> are the easiest to recognize of all improper integrals:

$$\boxed{\begin{array}{c}\text{AH}\\\text{HA!}\end{array}} \longrightarrow \int_a^{+\infty} f(x)\, dx \ , \qquad \int_{-\infty}^b f(x)\, dx \ , \qquad \int_{-\infty}^{+\infty} f(x)\, dx$$

$$\boxed{\begin{array}{c}\text{AH}\\\text{HA!}\end{array}} \qquad \boxed{\begin{array}{c}\text{AH}\\\text{HA!}\end{array}} \qquad \boxed{\ldots \text{and AH HA again!}}$$

Anton's Examples 1, 2 and 3 illustrate how to evaluate improper integrals with just one infinite limit of integration. Here is an additional example in which the "Replace - Integrate - Take the Limit" technique is emphasized. [This example will be important later when we discuss infinite series (§11.4)]:

<u>Example A</u> (Anton's Exercise 22). Show that

$$\int_1^{+\infty} \frac{dx}{x^p}$$

converges to $\frac{1}{p-1}$ if $p > 1$ and diverges if $p \leq 1$.

<u>Solution.</u> First we consider the case $p = 1$. Then the improper integral is $\displaystyle\int_1^{+\infty} \frac{dx}{x}$ which diverges (this is Anton's Example 2).

To evaluate the improper integral $\displaystyle\int_1^{+\infty} \frac{dx}{x^p}$ when $p \neq 1$, we:

(1) <u>Replace</u> the infinite limit of integration by ℓ

$$\int_1^\ell \frac{dx}{x^p}$$

(2)   <u>Integrate</u> the resulting definite integral

$$\int_1^\ell \frac{dx}{x^p} = \int_1^\ell x^{-p}\, dx = \left(\frac{1}{1-p}\right) \frac{1}{x^{p-1}} \Bigg|_1^\ell = \left(\frac{1}{1-p}\right)\left(\frac{1}{\ell^{p-1}} - 1\right)$$

(3)   <u>Take the limit</u> as $\ell \to +\infty$

$$\int_1^{+\infty} \frac{dx}{x^p} = \lim_{\ell \to +\infty} \int_1^\ell \frac{dx}{x^p} = \left(\frac{1}{1-p}\right)\left(\lim_{\ell \to +\infty} \frac{1}{\ell^{p-1}}\right) - \left(\frac{1}{1-p}\right)$$

When $p > 1$, then $p - 1 > 0$ so $\displaystyle\lim_{\ell \to +\infty} \frac{1}{\ell^{p-1}} = 0$

Hence $\displaystyle\int_1^{+\infty} \frac{dx}{x^p} = \left(\frac{1}{1-p}\right)(0) - \left(\frac{1}{1-p}\right) = \frac{1}{p-1}$

When $p < 1$, then $p - 1 < 0$ so $\displaystyle\lim_{\ell \to +\infty} \frac{1}{\ell^{p-1}} = \lim_{\ell \to +\infty} \ell^{1-p} = +\infty$

Hence $\displaystyle\int_1^{+\infty} \frac{dx}{x^p} = \left(\frac{1}{1-p}\right)\left(\lim_{\ell \to +\infty} \frac{1}{\ell^{p-1}}\right) - \frac{1}{1-p} = +\infty$

                          ↗   ↖

              [positive]   [approaches $+\infty$]

Therefore

$$\int_1^{+\infty} \frac{dx}{x^p} = \begin{cases} \dfrac{1}{p-1} & \text{if } p > 1 \\[2ex] \text{diverges} & \text{if } p \le 1 \end{cases}$$

□

            *   *   *   *   *

For an integral in which <u>both</u> limits of integration are infinite, the "Replace - Integrate - Take the Limit" technique must be applied <u>twice</u>. Simply take some convenient value of c and write

$$\int_{-\infty}^{+\infty} f(x)\,dx \;=\; \int_{-\infty}^{c} f(x)\,dx \;+\; \int_{c}^{+\infty} f(x)\,dx$$

Then the "Replace - Integrate - Take the Limit" technique can be used on <u>each</u> improper integral appearing on the right.

Anton's Example 4 provides a good illustration of this technique. Here is another example which focuses on the choice of the "division point" c :

<u>Example B.</u>   Evaluate

$$\int_{-\infty}^{+\infty} e^{-a\,|x|}\,dx\,, \qquad a > 0$$

if it converges.

<u>Solution.</u>   Since

$$|x| \;=\; \begin{cases} x & \text{if} \quad x \geq 0 \\ -x & \text{if} \quad x < 0 \end{cases}$$

we have

$$e^{-a\,|x|} \;=\; \begin{cases} e^{-ax} & \text{if} \quad x \geq 0 \\ e^{ax} & \text{if} \quad x < 0 \end{cases}$$

Thus the most appropriate choice for the "division point" is c = 0 , i.e., we split the improper integral as

$$\int_{-\infty}^{+\infty} e^{-a\,|x|}\,dx \;=\; \int_{-\infty}^{0} e^{-a\,|x|}\,dx \;+\; \int_{0}^{+\infty} e^{-a\,|x|}\,dx$$

$$=\; \int_{-\infty}^{0} e^{ax}\,dx \;+\; \int_{0}^{+\infty} e^{-ax}\,dx$$

Then we evaluate the two improper integrals:

(1)   <u>Replace</u> the infinite limit of integration by ℓ

$$\int_{\ell}^{0} e^{ax} \, dx \qquad \text{and} \qquad \int_{0}^{\ell} e^{-ax} \, dx$$

(2)   <u>Integrate</u> the resulting definite integrals

$$\int_{\ell}^{0} e^{ax} \, dx = \frac{1}{a} e^{ax} \Big|_{\ell}^{0} = \frac{1}{a} - \frac{1}{a} e^{a\ell}$$

$$\int_{0}^{\ell} e^{-ax} \, dx = -\frac{1}{a} e^{-ax} \Big|_{0}^{\ell} = \frac{1}{a} - \frac{1}{a} e^{-a\ell}$$

(3)   <u>Take the limit</u> of the first integral as $\ell \to -\infty$ and of the second as $\ell \to +\infty$

$$\int_{-\infty}^{0} e^{ax} \, dx = \lim_{\ell \to -\infty} \int_{\ell}^{0} e^{ax} \, dx$$

$$= \lim_{\ell \to -\infty} \left[\frac{1}{a} - \frac{1}{a} e^{a\ell} \right]$$

$$= \frac{1}{a} \qquad\qquad [\, e^{x} \to 0 \quad \text{as} \quad x \to -\infty \,]$$

and

$$\int_{0}^{+\infty} e^{-ax} \, dx = \lim_{\ell \to +\infty} \int_{0}^{\ell} e^{-ax} \, dx$$

$$= \lim_{\ell \to +\infty} \left[\frac{1}{a} - \frac{1}{a} e^{-a\ell} \right]$$

$$= \frac{1}{a} \qquad\qquad [\, e^{x} \to 0 \quad \text{as} \quad x \to -\infty \,]$$

Finally, adding these limits we get

$$\int_{-\infty}^{+\infty} e^{-a|x|}\,dx \;=\; \int_{-\infty}^{0} e^{ax}\,dx \;+\; \int_{0}^{+\infty} e^{-ax}\,dx \;=\; \frac{1}{a} + \frac{1}{a} \;=\; \frac{2}{a}$$

□

You may have noticed a shortcut for Example B: it is not hard to show that the two improper

integrals obtained by splitting at c = 0 must be equal (hence only one needs to be computed)!

The equality follows because the integrand

$f(x) = e^{-a|x|}$ is an <u>even</u> function of x ,

i. e. ,

$$f(-x) = f(x)$$

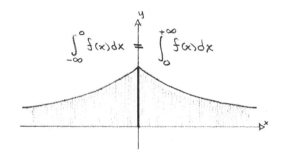

The diagram to the right shows how this

implies $\displaystyle\int_{-\infty}^{0} f(x)\,dx \;=\; \int_{0}^{+\infty} f(x)\,dx$.

The same trick applies to Anton's Example 4 !

3.   <u>Improper integrals in which the function becomes infinite</u> are more difficult to recognize than

those with infinite limits of integration.   Following Anton, we first consider integrals

$\displaystyle\int_{a}^{b} f(x)\,dx$ in which f approaches $+\infty$ or $-\infty$ <u>at one or both of the limits of integration</u>

a or b , but is continuous elsewhere on [a, b]. Anton illustrates this case in Examples 5

and 6. Here is one additional example:

Example C.   Evaluate $\displaystyle\int_{-1}^{1} \frac{1}{\sqrt{1 - x^2}}\, dx$ if it converges.

Solution.   If you are not observant, you might not recognize this as an improper integral.

However, although $f(x) = \dfrac{1}{\sqrt{1 - x^2}}$ is not infinite on the open interval $(-1, 1)$,   it becomes

infinite at both limits of integration!   Our solution technique will be to split the integral (by a

"division point" $x = c$)   into two integrals on each of which $f(x)$ becomes infinite at only

one limit of integration; then we will evaluate each of those improper integrals by the " Replace -

Integrate - Take the Limit" technique.   Hence

$$\int_{-1}^{1} \frac{1}{\sqrt{1 - x^2}}\, dx \; = \; \int_{-1}^{0} \frac{1}{\sqrt{1 - x^2}}\, dx \; + \; \int_{0}^{1} \frac{1}{\sqrt{1 - x^2}}\, dx$$

[$c = 0$ is a convenient choice for the "division point"]

Then

(1)   Replace the troublesome limits of integration by ℓ

$$\int_{\ell}^{0} \frac{1}{\sqrt{1 - x^2}}\, dx \quad \text{and} \quad \int_{0}^{\ell} \frac{1}{\sqrt{1 - x^2}}\, dx$$

(2)   Integrate the resulting definite integrals

$$\int_{\ell}^{0} \frac{1}{\sqrt{1 - x^2}}\, dx \; = \; \sin^{-1} x \,\Big|_{\ell}^{0} \; = \; \sin^{-1} 0 \, - \, \sin^{-1} \ell \; = \; - \sin^{-1} \ell$$

$$\int_{0}^{\ell} \frac{1}{\sqrt{1 - x^2}}\, dx \; = \; \sin^{-1} x \,\Big|_{0}^{\ell} \; = \; \sin^{-1} \ell \, - \, \sin^{-1} 0 \; = \; \sin^{-1} \ell$$

(3)   Take the Limit of the first integral as $\ell \to -1^{+}$ and of the second as $\ell \to 1^{-}$

$$\int_{-1}^{0} \frac{1}{\sqrt{1-x^2}}\, dx = \lim_{\ell \to -1^{+}} \int_{\ell}^{0} \frac{1}{\sqrt{1-x^2}}\, dx = \lim_{\ell \to -1^{+}} (-\sin^{-1}\ell) = -\left(\frac{-\pi}{2}\right) = \frac{\pi}{2}$$

$$\int_{0}^{1} \frac{1}{\sqrt{1-x^2}}\, dx = \lim_{\ell \to 1^{-}} \int_{0}^{\ell} \frac{1}{\sqrt{1-x^2}}\, dx = \lim_{\ell \to 1^{-}} (\sin^{-1}\ell) = \frac{\pi}{2}$$

Therefore, adding these two, we find

$$\int_{-1}^{1} \frac{1}{\sqrt{1-x^2}}\, dx = \int_{-1}^{0} \frac{1}{\sqrt{1-x^2}}\, dx + \int_{0}^{1} \frac{1}{\sqrt{1-x^2}}\, dx = \frac{\pi}{2} + \frac{\pi}{2} = \pi$$

$\square$

Note: As in Example B , it is not hard to see that

$$\int_{-1}^{0} \frac{1}{\sqrt{1-x^2}}\, dx = \int_{0}^{1} \frac{1}{\sqrt{1-x^2}}\, dx$$

so that we really have only to evaluate one of these improper integrals.

$$* \quad * \quad * \quad * \quad *$$

By far the most difficult type of improper integral to recognize is $\int_{a}^{b} f(x)\, dx$ where

f tends to $+\infty$ or $-\infty$ at an __interior point__ of $[a, b]$, i.e., at a point in the __open__ interval

(a, b) . Anton illustrates this case in his Example 7. Here is an additional example:

Example D.   Evaluate $\int_{-1}^{3} \frac{1}{x^{1/3}}\, dx$ if it converges.

Solution.   As in Example C , it would be very easy not to recognize this as an improper integral. However, observe that the integrand $f(x) = x^{-1/3}$ becomes infinite at $x = 0$, an interior point of $[-1, 3]$. Our solution technique is to split the original integral at the troublesome point $c = 0$ and evaluate the two resulting improper integrals by the " Replace - Integrate - Take the Limit" technique:

$$\int_{-1}^{3} x^{-1/3}\, dx \;=\; \int_{-1}^{0} x^{-1/3}\, dx \;+\; \int_{0}^{3} x^{-1/3}\, dx \qquad\qquad (*)$$

and

$$\int_{-1}^{0} x^{-1/3}\, dx = \lim_{\ell \to 0^-} \int_{-1}^{\ell} x^{-1/3}\, dx = \lim_{\ell \to 0^-} \frac{3}{2} x^{2/3}\Big|_{-1}^{\ell} = \lim_{\ell \to 0^-} \left[\frac{3}{2} \ell^{2/3} - \frac{3}{2} \right] \;\; -\frac{3}{2}$$

| Replace 0 | | Integrate | | Take the |
| by ℓ | | | | Limit |

$$\int_{0}^{3} x^{-1/3}\, dx = \lim_{\ell \to 0^+} \int_{\ell}^{3} x^{-1/3}\, dx = \lim_{\ell \to 0^+} \frac{3}{2} x^{2/3}\Big|_{\ell}^{3} = \lim_{\ell \to 0^+} \left[\frac{3^{5/3}}{2} - \frac{3}{2} \ell^{2/3} \right] = \frac{3^{5/3}}{2}$$

Thus Equation (*) becomes

$$\int_{-1}^{3} x^{-1/3}\, dx \;=\; -\frac{3}{2} + \frac{3^{5/3}}{2} \;=\; \frac{3}{2}\left(3^{2/3} - 1 \right) \qquad\qquad \Box$$

4.   <u>Look Before You Leap.</u>   Improper integrals occur frequently in applications (see Anton's Exercises 29 and 31, for example) and, of course, when you encounter them they rarely wear the label "THIS IS AN IMPROPER INTEGRAL!" As we have said, you will have no trouble recognizing improper integrals with infinite limits of integration, but improper integrals $\int_{a}^{b} f(x)\, dx$ in which the integrand f becomes infinite on $[a, b]$ are harder to spot.

You must be on the lookout for them. For instance, suppose you are confronted with ...

Example E.   Evaluate $\displaystyle\int_0^6 \frac{2x}{x^2 - 4}\, dx$.

Solution.   "Oh, that's easy," you might say! "This is a simple   u-substitution:"

$$\int_0^6 \frac{2x\, dx}{x^2 - 4} = \int_{-4}^{32} \frac{du}{u}$$

$\boxed{\begin{array}{l} u = x^2 - 4 \\ du = 2x\, dx \end{array}}$ $= \ln|u| \,\Big|_{-4}^{32}$ $\boxed{\begin{array}{c} \text{Appendix 3,} \\ \text{Table 3} \end{array}}$

$$= \ln(32) - \ln(4) = \ln\frac{32}{4} = \ln 8 \approx 2.0794$$

Uh, ... wait a minute! That denominator $x^2 - 4$ ... uh, it's zero when

$x = 2$ ... ?! ... AHA ! This is an improper integral since the integrand $f(x) = 2x/(x^2 - 4)$

becomes infinite at $x = 2$. Thus the correct solution goes as follows:

$$\int_0^6 \frac{2x\, dx}{x^2 - 4} = \int_0^2 \frac{2x\, dx}{x^2 - 4} + \int_2^6 \frac{2x\, dx}{x^2 - 4} \qquad (**)$$

$\boxed{\text{split the integral at }\ c = 2}$

and

$$\int_0^2 \frac{2x\, dx}{x^2 - 4} = \lim_{\ell \to 2^-} \int_0^\ell \frac{2x\, dx}{x^2 - 4} = \lim_{\ell \to 2^-} \ln|x^2 - 4|\,\Big|_0^\ell = \lim_{\ell \to 2^-} \ln(4 - \ell^2) - \ln 4 = -\infty$$

$\boxed{\begin{array}{c}\text{Replace } 2 \\ \text{by } \ell\end{array}}$ $\boxed{\text{Integrate}}$ $\boxed{\begin{array}{c}\text{Take the} \\ \text{Limit}\end{array}}$

$$\int_2^6 \frac{2x\, dx}{x^2 - 4} = \lim_{\ell \to 2^+} \int_\ell^6 \frac{2x\, dx}{x^2 - 4} = \lim_{\ell \to 2^+} \ln|x^2 - 4|\,\Big|_\ell^6 = \ln 32 - \lim_{\ell \to 2^+} \ln(\ell^2 - 4) = +\infty$$

Thus Equation (**) becomes

$$\int_0^6 \frac{2x\,dx}{x^2 - 4} = (-\infty) + (+\infty) = \underline{\text{undefined}}$$

since $(+\infty) + (-\infty)$ makes no sense. So $\int_0^6 \frac{2x\,dx}{x^2 - 4}$ is a $\underline{\text{divergent improper integral.}}$

We could have saved time and grief if we had checked for an improper integral $\underline{\text{before}}$ we worked the problem!! So look before you leap ... . $\square$

Section 10.2.   L'Hôpital's Rule (Indeterminant Forms of Type 0/0)

1.   L'Hôpital's Rule is a remarkably powerful technique for evaluating certain types of difficult (but important) limits. There are a number of versions of the Rule. In this section we will discuss the most basic version (for "indeterminant forms of type 0/0") of L'Hôpital's Rule; other versions will be considered in §10.3. Frankly, once you've mastered one version of L'Hôpital's Rule, you've mastered them all!

Anton gives L'Hôpital's Rule as Theorem 10.2.1; here is a slight rewording:

L'Hôpital's Rule for Form 0/0.

Suppose   (1) $\lim\limits_{x \to a} \dfrac{f(x)}{g(x)}$ is an $\underline{\text{indeterminant form of type } 0/0}$,

i.e., $\lim\limits_{x \to a} f(x) = 0$

and $\lim\limits_{x \to a} g(x) = 0$, and

(2) $\lim\limits_{x \to a} \dfrac{f'(x)}{g'(x)}$ has a finite limit L, or is $+\infty$ or $-\infty$.

Then L'Hôpital's Rule is valid:

$$\lim_{x \to a} \frac{f(x)}{g(x)} = \lim_{x \to a} \frac{f'(x)}{g'(x)}$$

Note: L'Hôpital's Rule is valid for other types of limits besides $\lim\limits_{x \to a}$, namely

$$\lim_{x \to a^+} , \quad \lim_{x \to a^-} , \quad \lim_{x \to +\infty} , \quad \text{and} \quad \lim_{x \to -\infty}$$

Example A. Evaluate $\lim\limits_{x \to 1} \dfrac{1 - e^{x-1}}{\ln x}$.

Solution. When asked to compute the limit of a quotient, you should first check (separately) the limits of the numerator and the denominator. After all, if these limits are both finite numbers, and if the denominator limit is non-zero, then we can use our old friend

"the limit of a quotient is the quotient of the limits"

(Theorem 2.5.1 (d))

So we'll try this approach:

$$\lim_{x \to 1} \frac{1 - e^{x-1}}{\ln x} = ? = \frac{\lim\limits_{x \to 1} (1 - e^{x-1})}{\lim\limits_{x \to 1} \ln x} = \frac{1 - e^0}{\ln 1} = \frac{0}{0} ? \quad \text{OOPS!}$$

We're hoping ...   ... no good!

As you can now see, we are faced with an indeterminant form of type $0/0$. If we try to evaluate this limit by the techniques of §§2.5 or 3.3 (e.g., cancelling like terms in the numerator and denominator, or using tricky geometric arguments) we will only end up frustrated. We thus turn to L'Hôpital's Rule ...

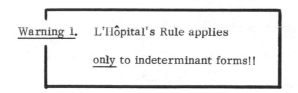

Well, we have already checked that we have an indeterminant form of type   0/0 ! So

now we apply the rule:

$$\lim_{x \to 1} \frac{1 - e^{x-1}}{\ln x} = ? = \lim_{x \to 1} \frac{\frac{d}{dx}[1 - e^{x-1}]}{\frac{d}{dx}[\ln x]} = \lim_{x \to 1} \frac{-e^{x-1}}{1/x}$$

We're hoping again...

$$= \frac{-e^{0}}{1/1} = \boxed{-1}$$

Since the quotient of the derivatives does have a finite limit, i. e. ,

$$\lim_{x \to 1} \frac{\frac{d}{dx}[1 - e^{x-1}]}{\frac{d}{dx}[\ln x]} = -1 \longleftarrow \boxed{\text{a finite number}}$$

then conditions (1) and (2) of L'Hôpital's Rule are both satisfied, proving that

our use of the rule was justified. Hence the "We're hoping again..." question

mark can be removed to give

$$\lim_{x \to 1} \frac{1 - e^{x-1}}{\ln x} = \boxed{-1} \quad .$$

$\square$

When applying L'Hôpital's Rule, it is rarely necessary to make the statement found in

the last paragraph of Example A (marked on the left with double lines). If applying L'Hôpital's

Rule to an indeterminate form yields a finite limit or $\pm \infty$, then applying the rule was legal to

begin with! This is the same observation that Anton makes in his Remark following Example 1.

Incidentally, this is a good spot to issue ...

Warning 2. L'Hôpital's Rule is

"the quotient of the derivatives" $\dfrac{\dfrac{d}{dx}[f(x)]}{\dfrac{d}{dx}[g(x)]}$

NOT

"the derivative of the quotient" $\dfrac{d}{dx}\left[\dfrac{f(x)}{g(x)}\right]$

Thus in Example A we used $\displaystyle\lim_{x\to 1}\ \dfrac{\dfrac{d}{dx}[1-e^{x-1}]}{\dfrac{d}{dx}[\ln x]}$,

NOT $\displaystyle\lim_{x\to 1}\ \dfrac{d}{dx}\left[\dfrac{1-e^{x-1}}{\ln x}\right]$

Example B. Evaluate $\displaystyle\lim_{x\to 1}\ \dfrac{\ln x}{e^{x}+1}$.

Solution. As in Example A we first check the separate limits of the numerator and the denominator:

$$\lim_{x\to 1}\ \frac{\ln x}{e^{x}+1}\ =\ ?\ =\ \frac{\displaystyle\lim_{x\to 1}\ \ln x}{\displaystyle\lim_{x\to 1}\ (e^{x}+1)}\ =\ \frac{\ln 1}{e^{1}+1}\ =\ \frac{0}{e+1}\ =\ 0\qquad \underline{\text{BINGO}}!$$

| We're hoping ... | ... It worked! |

In this case the elementary approach worked because we did <u>not</u> have an indeterminant form, and hence L'Hôpital's Rule was not needed. [Actually, we could NOT have applied L'Hôpital's Rule because "L'Hôpital's Rule applies <u>only</u> to indeterminant forms!!" (Warning 1).] □

2.   Tricks and further warnings.   Occasionally you must use L'Hôpital's Rule more than once

to obtain a solution. This is illustrated in Anton's Example 4. Here is an additional example:

Example C.   Evaluate $\lim\limits_{x \to 0} \dfrac{\sin x - e^x + 1}{x^2}$.

Solution.   As usual we first check the separate limits of the numerator and denominator:

$$\lim_{x \to 0} \frac{\sin x - e^x + 1}{x^2} = ? = \frac{\lim\limits_{x \to 0}(\sin x - e^x + 1)}{\lim\limits_{x \to 0} x^2} = \frac{0 - 1 + 1}{0} = \frac{0}{0} \; ? \; \text{OOPS!}$$

Optimistically...   ... no good!

So we have an indeterminant form of type 0/0 to which we apply L'Hôpital's Rule:

$$\lim_{x \to 0} \frac{\sin x - e^x + 1}{x^2} = \lim_{x \to 0} \frac{\frac{d}{dx}[\sin x - e^x + 1]}{\frac{d}{dx}[x^2]} = \lim_{x \to 0} \frac{\cos x - e^x}{2x}$$

Hmm... we are now faced with a new limit to compute, so we first check the separate limits

of the numerator and denominator:

$$\lim_{x \to 0} \frac{\cos x - e^x}{2x} = ? = \frac{\lim\limits_{x \to 0}(\cos x - e^x)}{\lim\limits_{x \to 0}(2x)} = \frac{0}{0} \; ? \; \text{OOPS!}$$

Optimistically...   ... but no luck again!

This result might at first seem discouraging: L'Hôpital's Rule has merely transformed our

original indeterminant form of type 0/0 ,

$$\lim_{x \to 0} \frac{\sin x - e^x + 1}{x^2}$$

into a <u>second</u> indeterminant form of type $0/0$,

$$\lim_{x \to 0} \frac{\cos x - e^x}{2x}$$

So do we give **up**? NO... because we take heed of

<u>Warning 3.</u> Often it is necessary to apply L'Hôpital's

Rule more than once!

This is such an occasion! Applying the rule to our second indeterminant form will yield

$$\lim_{x \to 0} \frac{\cos x - e^x}{2x} = \lim_{x \to 0} \frac{\frac{d}{dx}[\cos x - e^x]}{\frac{d}{dx}[2x]} = \lim_{x \to 0} \frac{-\sin x - e^x}{2}$$

$$= \frac{-0 - e^0}{2} = -\frac{1}{2} \qquad \underline{BINGO}!$$

Hence
$$\lim_{x \to 0} \frac{\sin x - e^x + 1}{x^2} = \boxed{-\frac{1}{2}}$$ □

*   *   *   *   *

Sometimes one is faced with a limit that must be rearranged to make it into an indeterminant form of type $0/0$. Here is an example similar to Anton's Exercise 25:

<u>Example D.</u>   Evaluate $\displaystyle\lim_{x \to \infty} x^2 \ln\left(\frac{x-1}{x+1}\right)$.

<u>Solution.</u>   Trying the basic rule "the limit of a product is the product of the limits" (Theorem 2.5.1c) won't work because one limit is infinite $\left(\displaystyle\lim_{x \to \infty} x^2 = \infty\right)$ while the other limit is zero:

$$\lim_{x \to \infty} \ln\left(\frac{x-1}{x+1}\right) = \lim_{x \to \infty} \ln\left(1 - \frac{2}{x+1}\right) = \ln 1 = 0$$

And $(\infty)(0)$ is undefined! Things look bleak ... until we realize ...

> **Warning 4.** Often a difficult limit can be rearranged into an indeterminant form of type $0/0$.

So we try this approach by moving x^2 into the denominator as x^{-2} :

$$\lim_{x \to \infty} x^2 \ln\left(\frac{x-1}{x+1}\right) = \lim_{x \to \infty} \frac{\ln\left(\frac{x-1}{x+1}\right)}{x^{-2}}$$

This is now an indeterminant form of type $0/0$ (check it!) and so we can try L'Hôpital's Rule:

$$\lim_{x \to \infty} \frac{\ln\left(\frac{x-1}{x+1}\right)}{x^{-2}} = \lim_{x \to \infty} \frac{\frac{d}{dx}\left[\ln(x-1) - \ln(x+1)\right]}{\frac{d}{dx}\left[x^{-2}\right]}$$

$$= \lim_{x \to \infty} \frac{\frac{1}{x-1} - \frac{1}{x+1}}{-2x^{-3}} = \lim_{x \to \infty} \left(-\frac{x^3}{2}\right)\left(\frac{2}{x^2-1}\right)$$

$$= \lim_{x \to \infty} (-x)\left(\frac{x^2}{x^2-1}\right) = \lim_{x \to \infty} (-x)\left(\frac{1}{1-(1/x^2)}\right)$$

Since the limit of the second term $\left(\dfrac{1}{1-(1/x^2)}\right)$ is 1 as $x \to \infty$, the first term $(-x)$

dominates the expression to give a limit of $-\infty$. Hence

$$\lim_{x \to \infty} x^2 \ln\left(\frac{x-1}{x+1}\right) = \boxed{-\infty}$$

*   *   *   *   *

* * * * *

Our final comment is simply to restate WARNING 1: (it is also found in Anton's Step 1):

$$\boxed{\begin{array}{c} \text{L'Hôpital's Rule applies } \underline{\text{only}} \\[4pt] \text{to indeterminant forms.} \end{array}}$$

Applying L'Hôpital's Rule to expressions which are NOT indeterminant forms will always yield garbage as answers, and is by far the biggest source of mistakes connected with L'Hôpital's Rule. For instance, ...

<u>Example E.</u> Evaluate $\lim\limits_{x \to 0} \dfrac{e^x - 1}{x^2 - x}$.

<u>Incorrect Solution.</u> This is an indeterminant form of type $0/0$ so if we apply L'Hôpital's Rule twice, we get

$$\lim_{x \to 0} \frac{e^x - 1}{x^2 - x} = \lim_{x \to 0} \frac{\frac{d}{dx}[e^x - 1]}{\frac{d}{dx}[x^2 - x]}$$

$$= \lim_{x \to 0} \frac{e^x}{2x - 1}$$

$$\boxed{\begin{array}{l}\text{Incorrect application of} \\ \text{L'Hôpital's Rule since} \\[4pt] \lim\limits_{x \to 0} \dfrac{e^x}{2x - 1} \text{ is } \underline{\text{not}} \text{ an} \\[4pt] \text{indeterminate form}\end{array}} \longrightarrow = \lim_{x \to 0} \frac{\frac{d}{dx}[e^x]}{\frac{d}{dx}[2x - 1]}$$

$$= \lim_{x \to 0} \frac{e^x}{2}$$

$$= \frac{1}{2}$$

Correct Solution. Apply L'Hôpital's Rule only once:

$$\lim_{x \to 0} \frac{e^x - 1}{x^2 - x} = \lim_{x \to 0} \frac{\frac{d}{dx}[e^x - 1]}{\frac{d}{dx}[x^2 - x]} = \lim_{x \to 0} \frac{e^x}{2x - 1} = \frac{\lim_{x \to 0} e^x}{\lim_{x \to 0} (2x - 1)} = \frac{1}{-1} = -1 \quad \square$$

Section 10.3.   Other Indeterminant Forms

There is just one new theorem in this section, L'Hôpital's Rule <u>for Form ∞/∞</u>, and that is only a slight variant of L'Hôpital's Rule for Form $0/0$ which was considered in the previous section. However, then we have a "bag of tricks" which show how to apply L'Hôpital's Rule to various types of indeterminant forms; these are most conveniently treated in three categories:

$$\text{Forms of type} \quad 0 \cdot \infty$$

$$\text{Forms of type} \quad 0^0, \quad \infty^0, \quad 1^\infty$$

$$\text{Forms of type} \quad \infty - \infty$$

First let's examine the new L'Hôpital's Rule (Theorem 10.3.1) :

1.   L'Hôpital's Rule for Form ∞/∞.

Suppose   (1) $\lim_{x \to a} \frac{f(x)}{g(x)}$ is an <u>indeterminant form of type ∞/∞</u>,

i.e., $\lim_{x \to a} f(x) = \infty$

and $\lim_{x \to a} g(x) = \infty$,   and

(2) $\lim_{x \to a} \frac{f'(x)}{g'(x)}$ has a finite limit L, or is $+\infty$ or $-\infty$.

Then <u>L'Hôpital's Rule</u> is valid:

$$\lim_{x \to a} \frac{f(x)}{g(x)} = \lim_{x \to a} \frac{f'(x)}{g'(x)}$$

Notice that <u>the only change</u> from the L'Hôpital's Rule of the previous section is in dealing with forms of <u>type ∞/∞</u>, i.e.,

$$\lim_{x \to a} f(x) = \infty \quad \text{and} \quad \lim_{x \to a} g(x) = \infty \qquad *$$

instead of forms of <u>type $0/0$</u>, i.e.,

$$\lim_{x \to a} f(x) = 0 \quad \text{and} \quad \lim_{x \to a} g(x) = 0 .$$

As with the earlier version, this L'Hôpital's Rule also applies to the limits

$$\lim_{x \to a^+} , \quad \lim_{x \to a^-} , \quad \lim_{x \to +\infty} \quad \text{and} \quad \lim_{x \to -\infty} .$$

This new version of L'Hôpital's Rule is used in exactly the same way as was the $0/0$ version. Anton illustrates this in Examples 1 and 2. Here is another illustration:

<u>Example A.</u>   Evaluate $\lim\limits_{x \to +\infty} \dfrac{\ln(e^x + 1)}{x}$.

<u>Solution.</u>   As we emphasized in the previous section, when asked to evaluate the limit of a quotient, you should begin by evaluating the limits of the numerator and denominator separately:

---

* Be sure you understand the meaning of $\lim\limits_{x \to a} f(x) = \infty$. Anton discusses this at the beginning of the section.

$$\lim_{x \to +\infty} \frac{\ln(e^x + 1)}{x} = ? = \frac{\displaystyle\lim_{x \to +\infty} \ln(e^x + 1)}{\displaystyle\lim_{x \to +\infty} x} = \frac{\infty}{\infty} ? \quad \text{NOPE!}$$

| Not likely in this case ... | ... no good! |

So we have an indeterminant form of type ∞/∞ to which we will apply L'Hôpital's Rule:

$$\lim_{x \to +\infty} \frac{\ln(e^x + 1)}{x} = \lim_{x \to +\infty} \frac{\dfrac{d}{dx}[\ln(e^x + 1)]}{\dfrac{d}{dx}[x]} = \lim_{x \to +\infty} \frac{\dfrac{e^x}{e^x + 1}}{1}$$

$$= \lim_{x \to +\infty} \frac{1}{1 + e^{-x}} = \frac{1}{1 + 0} = \boxed{1} \quad . \qquad \Box$$

2. <u>Indeterminant forms of type $0 \cdot \infty$</u> are limits of the form

$$\lim f(x)\, g(x)$$

where $\lim f(x) = 0$ and $\lim g(x) = \infty$. The "trick" in this situation is to rewrite the limit as an indeterminant form of either type $0/0$ or ∞/∞:

Type	
$0 \cdot \infty$	$\lim f(x)\, g(x) = \lim \dfrac{f(x)}{1/g(x)}$ (type $0/0$)
	$\lim f(x)\, g(x) = \lim \dfrac{g(x)}{1/f(x)}$ (type ∞/∞)

Then we can apply the appropriate version of L'Hôpital's Rule.

Anton's Examples 3 and 4 illustrate how this trick is used. Here is an additional example which illustrates that generally only one of the two possible conversions will work on a given limit:

Example B. Evaluate $\lim\limits_{x \to +\infty} x^2 e^{-3x}$.

Solution. This is an indeterminate form of type $0 \cdot \infty$, since

$$\lim_{x \to +\infty} e^{-3x} = 0 \quad \text{and} \quad \lim_{x \to +\infty} x^2 = \infty$$

We convert it to type ∞/∞ by writing

$$x^2 e^{-3x} = \frac{x^2}{e^{3x}}$$

Then, using L'Hôpital's Rule twice, we obtain

$$\lim_{x \to +\infty} x^2 e^{-3x} = \lim_{x \to +\infty} \frac{x^2}{e^{3x}} = \lim_{x \to +\infty} \frac{2x}{3e^{3x}} = \lim_{x \to +\infty} \frac{2}{9e^{3x}} = \boxed{0}$$

$\uparrow$ conversion $\uparrow$ L'H's Rule $\uparrow$ L'H's Rule again

Non-solution. This form could be converted to type $0/0$ by writing

$$x^2 e^{-3x} = \frac{e^{-3x}}{x^{-2}}$$

Then, using L'Hôpital's Rule repeatedly, we obtain

$$\lim_{x \to +\infty} x^2 e^{-3x} = \lim_{x \to +\infty} \frac{e^{-3x}}{x^{-2}} = \lim_{x \to +\infty} \frac{3e^{-3x}}{2x^{-3}} = \lim_{x \to +\infty} \frac{9e^{-3x}}{6x^{-4}} = \ldots \ !\&\$\#!$$

$\uparrow$ conversion $\uparrow$ L'H's Rule $\uparrow$ L'H's Rule again

The indeterminacy never disappears if we use this method! $\square$

> **The moral:** If one conversion doesn't work on a
> form of type $0 \cdot \infty$, then try the other!

3. <u>Indeterminant forms of type</u> 0^0, ∞^0 and 1^∞ are limits of the form

$$\lim f(x)^{g(x)}$$

where   (i) $\lim f(x) = 0$ and $\lim g(x) = 0$,   (type 0^0)

           (ii) $\lim f(x) = \infty$ and $\lim g(x) = 0$,   (type ∞^0)

or   (iii) $\lim f(x) = 1$ and $\lim g(x) = \infty$ (type 1^∞) .

The trick in each of these situations is to use the logarithm function to convert into an indeterminant form of type $0 \cdot \infty$:

Types
0^0
∞^0
1^∞

To compute $\lim f(x)^{g(x)}$, take the logarithm:

$$\ln\left(\lim f(x)^{g(x)}\right) = \lim\left(\ln f(x)^{g(x)}\right) \quad *$$

$$= \underbrace{\lim\,(g(x)\,\ln f(x))}_{} \qquad \text{(Equation (2) of §7.4)}$$

An indeterminant form
of type $0 \cdot \infty$; evaluate
as discussed previously
to obtain a value L .

Then $\ln\left(\lim f(x)^{g(x)}\right) = L$,

which gives $\lim f(x)^{g(x)} = e^L$.

Using this evaluation procedure is not as difficult as the above description might lead you to believe. Anton's Example 5 illustrates its use for an indeterminate form of type 1^∞ . Here are two additional examples, one for type 0^0 and one for type ∞^0 :

<u>Example C.</u>   Evaluate $\lim\limits_{x \to 0^+} x^x$.

---

* The limit can be pulled outside of the logarithm function because the logarithm function is continuous; this is an application of Theorem 3.7.6.

Solution. This is an indeterminate form of type 0^0, which we convert to type $0 \cdot \infty$ by taking $\ln$:

$$\ln \left(\lim_{x \to 0^+} x^x \right) = \lim_{x \to 0^+} (\ln x^x) = \lim_{x \to 0^+} (x \ln x)$$

Using our "trick" for type $0 \cdot \infty$, we convert this form to type ∞/∞ by writing

$$x \ln x = \frac{\ln x}{\frac{1}{x}} \quad *$$

Now L'Hôpital's Rule can be applied to yield

$$\lim_{x \to 0^+} (x \ln x) = \lim_{x \to 0^+} \frac{\ln x}{x^{-1}} = \lim_{x \to 0^+} \frac{x^{-1}}{-x^{-2}} = \lim_{x \to 0^+} (-x) = 0$$

This gives the logarithm of our desired limit:

$$\ln \left(\lim_{x \to 0^+} x^x \right) = 0$$

To get the limit itself, we have only to exponentiate both sides of this equation:

$$\lim_{x \to 0^+} x^x = e^0 = \boxed{1}$$

$\square$

Example D. Evaluate $\lim_{x \to +\infty} x^{1/x}$.

Solution. This is an indeterminate form of type ∞^0. We convert it to type $0 \cdot \infty$ by taking $\ln$:

---

* You can also convert to a form of type $0/0$ by writing $x \ln x = \dfrac{x}{1/\ln x}$. However, applying L'Hôpital's Rule to this form complicates, rather than simplifies matters.

$$\ln\left(\lim_{x \to +\infty} x^{1/x}\right) = \lim_{x \to +\infty}\left(\ln x^{1/x}\right) = \lim_{x \to +\infty}\left(\frac{1}{x}\ln x\right)$$

This form is of type $0 \cdot \infty$. However, it is easily converted into a form of type ∞/∞ by

writing $\frac{1}{x}\ln x = \ln x/x$. Thus, using L'Hôpital's Rule, we obtain

$$\lim_{x \to +\infty}\left(\frac{1}{x}\ln x\right) = \lim_{x \to +\infty}\frac{\ln x}{x} = \lim_{x \to +\infty}\frac{x^{-1}}{1} = \lim_{x \to +\infty}\frac{1}{x} = 0$$

This gives the logarithm of our desired limit:

$$\ln\left(\lim_{x \to +\infty} x^{1/x}\right) = 0$$

To get the limit itself, we have only to exponentiate both sides of this equation:

$$\lim_{x \to +\infty} x^{1/x} = e^0 = \boxed{1}$$

$\square$

4. <u>Indeterminant forms of type $\infty - \infty$</u> are limits of the form

$$\lim\,[\,f(x) - g(x)\,] \qquad \text{or} \qquad \lim\,[\,f(x) + g(x)\,]$$

which lead to one of the following expressions:

$$(+\infty) - (+\infty) \qquad\qquad (-\infty) - (-\infty)$$

$$(+\infty) + (-\infty) \qquad\qquad (-\infty) + (+\infty)$$

In contrast to the cases considered previously, no <u>precise</u> method can be given to handle

indeterminant forms of type $\infty - \infty$. However, any solution method will invariably involve the

following idea:

> Combine the terms $f(x)$ and $g(x)$ into a single term, and then "jiggle" the result into one of the previous indeterminant forms.

For example, if $\lim f(x) = \lim g(x) = 0$, then $\lim \left(\dfrac{1}{f(x)} - \dfrac{1}{g(x)} \right)$ is an indeterminate form of type $\infty - \infty$. We can combine the terms by putting the two quotients over a common denominator and adding:

$$\lim \left(\frac{1}{f(x)} - \frac{1}{g(x)} \right) = \lim \frac{g(x) - f(x)}{f(x)\, g(x)}$$

This is automatically an indeterminate form of type $0/0$ (no "jiggling" required) so L'Hôpital's Rule may be applied. Anton's Example 6 is an illustration of this exact situation. Here is a similar example, although at first it looks quite different:

Example E. Evaluate $\lim\limits_{x \to \pi/2} (\tan x - \sec x)$.

Solution. This is an indeterminate form of type $\infty - \infty$. We convert it to an indeterminate form of type $0/0$ by using a common denominator:

$$\lim_{x \to \pi/2} (\tan x - \sec x) = \lim_{x \to \pi/2} \left(\frac{\sin x}{\cos x} - \frac{1}{\cos x} \right) = \lim_{x \to \pi/2} \left(\frac{\sin x - 1}{\cos x} \right)$$

Thus L'Hôpital's Rule applies to yield

$$\lim_{x \to \pi/2} (\tan x - \sec x) = \lim_{x \to \pi/2} \left(\frac{\sin x - 1}{\cos x} \right) = \lim_{x \to \pi/2} \left(\frac{\cos x}{-\sin x} \right) = \frac{0}{-1} = \boxed{0}$$

5. Forms involving 0 and ∞ which are NOT indeterminate.   Not all forms involving 0 and ∞

are indeterminate. Anton's Exercise 27 gives a list of eight such forms which are NOT

indeterminate. Here is that list with the value of each:

	Form Type	Value of Limit	Intuitive Reason
(1)	$\dfrac{0}{\infty}$	0	A very small number divided by a very large number will yield a very small number!
(2)	$\dfrac{\infty}{0}$	$\pm\,\infty$ or undefined	A very large number divided by a very small number will yield a very large number! *
(3)	0^{∞}	0	Zero to any power, even a very large one, is zero.
(4)	$\infty \cdot \infty$	∞	Why?
(5)	$(+\infty) + (+\infty)$	$+\infty$	Why?
(6)	$(+\infty) - (-\infty)$	$+\infty$	Why?
(7)	$(-\infty) + (-\infty)$	$-\infty$	Why?
(8)	$(-\infty) - (+\infty)$	$-\infty$	Why?

DO NOT CONFUSE THESE FORMS WITH INDETERMINATE FORMS! They can be

evaluated "by inspection," with no need of L'Hôpital's Rule.   In fact, L'Hôpital's Rule cannot be

used in these situations.

Anton's Example 7 illustrates a form of type $(+\infty) - (-\infty)$. Here is another example:

---

* However, if the small number alternates in sign, then the sign of the quotient will also
    alternate and the limit will be undefined.

10.3.10

Example F. Evaluate $\lim\limits_{x \to +\infty} \left(\dfrac{1}{x}\right)^x$.

Solution. This is a form of type 0^∞ . It is <u>not</u> indeterminate, for "by inspection" we see that

$$\lim_{x \to +\infty} \left(\frac{1}{x}\right)^x = 0$$

The intuitive, "by inspection" reasoning goes as follows: as $\dfrac{1}{x}$ approaches zero, multiplying it by itself more and more times only makes it approach zero faster! □

6. <u>A useful formula.</u> In Example 5, Anton uses L'Hôpital's Rule to show that

$$\lim_{x \to 0} (1 + x)^{1/x} = e$$

Replacing x by $\dfrac{1}{h}$ and $x \to 0$ by $h \to +\infty$ yields

$$\boxed{\lim_{h \to +\infty} \left(1 + \frac{1}{h}\right)^h = e}$$

This formula provides our first method for <u>approximating the value of e</u> : we compute $\left(1 + \dfrac{1}{h}\right)^h$ for large values of h . The approximate values of e which result from various choices of h are given in Table 7.2.1 in Chapter 7.

In fact, we can generalize Anton's formula. By using the same Example 5 procedure on the limit

$$\lim_{x \to 0} (1 + ax)^{1/x}$$

where a is <u>any</u> real number, we can show

$$\lim_{x \to 0} (1 + a x)^{1/x} = e^{a}$$

or

$$\boxed{\lim_{h \to \infty} \left(1 + \frac{a}{h}\right)^{h} = e^{a}}$$

Hence we can approximate <u>any</u> value of the exponential function! (Approximation questions of this sort are <u>crucial</u> in applied mathematics. After all, what do you think your calculator does when you push the "exp" button?!?)

This method of estimating e^{a} is laborious and time consuming. Fortunately, we will develop a better one when we study Taylor Series in the next chapter ... .

Chapter 11:   Infinite Series

Introduction.   Have you ever wondered how your pocket calculator obtains a fact such as

the true value of   e   to five decimal places is $e = 2.71828$

or how the values in the tables in Anton's   Appendix 3   are obtained?   Well,   such results can

be obtained through the use of infinite series.   For example,   you will discover in this chapter

that

$$e^x = 1 + x + \frac{x^2}{2!} + \frac{x^3}{3!} + \frac{x^4}{4!} + \frac{x^5}{5!} + \cdots$$

This "infinite sum" is what is called an infinite series.   Once we have found it,   we can substitute

$x = 1$ and obtain

$$e = 1 + 1 + \frac{1}{2!} + \frac{1}{3!} + \frac{1}{4!} + \frac{1}{5!} + \frac{1}{6!} + \frac{1}{7!} + \frac{1}{8!} + \frac{1}{9!} + \frac{1}{10!} + \cdots$$

$$= 1 + 1 + \frac{1}{2} + \frac{1}{6} + \frac{1}{24} + \frac{1}{120} + \frac{1}{720} + \frac{1}{5040} + \frac{1}{40,320} + \frac{1}{362,880} + \frac{1}{3,628,800} + \cdots$$

$$= 1 + 1 + .5 + .1666666 + .0416666 + .0083333 + .0013888 + .0001984 + .0000248$$

$$+ .0000027 + .0000002 + \cdots$$

$$= 2.71828$$

(It can be shown that the addition of the remaining terms does not affect the   5-th   decimal place.)

Infinite series are useful in many other contexts as well.   Differential equations can

sometimes be solved by solutions which are infinite series and algorithms in numerical analysis

are devised by approximating functions with parts of infinite series.   The wealth of significant

applications is what makes the study of infinite series important.

Section 11. 1 :   Sequences

1.   The concept of a sequence.   This chapter on infinite series begins with a discussion of

sequences of real numbers.   Roughly speaking, a sequence is an infinite set of numbers which

has been listed in succession, i. e. , "counted" by the positive integers.

If you think about counting (for example, counting the words in this sentence) you realize

that what happens when you count is that you assign to each positive integer one of the objects

being counted   (1   is assigned "If, "   2   is assigned "you, "   3   is assigned "think, " etc.)

until you run out of objects.   In this way you have obtained a function   f   from a finite set of

positive integers to the objects you are counting:

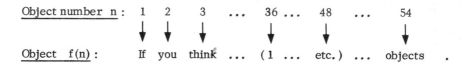

A sequence of real numbers is merely a function   f   from the infinite set of all positive

integers to a set of real numbers.   The "objects" in this case are real numbers, called the

terms of the sequence.   Thus the sequence

$$1 , \ 1/2 , \ 1/3 , \ 1/4 , \ 1/5 , \ \ldots \ , \ 1/n , \ \ldots$$

can be written as a function via the table

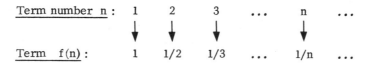

Since in this case we have a simple formula for the   n-th term   (or general term) of the

sequence,

$$[\,n\text{-th term}\,] \;=\; 1/n$$

we can employ a useful shorthand and express the sequence as

$$\{1/n\}_{n=1}^{\infty}$$

or

$$\{a_n\}_{n=1}^{\infty} \,, \quad \text{where} \quad a_n = 1/n$$

This shorthand is called <u>bracket notation</u> for a sequence. A natural question to ask is

"Given a sequence $\,a_1\,,\;a_2\,,\;a_3\,,\;a_4\,,\;\cdots\,,$

how can we write it in a bracket notation? "

This is equivalent to the question

"Given a sequence $\,a_1\,,\;a_2\,,\;a_3\,,\;a_4\,,\;\cdots\,,$

how can we determine a formula for $\;a_n\,,\;$ the $\,$n-th$\,$ term? "

To answer these questions it is best to write the sequence out as a function via the function table discussed above:

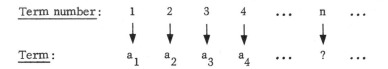

Then carefully examine the table for a <u>pattern,</u> i.e., some operation, or combination

of operations, which when performed on $\;1\;$ yields $\;a_1\,,\;$ when performed on $\;2\;$ yields $\;a_2\,,$

etc. ? Anton's Example 2 illustrates such techniques. Here are two more illustrations:

<u>Example A.</u>   Express the following sequences in bracket notation:

(a) 2 , $2/3$, $2/9$, $2/27$, ...

(b) 1 , $-3/4$, $5/7$, $-7/10$, ...

<u>Solution to (a)</u>.   We set up the table

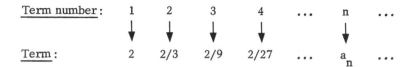

Term number:	1	2	3	4	...	n	...
Term:	2	2/3	2/9	2/27	...	a_n	...

We first notice that there is a 2 as the numerator in every term. Thus we start off with

$$a_n = 2/(?)$$

Then we notice that the denominators are all powers of 3 ; our table can therefore be rewritten as

Term number:	1	2	3	4	...	n	...
Term:	$2/3^0$	$2/3^1$	$2/3^2$	$2/3^3$	...	a_n	...

It is now evident that the denominator of each term is the power of 3 which is <u>one less than</u> the term number itself. Thus

$$a_n = 2/3^{n-1}$$

The bracket notation is therefore $\left\{ \dfrac{2}{3^{n-1}} \right\}_{n=1}^{\infty}$

<u>Solution to (b)</u>.   Our table is

Term number:	1	2	3	4	...	n	...
Term:	1	$-3/4$	5/7	$-7/10$	...	a_n	...

Leaving aside for the moment the fact that the signs alternate, notice that the numerators increase by 2 in each term (this is called a "linear increase" of 2). This should make you suspect that the numerator of a_n is of the form $2n + C$ for some constant C. To determine C, note that when $n = 1$ the numerator equals 1; thus

$$2(1) + C = 1, \text{ which implies } C = -1$$

Thus, the numerator of a_n is $2n - 1$, which does check out correctly for $n = 2, 3$ and 4.

As for the denominator, notice that it increases by 3 for each term (a linear increase of 3) and thus is of the form $3n + D$. We determine D as we did above for the numerator: when $n = 1$ we have $3(1) + D = 1$ so $D = -2$, and hence the denominator of a_n is $3n - 2$.

It only remains to determine the sign of a_n. The sign alternates and hence we have a choice between $(-1)^n$ and $(-1)^{n+1}$. However, a_1 is positive so that we need a positive sign when $n = 1$, proving that the sign of a_n is $(-1)^{n+1}$. Thus we have

$$a_n = (-1)^{n+1} (2n - 1)/(3n - 2)$$

so that in bracket notation the sequence is

$$\left\{ (-1)^{n+1} \frac{2n - 1}{3n - 2} \right\}_{n=1}^{\infty}$$

$\square$

As you can see, you have to be fairly clever to find the formula for the n-th term of a sequence. Although no sure-fire procedure can be given, our examples do illustrate some common tricks:

---

1. Look for <u>alternating signs</u> (use $(-1)^n$ or $(-1)^{n+1}$).

2. Look for <u>powers</u> (e.g., 2^n, $1/3^n$).

3. Look for <u>linear increases</u> (e.g., $2n-1$, $3n-2$).

   In particular, <u>even numbers</u> can often be written as $2n$

   while <u>odd numbers</u> can often be written as $2n-1$ or $2n+1$.

---

2. <u>Limits of Sequences: The Algebraic Operations Theorem.</u> The most important question to ask about any sequence is "what is happening to the n-th term as n becomes very large?" For example, in the sequence $\{1, 1/2, 1/3, 1/4, 1/5, \ldots, 1/n, \ldots\}$ it appears that the n-th term $1/n$ gets closer and closer to zero as n increases. In the sequence $\{2, 3/2, 4/3, 5/4, 6/5, \ldots, (n+1)/n, \ldots\}$, it appears that the n-th term gets closer and closer to 1 as n increases. A value that the n-th term approaches as n increases is called the <u>limit of the sequence.</u>

   <u>The formal definition of the limit of a sequence</u> (Anton's Definition 11. 1. 2) <u>is the most difficult part of this section.</u> We will give some hints for understanding it in the last subsection of this section. However, what we want to emphasize here is that <u>at this level of calculus you don't need to master the formal definition in order to "do" limits;</u> the calculation of most sequence limits can be accomplished by using rules which are much more intuitive than the limit definition itself. * Here's how this works:

   First learn two basic limits (and one "non-limit"):

---

* Of course these rules are <u>proven</u> by using the formal definition of the limit of a sequence.

1. $\lim_{n \to \infty} c = c$, when c is a constant.

    (The limit of a constant is the constant.)

2. $\lim_{n \to \infty} (1/n) = 0$

    (The limit of   1/n   is zero.)

3. $\lim_{n \to \infty} n = +\infty$

    (The limit of n <u>diverges</u> to positive infinity.)

The
Basic
Limits

Then the Algebraic Operations Theorem (Anton's Theorem 11.1.3) can often be used to break

down complicated sequence limit computations into combination of the two basic limits.

The
Algebraic
Operations
Theorem

<u>Theorem 11.1.3 in English.</u>

1. Constants move through limit signs.

2. The limit of the $\begin{cases} \text{sum} \\ \text{difference} \\ \text{product} \\ \text{quotient} \end{cases}$ of sequences is the $\begin{cases} \text{sum} \\ \text{difference} \\ \text{product} \\ \text{quotient} \end{cases}$ of their limits.

Anton illustrates the use of this theorem in Example 4 , parts (a) through (d). Here is

another illustration:

<u>Example D.</u>   Find the limit of the sequence

$$\left\{ \frac{2n^3 - 3n + 6}{4n^3 - n} \right\}_{n=1}^{\infty}$$

Solution.

$$\lim_{n \to \infty} \frac{2n^3 - 3n + 6}{4n^3 - n} = \lim_{n \to \infty} \frac{2n^3 - 3n + 6}{4n^3 - n} \cdot \frac{1/n^3}{1/n^3}$$

$$= \lim_{n \to \infty} \frac{2 - (3/n^2) + (6/n^3)}{4 - (1/n^2)}$$

divide numerator and denominator by n^3

$$= \frac{\lim_{n \to \infty} 2 - \lim_{n \to \infty} (3/n^2) + \lim_{n \to \infty} (6/n^3)}{\lim_{n \to \infty} (4) - \lim_{n \to \infty} (1/n^2)}$$

by the Algebraic Operations Theorem

$$= \frac{\lim_{n \to \infty} 2 - 3 \left[\lim_{n \to \infty} (1/n) \right]^2 + 6 \left[\lim_{n \to \infty} (1/n) \right]^3}{\lim_{n \to \infty} (4) - \left[\lim_{n \to \infty} (1/n) \right]^2}$$

by the Algebraic Operations Theorem

$$= \frac{2 - 3(0)^2 + 6(0)^3}{4 - (0)^2}$$

by the Basic Limits

$$= 1/2 \qquad \qquad \square$$

Note: You should recognize the similarities between the Algebraic Operations Theorem for limits of <u>sequences</u> and Theorem 2.5.1, the Algebraic Operations Theorem for limits of <u>functions.</u>

Our solution to Example B contained an important technique:

> If the n-th term is a rational function* of n ,
> then divide both the numerator and denominator
> by the highest power of n in the expression.

Thus in Example B we divided both numerator and denominator by n^3 , and in Anton's

_____

* Recall that a rational function is the quotient of two polynomials.

Example 4(a) he divides both numerator and denominator by n .

3. Finding limits of sequences by converting to ordinary function limits.   In many instances a

sequence limit $\lim\limits_{n \to \infty} a_n$ can be converted into an ordinary function limit $\lim\limits_{x \to \infty} f(x)$. In

this way the numerous techniques of §§2.4 - 2.6 and Chapter 10 (most notably L'Hôpital's Rule)

for evaluating function limits can be used to evaluate limits of sequences.   Anton uses this

technique in Example 4(e). Here is another illustration.

Example C.   Find the limit of the sequence $\left\{ \dfrac{n^{-4/3}}{\sin(1/n)} \right\}$.

Solution.   The n-th term of this sequence is $f(n)$ where $f(x) = x^{-4/3} / \sin(1/x)$. Since

$\lim\limits_{x \to +\infty} x^{-4/3} = \lim\limits_{x \to +\infty} \sin(1/x) = 0$, L'Hôpital's Rule (Theorem 10.2.1) applies and gives

$$\lim_{x \to +\infty} \frac{x^{-4/3}}{\sin(1/x)} = \lim_{x \to +\infty} \frac{(-4/3)\, x^{-7/3}}{(-1/x^2)\cos(1/x)}$$

$$= \lim_{x \to +\infty} \frac{(4/3)\, x^{-1/3}}{\cos(1/x)} = \frac{0}{1} = 0$$

Since $\lim\limits_{n \to \infty} f(n) = \lim\limits_{x \to +\infty} f(x)$, this says

$$\lim_{n \to \infty} \frac{n^{-4/3}}{\sin(1/n)} = \lim_{x \to +\infty} \frac{x^{-4/3}}{\sin(1/x)} = 0 \qquad\qquad \square$$

In general this conversion technique can be described as follows:

Suppose the expression a_n makes sense when the <u>integer</u> n is

replaced by any large <u>real number</u> x. In this case $a_n = f(n)$

for some function f(x), and we can evaluate $\lim_{n \to \infty} a_n$ as

follows:

i. Evaluate $\lim_{x \to \infty} f(x)$ by any technique (such as L'Hôpital's

Rule) which applies to function limits.

ii. Then $\lim_{n \to \infty} a_n = \lim_{x \to \infty} f(x)$.

<table><tr><td>Conversion
to
function
limits</td></tr></table>

<u>Warning: A Common Mistake.</u> The technique of converting the n-th term of a
sequence to a function of x cannot be used on every sequence. An example of a "non-
convertible" sequence is $\{1/n!\}_{n=1}^{\infty}$. The function f(x) such that f(n) = 1/n! would be
f(x) = 1/x!, but we have not defined x! for x's which are not positive integers (e.g.,
1/(1/2)! is not defined). Unless f(x) is defined for all large positive values of x, it
makes no sense to talk about the limit $\lim_{x \to +\infty} f(x)$. <u>People frequently make the mistake of</u>
<u>converting non-convertible sequences like these involving n! to functions of x, so ...</u>
<u>BEWARE!</u>

4. <u>Finding limits of sequences by graphing.</u> Sometimes the easiest way to check the convergence
or divergence of a sequence is to graph enough terms so that the behavior of the sequence becomes
obvious.

<u>Example D.</u> Does the following sequence converge or diverge?

$$\{-1/2, 2, -2/3, 3/2, -3/4, 4/3, -4/5, 5/4, -5/6, 6/5, ...\} \quad .$$

Solution.   The graph of the sequence is

From the graph, it is clear that the terms of the sequence are alternately clustering at two

different values   (1   and   - 1) .   Hence the sequence diverges.   □

5.   <u>The formal definition of the limit of a sequence (optional)</u>.   As we have said, Anton's

Definition 11. 1. 2 is the most difficult part of this section.   Most newcomers to calculus

find that mastering it requires going over it many times.

One way of understanding the definition is to proceed gradually through deepening levels

of understanding.   The following is a diagram of such a process:

The "That Means" Diagram

(the pointing finger should be read "that means")

The sequence $\{a_1, a_2, a_3, \ldots\}$ approaches L as a limit.

The terms a_n get very close to L (and stay very close to L).

The terms a_n get as close to L as anyone could want. That is, given any distance, the terms a_n eventually get within that distance of L (and they stay there! That is, after some term in the sequence, all the rest must be within the given distance of L).

If $\epsilon > 0$ is any given distance, then $|a_n - L| < \epsilon$ when a_n is far enough out in the sequence (and this is true for all the a_n's from then on).

If $\epsilon > 0$ is any given distance, then there is some N_ϵ so that $|a_n - L| < \epsilon$ for all the a_n past $n = N_\epsilon$, i.e., for all $n \geq N_\epsilon$.

The Formal
Definition of
the Limit of
a Sequence

Given $\epsilon > 0$, there is an integer N_ϵ such that $|a_n - L| < \epsilon$ whenever $n \geq N_\epsilon$.

Another way of approaching the formal definition of the limit of a sequence is to view it in terms of a graph as in Anton's Figure 11.1.3. Here is that figure with a few additional comments:

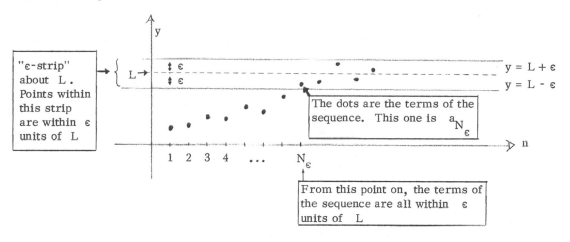

"ε-strip" about L. Points within this strip are within ε units of L

The dots are the terms of the sequence. This one is a_{N_ϵ}

From this point on, the terms of the sequence are all within ε units of L

Here are some important points to be kept in mind about the definition of the limit of the sequence:

1. ε is <u>given</u> and N_ϵ is then <u>chosen</u> to "work" for that ε. Different ε's may have different N_ϵ's which "work" for them.

2. The condition $|a_n - L| < \epsilon$ when $n \geq N_\epsilon$ must be satisfied for <u>every</u> ε > 0, no matter how small it may be. It is not enough for the condition to hold only for some ε's but not all.

Section 11.2: <u>Monotone Sequences.</u>

1. <u>Convergence properties of monotone sequences.</u> The primary reason for identifying montone sequences is that they are easy to test for convergence or divergence:

Monotone Convergence Principle

A monotone sequence will

(a) <u>converge</u> if it is bounded, or

(b) <u>diverge</u> to $+\infty$ or $-\infty$ if it is unbounded.

11. 2. 2

This is the essence of Theorems 11.2.2 and 11.2.3 , the first dealing with **nondecreasing** sequences, the second with **nonincreasing sequences.** Let's be more explicit in the non-decreasing situation:

A nondecreasing sequence $(a_1 \leq a_2 \leq \ldots \leq a_n \leq \ldots)$ will

   (a) <u>converge</u> if there exists a number M such that

        $a_n \leq M$ for <u>all</u> the terms a_n , or

   (b) <u>diverge to</u> $+\infty$ if there does not exist a number M

        such that $a_n \leq M$ for <u>all</u> the terms a_n .

Theorem
11. 2. 2

The <u>divergence</u> part of the Theorem is really very intuitive. For instance, most people would have no trouble seeing that the nondecreasing sequence

$$2 , 2 , 4 , 4 , 6 , 6 , 8 , 8 , \ldots .$$

diverges to $+\infty$. This is because the sequence has no <u>upper bound</u>, i. e. , there is no number M which is greater than or equal to <u>every</u> term in the sequence.

The <u>convergence</u> part of the Theorem, while still reasonably intuitive, is a bit more subtle. To understand it, perhaps you should think of Howard Ant crawling up your living room wall. Since your living room has a ceiling (we presume...), Howard's height on the wall, recorded at each second, is a nondecreasing sequence which is bounded above. The <u>lowest</u> point on the wall that Howard does <u>not rise above</u> will be his limit; to the

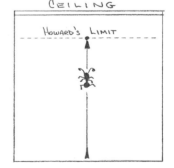

casual observer he will eventually appear to stop on this point. <u>There must be such a limit since</u>

the ceiling prevents Howard from crawling upward without bound!

Example A. Show that the sequence 1/2, 2/3, 3/4, 4/5, 5/6, ... converges.

Solution.   We first notice that our sequence is nondecreasing (in fact, increasing - see
Anton's Example 2) and that each term is less than 1. Thus Theorem 11.2.2a tells us
that the sequence converges, and that its limit will be less than or equal to 1. In fact the
limit must equal 1 since the sequence eventually exceeds any number which is less than 1,
leaving 1 as the only possible limit.

Alternate Solution.   For the sequence in Example A you can avoid the monotone convergence
principle and instead employ the techniques of §11.1.   To do this, first observe that the
sequence can be written in bracket notation as follows: $\{\frac{n}{n+1}\}_{n=1}^{\infty}$. Then

$$\lim_{n \to \infty} \frac{n}{n+1} = \lim_{n \to \infty} \frac{n}{n+1} \cdot \frac{(1/n)}{(1/n)}$$

$$= \lim_{n \to \infty} \frac{1}{1+(1/n)} = \frac{1}{1+0} = 1 \qquad \square$$

Do not let Example A deceive you, however; not all monotone sequences can be
handled by the techniques of §11.1 as done in the Alternate Solution. The convergence of
many monotone sequences can be tested only by using the results of this section, i.e., Theorems
11.2.2 and 11.2.3.

So why don't we give you an example of such a sequence? We will... later in the chapter,
when we develop the concept of infinite series. The methods for determining the convergence of
sequences that arise as infinite series (a topic of primary importance, as you will soon see!)
depend on the convergence of bounded, monotone sequences.

11.2.4

2. "Nondecreasing" verses "increasing." The distinction between a "nondecreasing sequence"
and an "increasing sequence" is really very slight: equality between terms is allowed in a nondecreasing sequence, while no equality between terms is allowed in an increasing sequence.
Thus

$$2, 2, 4, 4, 6, 6, 8, 8, \dots \quad \text{is nondecreasing},$$

$$2, 3, 4, 5, 6, 7, 8, 9, \dots \quad \text{is increasing}.$$

Frankly, as a practical matter, this distinction is not of great importance. Our primary
theorem (Theorem 11.2.2) is stated for nondecreasing sequences and hence also applies to
increasing sequences! There is no need for a special statement of the theorem for increasing
sequences. The same comments apply to the distinctions between

"nonincreasing" and "decreasing,"

"monotone" and "strictly monotone."

The moral of the story: it is important to determine if a sequence is monotone (i.e.,
nondecreasing or nonincreasing). However, unless specifically asked to do so, it is generally
unnecessary to check for the stronger condition of strictly monotone (i.e., increasing or
decreasing).

3. Testing to see if a sequence is monotone. There are essentially three techniques for testing
whether a sequence is monotone:

1. Examine $a_n - a_{n+1}$.

Difference between successive terms

If $a_n - a_{n+1} \leq 0$ for all n , then $\{a_n\}$ is nondecreasing.

If $a_n - a_{n+1} < 0$ for all n , then $\{a_n\}$ is increasing, etc.

2. <u>Examine a_{n+1}/a_n</u> . (Applicable only when all the a_n are positive.)

Ratio of
successive
terms

If $a_{n+1}/a_n \geq 1$ for all n, then $\{a_n\}$ is nondecreasing.

If $a_{n+1}/a_n > 1$ for all n, then $\{a_n\}$ is increasing, etc.

3. <u>Examine $f'(x)$ when $f(n) = a_n$</u> . (Applicable only when $a_n = f(n)$ for $f(x)$

a function defined for all $x \geq 1$.)

Derivative
Method

If $f'(x) \geq 0$ for all $x \geq 1$, then $\{a_n\}$ is nondecreasing

If $f'(x) > 0$ for all $x \geq 1$, then $\{a_n\}$ is increasing, etc.

Anton describes these three techniques completely and illustrates them in Examples 2, 3, 4, 5 and 7. Here are two additional examples:

<u>Example A.</u> Show that the sequence $a_n = \dfrac{n}{(n-1)!}$ converges.

$\underline{a_n - a_{n+1}}$ solution. We find

$$a_n - a_{n+1} = n/(n-1)! - (n+1)/n!$$

$$= \frac{n^2 - n - 1}{n!}$$ by multiplying the first fraction by n/n to get $n!$ as a common denominator.

Thus

for $n = 1$, $a_n - a_{n+1} < 0$

for $n \geq 2$, $a_n - a_{n+1} > 0$ since $n^2 - n - 1 > 0$ for $n \geq 2$

Hence $a_1 < a_2$ and $a_2 > a_3 > a_4 > \cdots$. That is, after the first term, the sequence is decreasing. But the first terms do not affect the convergence of the sequence (it's what happens for large values of n that matters), so we may regard this as a decreasing sequence. Hence, since all of the terms of the sequence are positive, 0 is a lower bound and, because bounded decreasing sequences converge (Theorem 11.2.3), the sequence converges.

a_{n+1}/a_n **solution.**   The terms of the sequence are all positive and

$$a_{n+1}/a_n = (n + 1)/n! \ / \ n/(n - 1)! = (n + 1)/n^2$$

Thus

    for $n = 1$, $a_{n+1}/a_n > 1$

    for $n \geq 2$, $a_{n+1}/a_n < 1$ since $n + 1 < n^2$ for $n \geq 2$

so $a_1 < a_2$ and $a_2 > a_3 > a_4 > \cdots$. Thus the sequence is, eventually, a decreasing sequence
of positive terms and thus converges as before.

$f'(x)$ **non-solution.**   If we let $f(n) = n/(n - 1)!$ and try to write $f(x) = x/(x - 1)!$, we have
a BIG problem! For $f(x)$ is <u>NOT</u> defined for all real $x \geq 1$ $((x - 1)!$ is defined only when
x is a positive integer ≥ 1). Thus the "examine $f'(x)$" technique won't work on this
sequence. (In general, it won't work on any sequence involving factorials.) $\square$

Example B.   Show that the sequence $a_n = \dfrac{n - \ln n}{\sqrt{n}}$ is increasing.

$a_n - a_{n+1}$ **non-solution.**   We find

$$a_n - a_{n+1} = \frac{n - \ln n}{\sqrt{n}} - \frac{(n + 1) - \ln (n + 1)}{\sqrt{n + 1}}$$

an expression so complicated that we despair of ever being able to show whether it is positive
or negative. We conclude that the wisest course of action is to abandon this approach!

a_{n+1}/a_n **non-solution.**   We find

$$a_{n+1}/a_n = [(n + 1) - \ln (n + 1)]/\sqrt{n + 1} \ / \ (n - \ln n)/\sqrt{n}$$

$$= (\sqrt{n} / \sqrt{n + 1}) ((n + 1) - \ln (n + 1))/(n - \ln n)$$

This too is such a complicated expression that we decide to abandon this approach.

$\underline{f'(x) \text{ solution.}}$ Letting $f(x) = \dfrac{x - \ln x}{\sqrt{x}}$,   we have a function which is defined for all $x \geq 1$

and is such that $f(n) = a_n$.   Its derivative is

$$f'(x) = \frac{\sqrt{x}\,(1 - 1/x) - (1/2\sqrt{x})(x - \ln x)}{x} = \frac{x - 2 + \ln x}{2\,x^{3/2}}$$

When $x \geq 1$, $x^{3/2}$ is positive so the derivative is positive if and only if its numerator

$x - 2 + \ln x$ is positive, i.e., $x + \ln x > 2$.   Now $1 + \ln 1 = 1 < 2$, but if $x \geq 2$,   then

$\ln x > 0$ and $x + \ln x > 2$.   Hence for $x \geq 2$ we have $f'(x) > 0$ and this says that,

after the first term, the sequence is increasing, i.e., $a_2 < a_3 < a_4 < \cdots$.   □

        Examples A and B illustrate

$\underline{\text{The Moral of The Story:}}$ Usually one of the three techniques for testing to see if a sequence is

monotone will be easier to use than the others. Frequently one (or more) of them will not work

at all! Do not stubbornly stick to a technique which does not seem to be working. Instead, try

one of the others.

        Another point made in Example A is worth repeating, for emphasis:

> The first terms of a sequence are not important
> in determining the convergence of the sequence.
> It's what happens for large values of   n   that
> matters.

Thus a sequence may be $-1, 1, -1, 1, -1, 1, -1, \ldots$ for, say, the first million terms

and still converge if it "settles down" after that and becomes a convergent sequence. For

instance, if

11. 2. 8

$$a_n = (-1)^n \quad \text{for} \quad 1 \le n \le 1,000,000$$

and

$$a_n = \frac{1}{n} \quad \text{for} \quad n \ge 1,000,001$$

the sequence is, eventually, a positive decreasing sequence and thus it converges by Theorem 11. 2. 3.

A more colorful way of putting this is to call the first terms of the sequence its "head" and all the rest of the sequence its "tail." Thus we have

> a sequence converges if and only if its "tail" converges.

(We will discuss the "tail" of an infinite series in §11. 4. 1 of The Companion.)

∴ Sequences and series: a bedtime story.   "... and when the three bears came home, Papa Bear said, 'somebody's been sleeping in my bed.' And Momma Bear said, "somebody's been sleeping in my bed.' But Baby Bear said, 'somebody's been sleeping in my bed, and here she is!'"

One can make either a Baby Bear or a Parent Bear statement about limits of sequences. In the Baby Bear approach, you say

"this sequence has a limit, and here it is. "

In the Parent Bear approach, you only say

"this sequence has a limit. "

There is no question that the Baby Bear statement is to be preferred because it gives more information. However, sometimes only Parent Bear statements are possible. Theorems 11. 2. 2 and 11. 2. 3 are examples. They say "bounded monotone sequences converge," but they do not give a method for determining the limits.

You may wonder why it is useful to know that a sequence converges if you do not also know its limit. Well, in most instances, <u>if you know that a given sequence converges (i. e. , a Parent Bear result) then there are ways to approximate the limit</u>! In fact, sequences are important in applied mathematics precisely because we can approximate their limits! Here is a typical application:

In Example 5 of §10.3 , Anton shows that the sequence $\left\{ (1 + \frac{1}{n})^n \right\}_{n=1}^{\infty}$ converges.

(Actually, he establishes a Baby Bear result and shows that this sequence converges <u>to e</u>). To determine an approximation for the limit we have only to compute $(1 + \frac{1}{n})^n$ for large values of n as Anton does in §10.3 . Note that his method of approximating the value of the limit e (as the limit of a convergent sequence) is valid even if you only know that there is a limit, but don't know exactly what that limit is.

In general, how do you think we approximate values for e^x , ln x , tan x , etc.? <u>Answer</u>: We show that these functions are the limits of very special types of convergent sequences called <u>infinite series</u> and then we just approximate the limits of these infinite series ...   [To be continued!]

Section 11.3 :   <u>Infinite Series</u>

1.   <u>Series convergence: a form of sequence convergence.</u>   Sometimes people new to the subject confuse infinite <u>sequences</u> with infinite <u>series.</u>   You can avoid this confusion by remembering that

> Infinite sequences are countable sets
>
> $$a_1, \ a_2, \ a_3, \ a_4, \ \cdots$$
>
> Infinite series are sums
>
> $$\sum_{k=1}^{\infty} a_k = a_1 + a_2 + a_3 + a_4 + \cdots$$

Or, to be more mundane about it,

> Infinite sequences have terms separated by <u>commas.</u>
>
> Infinite series have terms separated by <u>plus signs.</u>

The key question about an infinite series $\displaystyle\sum_{k=1}^{\infty} a_k$ is: does it add up to something

finite? (If it does, we say the series <u>converges</u> and call the finite number its <u>sum.</u>) We answer

this question by checking whether a certain <u>sequence</u> converges:

> | Definition of the Sum of a Series |
> | --- |

> The infinite series $\displaystyle\sum_{k=1}^{\infty} a_k$ converges to the
>
> sum $\ S \ $ if and only if the sequence $\ \{s_n\} \ $ of
>
> its partial sums converges to $\ S$.

That is, series convergence depends on sequence convergence (that's why we studied sequence

convergence first, folks!)

Let's get back to those <u>partial sums.</u> They are exactly what the name implies - sums

which are part of the infinite sum - but newcomers frequently find them confusing. Here are a

couple of examples which should help you understand the natural and critical role played by

partial sums:

Example A.   Does the series

$$\sum_{k=1}^{\infty} \frac{1}{2^k} = \frac{1}{2} + \frac{1}{4} + \frac{1}{8} + \cdots$$

converge? If so, find its sum.

Solution.   Let s_n denote the sum of the first n terms of the series or, in other words, the sum you have if you stop the infinite summation after n terms. (It is called the n-th partial sum.) That is,

$$s_1 = \frac{1}{2} \qquad\qquad = \frac{1}{2} \qquad \text{(the 1-st partial sum)}$$

$$s_2 = \frac{1}{2} + \frac{1}{4} \qquad\qquad = \frac{3}{4} \qquad \text{(the 2-nd partial sum)}$$

$$s_3 = \frac{1}{2} + \frac{1}{4} + \frac{1}{8} \qquad = \frac{7}{8} \qquad \text{(the 3-rd partial sum)}$$

$$s_4 = \frac{1}{2} + \frac{1}{4} + \frac{1}{8} + \frac{1}{16} = \frac{15}{16} \qquad \text{(the 4-th partial sum)}$$

$$s_5 = \frac{1}{2} + \frac{1}{4} + \cdots + \frac{1}{32} = \frac{31}{32} \qquad \text{(the 5-th partial sum)}$$

etc.

The partial sums $\{\frac{1}{2}, \frac{3}{4}, \frac{7}{8}, \frac{15}{16}, \frac{31}{32}, \ldots\}$ form a sequence, and the definition says that the original series converges if and only if this sequence converges.

The sequence $\{\frac{1}{2}, \frac{3}{4}, \frac{7}{8}, \frac{15}{16}, \frac{31}{32}, \ldots\}$ can be handled in different ways. If, as in §11.2.4 of The Companion, we are interested only in the Parent Bear question ("Does the sequence converge?"), we can observe that the sequence is increasing and that each term is less than 1 (for a proof, read on). Thus the sequence is a bounded monotone sequence which must converge by Theorem 11.2.2, proving that the series $\sum_{k=1}^{\infty} \frac{1}{2^k}$ also converges.

If we are also interested in answering the Baby Bear question ("To what does the series converge?") we can actually write out the n-th partial sum s_n in what is called "closed form"*: Since

$$s_n = \frac{1}{2} + \frac{1}{4} + \frac{1}{8} + \cdots + \frac{1}{2^n}$$

then

$$2 s_n = 1 + \frac{1}{2} + \frac{1}{4} + \cdots + \frac{1}{2^{n-1}}$$

$$= 1 + \left(\frac{1}{2} + \frac{1}{4} + \cdots + \frac{1}{2^{n-1}} + \frac{1}{2^n} \right) - \frac{1}{2^n}$$

$$= 1 + s_n - \frac{1}{2^n}$$

Thus

$$s_n = 1 - \frac{1}{2^n}$$

Hence

$$\lim_{n \to \infty} s_n = \lim_{n \to \infty} \left(1 - \frac{1}{2^n} \right) = 1$$

so the sequence of partial sums converges to 1 and hence the sum of the infinite series

$$\sum_{k=1}^{\infty} \frac{1}{2^k}$$ is 1. We write this as $$\sum_{k=1}^{\infty} \frac{1}{2^k} = 1.$$

□

Example B. (The Harmonic Series) Does the series $\sum_{k=1}^{\infty} \frac{1}{k} = 1 + \frac{1}{2} + \frac{1}{3} + \frac{1}{4} + \cdots$ converge? If so, find its sum.

_____

* A "closed form" is simply a formula which gives s_n as a function of n and does not involve an increasingly long summation.

Solution.   First we compute a few partial sums:

$$s_1 = 1 \qquad\qquad\qquad = 1 \qquad\qquad \text{(the \quad 1-st partial sum)}$$

$$s_2 = 1 + \frac{1}{2} \qquad\qquad = \frac{3}{2} = 1.500 \qquad \text{(the \quad 2-nd partial sum)}$$

$$s_3 = 1 + \frac{1}{2} + \frac{1}{3} \qquad = \frac{11}{6} \cong 1.833 \qquad \text{(the \quad 3-rd partial sum)}$$

$$s_4 = 1 + \frac{1}{2} + \frac{1}{3} + \frac{1}{4} = \frac{25}{12} \cong 2.083 \qquad \text{(the \quad 4-th partial sum)}$$

$$s_5 = 1 + \frac{1}{2} + \cdots + \frac{1}{5} = \frac{137}{60} \cong 2.283 \qquad \text{(the \quad 5-th partial sum)}$$

Continuing in this way, we find $s_6 = 147/60 = 2.450$, $s_7 = 1089/420 \cong 2.593$,
$s_8 = 2283/840 \cong 2.718$,   etc.

The series $\displaystyle\sum_{k=1}^{\infty} \frac{1}{k}$ converges if the sequence of these partial sums

$\{1, 1.500, 1.833, 2.083, 2.283, \ldots\}$ converges.

The sequence $\{1, 1.500, 1.833, 2.083, 2.283, \ldots, s_n, \ldots\}$ is an increasing sequence since each term is the previous term plus a positive number   (i.e., $s_n = s_{n-1} + \frac{1}{n}$) ,   but an increasing sequence converges only if there is an upper bound or its terms (there's the critical Theorem 11.2.2 again!) However, in Example 7 Anton shows that there is no upper bound! [This is an important argument and it is worth the effort to understand it.   Its essence is that the 2^n-th   term of the sequence is larger than the sum $\underbrace{\frac{1}{2} + \frac{1}{2} + \frac{1}{2} + \cdots + \frac{1}{2}}_{n+1}$.

Since an n can be found which makes the sum of $1/2$'s   larger than any potential upper bound, the terms of the sequence also eventually exceed any potential upper bound.]

Hence the increasing sequence of partial sums diverges.   This is exactly what is needed

to establish that the series $\displaystyle\sum_{k=1}^{\infty} \frac{1}{k}$ diverges.

$\square$

The Harmonic Series $\displaystyle\sum_{k=1}^{\infty}\frac{1}{k}$ of Example B is the most famous and basic example of a divergent series:

$$\boxed{\text{The harmonic series } \sum_{k=1}^{\infty}\frac{1}{k} \text{ diverges.}}$$

This is an important fact which you must remember; it will be used extensively to establish the divergence of other series. Here is an illustration:

Fix a number p in the interval $0 \le p \le 1$ and consider the corresponding p-series $\displaystyle\sum_{k=1}^{\infty}\frac{1}{k^p}$ ($p = 1$ gives the harmonic series). Since $0 \le p \le 1$, then $k^p \le k$ and $\dfrac{1}{k^p} \ge \dfrac{1}{k}$ for every $k \ge 1$. Thus the partial sums $s_n^{(p)}$ for the p-series,

$$s_n^{(p)} = 1 + \frac{1}{2^p} + \cdots + \frac{1}{n^p}$$

are larger than the corresponding partial sums s_n for the harmonic series,

$$s_n = 1 + \frac{1}{2} + \cdots + \frac{1}{n}$$

However, since these smaller harmonic series partial sums diverge to infinity (the harmonic series diverges!), then the larger p-series partial sums must also diverge to infinity. We have thus shown

$$\boxed{\text{The p-series } \sum_{k=1}^{\infty}\frac{1}{k^p} \text{ diverges for } 0 \le p \le 1.}$$

The principle used above to establish this result is known as The Comparison Test. It will be studied further in §§11.5-6.

Note that we have not said anything about what happens with the p-series when $p > 1$. Our comparison test breaks down: if $p > 1$, then $k^p \geq k$ and $\dfrac{1}{k^p} \leq \dfrac{1}{k}$ for any $k > 1$, so that $s_n^{(p)} \leq s_n$. But knowing that the partial sums of $\displaystyle\sum_{k=1}^{\infty} \dfrac{1}{k^p}$ are <u>smaller</u> than the partial sums of the divergent series $\displaystyle\sum_{k=1}^{\infty} \dfrac{1}{k}$ does not tell us whether the smaller series diverges or converges. Both are possible. We'll have to wait for §11.4 to prove that in fact the p-series

$$\sum_{k=1}^{\infty} \dfrac{1}{k^p} \quad \underline{\text{converges}} \text{ when } p > 1.$$

2. <u>Answering the Baby Bear Question: geometric and telescoping series.</u> Remember the Parent Bear and Baby Bear questions: "Does the series converge?" and "To <u>what</u> does the series converge?"? You should always know which one you are being asked. Instructions like "determine if the series converges" or "does the series converge?" are used to ask the Parent Bear question; instructions like "find the sum of the series" or "evaluate the series" are used to ask the Baby Bear question.

An important point to remember is that to be able to answer the Baby Bear question about an infinite series, that is, <u>to be able to find the sum of a convergent series, you must be able to write an expression for the n-th partial sum</u> s_n <u>from which it is absolutely clear what</u> $\displaystyle\lim_{n \to \infty} s_n$ <u>is</u>. For instance, in Example A we found $s_n = 1 - \dfrac{1}{2^n}$ so that we can establish that $\displaystyle\lim_{n \to \infty} s_n = 1$. However, series for which such nice expressions for s_n can be found are rare. Or, to put it in other words, for most series you will encounter, only the Parent Bear question ("Does...?") and not the Baby Bear question ("To what...?") can be answered.

There are two important types of series mentioned in this section whose sum <u>can</u> be found:

geometric and telescoping series. We will discuss each one:

3. Geometric Series.   A geometric series is a series in which each term is obtained from its predecessor by multiplying by a fixed number r , called the ratio. That is, the series has the form

$$\sum_{k=0}^{\infty} a r^k = a + ar + ar^2 + ar^3 + \cdots + ar^n + \cdots$$

where   a   is the first term of the series and   r   is the ratio. The situation for geometric series is very clear-cut:

The Geometric Series Theorem

$$\sum_{k=0}^{\infty} a r^k = a + ar + ar^2 + ar^3 + \cdots$$

(a) converges to $\dfrac{a}{1-r}$ if $|r| < 1$, and

(b) diverges if $|r| \geq 1$.

Anton's Examples 4 and 5 illustrate this theorem. Here are two more examples:

Example D.   Find the sum of the series $\displaystyle\sum_{k=2}^{\infty} (\tfrac{4}{7})^k$.

Solution.   Here the summation starts with $k = 2$, but that does not change the use of the Geometric Series Theorem at all. For $\displaystyle\sum_{k=2}^{\infty} (\tfrac{4}{7})^k = (\tfrac{4}{7})^2 + (\tfrac{4}{7})^3 + (\tfrac{4}{7})^4 + \cdots$ is a geometric series with first term $a = (\tfrac{4}{7})^2$ and ratio $r = \tfrac{4}{7}$. Since $|r| = \tfrac{4}{7} < 1$, the series converges and its sum is $\dfrac{a}{1-r} = \dfrac{(4/7)^2}{1 - (4/7)} = \dfrac{16/49}{3/7} = \dfrac{16}{21}$.   □

Note: Series which do not start with $k = 0$ are stumbling blocks for many newcomers to this

material, but, as Example D shows, they need not be. The first value of k simply determines

the first term of the series, whether it be $k = 0$ or $k = $ something else.

Example E.   Express the repeating decimal $.267757575\ldots$ as a fraction.

Solution.   This decimal has some nonrepeating terms at the beginning. When we break those

off, what remains is a geometric series:

$$.267757575 \ldots \; = \; .267 + .00075 + .0000075 + .000000075$$

$$+ \; .00000000075 \; + \; \cdots$$

$$= \; .267 + .00075 + (.00075)(.01)$$

$$+ \; (.00075)(.0001) + (.00075)(.000001) + \cdots$$

$$= \; .267 + \sum_{k=0}^{\infty} (.00075)(.01)^{k}$$

$$= \; .267 + \frac{.00075}{1 - (.01)} \qquad \text{by the Geometric Series Theorem}$$

$$= \; \frac{267}{1000} + \frac{75/100,000}{1 - (1/100)} = \frac{267}{1000} + \frac{75/100,000}{99/100}$$

$$= \; \frac{267}{1000} + \frac{75}{99,000} = \frac{26,508}{99,000} = \frac{6,627}{24,750} \qquad \Box$$

4.   Telescoping Series.   A telescoping series is a series

in which the n-th partial sum s_n collapses (like a

collapsing telescope) into a simpler, more manageable

expression. Anton's Example 6 is one example. Here

is another :

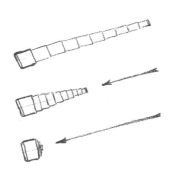

Example F. Find the sum of the series $\displaystyle\sum_{k=1}^{\infty} \frac{4}{(k+1)(k+2)}$.

Solution. Using the technique of partial fractions (see §9.7 if you need a review), we can write

$$\frac{4}{(k+1)(k+2)} = \frac{A}{(k+1)} + \frac{B}{(k+2)} = \frac{A(k+2) + B(k+1)}{(k+1)(k+2)} = \frac{(A+B)k + (2A+B)}{(k+1)(k+2)}$$

from which it follows that $0k + 4 = (A+B)k + (2A+B)$, or

$$\begin{cases} A + B = 0 \\ 2A + B = 4 \end{cases}$$

Solving these equations simultaneously yields $A = 4$ and $B = -4$ so

$$\frac{4}{(k+1)(k+2)} = \frac{4}{k+1} - \frac{4}{k+2}$$

Thus the series is $\displaystyle\sum_{k=1}^{\infty} \frac{4}{(k+1)(k+2)} = \sum_{k=1}^{\infty} \left(\frac{4}{k+1} - \frac{4}{k+2} \right)$. The n-th partial sum of this series is

$$s_n = \sum_{k=1}^{n} \left(\frac{4}{k+1} - \frac{4}{k+2} \right) = \left(\frac{4}{2} - \frac{4}{3} \right) + \left(\frac{4}{3} - \frac{4}{4} \right) + \left(\frac{4}{4} - \frac{4}{5} \right) + \cdots$$

$$\cdots + \left(\frac{4}{n} - \frac{4}{n+1} \right) + \left(\frac{4}{n+1} - \frac{4}{n+2} \right)$$

$$= 2 - \frac{4}{n+2} \qquad \text{since all the terms cancel except the first and last .}$$

Then we find $\displaystyle\lim_{n \to \infty} s_n = \lim_{n \to \infty} \left(2 - \frac{4}{n+2} \right) = 2$ so the sum of the series $\displaystyle\sum_{k=1}^{\infty} \frac{4}{(k+1)(k+2)}$

is 2 . $\qquad\qquad\qquad\qquad\qquad\qquad\qquad\qquad\qquad\qquad\qquad$ □

In Example F note that, once again, our ability to find the sum of the series depends upon our ability to find a "closed form" expression for the n-th partial sum s_n , in this case $s_n = 2 - \dfrac{4}{n+2}$. In telescoping series this is possible because the n-th partial sum "telescopes. " Series in which the terms are fractions with denominators which can be factored (such as Anton's Example 6 and Example F above) are frequently telescoping series. The old integration trick of partial fractions is the way to attack them.

Series in which the terms are logarithms are sometimes telescoping series. Here is an example:

Example G.   Does the series $\displaystyle\sum_{k=2}^{\infty} \ln\left(1 - \frac{1}{k+1}\right)$ converge? If so, find its sum.

Solution.   Since $\ln\left(1 - \dfrac{1}{k+1}\right) = \ln\left(\dfrac{(k+1)-1}{k+1}\right) = \ln\left(\dfrac{k}{k+1}\right) = \ln k - \ln(k+1)$, the

series is $\displaystyle\sum_{k=2}^{\infty} \ln\left(1 - \frac{1}{k+1}\right) = \sum_{k=2}^{\infty} [\ln k - \ln(k+1)]$. Thus the n-th partial sum s_n

is

$$s_n = \sum_{k=2}^{n} [\ln k - \ln(k+1)] = (\ln 2 - \ln 3) + (\ln 3 - \ln 4) + (\ln 4 - \ln 5)$$
$$+ \cdots + (\ln(n-1) - \ln n) + (\ln n - \ln(n+1))$$
$$= \ln 2 - \ln(n+1) \qquad \text{(cancellation)}$$

Thus $\displaystyle\lim_{n \to \infty} s_n = \lim_{n \to \infty} [\ln 2 - \ln(n+1)] = \infty$, so the series $\displaystyle\sum_{k=2}^{\infty} \ln(1 - \frac{1}{k+1})$ diverges.

$\square$

Note that in Example G   we have a series which "telescopes" into a closed form $s_n = \ln 2 - \ln(n+1)$ for the   n-th   partial sum s_n , but that alone does not guarantee that the series converges. Telescoping series can diverge as well as converge! You must always check the limit of partial sums, $\lim\limits_{n \to \infty} s_n$.

Section 11.4: Convegence; The Integral Test

1. <u>The tail of a series; it affects the sum but not the convergence of a series.</u>   It is sometimes useful to think of series as objects with "tails" and "heads" (confirming, for the weak-at-heart, that series are frightful monsters! Actually they're not!). For example, consider the infinite series

$$\underbrace{1 + 2 + 3 + 4 + 5}_{\text{head}} + \underbrace{\frac{1}{6} + \frac{1}{7} + \frac{1}{8} + \frac{1}{9}}_{\text{tail}} + \cdots$$

As the brackets indicate, we will choose to regard the first five terms as a "head" of the series and all the other terms as the corresponding "tail." <u>That is, a "head" of a series consists of the first terms up through some finite number; the corresponding "tail" consists of everything else.</u>

The point is that, because a "head" of a series is just a <u>finite</u> sum of terms, it is a finite number (in the illustration above, the "head" = $1 + 2 + 3 + 4 + 5 = 15$) and hence it does not affect the convergence or divergence of the series. That is,

<table>
<tr><td>

The Tail
Principle
</td><td>

(1) An infinite series converges if and only if its "tail" converges.

(2) An infinite series diverges if and only if its "tail" diverges.
</td></tr>
</table>

Example A.   Determine whether the series $\displaystyle\sum_{k=6}^{\infty} \frac{1}{k}$ converges or diverges.

Solution.   Since $\displaystyle\sum_{k=1}^{\infty} \frac{1}{k} = 1 + \frac{1}{2} + \frac{1}{3} + \frac{1}{4} + \frac{1}{5} + \sum_{k=6}^{\infty} \frac{1}{k}$, the series $\displaystyle\sum_{k=6}^{\infty} \frac{1}{k}$ is the

"tail" of the Harmonic Series corresponding to the "head" $\ 1 + \frac{1}{2} + \frac{1}{3} + \frac{1}{4} + \frac{1}{5}$. Since the

Harmonic Series diverges, The Tail Principle says that its "tail" $\displaystyle\sum_{k=6}^{\infty} \frac{1}{k}$ also diverges.   □

Example B.   Determine whether the series

$$1 + 2 + 3 + 4 + 5 + \frac{1}{6} + \frac{1}{7} + \frac{1}{8} + \frac{1}{9} + \cdots$$

converges or diverges.

Solution.   This series is $\ 1 + 2 + 3 + 4 + 5 + \displaystyle\sum_{k=6}^{\infty} \frac{1}{k}$, i.e., $\displaystyle\sum_{k=6}^{\infty} \frac{1}{k}$ is the "tail"

corresponding to the "head" $\ 1 + 2 + 3 + 4 + 5$. This "tail" is the divergent series of

Example A so, by the Tail Principle, the whole series $\ 1 + 2 + 3 + 4 + 5 + \frac{1}{6} + \frac{1}{7} + \frac{1}{8} + \frac{1}{9} + \cdots$

diverges.   □

Example C.   Express the repeating decimal $.267757575\ldots$ as a fraction.

Solution.   We will regard the nonrepeating terms at the beginning as the "head" of a series:

$$.267757575\ldots = \underbrace{.2 + .06 + .007}_{\text{head}} + \underbrace{\sum_{k=0}^{\infty} (.00075)(.01)^{k}}_{\text{tail}}$$

The "tail" of the series is a geometric series with sum $75/99{,}000$ and thus the series above

converges to "head" + "tail" = $267/1{,}000 + 75/99{,}000 = 6{,}627/24{,}750.$ (See Example E of §11.3 of <u>The Companion</u> for the details.) □

In Theorem 11.4.3(c), Anton states the Tail Principle in a slightly different way: a finite number of terms from the beginning of a series may be deleted without affecting its convergence or divergence. But WATCH OUT! While deleting a finite number of terms at the beginning of a series will not affect its convergence ("Does it converge?"), <u>it will affect its sum.</u> ("To <u>what</u> does it converge?"). This is the point Anton makes in the Remark following Theorem 11.4.3 and newcomers to series sometimes overlook it. Here is the new principle:

> The Head Principle

> "Chopping off the head" of a series
>
> affects its sum but not its convergence.

<u>Example D.</u> Does the series $4 - 3 + \frac{1}{2} + 6 + \left(\frac{2}{3}\right)^3 + \left(\frac{2}{3}\right)^4 + \left(\frac{2}{3}\right)^5 + \cdots$ converge? If so, find its sum.

<u>Solution.</u> If we "chop off the head" of this series, we are left with the "tail" $\sum_{k=3}^{\infty} \left(\frac{2}{3}\right)^k = \left(\frac{2}{3}\right)^3 + \left(\frac{2}{3}\right)^4 + \cdots$. The "tail" is a geometric series with first term $a = \left(\frac{2}{3}\right)^3$ and ratio $r = \frac{2}{3}$ so, by the Geometric Series Theorem of §11.3, the "tail"

converges to $\dfrac{\left(\frac{2}{3}\right)^3}{1 - \frac{2}{3}} = \frac{4}{3}$. Thus we know that the original series converges. It converges to

$$\text{"head"} + \text{"tail"} = \left(4 - 3 + \frac{1}{2} + 6\right) + \frac{4}{3}$$

$$= \frac{15}{2} + \frac{4}{3} = \frac{53}{6}$$

□

2. A Preview.   In §11.3 , we listed two types of series (geometric and telescoping) for which you

can answer the Baby Bear question ("To what does a series converge?").   Other series (Taylor

and Maclaurin series) for which that question can usually be answered will come up later.

For the next few sections (through §11.7), we will be concerned with practical tests

for answering the Parent Bear question ("Does a series converge?").   The tests we will consider

are all based on the   partial sums   definition of convergence, but they are much easier to use.

We look first at series with positive terms; the tests we derive all have the important

Theorem 11.2.2 on bounded monotone sequences (and its series version,   Theorem 11.4.4)

as their basis.   After that, we will consider the more general case: series whose terms are

not necessarily positive.

But we begin with a test which applies to all series: the divergence test...

3. The divergence test: the easiest test of all, but its converse is not true!   The first test you

should use on any series is the divergence test, Anton's Theorem 11.4.2:

> The Divergence Test.   A series diverges if the
>
> limit of its individual terms is not zero.

The reason you should turn to it first is that it is very easy to use and hence it can give you a

lot of information for very little effort.   Even if it proves to be inconclusive, you have lost only

a few seconds.

But beware of A COMMON MISTAKE ! People sometimes try to use the converse of the

divergence test, but the converse of the divergence test is false.   To be specific,

> FALSE Converse: A series converges if the limit
>
> of its individual terms is zero.
>
> NO!   NOT TRUE IN ALL CASES!   A BLUNDER!

The Divergence Test gives information about <u>divergence</u> (hence the name), but <u>not</u> about convergence. Keep this important point in mind!

Example E.   Does the series $\displaystyle\sum_{k=1}^{\infty} \frac{2k^2 + 3k - 6}{k^2 - k + 1}$ converge?

Solution.   Taking the limit of the terms, we get

$$\lim_{k \to \infty} \frac{2k^2 + 3k - 6}{k^2 - k + 1} = \lim_{k \to \infty} \frac{2 + \dfrac{3}{k} + \dfrac{6}{k^2}}{1 - \dfrac{1}{k} + \dfrac{1}{k^2}} = 2$$

(we divided numerator and denominator by k^2). Thus, since the limit is $2 \neq 0$, we know that the series diverges by The Divergence Test.   □

Example F.   Does the series $\displaystyle\sum_{k=1}^{\infty} \frac{1}{\sqrt{k}}$ converge?

Solution.   Taking the limit of the terms, we get $\displaystyle\lim_{k \to \infty} \frac{1}{\sqrt{k}} = 0$. It is tempting, but WRONG, to conclude that this means that the series $\displaystyle\sum_{k=1}^{\infty} \frac{1}{\sqrt{k}}$ converges. In fact, this series <u>diverges</u> since it is a p-series with $p = \frac{1}{2}$ (as discussed in §11.3.2 of <u>The Companion</u>).   □

Example G.   Does the series $\displaystyle\sum_{k=1}^{\infty} \frac{1}{k^2}$ converge?

Solution.   Taking the limit of the terms, we get $\lim\limits_{k \to \infty} \dfrac{1}{k^2} = 0$. Once again, it is tempting,

but WRONG, to use this fact to conclude that the series $\sum\limits_{k=1}^{\infty} \dfrac{1}{k^2}$ converges. In fact, this

series does converge since it is a p-series with p = 2 (see §11.4.3 below), but saying

that it converges because $\lim\limits_{k \to \infty} \dfrac{1}{k^2} = 0$ would be getting the right answer for the wrong

reason!   □

As Example F and G show,

> If the limit of the individual terms of a series equals zero,
>
> the series may converge or it may diverge.

4.   The integral test: you can't use it to evaluate series.   The essence of the integral test (Anton's

Theorem 11.4.5) is that the convergence of a series may be tested by checking the convergence

of an improper integral:

The Integral Test

> $\sum\limits_{k=1}^{\infty} u_k$ converges if and only if $\displaystyle\int_{1}^{\infty} f(x)\,dx$ converges
>
> where $f(x)$ is the function obtained by replacing k by x in u_k.

(Note from Anton's statement of the theorem that the u_k must be positive and that f must

be decreasing and continuous. As a practical matter, these conditions will usually hold in the

exercises you are asked to do.) An important point to remember is that the integral test is a

11.4.7

"Parent Bear" test and not a "Baby Bear" test; that is, it cannot be used to evaluate series.

It does NOT say

<table>
<tr><td>

What the
Integral Test
Does NOT Say

</td><td>

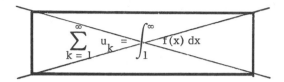

</td></tr>
</table>

Example H.   Does the series $\displaystyle\sum_{k=1}^{\infty} \frac{k}{(4+k^2)^{3/4}}$ converge?

Solution.   The limit of the individual terms is zero (check it!) so The Divergence Test is inconclusive. We thus turn to The Integral Test and replace k by x in $u_k = \dfrac{k}{(4+k^2)^{3/4}}$

to obtain $f(x) = \dfrac{x}{(4+x)^{3/4}}$. Then we evaluate the improper integral

$$\int_1^\infty f(x)\,dx = \int_1^\infty \frac{x\,dx}{(4+x^2)^{3/4}} = \lim_{b\to\infty}\int_1^b \frac{x\,dx}{(4+x^2)^{3/4}}$$

$$= \lim_{b\to\infty}\frac{1}{2}\int_5^{4+b^2} u^{-3/4}\,du$$

Let $u = 4 + x^2$

$\dfrac{du}{dx} = 2x$

$\dfrac{1}{2}\,du = x\,dx$

$$= \lim_{b\to\infty}\frac{1}{2}\left.\frac{u^{1/4}}{1/4}\right|_5^{4+b^2}$$

$$= \lim_{b\to\infty}\left(2\sqrt[4]{4+b^2} - 2\sqrt[4]{5}\right)$$

$$= \infty$$

Thus the improper integral diverges and hence, by The Integral Test, the series

$$\sum_{k=1}^{\infty} \frac{k}{(4 + k^2)^{3/4}} \quad \text{also diverges.}$$ □

Example I.   Does the series $\displaystyle\sum_{k=1}^{\infty} \frac{1}{(k + 1)^3}$ converge?

Solution.   The limit of the individual terms is zero so The Divergence Test is inconclusive.

Hence we turn to The Integral Test: Replacing k by x in $u_k = \dfrac{1}{(k + 1)^3}$ gives

$f(x) = \dfrac{1}{(x + 1)^3}$. Then we evaluate the improper integral

$$\int_1^{\infty} f(x)\, dx = \int_1^{\infty} \frac{1}{(x + 1)^3}\, dx = \lim_{b \to \infty} \int_1^b \frac{1}{(x + 1)^3}\, dx$$

$$= \lim_{b \to \infty} \left. -\frac{1}{2}\, \frac{1}{(x + 1)^2} \right|_1^b \qquad (\text{let}\quad u = x + 1 \quad \text{and integrate})$$

$$= \lim_{b \to \infty} \left[\frac{-1}{2(b + 1)^2} + \frac{1}{8} \right]$$

$$= 0 + \frac{1}{8} = \frac{1}{8}$$

Thus the improper integral converges to $\dfrac{1}{8}$ and hence, by The Integral Test, the series

$\displaystyle\sum_{k=1}^{\infty} \frac{1}{(k + 1)^3}$ also converges. However, we do NOT know what its sum is (it is NOT $\frac{1}{8}$).   □

There is nothing sacred about starting with $k = 1$ in The Integral Test. If a series

starts with $k = m$, take the lower limit of the integral to be m :

11. 4. 9

The Integral Test when k does not start at k = 1

$$\sum_{k=m}^{\infty} u_k \quad \text{converges if and only if} \quad \int_m^{\infty} f(x)\, dx \quad \text{converges}$$

where $f(x)$ is the function obtained by replacing k by x

in u_k and m is a positive integer.

Example J. Does the series $\sum_{k=6}^{\infty} \dfrac{1}{(k-4)^2}$ converge?

Solution. Since the Divergence Test is inconclusive (why?), we turn to The Integral Test: Replacing k by x in $u_k = 1/(k-4)^2$, we obtain $f(x) = 1/(x-4)^2$. Then we evaluate the improper integral

$$\int_6^{\infty} \frac{1}{(x-4)^2}\, dx = \lim_{b \to \infty} \int_6^b \frac{1}{(x-4)^2}\, dx = \lim_{b \to \infty} \left. \frac{-1}{(x-4)} \right|_6^b$$

$$\boxed{\text{NOTE THE } 6}$$

$$= \lim_{b \to \infty} \left[\frac{-1}{b-4} + \frac{1}{2} \right] = \frac{1}{2}$$

Thus by The Integral Test with $m = 6$, the series $\sum_{k=6}^{\infty} \dfrac{1}{(k-4)^2}$ converges (but not necessarily to $\frac{1}{2}$). $\qquad \square$

5. The p-series. If you recall, we jumped the gun on Anton and introduced the p-series in §11.3.1 of The Companion:

$$\sum_{k=1}^{\infty} \frac{1}{k^p} = 1 + \frac{1}{2^p} + \frac{1}{3^p} + \cdots$$

In that section we showed that the p-series diverges for $0 < p \le 1$. However, we did not (could not) establish anything about the case when p is greater than one. Now that he has The Integral Test, Anton can handle that case too:

<table>
<tr><td>Convergence
of p-series</td><td>Suppose p is any fixed, positive number. Then

$$\sum_{k=1}^{\infty} \frac{1}{k^p} \quad \text{converges if } p > 1, \text{ and}$$

diverges if $0 < p \le 1$.</td></tr>
</table>

To prove this result, Anton considers three cases: (i) $p > 1$, (ii) $0 < p < 1$ and (iii) $p = 1$. In the last case $(p = 1)$, the p-series becomes the Harmonic Series, a series already known to diverge (from §11.3). In the first two cases $(p \ne 1)$, Anton applies The Integral Test by showing

$$\int_1^{+\infty} \frac{1}{x^p} \, dx = \lim_{\ell \to +\infty} \left[\frac{\ell^{1-p}}{1-p} - \frac{1}{1-p} \right]$$

Anton then shows this limit is (i) finite when $p > 1$ and (ii) infinite when $0 < p < 1$. If you have trouble following the details, simply plug in specific values of p and observe what happens. For example, in the $p > 1$ case, try $p = 2$:

$$\lim_{\ell \to +\infty} \left[\frac{\ell^{1-2}}{1-2} - \frac{1}{1-2} \right] = \lim_{\ell \to +\infty} \left[-\frac{1}{\ell} + 1 \right] = 1$$

Then try $p = 3$ and get

$$\lim_{\ell \to +\infty} \left[\frac{\ell^{1-3}}{1-3} - \frac{1}{1-3} \right] = \lim_{\ell \to +\infty} \left[-\frac{1}{2\ell^2} + \frac{1}{2} \right] = \frac{1}{2}$$

Soon you will see the general point: when $p > 1$ the exponent of ℓ will be negative and hence the term containing ℓ will approach zero as $\ell \to +\infty$.

At first glance, you might think the p-series test leaves out a major case: the case when $p \leq 0$. But that's an easy case. For when $p = 0$, $\frac{1}{k^p} = \frac{1}{k^0} = 1$ and when $p < 0$, $\frac{1}{k^p}$ is k to a positive exponent. In both of these cases, the limit of the terms of the series is not zero so the series diverges by The Divergence Test, i.e.,

$$\sum_{k=1}^{\infty} \frac{1}{k^p} \text{ diverges if } p \leq 0$$

Example K. Does the series $\sum_{k=1}^{\infty} \sqrt[4]{k}$ converge?

Solution. Written in the form of a p-series (i.e., with k in the denominator), this series is $\sum_{k=1}^{\infty} \frac{1}{k^{-1/4}}$. Thus $p = -\frac{1}{4} < 0$ so the series diverges. □

Example L. Does the series $\sum_{k=1}^{\infty} \frac{1}{\sqrt[4]{k}}$ converge?

Solution. This series may be written $\sum_{k=1}^{\infty} \frac{1}{k^{1/4}}$ so it is a p-series with $p = \frac{1}{4} < 1$.

Thus the series diverges by the p-series result. □

Example M. Does the series $\sum_{k=1}^{\infty} \frac{1}{k^4}$ converge?

Solution. This is a p-series with $p = 4$ so it converges by the p-series result. □

Section 11.5:   Additional Convergence Tests

For a discussion of The Comparison Test, see the next section (§11.6) of The Companion. Keep in mind that in this section we consider only series with positive terms.

1.   The Ratio and Root Tests: sometimes they are inconclusive.   You should notice how similar Anton's Theorems 11.5.2 and 11.5.3 are:

Ratio and Root Tests

Let $\sum u_k$ be a series with positive terms and suppose

$$\rho = \lim_{k \to \infty} \frac{u_{k+1}}{u_k} \quad \text{(Ratio Test)} \quad \text{or} \quad \rho = \lim_{k \to \infty} \sqrt[k]{u_k} \quad \text{(Root Test)} .$$

Then

a)   If $\rho < 1$, the series converges.

b)   If $\rho > 1$ or $\rho = +\infty$, the series diverges.

c)   If $\rho = 1$, the test is inconclusive.

An important thing to note is that there are cases in which each test is inconclusive.   Under those circumstances, another test must be tried.

Example A.   Does the series $\displaystyle\sum_{k=1}^{\infty} \frac{1}{4k+6}$ converge?

Solution.   Applying The Ratio Test, we find

$$\rho = \lim_{k \to \infty} \frac{u_{k+1}}{u_k} = \lim_{k \to \infty} \frac{1/(4(k+1)+6)}{1/(4k+6)} = \lim_{k \to \infty} \frac{4k+6}{4k+10} = \frac{4}{4} = 1$$

so The Ratio Test is inconclusive.   However, using The Integral Test, we find

$$\int_1^\infty \frac{1}{4x + 6}\, dx = \lim_{b \to \infty} \int_1^b \frac{1}{4x + 6}\, dx = \lim_{b \to \infty} \frac{1}{4} \int_{10}^{4b+6} \frac{du}{u}$$

$$= \frac{1}{4} \lim_{b \to \infty} (\ln |4b + 6| - \ln 10) = \infty$$

so the series $\displaystyle\sum_{k=1}^\infty \frac{1}{4k + 6}$ diverges. □

Example B. Does the series $\displaystyle\sum_{k=1}^\infty \frac{1}{k^3}$ converge?

Solution. This is a convergent p-series (i.e., with p = 3). We could stop there, but this series provides a good illustration of how both The Ratio Test and The Root Test can be inconclusive. Applying The Root Test, we find

$$\rho = \lim_{k \to \infty} \sqrt[k]{u_k} = \lim_{k \to \infty} \sqrt[k]{k^{-3}} = \lim_{k \to \infty} k^{-3/k}$$

L'Hôpital's Rule (§10.2) then will show

$$\ln \rho = \ln \left(\lim_{k \to \infty} k^{-3/k} \right) = \lim_{k \to \infty} \left(\ln k^{-3/k} \right) = \lim_{k \to \infty} \left(\frac{-3 \ln k}{k} \right)$$

$$= \lim_{k \to \infty} \left(\frac{-3/k}{1} \right) = 0$$

So $\rho = \displaystyle\lim_{k \to \infty} k^{-3/k} = e^0 = 1$. Thus The Root Test is inconclusive.

Applying The Ratio Test, we find

$$\rho = \lim_{k \to \infty} \frac{u_{k+1}}{u_k} = \lim_{k \to \infty} \frac{1/(k + 1)^3}{1/k^3} = \lim_{k \to \infty} \left(\frac{k}{k + 1} \right)^3 = \left(\lim_{k \to \infty} \frac{1}{1 + (1/k)} \right)^3 = 1$$

Thus The Ratio Test is also inconclusive. □

In Examples 1, 2, 3, 5, 6 and 7, Anton amply illustrates situations in which The Ratio and Root Tests are conclusive.

2. **Three small but frequently occurring results.**   In using The Ratio and Root Tests, three specific results occur frequently (see Anton's Examples as well as those above):

$$
\begin{array}{ll}
\text{i)} & \dfrac{(k+1)!}{k!} = k + 1 \\[2em]
\text{ii)} & \lim_{k \to \infty} \left(1 + \dfrac{1}{k}\right)^{k} = e \\[2em]
\text{iii)} & \lim_{k \to \infty} \dfrac{\ln k}{k} = 0
\end{array}
$$

The first result follows from the fact that $(k + 1)! = (k + 1)(k!)$, the second may be found in §10.3 and the third may be proved using L'Hôpital's Rule as in Example B above. Keeping these results in mind should facilitate your use of The Ratio and Root Tests.

3. **"Which test do I use?" (Version 1).**   When confronted with a series and asked to determine whether it converges or diverges, you have to decide which test to use first (you hope that your first test will also be your last test, but you may not be so lucky!). After you have more experience with the various tests, you'll get better at deciding which test is most appropriate for a given series. Until then, however, we recommend trying the tests in the following order until you find one which works:

1. <u>Direct recognition</u>:   geometric series
                                  p-series
                                  harmonic series

    There is no need to apply special tests to the series just listed: you already

    know when they converge and when they diverge.

2. <u>Divergence Test</u>:   a fast test that will dispense with many divergent series. But

    remember: the test can only prove divergence, not convergence.

Restricting to series with positive terms...

3. <u>Ratio Test</u>:   an easy test to apply to just about any series, it is particularly good

    with <u>factorials</u> and simple expressions raised to the   <u>k-th power</u>, e.g.

    $$2^k / k! \quad , \quad (2k + 1)/3^k$$

    The major problem: the inconclusive case $\rho = 1$ occurs quite often (which

    means you must on on to another test...).

4. <u>Root Test</u>:   a test that is useful with a fairly restricted class of series, i.e., those

    with   <u>k-th powers</u> such as

    $$\left(\frac{k + 1}{4k - 2} \right)^k \quad \text{or} \quad \left(1 - \frac{1}{k} \right)^{k^2}$$

    The major problem is that the expression $\sqrt[k]{u_k}$ can be very unpleasant unless

    u_k is a   k-th power of something reasonable.

[5.   <u>Limit Comparison Test</u>] - To be covered in the next section.

6. <u>Integral Test</u>:   a test that is applicable only to series whose terms become

    <u>integrable</u> functions, e.g.,

    $$\frac{1}{k^2 + 1} \quad , \quad \frac{1}{k \ln k}$$

The major problem is that many series have terms which do not yield integrable

functions (e. g. , any series with a factorial).

[7.   Basic Comparison Test] - To be covered in the next section.

Note that we have listed two tests which have not yet been covered.   Those blanks will

be filled in later.

We emphasize that following this order is not mandatory. With practice, you will develop

a "feel" for the various tests which will lead you to try a particular test without considering

tests previous to it on the list.   However, until you develop the confidence to "skip around, " the

order we suggest should prove helpful.

Example C.   Use any appropriate test to determine whether the series $\displaystyle\sum_{k=1}^{\infty} \frac{\ln k}{k}$ converges

or diverges.

Solution.   We try the tests in the suggested order:

1.   Direct recognition.   This series is not a geometric series or the harmonic series or

   a p-series . So on we go to the ...

2.   Divergence Test. $\displaystyle\lim_{k \to \infty} \frac{\ln k}{k} = \lim_{k \to \infty} \frac{1/k}{1} = \lim_{k \to \infty} \frac{1}{k} = 0$

   $\uparrow$

   L'Hôpital's Rule (§10. 3)

so The Divergence Test is inconclusive. Since the terms of the series are positive,

on we go to the ...

3. <u>Ratio Test.</u> $\rho = \lim\limits_{k \to \infty} \dfrac{u_{k+1}}{u_k} = \lim\limits_{k \to \infty} \dfrac{\frac{\ln(k+1)}{k+1}}{\frac{\ln k}{k}} = \lim\limits_{k \to \infty} \dfrac{k}{k+1} \; \dfrac{\ln(k+1)}{\ln k}$

$$= \left(\lim\limits_{k \to \infty} \frac{k}{k+1} \right) \left(\lim\limits_{k \to \infty} \frac{\ln(k+1)}{\ln k} \right)$$

$$= \left(\lim\limits_{k \to \infty} \frac{k}{k+1} \right) \left(\lim\limits_{k \to \infty} \frac{1/(k+1)}{1/k} \right)$$

$\boxed{\text{L'Hôpital's Rule } (\S 10.3)}$

$$= \left(\lim\limits_{k \to \infty} \frac{1}{1+(1/k)} \right) \left(\lim\limits_{k \to \infty} \frac{1}{1+(1/k)} \right)$$

$$= (1)(1) = 1$$

so The Ratio Test in inconclusive. So on we go to the ...

4. <u>Root Test.</u> $\rho = \sqrt[k]{u_k} = \sqrt[k]{\dfrac{\ln k}{k}} = ????$

so we abandon The Root Test and go on to the ...

5. <u>Limit Comparison Test</u> - which we don't have yet. So on we go to the ...

6. <u>Integral Test.</u> Replacing k by x in $u_k = \dfrac{\ln k}{k}$ gives $f(x) = \dfrac{\ln x}{x}$. Then we evaluate the improper integral

$$\int_1^{\infty} \frac{\ln x}{x} \, dx = \lim\limits_{b \to \infty} \int_1^{b} \frac{\ln x}{x} \, dx = \lim\limits_{b \to \infty} \left. (\ln x)^2 \right|_1^{b} \quad (\text{let } u = \ln x \text{ so } du = \frac{1}{x} \, dx)$$

$$= \lim\limits_{b \to \infty} [(\ln b)^2 - (\ln 1)^2] = \lim\limits_{b \to \infty} [(\ln b)^2 - 0]$$

$$= \lim\limits_{b \to \infty} (\ln b)^2 = \infty$$

BINGO! We finally found a test which works! The series diverges by The Integral Test. $\square$

Example D.   Use any appropriate test to determine whether the series $\displaystyle\sum_{k=3}^{\infty} \frac{2k-1}{2^k}$ converges or diverges.

Solution.   We try the tests in the suggested order:

1.   **Direct recognition.**   This series is not a geometric, harmonic or p-series. So on we go to the . . .

2.   **Divergence Test.**

$$\lim_{k \to \infty} \frac{2k-1}{2^k} = \lim_{k \to \infty} \frac{2}{2^k \ln 2} = \frac{1}{\ln 2} \lim_{k \to \infty} \frac{2}{2^k} = 0$$

$$\boxed{\text{L'Hôpital's Rule } (\S 10.3)}$$

so The Divergence Test is inconclusive. Since the terms of the series are positive, on we go to the . . .

3.   **Ratio Test.**

$$\rho = \lim_{k \to \infty} \frac{u_{k+1}}{u_k} = \lim_{k \to \infty} \frac{\dfrac{2(k+1)+1}{2^{k+1}}}{\dfrac{2k+1}{2^k}} = \lim_{k \to \infty} \frac{1}{2} \frac{2k+3}{2k+1}$$

$$= \frac{1}{2} \lim_{k \to \infty} \frac{2+(3/k)}{2+(1/k)} = \frac{1}{2} < 1$$

BINGO! The series converges by The Ratio Test. (Notice that the fact that the series starts with $k = 3$ did not need to be considered because of the "Tail Principle" ($\S 11.4.1$ of The Companion).)   ☐

Example E.   Use any appropriate test to determine whether the series $\displaystyle\sum_{k=1}^{\infty} \left(\frac{3k}{4k-1}\right)^{2k}$ converges.

Solution.   We try the tests in the suggested order:

1.   <u>Direct recognition.</u>   The series is not a geometric, harmonic or p-series. So on

we go to the...

2.   <u>Divergence Test.</u>

$$\lim_{k \to \infty} \left(\frac{3k}{4k - 1} \right)^{2k} = \lim_{k \to \infty} \left(\frac{3}{4 - (1/k)} \right)^{2k} = \lim_{k \to \infty} \left(\frac{3}{4} \right)^{2k} = 0$$

$$\boxed{\text{since } \frac{1}{k} \to 0} \qquad \boxed{\text{since } \frac{3}{4} < 1}$$

so The Divergence Test is inconclusive. Since the terms of the series are positive,

on we go to the...

3.   <u>Ratio Test.</u>

$$\rho = \lim_{k \to \infty} \frac{u_{k+1}}{u_k} = \lim_{k \to \infty} \frac{\left(\dfrac{3(k+1)}{4(k+1) - 1} \right)^{2(k+1)}}{\left(\dfrac{3k}{4k - 1} \right)^{2k}}$$

$$= \lim_{k \to \infty} \frac{(3k + 3)^{2k+2}}{(3k)^{2k}} \cdot \frac{(4k - 1)^{2k}}{(4k + 3)^{2k+2}} = \; ????$$

so we abandon The Ratio Test and go on to the...

4.   <u>Root Test.</u>

$$\rho = \lim_{k \to \infty} \sqrt[k]{u_k} = \lim_{k \to \infty} \sqrt[k]{\left(\frac{3k}{4k - 1} \right)^{2k}} = \lim_{k \to \infty} \left(\frac{3k}{4k - 1} \right)^2$$

$$= \left(\lim_{k \to \infty} \frac{3k}{4k - 1} \right)^2 = \left(\lim_{k \to \infty} \frac{3}{4 - (1/k)} \right)^2 = \left(\frac{3}{4} \right)^2 = \frac{9}{16} < 1$$

BINGO! The series converges by The Root Test. ☐

Example F.   Use any appropriate test to determine whether the series $\displaystyle\sum_{k=1}^{\infty}\left(\frac{3\,k^2}{2\,k^2+5}\right)^{k}$ converges or diverges.

Solution.   We try the tests in the suggested order:

1.   Direct recognition.   The series is not a geometric, harmonic or p-series. So on we go to the ...

2.   Divergence Test.

$$\lim_{k\to\infty}\left(\frac{3\,k^2}{2\,k^2+5}\right)^{k} = \lim_{k\to\infty}\left(\frac{3}{2+(5/k^2)}\right)^{k} = \lim_{k\to\infty}\left(\frac{3}{2}\right)^{k} = +\infty$$

$$\boxed{\text{since } \frac{5}{k^2}\to 0}\qquad\boxed{\text{since } \frac{3}{2}>1}$$

        BINGO! The series diverges by The Divergence Test.   □

Example G.   Use any appropriate test to determine whether the series $\displaystyle\sum_{k=1}^{\infty}\frac{2^{k+1}}{3^{k-1}}$ converges or diverges.

Solution.   We try the tests in the suggested order:

1.   Direct recognition.   BINGO! The series is

$$\sum_{k=1}^{\infty}\frac{2^{k+1}}{3^{k-1}} = \frac{2^2}{3^0} + \frac{2^3}{3^1} + \frac{2^4}{3^2} + \frac{2^5}{3^3} + \cdots \; ,$$

        a geometric series with first term $a = \dfrac{2^2}{3^0} = 4$ and ratio $r = \dfrac{2}{3}$. Since

$|r| = \dfrac{2}{3} < 1$, the series converges to $\dfrac{a}{1-r} = \dfrac{4}{1-(2/3)} = 12$.   □

Section 11.6: <u>Applying the Comparison Test</u>

As was the case with The Ratio, Root and Integral Tests, <u>in this section we consider only</u> <u>series with positive terms.</u>

1. <u>The Basic Comparison Test: the test of last resort.</u> Here is an informal (and hopefully more memorable) statement of Anton's Theorem 11.5.1:

Basic
Companion
Test

> The convergence of a positive series $\sum a_k$ can be tested as follows:
>
> a) If $\sum a_k$ is "smaller" than a convergent positive series $\sum b_k$,
>
> then $\sum a_k$ <u>converges.</u>
>
> b) If $\sum a_k$ is "larger" than a divergent positive series $\sum b_k$,
>
> then $\sum a_k$ <u>diverges.</u>

More loosely

> If a series converges, any "smaller" one also converges.
>
> If a series diverges, any "larger" one also diverges.

Stated in this form, The Basic Comparison Test is reminiscent of the Pinching Theorem for functions (Anton's Theorem 3.3.1); it says that one series "pushes" another toward convergence or divergence.

The Basic Comparison Test is by far the hardest series convergence test to use for a very simple reason: not only do you have to apply the test, you have to <u>choose</u> what to apply it to! That is, you have to compare a given series to one you choose. <u>It's the choosing of the comparison</u> <u>series that's the hard part</u>! For this reason, The Basic Comparison Test should be used only

a) when the comparison series is absolutely obvious to you or b) as a last resort, after all

other tests have been tried with no success.

When the comparison series is "obvious" to you, The Basic Comparison Test can save

a lot of time. For example, in Example C of §11.5 in The Companion, we went all the way

down our list to The Integral Test to show that $\displaystyle\sum_{k=1}^{\infty} \frac{\ln k}{k}$ diverges. But if we had noticed that

$\ln k \geq 1$ for $k \geq 3$, then we could have said:

$$\frac{\ln k}{k} \geq \frac{1}{k} \qquad \text{for} \qquad k \geq 3$$

so $\displaystyle\sum_{k=3}^{\infty} \frac{\ln k}{k}$ is "larger" than the divergent Harmonic Series $\displaystyle\sum_{k=3}^{\infty} \frac{1}{k}$ and hence diverges

by The Basic Comparison Test. (Actually, we're also using the "Tail Principle" of §11.4 of

The Companion.) Much faster!

If a comparison series is not "obvious" to you, however, it really makes very little sense

to thrash about looking for one, except of course as a last resort. Remember that The

Basic Comparison Test is listed last in the suggested order of tests in §11.5.3 of The

Companion.

2.   A two-step procedure for choosing a comparison series.   Anton gives what amounts to a two-

step procedure for finding an appropriate comparison series $\sum b_k$ for the given series $\sum a_k$:

Step I.   Use the Informal Principles to determine the "basic form" of the comparison

series $\sum b_k$, i.e.,

drop terms containing lower powers of k (including

constant terms) out of a_k to obtain b_k.

If the new series $\sum b_k$ is a series you recognize (e. g. , a geometric, harmonic or p-series) or a series whose convergence or divergence you can determine using one of the other tests, go on to Step II. <u>If you cannot determine the convergence or divergence of</u> $\sum b_k$, <u>using it in The Basic Comparison Test accomplishes nothing.</u>

<u>Step II.</u> See whether the series $\sum b_k$ obtained in Step I compares properly with the original series $\sum a_k$, i. e. ,

 (i) If $\sum b_k$ converges, then we want $a_k \leq b_k$ (conclusion: $\sum a_k$ converges), or

 (ii) If $\sum b_k$ diverges, then we want $a_k \geq b_k$ (conclusion: $\sum a_k$ diverges).

If the series $\sum b_k$ does NOT compare properly with the original series $\sum a_k$, try to modify the $\sum b_k$ series in such a way that it will compare properly. <u>If it cannot be made to compare properly, using the series</u> $\sum b_k$ <u>in The Basic Comparison Test accomplishes nothing.</u>

Step I in the procedure is straightforward and poses no serious problems; the difficulties lie in Step II. Here are two examples illustrating some of the problems:

<u>Example A.</u> Does the series $\displaystyle\sum_{k=1}^{\infty} \frac{1}{k + \sqrt{k}}$ converge or diverge?

<u>Solution.</u> Step I gives us no trouble: since $k + \sqrt{k}$ equals $k^1 + k^{1/2}$, we drop the lower power $(k^{1/2})$ of k to obtain the harmonic series $\displaystyle\sum_{k=1}^{\infty} \frac{1}{k}$ as the "basic form" of our

comparison series.   Therefore we strongly suspect that the original series diverges.

Step II is more troublesome: Since $k + \sqrt{k} \geq k$, taking reciprocals gives

$\dfrac{1}{k + \sqrt{k}} \leq \dfrac{1}{k}$ (note that the inequality is reversed). So $\sum \dfrac{1}{k + \sqrt{k}}$ is a _smaller_ series

than the divergent harmonic series $\sum \dfrac{1}{k}$. Oops! The Basic Comparison Test gives _no_

information about what happens when a smaller series is compared with larger divergent series.

[If you wrote on a test "$\sum \dfrac{1}{k + \sqrt{k}}$ diverges by The Basic Comparison Test with the Harmonic

Series $\sum \dfrac{1}{k}$," you would probably lose credit for "insufficient reason."]

So we must tinker with $\sum \dfrac{1}{k}$: we must do something to it which makes $\sum \dfrac{1}{k + \sqrt{k}}$

greater (not less) than the resulting divergent series. Taking reciprocals reverses inequalities

so we want to make the denominator $k + \sqrt{k}$ _less_ than the denominator of the result. Well,

instead of dropping $\sqrt{k}$ from $k + \sqrt{k}$, i.e., replacing it with the smaller term 0, let's

replace it with a _larger_ term, say k. Then $k + \sqrt{k} \leq k + k = 2k$, so that taking recip-

rocals gives

$$\dfrac{1}{k + \sqrt{k}} \geq \dfrac{1}{2k}$$

Thus $\sum \dfrac{1}{k + \sqrt{k}}$ is a larger series than the divergent series $\dfrac{1}{2}\left(\sum \dfrac{1}{k}\right)$. Ahh, The Basic

Comparison Test now proves that $\sum \dfrac{1}{k + \sqrt{k}}$ diverges. Whew!   □

Example B.   Does the series $\displaystyle\sum_{k=2}^{\infty} \dfrac{1}{k^3 - k}$ converge or diverge?

Solution.   Step I is easy: we drop the lower power (k^1) of the denominator and obtain the convergent p-series $\displaystyle\sum_{k=2}^{\infty} \frac{1}{k^3}$ as the "basic form" of our comparison series. Thus we

stongly suspect that the original series converges.

    Step II is not so easy: Since $k^3 - k < k^3$, we have $\dfrac{1}{k^3 - k} > \dfrac{1}{k^3}$ for all $k \geq 2$.

So $\displaystyle\sum \frac{1}{k^3 - k}$ is a <u>larger</u> series than the convergent series $\displaystyle\sum \frac{1}{k^3}$, a situation about which

The Basic Comparison Test gives <u>no information.</u>

    Thus we must tinker with $\displaystyle\sum \frac{1}{k^3}$: we must do something to it which makes $\displaystyle\sum \frac{1}{k^3 - k}$

<u>less</u> (not greater) than the resulting convergent series. Taking reciprocals reverses inequalities

so we want to make the denominator $k^3 - k$ <u>greater</u> than the denominator of the result. Well,

$k^3 - k$ will be greater than an expression in which something larger than k is subtracted

from k^3 . We note that $\frac{1}{2} k^3 \geq k$ for $k \geq 2$ so $k^3 - \frac{1}{2} k^3 \leq k^3 - k$ for $k \geq 2$. Since

$k^3 - \frac{1}{2} k^3 = \frac{1}{2} k^3$, taking reciprocals yields

$$\frac{1}{k^3 - k} \leq \frac{2}{k^3} \qquad \text{for} \quad k \geq 2$$

Thus $\displaystyle\sum \frac{1}{k^3 - k}$ is a smaller series than the convergent series $2 \displaystyle\sum \frac{1}{k^3}$ and therefore it

converges by The Basic Comparison Test. Whew again! $\square$

    Anton's Examples 3, 4, 5 and 6 also illustrate the Step I - II procedure for finding

the comparison series. In all four Examples, Step I is an easy application of The Informal

Principles. In Examples 3 and 5 , there is no need to perform Step II modifications; the

series $\displaystyle\sum b_k$ compares properly without modification with the original series. In Examples 4

and 6 , however, Step II modifications are necessary.

You will become more skilled at making Step II modifications (i. e. , at "tinkering") as you acquire more experience working with series, but, fundamentally, the modification process will still be a "fishing expedition" : you cast about (in an informed, intelligent way of course) for a comparison series which "works. " However, there is...

3. A better way: avoid Step II by using The Limit Comparison Test.   Fortunately, there is an easier-to-use version of the Comparison Test which avoids Step II! It is Anton's Theorem 11.6.3:

The Limit
Comparison
Test

> If two series $\sum a_k$ and $\sum b_k$ are such that
>
> $$\rho = \lim_{k \to \infty} \frac{a_k}{b_k}$$ is positive and finite, then both
>
> series converge or both series diverge.

We will usually denote "ρ is positive and finite" by writing $0 < \rho < \infty$.

Here's how to use The Limit Comparison Test: Given a positive series $\sum a_k$, apply Step I (The Informal Principles) as before to obtain $\sum b_k$. However, we no longer have to obtain an inequality between $\sum a_k$ and $\sum b_k$ so we drop Step II, and in its place perform

> Step III.   Compute the limit
>
> $$\rho = \lim_{k \to \infty} \frac{a_k}{b_k}$$
>
> If $0 < \rho < \infty$, then (i) $\sum a_k$ converges if $\sum b_k$ converges,
>
>   (ii) $\sum a_k$ diverges if $\sum b_k$ diverges.

Step III is usually easier to use than Step II, and hence The Limit Comparison Test is generally to be preferred over The Basic Comparison Test. Anton's Example 7 illustrates the Step I / III procedure. Here are other examples:

Example A (Revisited). Does the series $\displaystyle\sum_{k=1}^{\infty} \frac{1}{k + \sqrt{k}}$ converge or diverge?

Solution. Step I: Applying The Informal Principles as before yields $\displaystyle\sum_{k=1}^{\infty} \frac{1}{k}$, the divergent Harmonic Series, as the comparison series. Step III: Using The Limit Comparison Test, we find

$$\rho = \lim_{k \to \infty} \frac{a_k}{b_k} = \lim_{k \to \infty} \frac{1/(k + \sqrt{k})}{1/k} = \lim_{k \to \infty} \frac{k}{k + \sqrt{k}} = \lim_{k \to \infty} \frac{1}{1 + (1/\sqrt{k})} = 1$$

Since $\rho = 1$ is positive and finite, we know that $\displaystyle\sum_{k=1}^{\infty} \frac{1}{k + \sqrt{k}}$ diverges (since the comparison series $\displaystyle\sum_{k=1}^{\infty} \frac{1}{k}$ diverges). $\qquad\square$

Example B (Revisited). Does the series $\displaystyle\sum_{k=1}^{\infty} \frac{1}{k^3 - k}$ converge or diverge?

Solution. Step I: As before The Informal Principles yield $\displaystyle\sum_{k=1}^{\infty} \frac{1}{k^3}$, a convergent p-series with $p = 3 > 1$, as the comparison series. Step III: We find

$$\rho = \lim_{k \to \infty} \frac{a_k}{b_k} = \lim_{k \to \infty} \frac{1/(k^3 - 1)}{1/k^3} = \lim_{k \to \infty} \frac{k^3}{k^3 - 1} = 1$$

Since $\rho = 1$ is positive and finite, we know that $\displaystyle\sum_{k=1}^{\infty} \frac{1}{k^3 - 1}$ converges (since the comparison series $\displaystyle\sum_{k=1}^{\infty} \frac{1}{k^3}$ converges) by The Limit Comparison Test. $\qquad\square$

You can see that Examples A and B were easier with The Limit Comparison Test than with The Basic Comparison Test.

Example C.   Does the series $\displaystyle\sum \frac{4k^2 - 2k + 7}{6k^3 + k - 6}$ converge or diverge?

Solution.   Step I: Ignoring all the terms containing powers smaller than k^2 in the numerator and k^3 in the denominator, we obtain $\displaystyle\sum_{k=1}^{\infty} \frac{4k^2}{6k^3} = \frac{2}{3} \sum_{k=1}^{\infty} \frac{1}{k}$ (a divergent series since it is a multiple of the divergent Harmonic Series) as the comparison series.

Step III: Then

$$\rho = \lim_{k \to \infty} \frac{a_k}{b_k} = \lim_{k \to \infty} \frac{(4k^2 - 2k + 7)/(6k^3 + k - 6)}{2/3\,k}$$

$$= \frac{3}{2} \lim_{k \to \infty} \frac{4k^3 - 2k^2 + 7k}{6k^3 + k - 6} = \frac{3}{2} \cdot \frac{4}{6} = 1$$

so by The Limit Comparison Test we know that $\displaystyle\sum_{k=1}^{\infty} \frac{4k^2 - 2k + 7}{6k^3 + k - 6}$ diverges (since the comparison series $\displaystyle\frac{2}{3} \sum_{k=1}^{\infty} \frac{1}{k}$ diverges).

Example D.   Does the series $\displaystyle\sum_{k=1}^{\infty} \frac{6}{4(2^k) - 7}$ converge?

Solution.   Step I: Ignoring the constant -7, we obtain $\displaystyle\frac{3}{2} \sum_{k=1}^{\infty} \frac{1}{2^k}$ as the comparison series.   This is a geometric series with ratio $r = \frac{1}{2}$, and hence it converges.

Step III: Then

$$\rho = \lim_{k \to \infty} \frac{a_k}{b_k} = \lim_{k \to \infty} \frac{6/(4(2^k) - 7)}{(\frac{3}{2})(\frac{1}{2^k})} = \left(\frac{2 \cdot 6}{3}\right) \lim_{k \to \infty} \frac{2^k}{4(2^k) - 7}$$

$$= 4 \lim_{k \to \infty} \frac{1}{4 - 7/2^k} = 4\left(\frac{1}{4}\right) = 1$$

so by The Limit Comparison Test we know that $\displaystyle\sum_{k=1}^{\infty} \frac{6}{4(2^k) - 7}$ converges (since the comparison

series $\displaystyle\frac{3}{2} \sum_{k=1}^{\infty} \frac{1}{(2^k)}$ converges). $\square$

4. <u>Some common mistakes in using The Limit Comparison Test.</u> When using the Step I / III

procedure, watch out for the following <u>COMMON MISTAKES</u>:

i.) Do not confuse the ρ of The Limit Comparison Test with the ρ's of The Ratio and

Root Tests:

Limit Comparison Test: $\rho = \lim_{k \to \infty} a_k / b_k$

If $0 < \rho < \infty$, then $\sum a_k$ and $\sum b_k$ either both converge or

both diverge.

Ratio and Root Tests: $\rho = \lim_{k \to \infty} \frac{a_{k+1}}{a_k}$ <u>or</u> $\rho = \lim_{k \to \infty} \sqrt[k]{a_k}$

If $0 \leq \rho < 1$, then $\sum a_k$ converges,

$\rho = 1$, then the test is inconclusive,

$1 < \rho$, then $\sum a_k$ diverges.

Thus, for example, obtaining $\rho = 1$ in The Limit Comparison Test does <u>not</u> imply that "the test is inconclusive" and obtaining $\rho < 1$ in The Limit Comparison Test does not imply that "the series converges."

ii.)   Do not attempt to <u>evaluate</u> series using The Limit Comparison Test. Remember that if $\rho = \lim\limits_{k \to \infty} \dfrac{a_k}{b_k}$ is finite and positive, the series $\sum a_k$ and $\sum b_k$ both converge or

both diverge, but <u>they do not necessarily have the same sum.</u> Thus you <u>cannot</u> say that

$\sum a_k = \sum b_k$. For example, you cannot conclude that $\displaystyle\sum_{k=1}^{\infty} \dfrac{6}{4(2^k) - 7} = \dfrac{3}{2}$ in

Example D even though the comparison series $\dfrac{3}{2} \displaystyle\sum_{k=1}^{\infty} \dfrac{1}{2^k}$ converges to $\dfrac{3}{2}$.

iii.)   Do not continue to use a bad comparison series out of stubbornness or lack of alertness. Recognize bad choices of comparison series: if $\rho = 0$ or $\rho = \infty$ in The Limit Comparison Test, then that test is inconclusive and you have made a poor choice of comparison series (you should try again until you find one which makes ρ positive and finite). You have also made a poor choice of comparison series if you discover that you do not know that the comparison series converges or diverges.

iv.)   Do not use The Informal Principles (Step I) alone to make a statement such as

" $\displaystyle\sum_{k=1}^{\infty} \dfrac{4k^2 - 2k + 7}{6k^3 + k - 6}$ diverges by The Informal Principles since it behaves like $\dfrac{2}{3} \displaystyle\sum_{k=1}^{\infty} \dfrac{1}{k}$. "

The Informal Principles are just that, <u>informal</u> principles. They can be used to choose a comparison series, but not to justify formally a result. The Limit Comparison Test is needed for that formal justification.

5.   <u>Which test do I use? (an up-date).</u> In §11.5.3 of <u>The Companion</u> we gave a preferred order for using the various convergence tests. In that list, we left room for The Limit Comparison Test as #5 and The Basic Comparison Test as #7. Their descriptions

should be:

5. <u>Limit Comparison Test</u>: a test that applies well to any series which can be "simplified" (using The Informal Principles) into a series whose convergence properties are known. Problems arise if it is not known whether the "simplified" series converges or diverges.

7. <u>Basic Comparison Test</u>: a test that applies well to any series which can be "simplified" into a series whose convergence properties are known. Problems arise if (i) it is not known whether the "simplified" series converges or diverges or (ii) the "simplified" series does not compare properly with the original series.

<u>Section 11.7</u>: Alternating Series; Conditional Convergence

1. <u>Alternating Series: an easy-to-recognize type.</u> Alternating series are series whose terms alternate between being positive and being negative. For this reason, they are extremely easy to recognize. You can spot them instantly by recognizing

i) alternating + and - signs in the "written-out" version $a_1 - a_2 + a_3 - a_4 + \cdots$ (or $-a_1 + a_2 - a_3 + \cdots$), e.g.,

$$1 - \frac{1}{2} + \frac{1}{3} - \frac{1}{4} + \cdots \, , \qquad \text{or}$$

ii) an "alternating sign term" such as $(-1)^k$ or $(-1)^{k+1}$ in the $\sum$ version, e.g.,

$$\sum_{k=1}^{\infty} (-1)^{k+1} \frac{1}{k}$$

Once you spot an alternating series, your reaction should be "whoopee!" because you do not have to "shop around" for the correct convergence test. <u>Use The Alternating Series Test;</u>

<u>it's as simple as that.</u> The test is Anton's Theorem 11.7.1 :

<table>
<tr><td>Alternating
Series
Test</td><td>An alternating series converges if

a) the absolute values of its terms are nonincreasing, and

b) the absolute values of its terms have limit zero as $k \to \infty$.</td></tr>
</table>

(Note that in Anton the <u>terms</u> of an alternating series are designated $(-1)^{k+1} a_k$ or $(-1)^k a_k$

and <u>their absolute values</u> are designated a_k .)

      Anton's Examples 1 and 2 illustrate the use of The Alternating Series Test. Here is a

more complicated example, one which does not appear initially to be an alternating series:

<u>Example A.</u>   Does the series $\displaystyle\sum_{k=0}^{\infty} \frac{\sin\left(\frac{\pi}{2} + k\pi\right)(4k+3)}{(2k+1)(2k+2)}$ converge or diverge?

<u>Solution.</u>   This does not appear to be an alternating series because the tell-tale $(-1)^k$ sign

is not there. However, using a trigonometric identity for $\sin(A + B)$ (see Appendix G) , we

find

$$\sin\left(\frac{\pi}{2} + k\pi\right) = \sin\frac{\pi}{2} \cos k\pi + \cos\frac{\pi}{2} \sin k\pi$$

$$= (1)(-1)^k + (0)(0)$$

$$= (-1)^k$$

Thus

$$\sum_{k=0}^{\infty} \frac{\sin\left(\frac{\pi}{2} + k\pi\right)(4k+3)}{(2k+1)(2k+2)} = \sum_{k=0}^{\infty} (-1)^k \frac{(4k+3)}{(2k+1)(2k+2)}$$

Now we have an alternating series so we apply The Alternating Series Test:

a)   The absolute values of the k and (k + 1) terms are just those terms without the

alternating sign term: $a_k = \dfrac{4k + 3}{(2k + 1)(2k + 2)}$ and

$a_{k+1} = \dfrac{4(k + 1) + 3}{[2(k + 1) + 1][2(k + 1) + 2]} = \dfrac{4k + 7}{(2k + 3)(2k + 4)}$. We want to prove that

$a_k \geq a_{k+1}$, i.e., that

$$\frac{4k + 3}{(2k + 1)(2k + 2)} \geq \frac{4k + 7}{(2k + 3)(2k + 4)} \qquad (*)$$

Cross multiplying (everything is positive so there is no change in the direction of the

inequality) and then cancelling out similar terms results in the equivalent inequality

$$16k^2 + 40k + 22 \geq 0 \qquad (**)$$

Since (**) is always true for $k \geq 0$, then (*) is always true for $k \geq 0$. Thus

the absolute values of the terms of the series are nonincreasing, as desired.

b)   The limit of the absolute values of the terms is

$$\lim_{k \to \infty} \frac{4k + 3}{(2k + 1)(2k + 2)} = \lim_{k \to \infty} \frac{4k + 3}{4k^2 + 6k + 2} = \lim_{k \to \infty} \frac{4/k + 3/k^2}{4 + 6/k + 2/k^2} = 0$$

Hence the series converges by The Alternating Series Test.   □

2. <u>Error bounds in series approximations: the general case.</u>   The determination of error bounds

in approximating the sum of an infinite series is of crucial importance in applications. As you

will see in later sections, the values for all the functions e^x , ln x , sin x , etc. are

computed by the use of infinite series, and hence we must have reliable methods to control the

errors in our approximations.

There are two important types of error bound questions which can be asked for any

series $\sum u_k$. Suppose

$$S \quad \text{is the sum of the series,} \quad S = \sum_{k=1}^{\infty} u_k , \quad \text{and}$$

$$s_n \quad \text{is the} \quad \text{n-th} \quad \text{partial sum,} \quad s_n = \sum_{k=1}^{n} u_k$$

<table>
<tr><td rowspan="2">Error
Bound
Questions</td><td>Type I Question: For a <u>given value of</u> n , what is the maximum possible error ε obtained when approximating S by s_n ?</td></tr>
<tr><td>Type II Question: For a <u>given accuracy</u> ε > 0 , what is the smallest value of n for which s_n approximates S to within the accuracy ε ?</td></tr>
</table>

In the first question, we are given n and want ε ; in the second, we are given ε and

want n .

We can illustrate these questions with a simple but important example: In §11.10 we

will show that

$$e = \sum_{k=0}^{\infty} \frac{1}{k!} = 1 + 1 + \frac{1}{2!} + \frac{1}{3!} + \frac{1}{4!} + \cdots$$

This will give us a way to calculate the value of e ... well, to <u>approximate</u> its value, anyway!

Specific examples of Type I and Type II error bound questions we could ask about this

approximation are:

Type I: What is the largest possible error we might obtain if we approximate

e   by the first five terms in the series expansion, i. e. ,

$$e \cong 1 + 1 + \frac{1}{2!} + \frac{1}{3!} + \frac{1}{4!} \quad ?$$

Type II: How many terms in the series

$$\sum_{k=0}^{\infty} \frac{1}{k!} = 1 + 1 + \frac{1}{2!} + \frac{1}{3!} + \frac{1}{4!} + \cdots$$

must we add up in order to be guaranteed an approximation of   e   which

is in error by no more than $\pm .00005$?

As critical as these questions are, they are not, in general, easy to answer; we will discuss

them in depth in §§11. 10 and 11. 11. However, in one important situation there <u>is</u> an easy

answer: alternating series whose terms decrease to zero in absolute value. We will now

discuss this case:

3. <u>Error bounds in series approximations: the alternating case.</u>   Alternating series have a very

important property which Anton states as Theorem 11. 7. 2 :

Error in
Alternating
Series
Approximations

Suppose $\sum (-1)^k a_k$ is an alternating series, the terms of which

decrease to zero in absolute value (i. e. , which converges by The

Alternating Series Test).

Then the error in approximating the sum of the series by stopping

the series after its   n-th   term is less than the absolute value

of the   (n + 1) - st   term.

In a more pictorial manner:

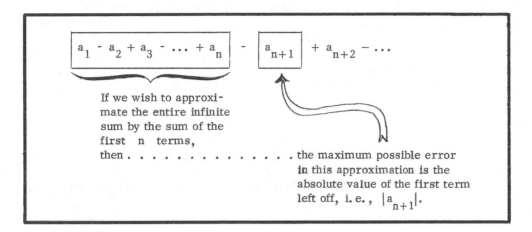

This result allows us to make quick work of both types of error bound questions. Here is an example of each kind:

<u>Example B.</u>   Find the maximum possible error if the value of $\displaystyle\sum_{k=0}^{\infty} \frac{(-1)^k (4k+3)}{(2k+1)(2k+2)}$

is approximated by its partial sum $\displaystyle\sum_{k=0}^{3} \frac{(-1)^k (4k+3)}{(2k+1)(2k+2)}$.

<u>Solution.</u>   This is a typical Type I error bound question, and by the error property of alternating series approximations, the answer is easy: the error is less than the absolute value of the $k=4$ term. That is,

$$\text{error} \quad < \quad \left| \frac{(-1)^4 (4(4)+3)}{(2(4)+1)(2(4)+2)} \right| \quad = \quad \frac{19}{90} \quad < \quad .22 \qquad \square$$

<u>Example C.</u>   Determine the value of $\displaystyle\sum_{k=0}^{\infty} \frac{(-1)^k (4k+3)}{(2k+1)(2k+2)}$ so that the error is less than .01.

<u>Solution.</u>   This is a Type II question: how many terms must we add up to approximate the

11.7.7

entire sum to within an accuracy of .01? The error property of alternating series tells us

what to do: find the first n such that the absolute value of the (n + 1)-st term of the

series is less than .01 and then compute $\sum_{k=0}^{n} \frac{(-1)^k (4k+3)}{(2k+1)(2k+2)}$. The error in using

that number will be less than the absolute value of the (n + 1)-st term which, in turn, will

be less than .01.

Thus we want to find n such that $\left| \frac{(-1)^{n+1} (4(n+1)+3)}{(2(n+1)+1)(2(n+1)+2)} \right| =$

$\frac{4n+7}{(2n+3)(2n+4)} < .01$. Now is a good time to have a calculator! For the approach is to

calculate the expression $\frac{4n+7}{(2n+3)(2n+4)}$ for various values of n to find the smallest one

giving a value less than .01. (Remember that we said in Example A that the terms decrease.)

If you do this, you find that when n = 98 , the value is .0100... and when n = 99 , the

value is .0099.... . Thus n = 99 yields the first value less than .01. Hence the exact

value of $\sum_{k=0}^{\infty} \frac{(-1)^k (4k+3)}{(2k+1)(2k+2)}$ is $\sum_{k=0}^{99} \frac{(-1)^k (4k+3)}{(2k+1)(2k+2)} = \frac{3}{2} - \frac{7}{2} + \frac{11}{30} - \frac{15}{56} + \cdots - \frac{399}{39,800}$

with an error less than .01. (If you want to calculate this number, go ahead. Here's where

a computer would be mighty handy!) □

4. Absolute Convergence. A series is said to converge absolutely if the series of its absolute

values converges. The primary importance of this concept lies in Anton's Theorem 11.7.4:

> If a series converges absolutely, it converges.

In other words,

> If $\sum |a_k|$ converges, then $\sum a_k$ converges.

<u>Now</u> you can understand why we were not really restricting ourselves very much in the previous convergence tests when we "restrict" ourselves to series with <u>positive</u> terms. If a series does <u>not</u> have positive terms, all we need to do is to consider the series of its absolute values (which <u>will</u> of course have positive terms); if we can show by using one of those tests that the series of absolute values converges, then the original series will also converge.

If you understand the concept of absolute convergence, it is not necessary to memorize The Ratio Test for Absolute Convergence (Anton's Theorem 11. 7. 5). That test is simply The Ratio Test applied to a series of absolute values. The only additional piece of information contained in the theorem is the following: if $\rho > 1$, then not only does $\sum |u_k|$ diverge, but $\sum u_k$ also diverges. Said another way, if $\rho > 1$, then the series not only fails to converge absolutely, but it also fails to converge conditionally (a subject we now discuss...).

5. <u>Conditional Convergence: when a series converges but does not converge absolutely.</u>   A concept intimately related to absolute convergence is conditional convergence:

Definition of Conditional Convergence	A series is said to <u>converge conditionally</u> if it converges, but does not converge absolutely.

It is convenient to think of conditional convergence as being "between" absolute convergence and divergence:

Absolute convergence: $\sum u_k$ and $\sum |u_k|$ both converge

Conditional convergence: $\sum u_k$ converges but $\sum |u_k|$ diverges

Divergence: $\sum u_k$ and $\sum |u_k|$ both diverge.

The fourth possibility

$$\sum u_k \quad \text{diverges but} \quad \sum |u_k| \quad \text{converges}$$

cannot happen; that is the meaning of the result that absolute convergence implies convergence! Thus, for any series $\sum u_k$, exactly one of the above three situations must hold true: it converges absolutely, it converges conditionally, or it diverges.

Thus, given a series $\sum u_k$, a very common question is: classify $\sum u_k$ as absolutely convergent, conditionally convergent, or divergent. The following is an orderly three-step procedure for making that classification:

1. Test for <u>divergence</u> of $\sum u_k$ by using The Divergence Test.

    If $\sum u_k$ diverges, conclude DIVERGENCE.

    If the test is inconclusive *, then proceed to Step 2 .

2. Test for <u>absolute convergence</u> of $\sum |u_k|$ by using any of the positive series tests described in previous sections (see the Summary in §11.5.3 of <u>The Companion</u>).

    If $\sum |u_k|$ converges, conclude ABSOLUTE CONVERGENCE.

    If $\sum |u_k|$ diverges, then proceed to Step 3 .

---

* <u>Remember:</u> The Divergence Test can only prove divergence or be inconclusive. It can <u>never</u> prove convergence.

3. Test for <u>conditional convergence</u> of $\sum u_k$. As a practical matter, almost all the series you will have at this stage are alternating, and thus The Alternating Series Test will be the test to use.

If $\sum u_k$ converges, conclude CONDITIONAL CONVERGENCE.

If $\sum u_k$ diverges, conclude DIVERGENCE.

Here are some examples:

<u>Example D.</u>   Determine whether the series $\sum_{k=1}^{\infty} \frac{(-1)^k}{7k}$ converges absolutely, converges conditionally or diverges.

<u>Solution.</u> First, we apply The Divergence Test. Since $\lim_{k \to \infty} \frac{(-1)^k}{7k} = 0$, the test is inconclusive and we move on to test for absolute convergence. The series of absolute values

is $\sum_{k=1}^{\infty} \left| \frac{(-1)^k}{7k} \right| = \sum_{k=1}^{\infty} \frac{1}{7k} = \frac{1}{7} \sum_{k=1}^{\infty} \frac{1}{k}$ which diverges since the Harmonic Series

diverges. Thus we know that the series does <u>not</u> converge absolutely and we move on to test

for conditional convergence. That is, we want to know whether $\sum_{k=1}^{\infty} \frac{(-1)^k}{7k}$ converges. It

is an alternating series, so we apply The Alternating Series Test:

a)   The absolute values of the k and k + 1 terms are $\frac{1}{7k}$ and $\frac{1}{7(k+1)}$, respectively. Since

$$\frac{1}{7k} > \frac{1}{7k+7}$$

we know that the absolute values of the terms of the series are <u>decreasing</u>.

11.7.11

b)   The limit of the absolute values of the terms is

$$\lim_{k \to \infty} \left| \frac{(-1)^k}{7k} \right| = \lim_{k \to \infty} \frac{1}{7k} = 0$$

Hence the series $\sum_{k=1}^{\infty} \frac{(-1)^k}{7k}$ converges by The Alternating Series Test. In view of the above,

this means that it <u>converges conditionally.</u>   ☐

<u>Example E.</u>   Determine whether the series $\sum (-1)^k \frac{k^2}{e^k}$ converges absolutely, converges

conditionally or diverges.

<u>Solution.</u>   First, we apply The Divergence Test:

$$\lim_{k \to \infty} (-1)^k \frac{k^2}{e^k} = \lim_{k \to \infty} (-1)^k \frac{2k}{e^k} = \lim_{k \to \infty} (-1)^k \frac{2}{e^k} = 0$$

(using L'Hôpital's Rule twice). Thus The Divergence Test is inconclusive and we move on to

test for absolute convergence. The series of absolute values is $\sum_{k=1}^{\infty} \left| (-1)^k \frac{k^2}{e^k} \right| =$

$\sum_{k=1}^{\infty} \frac{k^2}{e^k}$. Applying The Ratio Test to this series we get

$$\rho = \lim_{k \to \infty} \frac{u_{k+1}}{u_k} = \lim_{k \to \infty} \frac{(k+1)^2/e^{k+1}}{k^2/e^k} = \lim_{k \to \infty} \frac{e^k}{e^{k+1}} \cdot \frac{(k+1)^2}{k^2}$$

$$= \lim_{k \to \infty} \frac{e^k}{e(e^k)} \cdot \left(\frac{k+1}{k} \right)^2 = \lim_{k \to \infty} \frac{1}{e} \left(1 + \frac{1}{k} \right)^2 = \frac{1}{e} < 1$$

Thus the series $\sum_{k=1}^{\infty} \frac{k^2}{e^k}$ converges by The Ratio Test so the series $\sum_{k=1}^{\infty} (-1)^k \frac{k^2}{e^k}$

<u>converges absolutely.</u>   ☐

Example F.   Determine whether the series $\sum\limits_{k=1}^{\infty} (-1)^k \dfrac{k^2 - 1}{2k^2 + 2}$ converges absolutely, converges

conditionally or diverges.

Solution.   First, we apply The Divergence Test:

$$\lim_{k \to \infty} (-1)^k \frac{k^2 - 1}{2k^2 + 2} = \lim_{k \to \infty} (-1)^k \frac{1 - (1/k^2)}{2 + (2/k^2)}$$

$$= \left(\lim_{k \to \infty} (-1)^k \right) \left(\lim_{k \to \infty} \frac{1 - (1/k^2)}{2 + (2/k^2)} \right)$$

$$= \frac{1}{2} \lim_{k \to \infty} (-1)^k \neq 0$$

Actually, this limit is undefined since it oscillates between $\dfrac{1}{2}$ and $-\dfrac{1}{2}$. However,

the significant point is that it is not zero. Hence the series diverges by The Divergence Test.   □

Comment: Example F makes it clear why we recommend applying The Divergence Test

first. It takes very little time to apply, and when it is conclusive it saves considerable time!

6.   Which test do I use? (A third look).   In the last few sections (see §§11.5.3 and 11.6.5) we

have slowly built up the collection of convergence tests, and have tried to indicate the proper

order of their use. * Here is a brief recap:

---

*
We emphasize again that there is no one "proper order" for applying the tests, but our list pre-
sents a solid procedure that will serve you well, at least until you develop the confidence to "go
it alone."

FOR STARTERS ...

    (1) Direct recognition

TO TEST FOR DIVERGENCE

    (2) The Divergence Test

TO TEST FOR ABSOLUTE CONVERGENCE

    (3) Ratio Test

    (4) Root Test

    (5) Limit Comparison Test

    (6) Integral Test

    (7) Basic Comparison Test

TO TEST FOR CONDITIONAL CONVERGENCE

    (8) Alternating Series Test

This is just the skeleton of a list; you should expand upon it (adding details, hints, etc.) to make your own list. If you do so, you will understand and "own" the list in a way that is not possible if you simply read or copy a list from a book.

Section 11. 8: Power Series

1. Power Series: very long polynomials. The power series is an extension of the familiar polynomial. For example, suppose we have the polynomial

$$1 + x + \frac{x^2}{2} + \frac{x^3}{6} + \frac{x^4}{24}$$

Since there is a pattern to the terms (i. e., $x^k/k!$ for $k = 0, 1, 2, 3, 4$), it is natural to wonder what happens if this polynomial is extended indefinitely:

$$1 + x + \frac{x^2}{2} + \frac{x^3}{6} + \frac{x^4}{24} + \frac{x^5}{5!} + \cdots + \frac{x^k}{k!} + \cdots = \sum_{k=0}^{\infty} \frac{x^k}{k!}$$

This is an infinite series! We call it a "power series" because it is a series in powers of x. The series we have dealt with to this point have been series of <u>constants.</u> Power series, however, contain a <u>variable</u>, usually x; you get a series of constants for each value of x by plugging the specific value of x into the power series. For example, letting $x = 2$ in the power series discussed above yields the series (of constants)

$$\sum_{k=0}^{\infty} \frac{2^k}{k!} = 1 + 2 + \frac{2^2}{2} + \frac{2^3}{3!} + \cdots + \frac{2^k}{k!} + \cdots$$

Letting $x = -3$, we get a different series (of constants)

$$\sum_{k=0}^{\infty} \frac{(-3)^k}{k!} = 1 - 3 + \frac{3^2}{2} - \frac{3^3}{3!} + \cdots + \frac{(-1)^k 3^k}{k!} + \cdots$$

You may remember that the general form of a polynomial is

$$c_0 + c_1(x - a) + c_2(x - a)^2 + c_3(x - a)^3 + \cdots + c_n(x - a)^n$$

Here the c_i's are coefficients and a is a number which is sometimes called the <u>center</u> of the polynomial. We say that this is a polynomial "about $x = a$" or "in power of $x - a$." For example,

$$2 + 3(x - 3) + 4(x - 3)^2 + 5(x - 3)^3 + 6(x - 3)^4$$

is a polynomial of degree 4 about $x = 3$. (We could multiply it out and rewrite it as a polynomial in powers of x, but there is generally no need to do so.)

Accordingly, the general form of a <u>power series</u> is

11. 8. 3

General power series	$$\sum_{k=0}^{\infty} c_k (x - a)^k = c_0 + c_1 (x - a) + c_2 (x - a)^2 + \cdots + c_k (x - a)^k + \cdots$$

The number a is called the <u>center</u> of the power series and we say this is a power series "about $x = a$" or "in powers of $(x - a)$." In the <u>special case</u> when $a = 0$, the general power series is

$$\sum_{k=0}^{\infty} c_k x^k = c_0 + c_1 x + c_2 x^2 + \cdots + c_k x^k + \cdots$$

For example,

$$\sum_{k=0}^{\infty} (k + 2) (x - 3)^k = 2 + 3 (x - 3) + 4 (x - 3)^2 + \cdots + (k + 2) (x - 3)^k + \cdots$$

is a power series about $x = 3$, while

$$\sum_{k=0}^{\infty} \frac{x^k}{k!} = 1 + x + \frac{x^2}{2} + \frac{x^3}{6} + \cdots + \frac{x^k}{k!} + \cdots$$

is a power series about $x = 0$.

2. "<u>Why do we spend so much time on power series?</u>" A good question, with a good answer: because they comprise the most commonly occurring and most important class of infinite series. In the following sections you will see that most of our well known functions can be written as (the sums of) power series. For example, the power series discussed at the start of the previous subsection is merely the expansion for e^x about $x = 0$:

$$e^x = \sum_{k=0}^{\infty} \frac{x^k}{k!} = 1 + x + \frac{x^2}{2!} + \frac{x^3}{3!} + \cdots$$

This is called the <u>Maclaurin expansion</u> for e^x, and its importance cannot be overemphasized.

When you calculate $e^{.324}$ on a hand calculator, what do you think is occurring? That's right, the Maclaurin expansion

$$e^{.324} = 1 + .324 + \frac{(.324)^2}{2!} + \frac{(.324)^3}{3!} + \cdots$$

is being used to <u>approximate</u> $e^{.324}$ to however many decimal places of accuracy the calculator will print out. So values of e^x are obtained by the use of power series.

This is simply one application of power series; others abound in applications, and we will discuss them in later sections. However, in order to do so, we must familarize ourselves with the theory of power series. So let's get on with it...

3. <u>Interval of Convergence and Radius of Convergence.</u> The "Fundamental Problem" (as Anton calls it) we consider is

<table>
<tr>
<td>The
Fundamental
Problem</td>
<td>For what values of x does a given power series

$\displaystyle\sum_{k=0}^{\infty} c_k(x-a)^k$ converge?</td>
</tr>
</table>

That question can be rephrased. When $x = a$, we see that all the terms in our series are zero except for the first one c_0. This shows that <u>a power series about $x - a$ will always converge at $x = a$</u>. That gives part of the answer to the "Fundamental Problem" and hence we can re-phrase it as follows: for what values of x <u>other than</u> a does $\displaystyle\sum_{k=0}^{\infty} c_k(x-a)^k$ converge?

Or

<table>
<tr>
<td>The Fundamental
Problem
(rephrased)</td>
<td>How far away from $x = a$ can you go and

still have convergence of $\displaystyle\sum_{k=0}^{\infty} c_k(x-a)^k$?</td>
</tr>
</table>

The answer to the Fundamental Problem will vary from series to series, but Anton's Theorem 11. 8. 2 (and Theorem 11. 8. 1, the special case when $a = 0$) gives us the answer's basic form:

Interval of Convergence Theorem

Exactly one of the following happens for the power series $\displaystyle\sum_{k=0}^{\infty} c_k (x - a)^k$:

case a)   Convergence only at $x = a$.

case b)   Absolute convergence for all x .

case c) $\begin{cases} \text{Absolute convergence in an open interval } (a - R, a + R). \\ \text{Divergence in the open intervals } (-\infty, a - R) \text{ and } (a + R, +\infty). \\ \text{Anything is possible at the endpoints } x = a + R \text{ and } x = a - R. \end{cases}$

| divergence | ? | absolute convergence | ? | divergence |

$\longleftarrow$ ———○————————|————————○———— $\longrightarrow$

   $a - R$ a $a+R$

Case (c) is the interesting situation: there is some number R, $0 < R < \infty$, called the <u>radius of convergence</u>, such that the series converges absolutely for every x in the open interval $(a - R, a + R)$, with any type of convergence or divergence possible at the two endpoints $a + R$ and $a - R$. If you like, you can consider cases (a) and (b) as the two "extremes" of case (c): $R = 0$ gives case (a), while $R = \infty$ gives case (b).

The corresponding <u>interval of convergence</u> consists of all those x values for which $\sum c_k (x - a)^k$ converges. We see from our theorem that this interval must contain the open interval $(a - R, a + R)$, and that the only other possible points in it are the two endpoints $a \pm R$. Thus the interval of convergence must be one of the four forms

<table>
<tr><td>(a - R , a + R)</td><td>open</td></tr>
<tr><td>[a - R , a + R)</td><td>closed-open</td></tr>
<tr><td>(a - R , a + R]</td><td>open-closed</td></tr>
<tr><td>[a - R , a + R]</td><td>closed</td></tr>
</table>

Forms for the interval of convergence

Notice that all four forms have $x = a$ as the "center" and R as the "radius" of the interval; the only differences occur at the endpoints. Thus R provides an answer to the (rephrased) Fundamental Problem: you can go R units away from $x = a$ and still have convergence, but you have to check convergence at the endpoints $a \pm R$.

Let's summarize. The Fundamental Problem,

$$\text{"for what } x \text{ values does } \sum c_k (x - a)^k \text{ converge?"}$$

becomes

$$\text{"what is the } \underline{\text{interval of convergence}} \text{ for } \sum c_k (x - a)^k ?"$$

To determine this, we must

Determining the interval of convergence

1. determine the <u>radius of convergence</u> R , and

2. check the convergence at the endpoints $x = a \pm R$.

Then our interval of convergence will be specified as one of the four forms given above, and the Fundamental Problem will be solved!

Given the series $\sum c_k (x - a)^k$, here's how to execute our two-step method for finding the interval of convergence:

1)   Determine the <u>radius of convergence</u>   R : Apply The Ratio Test or The Root Test to

$$\sum |c_k| \, |x - a|^k$$ to obtain an expression for ρ which contains $|x - a|$.

Set this expression less than   1 , i.e.,

$$\rho < 1$$

and solve for $|x - a|$; what you obtain will be   R , i.e., you'll get

$$|x - a| < [\text{something}]$$

and the "something" will be   R .

2)   Check the convergence at the <u>endpoints</u> $x = a \pm R$:

You have two series of constants to check for convergence,

$$\sum c_k R^k \quad \text{and} \quad \sum (-1)^k c_k R^k$$

Attack each one with the tests of the previous sections. <u>However</u> ... don't

bother with The Ratio or Root Tests, for they will always be inconclusive!

(That's how you got the endpoints to begin with).

Anton gives 6 examples in Section 11. 8 , the last five of which are examples of this

procedure. The first, Example 1, uses Direct Recognition because the series is simply a

geometric series. However, we can also use the above procedure on the series of Example 1;

it gives a simple illustration of our two steps:

<u>Example A.</u>   Find the interval of convergence and the radius of convergence of the power series

$$\sum_{k=0}^{\infty} x^k .$$

<u>Solution.</u>   1)   The series of absolute values is $\sum_{k=0}^{\infty} |x^k| = \sum_{k=0}^{\infty} |x|^k$. The ρ of The

Ratio Test is

$$\rho = \lim_{k \to \infty} \frac{u_{k+1}}{u_k} = \lim_{k \to \infty} \frac{|x|^{k+1}}{|x|^k} = \lim_{k \to \infty} |x| = |x|$$

The Ratio Test says the series of absolute values converges for $\rho = |x| < 1$, i.e., for $-1 < x < 1$. Thus, since $a = 0$, the radius of convergence is $R = 1$. (It is the "radius" of the interval $(-1, 1)$ which has center $a = 0$.)

2)   Plugging the endpoints $a - R = 0 - 1 = -1$ and $a + R = 0 + 1 = 1$ into the series, we get

$$\sum_{k=0}^{\infty} (-1)^k \quad \text{and} \quad \sum_{k=0}^{\infty} (1)^k$$

Both of these series diverge by The Divergence Test since the limits of the individual terms are $\lim_{k \to \infty} (-1)^k =$ "undefined" and $\lim_{k \to \infty} (1)^k = 1$, respectively, and neither is zero. Thus the interval of convergence is the open interval $(-1, 1)$.   ☐

Here is a more complicated example:

Example B.   Find the interval of convergence and the radius of convergence of the power series

$$\sum_{k=1}^{\infty} \frac{(x+3)^{2k}}{k^2 2^k} .$$

Solution.   1)   The series of absolute values is $\displaystyle\sum_{k=1}^{\infty} \frac{|x+3|^{2k}}{k^2 2^k}$. The ρ of The Ratio Test is

$$\rho = \lim_{k \to \infty} \frac{u_{k+1}}{u_k} = \lim_{k \to \infty} \frac{|x+3|^{2k+2}/(k+1)^2 2^{k+1}}{|x+3|^{2k}/k^2 2^k} = \lim_{k \to \infty} \frac{k^2}{2(k+1)^2} |x+3|^2$$

$$= \frac{1}{2} |x+3|^2$$

The Ratio Test says the series of absolute values converges for $\rho = \frac{1}{2}|x+3|^2 < 1$, i.e.,

for $|x+3| < \sqrt{2}$ or $-3 - \sqrt{2} < x < -3 + \sqrt{2}$. Thus, since $a = -3$, the radius of

convergence is $R = \sqrt{2}$. (It is the "radius" of the interval $(-3 - \sqrt{2}, -3 + \sqrt{2})$ which

has center $a = -3$.)

2) Plugging the endpoint $a - R = -3 - \sqrt{2}$ into the series yields $\displaystyle\sum_{k=1}^{\infty} \frac{(-\sqrt{2})^{2k}}{k^2 2^k} =$

$\displaystyle\sum_{k=1}^{\infty} \frac{1}{k^2}$ which is a convergent p-series $(p = 2)$. Plugging in the other endpoint $a + R =$

$-3 + \sqrt{2}$ yields the same convergent p-series. Thus the interval of convergence is the

closed interval $[-3 - \sqrt{2}, -3 + \sqrt{2}]$. $\qquad\qquad\qquad\qquad\qquad\qquad\qquad \square$

4. <u>An important limit.</u> Anton's Equation (1) is a useful and important result which will be needed

later:

$$\lim_{k \to \infty} \frac{x^k}{k!} = 0 \qquad \text{for } \underline{any} \quad x$$

Anton proves it as a "byproduct" of another computation, but here is another justification which

may help you to understand and remember it better:

Let x be any real number. Then for any k

$$\frac{x^k}{k!} = \overbrace{\frac{x \cdot x \cdot x \ldots x \cdot x \cdot x}{\underbrace{k(k-1)(k-2) \ldots 3 \cdot 2 \cdot 1}_{k}}}^{k} = \left(\frac{x}{k}\right)\left(\frac{x}{k-1}\right)\left(\frac{x}{k-2}\right) \cdots \left(\frac{x}{3}\right)\left(\frac{x}{2}\right)\left(\frac{x}{1}\right)$$

As $k \to \infty$, the number of terms in this product increases indefinitely

$$\lim_{k \to \infty} \frac{x^k}{k!} = \left(\frac{x}{1}\right)\left(\frac{x}{2}\right)\left(\frac{x}{3}\right)\left(\frac{x}{4}\right)\left(\frac{x}{5}\right) \cdots$$

Since x is a fixed number, as $k \to \infty$ sooner or later k passes x (no matter how big x is!). When that happens, the fractions $\frac{x}{k}$ became <u>less than 1</u> and, from that point on, the fractions become smaller and smaller and approach zero. So the infinite product

$$\left(\frac{x}{1}\right)\left(\frac{x}{2}\right)\left(\frac{x}{3}\right)\left(\frac{x}{4}\right)\left(\frac{x}{5}\right) \cdots \left(\frac{x}{k}\right)\left(\frac{x}{k+1}\right) \cdots$$

must be zero. That is $\lim_{k \to \omega} \dfrac{x^k}{k!} = 0$.

Section 11.9: Taylor and Maclaurin Series

1. <u>Taylor and Maclaurin polynomials.</u> First we talk about <u>polynomials</u>, not just plain old polynomials, but special polynomials:

<u>The n-th Taylor Polynomial for $f(x)$ about $x = a$.</u>

$$p_n(x) = \sum_{k=0}^{n} \frac{f^{(k)}(a)}{k!}(x-a)^k$$

$$= f(a) + f'(a)(x-a) + \frac{f''(a)}{2!}(x-a)^2 + \cdots + \frac{f^{(n)}(a)}{n!}(x-a)^n$$

When $a = 0$, we call the Taylor polynomial the Maclaurin polynomial:

The   n-th   Maclaurin Polynomial for $f(x)$.

$$p_n(x) = \sum_{k=0}^{n} \frac{f^{(k)}(0)}{k!} x^k$$

$$= f(0) + f'(0)x + \frac{f''(0)}{2!} x^2 + \cdots + \frac{f^{(n)}(0)}{n!} x^n$$

Remember that $f^{(k)}(a)$ is the   k-th   derivative of $f(x)$ evaluated at $x = a$;   it is <u>not</u> the

k-th   power of $f(a)$!!   Also, when $k = 0$,   then $f^{(0)}(a) = f(0)$,   i.e.,   the   "0-th   derivative"

is the function itself.

Do **not** lose sight of the process that generated $p_n(x)$,   the   n-th   Taylor polynomial

for $f(x)$ about $x = a$:

$p_n(x)$ is an (in fact, the only)   <u>n-th degree</u>   polynomial such

that, at $x = a$,   the values of $p_n(x)$ and its first   n

derivatives equal the corresponding quantities for $f(x)$,   i.e.,

$$p_n(a) = f(a), \quad p_n'(a) = f'(a), \ldots, \quad p_n^{(n)}(a) = f^{(n)}(a)$$

Thus, from the standpoint of derivatives, $p_n(x)$ is the "best"   n-th degree   polynomial

approximation for $f(x)$ <u>at $x = a$</u> .   (Remember that the function $f(x)$ need not be a poly-

nomial itself, so we are approximating a function $f(x)$ with an easier-to-understand function,

the polynomial $p_n(x)$.)   This leads to the key question concerning Taylor polynomials:

The Taylor Approximation Question	How good an approximation of $f(x)$ is the n-th Taylor polynomial for $f(x)$ about $x = a$ at values of x <u>near a</u> ?

For many functions $f(x)$ the answer is "very good," and this has profound consequences in applied mathematics.

The approximation question will be taken up in detail in §11.11. For now we consider a more basic and mundane concern:

> How do we calculate a Taylor or
>
> Maclaurin polynomial for a given
>
> function f(x) ?

Such a computation can be time-consuming, but it is usually not difficult. Anton's first five examples all illustrate the technique: you simply make the calculations specified in the Taylor (or Maclaurin) formula. The only tricky part comes when you try to generalize to say what the n-th Taylor or Maclaurin polynomial is (for example, see Anton's Examples 3 and 4 where he generalizes to say what $p_{2n+1}(x)$ and $p_{2n}(x)$ are). In general, the method is to compute enough terms so that a pattern becomes "obvious." Even as a newcomer to Taylor and Maclaurin series, the pattern will usually become clear to you if you compute enough specific cases. *

Example A. Find the n-th Taylor polynomial for $\ln x$ about $x = 2$.

Solution. The Taylor polynomial formula is $p_n(x) = \sum_{k=0}^{n} \frac{f^{(k)}(a)}{k!} (x - a)^k$. Here $f(x) = \ln x$ and we must compute the derivatives:

---

* In later sections we will develop techniques that allow you to obtain Taylor polynomials indirectly, without going through the tedious direct calculations.

$$f(x) = \ln x$$

$$f'(x) = x^{-1}$$

$$f''(x) = -x^{-2}$$

$$f^{(3)}(x) = 2x^{-3} \qquad = 2!\,x^{-3}$$

$$f^{(4)}(x) = -3 \cdot 2x^{-4} \qquad = -3!\,x^{-4}$$

$$f^{(5)}(x) = 4 \cdot 3 \cdot 2x^{-5} = 4!\,x^{-5}$$

etc.

By now, the pattern is "obvious" (if it isn't, compute a few more terms and it will be):

$$f^{(k)}(x) = (-1)^{k-1}(k-1)!\,x^{-k} \qquad \text{for} \qquad k = 1, 2, 3, \ldots$$

(The $k = 0$ case does not fit this pattern so we have to consider it separately.) The formula tells us to evaluate these derivatives at $a = 2$, so we find

$$f^{(k)}(2) = (-1)^{k-1}(k-1)!\,2^{-k} \qquad \text{for} \qquad k = 1, 2, 3, \ldots$$

(and $f^{(0)}(2) = f(2) = \ln 2$). Thus the n-th Taylor polynomial for $\ln x$ about $x = 2$ is

$$p_n(x) = \sum_{k=0}^{n} \frac{f^{(k)}(2)}{k!} (x-2)^k$$

$$= \frac{\ln 2}{0!}(x-2)^0 + \sum_{k=1}^{n} \frac{(-1)^{k-1}(k-1)!}{k!\,2^k}(x-2)^k$$

$$= \ln 2 + \sum_{k=1}^{n} \frac{(-1)^{k-1}}{k\,2^k}(x-2)^k \qquad \text{since} \qquad \frac{(k-1)!}{k!} = \frac{(k-1)!}{k(k-1)!} = \frac{1}{k}$$

$$= \ln 2 + \frac{1}{2}(x-2) - \frac{1}{2 \cdot 2^2}(x-2)^2 + \cdots + \frac{(-1)^{n-1}}{n\,2^n}(x-2)^n \qquad \Box$$

2. <u>Taylor and Maclaurin Series.</u>   Once you understand Taylor and Maclaurin polynomials, it is natural to "extend them" (by letting $n \to \infty$) to power series:

The Taylor Series for $f(x)$ about $x = a$.

$$\sum_{k=0}^{\infty} \frac{f^{(k)}(a)}{k!} (x-a)^k = f(a) + f'(a)(x-a) + \frac{f''(a)}{2!}(x-a)^2$$

$$+ \cdots + \frac{f^{(k)}(a)}{k!}(x-a)^k + \cdots$$

When $a = 0$, we call the Taylor series the Maclaurin series.

The Maclaurin Series for $f(x)$.

$$\sum_{k=0}^{\omega} \frac{f^{(k)}(0)}{k!} x^k = f(0) + f'(0)x + \frac{f''(0)}{2!}x^2 + \cdots + \frac{f^{(k)}(0)}{k!}x^k + \cdots$$

The practical motivation for considering Taylor and Maclaurin series is simple: since Taylor <u>polynomials</u> for $f(x)$ about $x = a$ give a good <u>approximation</u> for $f(x)$ near a , then we hope that the limits of these polynomials (i. e. , the Taylor <u>series</u>) would <u>equal</u> $f(x)$ near a . This is the key question concerning Taylor series:

The Taylor
Convergence
Question

Does the Taylor series for $f(x)$ about $x = a$

converge to $f(x)$ at values of x <u>near a</u> ?

For many functions $f(x)$ the answer is "yes, " as we shall see in § 11. 10.

For now we consider a more elementary question:

How do we calculate a Taylor or

Maclaurin series for a given function $f(x)$?

The answer at this stage in our development is the same as for Taylor and Maclaurin polynomials: you simply make the calculations specified in the Taylor or Maclaurin formulas. * Again, the only tricky part comes when you try to find a general formula for the n-th term of the series; this is exactly the same problem as finding the general term of the n-th Taylor polynomial for $f(x)$ about $x = a$. Anton's Examples 6 and 7 illustrate this technique. Here are two more examples:

<u>Example B.</u>   Find the Maclaurin series for $\ln (1 - x)$.

<u>Solution.</u>   The Maclaurin series formula is $\displaystyle\sum_{k=0}^{\infty} \frac{f^{(k)}(0)}{k!} x^k$.   Here $f(x) = \ln (1 - x)$ and we must compute the derivatives:

$$f(x) = \ln (1 - x)$$

$$f'(x) = - (1 - x)^{-1}$$

$$f''(x) = - (1 - x)^{-2}$$

$$f^{(3)}(x) = - 2 (1 - x)^{-3} \qquad = - 2! (1 - x)^{-3}$$

$$f^{(4)}(x) = - 3 \cdot 2 (1 - x)^{-4} \qquad = - 3! (1 - x)^{-4}$$

$$f^{(5)}(x) = - 4 \cdot 3 \cdot 2 (1 - x)^{-5} = - 4! (1 - x)^{-5}$$

etc.

By now, the pattern is becoming apparent (is it?) :

$$f^{(k)}(x) = - (k - 1)! (1 - x)^{-k} \qquad \text{for} \qquad k = 1, 2, 3, \ldots$$

(The $k = 0$ case does not fit this pattern, so we have to consider it separately.) The formula says we want to evaluate these at $x = 0$, so we find

---

*
   In the following sections we will develop techniques that allow you to obtain Taylor series indirectly, without going through the tedious direct calculations.

$$f^{(k)}(0) = -(k-1)! \quad \text{for} \quad k = 1, 2, 3, \ldots$$

(and $f^{(0)}(0) = f(0) = \ln(1-0) = \ln 1 = 0$). Thus the $k = 0$ term is zero and the Maclaurin

series is

$$\sum_{k=0}^{\infty} \frac{f^{(k)}(0)}{k!} x^k = \sum_{k=1}^{\infty} \frac{-(k-1)!}{k!} x^k$$

$$= \sum_{k=1}^{\infty} \frac{-1}{k} x^k \quad \text{since} \quad \frac{(k-1)!}{k!} = \frac{1}{k} \quad \text{as in Example A}$$

$$= -x - \frac{x^2}{2} - \frac{x^3}{3} - \frac{x^4}{4} - \frac{x^5}{5} - \frac{x^6}{6} - \cdots \qquad \square$$

Example C. Find the Taylor series for $\sqrt{x}$ about $a = 4$.

Solution. The Taylor series formula about $a = 4$ is $\displaystyle\sum_{k=0}^{\infty} \frac{f^{(k)}(4)}{k!} (x-4)^k$. Here

$f(x) = \sqrt{x}$ and we must compute the derivatives:

$$f(x) = \sqrt{x} = x^{1/2}$$

$$f'(x) = \frac{1}{2} x^{-1/2}$$

$$f''(x) = -\frac{1}{2^2} x^{3/2} \qquad = -\frac{1}{4} x^{-3/2}$$

$$f^{(3)}(x) = \frac{1 \cdot 3}{2^3} x^{-5/2} \qquad = \frac{3}{8} x^{-5/2}$$

$$f^{(4)}(x) = \frac{-1 \cdot 3 \cdot 5}{2^4} x^{-7/2} \qquad = -\frac{15}{16} x^{-7/2}$$

$$f^{(5)}(x) = \frac{1 \cdot 3 \cdot 5 \cdot 7}{2^5} x^{-9/2} = \frac{105}{32} x^{-9/2}$$

etc.

The pattern can be determined as follows: Leaving aside the $k = 0$ case $(f(x) = x^{1/2})$ because it is different, note that the x-exponent in $f^{(k)}(x)$ is $-(2k-1)/2$, that the denominator in $f^{(k)}(x)$ is 2^k and that the signs of the derivatives alternate so that $f^{(k)}(x)$ has sign $(-1)^{k+1}$. The numerator of the coefficient in $f^{(k)}(x)$ <u>from $k = 2$ on</u> is a product of the <u>odd</u> integers from 1 up through $2k-3$; that is, it is $1 \cdot 3 \cdot 5 \cdot 7 \cdot \ldots \cdot (2k-3)$. Putting this all together, we see that the k-th derivative is

$$f^{(k)}(x) = (-1)^{k+1} \frac{1 \cdot 3 \cdot 5 \cdot 7 \cdot \ldots \cdot (2k-3)}{2^k} x^{-(2k-1)/2} \quad \text{for } k \geq 2$$

(The $k = 0$ and $k = 1$ derivatives do not fit this pattern so they will be dealt with separately.) Evaluating the derivatives at $a = 4$ yields

$$f(4) = (4)^{1/2} = 2$$

$$f'(4) = \frac{1}{2}(4)^{-1/2} = \frac{1}{4}$$

and for $k \geq 2$, $\quad f^{(k)}(4) = (-1)^{k+1} \dfrac{1 \cdot 3 \cdot 5 \cdot 7 \cdot \ldots \cdot (2k-3)}{2^k} (4)^{-(2k-1)/2}$

$$= (-1)^{k+1} \frac{1 \cdot 3 \cdot 5 \cdot 7 \cdot \ldots \cdot (2k-3)}{2^{3k-1}}$$

$$\left(\text{since} \qquad (4)^{-(2k-1)/2} = 2^{-(2k-1)} = \frac{1}{2^{2k-1}} \right)$$

Thus the Taylor series about $a = 4$ is

$$\sum_{k=0}^{\infty} \frac{f^{(k)}(4)}{k!} (x-4)^k = \frac{2}{0!} (x-4)^0 + \frac{1/4}{1!} (x-4)^1$$

$$+ \sum_{k=2}^{\infty} \frac{(-1)^{k+1}}{2^{3k-1}} \frac{1 \cdot 3 \cdot 5 \cdot 7 \cdot \ldots \cdot (2k-3)}{k!} (x-4)^k$$

$$- 2 + \frac{1}{4} (x-4) - \frac{1}{2^5} \frac{1}{2} (x-4)^2$$

$$+ \frac{1}{2^8} \frac{1 \cdot 3}{6} (x-4)^3 - \frac{1}{2^{11}} \frac{1 \cdot 3 \cdot 5}{24} (x-4)^4 + \ldots$$

$$= 2 + \frac{1}{4} (x-4) - \frac{1}{64} (x-4)^2 + \frac{1}{512} (x-4)^3 - \frac{5}{16,384} (x-4)^4 + \ldots$$

$\square$

Section 11. 10: Taylor Formula with Remainder; Convergence of Taylor Series

1. <u>Taylor's Theorem and the remainder term.</u> Given a function $f(x)$ with sufficiently many derivatives defined at $x = a$, we can form the n-th Taylor polynomial for $f(x)$ about $x = a$,

$$p_n(x) = \sum_{k=0}^{n} \frac{f^{(k)}(a)}{k!} (x-a)^k$$

and the (infinite) Taylor series expansion for $f(x)$ about $x = a$,

$$\lim_{n \to \infty} p_n(x) = \sum_{k=0}^{\infty} \frac{f^{(k)}(a)}{k!} (x-a)^k$$

The two key questions (see §11. 9) are

| The Taylor Approximation Question | How good an approximation for $f(x)$ is $p_n(x)$ at values of x <u>near</u> a ? |

and

The Taylor Convergence Question

Does the Taylor series for $f(x)$ about $x = a$

converge to $f(x)$ at values of x <u>near a</u> ?

Both of these questions can be answered by using <u>Taylor's Theorem.</u> Suppose $f(x)$ has $n + 1$ derivatives on an open interval containing the point a . Then for every x in that interval we have

Taylor's Theorem

$$f(x) = p_n(x) + R_n(x)$$

$$\text{where} \quad R_n(x) = \frac{f^{(n+1)}(c)}{(n+1)!}(x-a)^{n+1}$$

where c is some number between a and x .

<u>Lagrange's form</u> of the <u>remainder term</u> (or <u>error term</u>) $R_n(x)$ is easy to remember since it is almost identical to the $(n+1)$-st term in the Taylor series of $f(x)$ about $x = a$. The only difference is that the $(n+1)$-st derivative is evaluated at some mysterious number c , not at a :

$$\left[\begin{array}{c} (n+1)\text{-st term of} \\ \text{Taylor series} \end{array}\right] = \frac{f^{(n+1)}(a)}{(n+1)!}(x-a)^{n+1} \quad \boxed{\text{Note the } a}$$

$$\left[\begin{array}{c} \text{Remainder term} \\ R_n(x) \end{array}\right] = \frac{f^{(n+1)}(c)}{(n+1)!}(x-a)^{n+1} \quad \boxed{\text{Note the } c}$$

Keep the following points in mind concerning the number c ; if its nature is not clearly understood, then the subsequent applications of Taylor's Theorem will be quite unintelligible:

1. The number c varies with x ; if you change the x value you will almost certainly change the c value.

2. The phrase   "c is between   a   and   x"   covers two possibilities: if
    a   is less than   x ,   then $a < c < x$,   while if   a   is greater than
    x ,   then $x < c < a$.   Certain applications of Taylor's Theorem
    require the handling of each case separately.

3. Taylor's Theorem does not tell us how to compute a specific value for
    c .   In fact, <u>we do not need to know the specific value for   c   in order
    to use Taylor's Theorem!</u> Just knowing that   "c is between   a   and
    x "   is generally sufficient for our needs, as we will demonstrate in
    this section.

Here is an example that illustrates some of these properties in a specific case:

<u>Example A.</u>   For the function $f(x) = \dfrac{1}{(x+2)^2}$ with   a = - 3   and   n = 4 ,   find (a) Lagrange's

form of the remainder $R_n(x)$, (b) the largest open interval on which the remainder formula
is valid, (c) the interval of possible values for the number   c   in the remainder formula.

<u>Solution.</u>   (a) Lagrange's form of $R_4(x)$ is

$$R_4(x) = \frac{f^{(4+1)}(c)}{(4+1)!} (x+3)^{4+1} = \frac{f^{(5)}(c)}{5!} (x+3)^5$$

for some   c   between   a = - 3   and   x .   Thus we need to compute the first five derivatives
of   f (x)   in order to get the fifth derivative needed in the formula:

$$f(x) = (x + 2)^{-2}$$

$$f'(x) = -2(x + 2)^{-3}$$

$$f''(x) = 6(x + 2)^{-4}$$

$$f^{(3)}(x) = -24(x + 2)^{-5}$$

$$f^{(4)}(x) = 120(x + 2)^{-6}$$

$$f^{(5)}(x) = -720(x + 2)^{-7}$$

Thus $R_4(x) = \dfrac{-720(c + 2)^{-7}}{5!}(x + 3)^5 = \dfrac{-6}{(c + 2)^7}(x + 3)^5$ for some c between -3

and x.

(b) The remainder formula is valid on the largest interval containing $a = -3$ for which the first five derivatives of $f(x)$ exist. It is clear from the computations above that those derivatives exist everywhere except at $x = -2$ (that value makes the denominators zero). Therefore the derivatives exist everywhere on each of the open intervals $(-\infty, -2)$ and $(-2, +\infty)$. Since $(-\infty, -2)$ contains $a = -3$, then it is the largest open interval on which the remainder formula is valid.

(c) The interval of possible values for the number c is "between -3 and x." This means either $(-3, x)$ or $(x, -3)$, depending on whether $x > -3$ or $x < -3$. □

2. <u>Answering the Taylor convergence question: determine the interval of convergence.</u> The answer to the Taylor convergence question ("Does the Taylor series for $f(x)$ about $x = a$ converge to $f(x)$ at values of x near a?") is provided by Anton's Theorem 11. 10. 2, a simple corollary of Taylor's theorem:

Taylor Convergence Criterion

The Taylor series for $f(x)$ about $x = a$ converges to $f(x)$ at precisely those x-values for which

$$\lim_{n \to \infty} R_n(x) = 0$$

This should be a very believable result: the Taylor series converges to $f(x)$ whenever the remainder term tends to zero.

As the wording of this result indicates, given a function f, the Taylor series for f about $x = a$ might converge to $f(x)$ for some values of x but might not converge for others. The largest interval containing $x = a$ for which the Taylor series converges to $f(x)$ is called the <u>interval of convergence</u>, or <u>interval of validity</u>, for the Taylor series, i.e.,

if x is in the interval of convergence,

$$\text{then} \quad f(x) = \sum_{k=0}^{\infty} \frac{f^{(k)}(a)}{k!} (x - a)^k$$

Thus we answer the convergence question by determining the interval of convergence.

<u>Determining the Interval of Convergence of the Taylor Series for f about $x = a$</u>

1. Determine (as in Example A) the Lagrange form of the remainder

$$R_n(x) = \frac{f^{(n+1)}(c)}{(n+1)!} (x - a)^{n+1}$$

where c is a number between a and x.

2. Determine an upper bound for $\left| f^{(n+1)}(c) \right|$ which is <u>independent of</u> c (but may depend on a and x). This will give an upper bound for $\left| R_n(x) \right|$ which is independent of c.

3. Determine those x-values for which $\lim\limits_{n \to \infty} R_n(x) = 0$, i.e., those

    for which the upper bound in Step 2 goes to 0 as $n \to \infty$.

4. Determine the largest interval containing x = a which is contained

    in the set of x-values found in Step 3. This is the desired interval

    of convergence.

To illustrate the details of this procedure, let us see how Anton's Example 4 fits the pattern.

Anton's Example 4. Show that the Maclaurin series for sin x converges to sin x for all x .

Solution. Here we are dealing with the Taylor series for sin x about a = 0 .

Step 1: Since the (n + 1)-st derivative of sin x is $\pm \cos x$ or $\pm \sin x$, then

$$R_n(x) = \pm \frac{\cos c}{(n+1)!} x^{n+1} \quad \text{or} \quad \pm \frac{\sin c}{(n+1)!} x^{n+1}$$

where c is between 0 and x .

Step 2: Since $|\sin c| \le 1$ and $|\cos c| \le 1$, we know

$|f^{(n+1)}(c)| \le 1$. Hence $|R_n(x)| \le \frac{|x|^{n+1}}{(n+1)!}$ for all x .

The purpose of Step 2 is evident: we remove the vagueness of dealing with the

unspecified value of c in $f^{(n+1)}(c)$ by going to an upper bound which does

not contain c !

Step 3: To determine those x-values for which $\lim\limits_{n \to \infty} R_n(x) = 0$ we look at the limit

as n tends to ∞ of the upper bound found in Step 2:

$$\lim_{n \to \infty} \frac{|x|^{n+1}}{(n+1)!} = 0 \quad \underline{\text{for all } x} \quad \text{by Anton's (9)} \quad (\text{An important result!}) \quad .$$

Hence, since $R_n(x)$ is "pinched" between $\dfrac{-|x|^{n+1}}{(n+1)!}$ and $\dfrac{|x|^{n+1}}{(n+1)!}$, both of

which go to 0 as $n \to \infty$, the Pinching Theorem (3.3.1) guarantees that

$$\lim_{n \to \infty} R_n(x) = 0 \quad \underline{\text{for all } x}$$

Step 3 almost always uses the Pinching Theorem: If $M_n(x)$ is the upper bound for

$|R_n(x)|$ found in Step 2 , then $-M_n(x) \le R_n(x) \le M_n(x)$. The Pinching Theorem

(3.3.1) then gives us that $\lim\limits_{n \to \infty} R_n(x) = 0$ whenever $\lim\limits_{n \to \infty} M_n(x) = 0$. $\underline{\text{Hence}}$

$\underline{\text{we have only to determine those}}$ $\underline{x\text{-values}}$ $\underline{\text{for which}}$ $\lim\limits_{n \to \infty} M_n(x) = 0$.

Step 4 :   Since $\lim\limits_{n \to \infty} R_n(x) = 0$ $\underline{\text{for all } x}$, the largest interval in this set which contains

$x = 0$ is simply $(-\infty, \infty)$. This is the interval of convergence for the Maclaurin series

for $\sin x$. $\square$

Anton's Example 3 (convergence of the Maclaurin series for e^x) is harder than his

Example 4 which we just analyzed. The reason is that in Step 2 you do not obtain one upper

bound for $|R_n(x)|$, but $\underline{\text{three}}$, depending on the sign of x :

    i.   if $x > 0$, then $0 < c < x$, proving $e^c < e^x$ and

$$|R_n(x)| \le \frac{e^x}{(n+1)!} x^{n+1}$$

    ii.   if $x < 0$, then $x < c < 0$, proving $e^c < e^0 = 1$ and

$$|R_n(x)| \le \frac{1}{(n+1)!} |x|^{n+1}$$

iii.   if x = 0 , then c = 0 , proving $e^c = 1$ and

$$\left| R_n(x) \right| = 0$$

The third case is trivial: a Maclaurin series always converges for x = 0 . However, cases

i and ii must each be handled separately with our procedures (i. e. , perform Steps 2 , 3

and 4 for i and then for ii). This is what Anton does in his solution.

Here is another example of determining an interval of convergence to complement

Anton's Examples 3 , 4 and 5 :

Example B.   Find the Maclaurin series for cos x and prove that it converges to cos x for

all x .

Solution.   The Maclaurin series formula is $\displaystyle\sum_{k=0}^{\infty} \frac{f^{(k)}(0)}{k!} x^k$ so we need to find the derivatives

of cos x :

$$f(x) = \cos x$$

$$f'(x) = -\sin x$$

$$f''(x) = -\cos x$$

$$f^{(3)}(x) = \sin x$$

$$f^{(4)}(x) = \cos x$$

etc.

(From this point, the derivatives begin to repeat - they repeat the cycle $\cos x$, $-\sin x$, $-\cos x$,

$\sin x$ over and over.) The pattern is: when k is even, $f^{(k)}(x) = \pm \cos x$ and when k

is odd, $f^{(k)}(x) = \pm \sin x$. Moreover, the signs are determined as follows:

$$f^{(2k)}(x) = (-1)^k \cos x$$

for   k = 0, 1, 2, 3, ...

$$f^{(2k+1)}(x) = (-1)^{k+1} \sin x$$

Thus when we evaluate the derivatives at 0 , we find

$$f^{(2k)}(0) = (-1)^k \cos 0 = (-1)^k$$

$$\text{for} \quad k = 0, 1, 2, 3, \ldots$$

$$f^{(2k+1)}(0) = (-1)^{k+1} \sin 0 = 0$$

That is, the value of every other derivative is zero! Hence the Maclaurin series for $\cos x$

is

$$\frac{f(0)}{0!} x^0 + \frac{f'(0)}{1!} x^1 + \frac{f''(0)}{2!} x^2 + \frac{f^{(3)}(0)}{3!} x^3 + \cdots$$

$$= 1 + 0 + \frac{(-1)}{2} x^2 + 0 + \cdots$$

$$= \sum_{k=0}^{\infty} \frac{(-1)^k}{(2k)!} x^{2k}$$

To prove that the Maclaurin series converges to $\cos x$ for all x , we use the

procedure given above :

Step 1:   Since $a = 0$ in a Maclaurin series, the Lagrange form of the remainder is

$$R_n(x) = \frac{f^{(n+1)}(c)}{(n+1)!} (x-a)^{n+1} = \frac{f^{(n+1)}(c)}{(n+1)!} x^{n+1}$$

where c is some number between 0 and x . By the calculations above

$f^{(n+1)}(c) = \pm \cos c$ or $\pm \sin c$, depending on whether n is odd or even. Thus

$$R_n(x) = \frac{\pm \cos c}{(n+1)!} x^{n+1} \quad \text{or} \quad \frac{\pm \sin c}{(n+1)!} x^{n+1}$$

Step 2:   Since $|\sin c| \leq 1$ and $|\cos c| \leq 1$, then $|f^{(n+1)}(c)| \leq 1$, proving

$$|R_n(x)| \leq \frac{|x|^{n+1}}{(n+1)!} \qquad \underline{\text{for all } x}$$

11. 10. 10

Step 3:   From Step 2 we have

$$-\frac{|x|^{n+1}}{(n+1)!} \leq R_n(x) \leq \frac{|x|^{n+1}}{(n+1)!}$$

Since $\lim\limits_{n \to \infty} \dfrac{|x|^{n+1}}{(n+1)!} = 0$ for all x (there's that crucial result from (1) of §11.8

again!), the Pinching Theorem says that

$$\lim_{n \to \infty} R_n(x) = 0 \quad \underline{\text{for all } x}$$

Step 4:   Note that in 2), it was not necessary to put any restrictions on x . That is, we

found $\lim\limits_{n \to \infty} R_n(x) = 0$ $\underline{\text{for all } x}$. Thus the interval of convergence of the Maclaurin

series of cos x is $(-\infty, +\infty)$.   □

Anton's Table 11. 10. 1 summarizes some important Maclaurin series and their intervals

of convergence. (Note that the series of Example B appears in that table.) As with most

tables, the more of it you know, the better off you will be, but it is probably enough if you

memorize the series for e^x , sin x , cos x and, perhaps, $\ln(1+x)$ and $\dfrac{1}{1-x}$ (this last

one is simply the geometric series!).

3. Using one Maclaurin series to find another: a useful technique.   Sometimes it is possible to

avoid all the calculations involved in writing down a Maclaurin series directly from the formula

by using another, known Maclaurin series. Moreover, the interval of convergence of the new

series can often be determined from the old series! Anton's Examples 6 and 7 illustrate this

extremely valuable technique. Here is another example:

Example C. Find the Maclaurin series for $\dfrac{x^3}{1+2x}$ and determine its interval of convergence

Solution. The Maclaurin series for $\dfrac{1}{1-x}$ is (see Anton's Table 11. 10. 1)

$$\frac{1}{1-x} = \sum_{k=0}^{\infty} x^k = 1 + x + x^2 + x^3 + \cdots$$

If x is replaced by $-2x$, we obtain

$$\frac{1}{1+2x} = \frac{1}{1-(-2x)} = \sum_{k=0}^{\infty} (-2x)^k = \sum_{k=0}^{\infty} (-1)^k \, 2^k \, x^k$$

$$= 1 - 2x + 4x^2 - 8x^3 + 16x^4 - 32x^5 + \cdots$$

Finally, we multiply by x^3 :

$$\frac{x^3}{1+2x} = \sum_{k=0}^{\infty} (-1)^k \, 2^k \, x^{k+3} = x^3 - 2x^4 + 4x^5 - 8x^6 + \cdots$$

This is the desired Maclaurin series.

The interval of convergence is determined as follows: The Maclaurin series for $\dfrac{1}{1-x}$ converges for $-1 < x < 1$ (see Anton's Table 11. 10. 1) so the Maclaurin series for $\dfrac{1}{1+2x} = \dfrac{1}{1-(-2x)}$ converges for $-1 < -2x < 1$ or $-\dfrac{1}{2} < x < \dfrac{1}{2}$. Multiplication by x^3 does not affect the interval of convergence further, so the Maclaurin series for $\dfrac{x^3}{1+2x}$ converges to $\dfrac{x^3}{1+2x}$ for $-\dfrac{1}{2} < x < \dfrac{1}{2}$.   ⊓

Notice how much simpler this technique is than a direct calculation (as in Example B, for instance), both in determining the Maclaurin series and in determining the interval of convergence. There is, of course, one major drawback: we must already have an appropriate Maclaurin series to use these tricks on! That is, we must already have determined by direct calculation several Maclaurin series and their corresponding intervals of convergence before these techniques will be useful.

4. **A common mistake.** Sometimes when asked to determine the interval on which a Maclaurin or Taylor series converges to the original function, you might be tempted to think

> "Instead of going through all this stuff with the remainder
> term $R_n(x)$, I'll just test my series for convergence
> by The Ratio or Root Test (after all, the series is just a
> power series). The techniques of §11.9 will give me
> the interval of convergence very easily!"

It's a clever thought, but **it won't work**! There exist perverse functions (such as that in Anton's Exercise 40) whose Taylor or Maclaurin series converge... but **to something different than** $f(x)$! To prove convergence <u>to $f(x)$</u> you must use the Remainder formula or one of the techniques discussed above in subsection 3.

Section 11.11 : Computations Using Taylor Series

1. **Accuracy to n decimal places.** No doubt you are familiar with the process for rounding off a number to 2 decimal places. For example,

$$\pi \cong 3.14159 \quad \text{rounds off to} \quad 3.14$$
$$e \cong 2.71828 \quad \text{rounds off to} \quad 2.72$$
$$\sqrt{2} \cong 1.41421 \quad \text{rounds off to} \quad 1.41$$
$$\sqrt{5} \cong 2.23607 \quad \text{rounds off to} \quad 2.24$$

Observe that in all these cases, rounding off the given number to 2 decimal places changes the original number by less than $.005 = .5 \times 10^{-2}$:

$$|3.14159 - 3.14| = .00159$$

$$|2.71828 - 2.72| = .00172$$

$$|1.41421 - 1.41| = .00421$$

$$|2.23607 - 2.24| = .00393$$

In general, rounding a number off to n decimal places will change the original number by less than $.5 \times 10^{-n}$, i. e., the error introduced by rounding off will be less than $.5 \times 10^{-n}$. This is the motivation for the following definition:

> An approximation is accurate to n decimal places
>
> if the absolute value of the error (i. e., the
>
> difference between the approximation and the true
>
> value) is less than $.5 \times 10^{-n}$.

In those terms, what we have shown above is that rounding off a number to n decimal places is one way to obtain an approximation which is accurate to n decimal places.

There is only one major misconception that can be held concerning this definition: to say that $\bar{x}$ is an approximation of x which is accurate to n decimal places does not mean that x and $\bar{x}$ have the same digits in the first n decimal places! For example, we saw above that an approximation of $x = 2.71828...$ which is accurate to 2 decimal places is $\bar{x} = 2.72$; however, x and $\bar{x}$ differ in the 2nd decimal place. As a more dramatic example, an approximation (by rounding off) of $x = 2.99999$ which is accurate to four decimal places is $\bar{x} = 3.00003$; however, x and $\bar{x}$ differ in all their digits !

2. <u>The Taylor approximation question.</u> Using Taylor's Theorem, we are now in a position to answer the approximation question about $p_n(x)$, the n-th Taylor polynomial of a function $f(x)$ about $x = a$:

How good an approximation for $f(x)$ is $p_n(x)$ at values of x near a ?

By Taylor's Theorem we know that the absolute value of the error term (or Remainder term) is given by

$$\left| R_n(x) \right| = \frac{\left| f^{(n+1)}(c) \right|}{(n+1)!} \left| x - a \right|^{n+1}$$

for some value of c between x and a . The vagueness stemming from the unspecified value c can be removed if we can find an <u>upper bound</u> M for the values of $\left| f^{(n+1)}(c) \right|$ as c varies between x and a , i.e., a number M such that $\left| f^{(n+1)}(c) \right| \leq M$ for all c between x and a . Then

$$\left| R_n(x) \right| = \frac{\left| f^{(n+1)}(c) \right|}{(n+1)!} \left| x - a \right|^{n+1} \leq \frac{M}{(n+1)!} \left| x - a \right|^{n+1}$$

Putting this together with the definition of accuracy to n decimal places, we obtain a practical answer to the Taylor approximation question:

The Taylor Approximation Principle

The approximation of $f(x_0)$ by the value of the n-th Taylor polynomial about $x = a$,

$$p_n(x_0) = \sum_{k=0}^{n} \frac{f^{(k)}(a)}{k!} (x_0 - a)^n$$

is accurate to m decimal places whenever

$$\frac{M}{(n+1)!} \left| x_0 - a \right|^{n+1} < .5 \times 10^{-m}$$

Here M is a number (independent of c) such that $\left| f^{(n+1)}(c) \right| \leq M$ for all c between a and x_0

We use this principle most often as follows:. When we are asked to approximate $f(x)$

to a given decimal place accuracy, the approximation principle gives us a method to determine

which Taylor polynomial $p_n(x)$ will achieve the desired accuracy. Here's a detailed

description (which looks more complicated than it really is...):

Taylor series approximation of $f(x_0)$ to m decimal place   accuracy

---

Step 1:   Choose a convenient value of $x = a$ which is close to x_0 ,   i.e., choose   a

so that $|x_0 - a|$ is small. Then calculate the Taylor series for $f(x)$ about

$x = a$,

$$\sum_{k=0}^{\infty} \frac{f^{(k)}(a)}{k!} (x - a)^k$$

Step 2:   Determine the Lagrange form of the remainder,

$$R_n(x_0) = \frac{f^{(n+1)}(c)}{(n+1)!} (x_0 - a)^{n+1}$$

where   c   is some value between x_0 and   a ,   and   n   is any positive integer.

Step 3:   Determine an upper bound   M   for $|f^{(n+1)}(c)|$ which is valid for all   c

between x_0 and   a .   This gives an upper bound for $|R_n(x_0)|$ of the form

$$\frac{M}{(n+1)!} |x_0 - a|^{n+1}$$

Step 4:   Determine the smallest value of   n   for which

$$\frac{M}{(n+1)!} |x_0 - a|^{n+1} < .5 \times 10^{-m}$$

This is done by simply computing the left-hand side of the inequality for various

values of   n .   (Remember: x_0 , a   and   m   are given, while   M   was

obtained in Step 3. In general,   M   will depend on   n .)

Step 5: Compute $p_n(x_0) = \sum_{k=0}^{n} \frac{f^{(k)}(a)}{k!}(x_0 - a)^k$ for the value of n found in

Step 4. This is an approximation for $f(x_0)$ which is accurate to m decimal

places.

The reason for Step 1 (choose a convenient value of $x = a$ which is close to x_0)

is that if $|x_0 - a|$ is small, then we'll have fast convergence of the Taylor series to $f(x_0)$

(i. e. , only a small number of terms will be needed to obtain a very accurate answer). If

$|x_0 - a|$ is large, we'll have slow convergence (i. e. , a large number of terms will be needed).

Example A. Approximate $\sin 33^\circ$ to three decimal place accuracy.

Step 1: Since 33° corresponds to $x_0 = \frac{33}{180}\pi = \frac{11}{60}\pi$ radians, we choose $a = \frac{\pi}{6}$

(since $a = \frac{\pi}{6}$ is close to $x_0 = \frac{11}{60}\pi$). Then the Taylor series for $\sin x$ about

$a = \frac{\pi}{6}$ is

$$\sum_{k=0}^{\infty} \frac{f^{(k)}\left(\frac{\pi}{6}\right)}{k!}\left(x - \frac{\pi}{6}\right)^k$$

The derivatives of $f(x) = \sin x$ are

$$f^{(2k)}(x) = (-1)^k \sin x$$
$$\text{for} \quad k = 0, 1, 2, \ldots$$
$$f^{(2k+1)}(x) = (-1)^k \cos x$$

(cf. Example B of §11. 10 of <u>The Companion</u>) so for $k = 0, 1, 2, \ldots$

$$f^{(2k)}\left(\frac{\pi}{6}\right) = (-1)^k \sin\frac{\pi}{6} = (-1)^k\left(\frac{1}{2}\right)$$

$$f^{(2k+1)}\left(\frac{\pi}{6}\right) = (-1)^k \cos\frac{\pi}{6} = (-1)^k\left(\frac{\sqrt{3}}{2}\right)$$

Thus the Taylor series for $\sin x$ about $a - \dfrac{\pi}{6}$ is

$$\sum_{k=0}^{\infty} \frac{f^{(k)}\left(\frac{\pi}{6}\right)}{k!}\left(x - \frac{\pi}{6}\right)^k = \frac{1}{2} + \frac{\sqrt{3}}{2}\left(x - \frac{\pi}{6}\right) - \frac{1}{2}\frac{1}{2!}\left(x - \frac{\pi}{6}\right)^2 - \frac{\sqrt{3}}{2}\frac{1}{3!}\left(x - \frac{\pi}{6}\right)^3$$

$$= \frac{1}{2} + \frac{\sqrt{3}}{2}\left(x - \frac{\pi}{6}\right) - \frac{1}{4}\left(x - \frac{\pi}{6}\right)^2 - \frac{\sqrt{3}}{12}\left(x - \frac{\pi}{6}\right)^3 + \cdots$$

Step 2:   Using $f^{(n+1)}(x) = \pm \sin x$ or $\pm \cos x$, $a = \pi/6$ and $x_0 = 11\pi/60$

in the Lagrange Remainder formula yields

$$R_n\left(\frac{11\pi}{60}\right) = \frac{f^{(n+1)}(c)}{(n+1)!}\left(\frac{11}{60}\pi - \frac{1}{6}\pi\right)^{n+1} = \frac{f^{(n+1)}(c)}{(n+1)!}\left(\frac{\pi}{60}\right)^{n+1}$$

where c is a number between $\pi/6$ and $11\pi/60$.

Step 3:   Since $f^{(n+1)}(c) = \pm \sin c$ or $\pm \cos c$, then $\left|f^{(n+1)}(c)\right| < 1$ for all c and

hence, in particular, for all c between $\pi/6$ and $11\pi/60$. Then $M = 1$ is an

upper bound for $\left|f^{(n+1)}(c)\right|$ which yields

$$\left|R_n\left(\frac{11}{60}\pi\right)\right| \leq \frac{1}{(n+1)!}\left(\frac{\pi}{60}\right)^{n+1}$$

Step 4:   We wish to make $\left|R_n\left(\frac{11}{60}\pi\right)\right|$ less than $.5 \times 10^3$; from Step 3 this means finding

the smallest n such that

$$\frac{1}{(n+1)!}\left(\frac{\pi}{60}\right)^{n+1} < .5 \times 10^{-3}$$

Evaluating the expression on the left for $n - 0, 1, 2, \ldots$ (either by hand or with

a calculator) we find

n	$\dfrac{1}{(n+1)!}\left(\dfrac{\pi}{60}\right)^{n+1}$
0	.05236
1	.00137
2	.00002

Thus $n = 2$ is the smallest integer for which the desired inequality holds.

Step 5: Since $\left|R_n\left(\dfrac{11}{60}\pi\right)\right| < .5 \times 10^{-3}$ when $n = 2$, then $p_2\left(\dfrac{11}{60}\pi\right)$ is an approximation

for $\sin 33^{\circ}$ which is accurate to 3 decimal places. We compute $p_2\left(\dfrac{11}{60}\pi\right)$ directly

from its formula (the first three terms of the Taylor series for $\sin x$ about $x = \dfrac{\pi}{6}$):

$$p_2\left(\frac{11}{60}\pi\right) = \frac{1}{2} + \frac{\sqrt{3}}{2}\left(\frac{11}{60}\pi - \frac{1}{6}\pi\right) - \frac{1}{4}\left(\frac{11}{60}\pi - \frac{1}{6}\pi\right)^2$$

$$= \frac{1}{2} + \frac{\sqrt{3}}{2}\left(\frac{\pi}{60}\right) - \frac{1}{4}\left(\frac{\pi}{60}\right)^2$$

$$\cong .5 + (.8660)(.0524) - (.25)(.0027)$$

$$\cong .5 + .0454 - .0007$$

$$= .5447$$

$$\cong .545 \qquad \text{(rounding off to 3 decimal places)}$$

Thus $\sin 33^{\circ} \cong .545$ with an accuracy of 3 decimal places. * $\qquad\qquad \square$

Anton's Examples 1, 2 and 3 also use our 5-step procedure, although the steps are not labeled. For instance, here is a summary of Anton's solution to Example 1 using the 5-step procedure:

---

* Note Anton's Remark concerning round-off errors preceding Example 3. In our computations, we followed the procedure he describes.

Anton's Example 1.   Approximate e to four decimal place accuracy.

Solution.   Using $f(x) = e^x$, we wish to approximate $e = f(1)$ to $m = 4$ decimal

place accuracy.

Step 1:   Choose $a = 0$ as a convenient number which is close to $x_0 = 1$. Thus we will use

the Maclaurin series for e^x : $1 + x + \frac{x^2}{2!} + \frac{x^3}{3!} + \cdots$.

Step 2:   Using $f^{(n+1)}(x) = e^x$, $a = 0$, and $x_0 = 1$ in the Remainder formula yields

$$R_n(1) = \frac{e^c}{(n+1)!}$$

where c is a number between 0 and 1 .

Step 3:   Since $\left| f^{(n+1)}(c) \right| = \left| e^c \right| \le e^1 < 3$ for $0 < c < 1$, then $M = 3$ is an upper

bound for $\left| f^{(n+1)}(c) \right|$ which yields

$$\left| R_n(1) \right| < \frac{3}{(n+1)!}$$

Step 4:   By simple computations, $n = 8$ is the smallest integer such that

$$\frac{3}{(n+1)!} < .5 \times 10^{-4}$$

Step 5:   Thus $p_8(1) \cong 2.7183$ is an approximation for $e - f(1)$ which is accurate to 4

decimal places.   □

Here is an illustration of a slightly different question which can be asked regarding

Taylor series approximations. (This is like Anton's Exercise 17):

Example B. (Optional) For what range of x-values can e^x be approximated by

$1 + x + x^2/2! + x^3/3!$ if the allowable error is at most $.5 \times 10^{-6}$?

<u>Solution.</u> The given four term summation is $p_3(x)$, the third Maclaurin polynomial for e^x .

Hence, since the error in approximating $f(x) = e^x$ by $p_3(x)$ is given by the absolute value

of the remainder term $|R_3(x)|$, the question can be rephrased as

<blockquote>
"For what range of x-values will

$|R_3(x)|$ be less than $.5 \times 10^{-6}$?"
</blockquote>

Since $f^{(n+1)}(x) = e^x$, then

$$R_3(x) = \frac{e^c}{(n+1)!} x^{n+1} = \frac{e^c}{4!} x^4 = \frac{e^c}{24} x^4 \quad \text{for} \quad c \quad \text{between} \quad 0 \quad \text{and} \quad x \quad .$$

In order to continue we must consider two cases:

Case i. Suppose $x \le 0$. Then

$$|R_3(x)| \le \frac{|x|^4}{24} \quad \text{since} \quad e^c \le e^0 = 1$$

Thus $|R_3(x)|$ will be less than $.5 \times 10^{-6}$ whenever

$$|x|^4 = x^4 \le 24 \, (.5 \times 10^{-6}) \cong 1.2 \times 10^{-5}$$

or $\boxed{-.0586 \le x}$

Case ii. Suppose $x > 0$. Then

$$|R_3(x)| \le \frac{x^4}{24} e^x \quad \text{since} \quad e^c \le e^x$$

Thus $|R_3(x)|$ will be less than $.5 \times 10^{-6}$ whenever

$$x^4 e^x \le 1.2 \times 10^{-5}$$

However, because of the e^x term, this inequality must be treated carefully. Since

$x^4 e^x$ is supposed to be very small, then we know that x will be close to zero. Since

this is the case, we can begin by assuming $e^x < 2$. In that case

$$x^4 \leq .6 \times 10^{-5}$$

so that $x \leq .0495$

However, if $x = .0495$,   then $e^x \cong 1.0050$,   so that we were much too generous in

assuming $e^x < 2$. Instead, let's assume $e^x < 1.1$. In that case

$$x^4 \leq 1.091 \times 10^{-5}$$

so that $\boxed{x \leq .0575}$

Now, if $x = .0575$,   then $e^x \cong 1.059$ which is close to, but still less than, the   1.1

bound on e^x .

Combining Cases i and ii we see that if

$$\boxed{-.0586 \leq x \leq .0575}$$

then $1 + x + x^2/2! + x^3/3!$ approximates e^x to within an allowable error of

$.5 \times 10^{-6}$.   □

3.   Use of The Alternating Series Test.   Suppose you are computing a Taylor series approximation

of $f(x_0)$ to m decimal place accuracy using our 5-step procedure, and in Step 1 you end

up with an alternating series whose terms decrease to zero in absolute value, i. e.,

$$\sum_{k=0}^{\infty} \frac{f^{(k)}(a)}{k!} (x - a)^k$$

is an alternating series when x_0 is substituted for x . Then rejoice (!), for Steps 2, 3 and 4 can be greatly simplified. Why? Because you no longer need to use the Remainder formula* to estimate the error between $f(x_0)$ and

$$P_n(x_0) = \sum_{k=0}^{n} \frac{f^{(k)}(a)}{k!} (x_0 - a)^k$$

Instead use The Alternating Series Test to establish that the error between $f(x_0)$ and $P_n(x_0)$ is less than the magnitude of the first term left out of $P_n(x_0)$. Thus we have only to determine the smallest integer n so that the first term left out of $P_n(x_0)$ is less than $.5 \times 10^{-m}$ in absolute value.

Anton uses this important technique to give an alternate solution to his Example 2. Here is another illustration:

Example C. Approximate $1/\sqrt{e}$ to five decimal place accuracy.

Solution. Using $f(x) = e^x$, we will approximate $1/\sqrt{e} = e^{-1/2} = f(-\frac{1}{2})$ to $m = 5$ decimal place accuracy.

Step 1: Choose $a = 0$ as a convenient number which is close to $x_0 = -1/2$. Thus we will use the Maclaurin series for e^x:

$$e^x = 1 + x + x^2/2! + x^3/3! + \cdots$$

You might at first not realize that this is an alternating series.(!?!)... well, it is when $x < 0$! In particular,

$$e^{-1/2} = 1 - \frac{1}{2} + \frac{1}{2^2 \cdot 2!} - \frac{1}{2^3 \cdot 3!} + \cdots + \frac{(-1)^n}{2^n \cdot n!} + \cdots$$

---

* Assuming, of course, that you have already established the convergence of the Taylor series to $f(x_0)$. This is almost always the case with the examples we consider.

Steps 2, 3 & 4:   Since the terms in our alternating series clearly decrease to zero in absolute

value, then the magnitude of the error between $f(-1/2)$ and $p_n(-1/2)$ will be at

most

$$\frac{1}{2^{n+1}(n+1)!}$$ (the first term left out of $p_n(-1/2)$)

Thus, for five decimal place accuracy, we look for the first positive integer n such

that

$$\frac{1}{2^{n+1}(n+1)!} < .5 \times 10^{-5}$$

By trial and error, $n = 6$ is the first such integer.

Step 5:   Thus to five decimal place accuracy we have

$$e^{-1/2} \approx 1 - \frac{1}{2} + \frac{1}{2^2 \cdot 2!} - \cdots + \frac{1}{2^6 \cdot 6!}$$

$$\approx 1 - .5 + .125 - .020833 + .002604 - .000260 + 000022$$

$$= .606533$$

$$\approx .60653 \quad \text{to } 5 \text{ decimal place accuracy.}$$

Section 11. 12.   Differentiation and Integration of Power Series

Introduction.   This section gathers together a number of "theoretical" results concerning

power series, all of which have important "practical" applications for the computation of Taylor

and Maclaurin series. As we have seen in earlier sections, obtaining a Taylor series expansion

for a given function   f   directly from the formula involves two steps, both of which can be

difficult:

1. Obtain an expression for the general term (i.e., the n-th term)

$$\frac{f^{(n)}(a)}{n!} (x - a)^n$$ of the series, and

2. Prove that the series so obtained does converge on some interval to f.

However, at the end of §11.10 we indicated that sometimes a Taylor series for one function can be obtained from the known Taylor series of another function by "substitution," e.g.,

since $\dfrac{1}{1 - x} = 1 + x + x^2 + x^3 + \cdots + x^n + \cdots$ for $-1 < x < 1$,

then $\dfrac{1}{1 + 2x^2} = 1 - 2x^2 + 4x^4 - 8x^6 + \cdots + (-1)^n 2^n x^{2n} + \cdots$ for $-\dfrac{\sqrt{2}}{2} < x < \dfrac{\sqrt{2}}{2}$.

Thus, in one step, we have obtained the formula for the Maclaurin series of $1/(1 + 2x^2)$ and have proven convergence on an open interval about zero. Quite a savings in time and effort!

In this section we will give the justification for this technique (Theorem 11. 12. 2 concerning the "uniqueness" of Taylor series) and will develop other useful procedures for obtaining Taylor series expansions "indirectly."

1. <u>Term-by-term differentiation and integration of power series.</u> The first key theorem of this section (Theorem 11. 12. 1) may be summarized quite simply:

Differentiation and Integration of Power Series

Power series can be differentiated and integrated term-by-term on the <u>open</u> interval of convergence $(a - R, a + R)$.

Do not let Anton's more detailed statement of the theorem intimidate you; our one-sentence summary really says everything you need to know. Anton's Examples 1, 2 and 3 illustrate how this theorem can be used to evaluate integrals and obtain Maclaurin series for new functions from known Maclaurin series. Here are three more examples:

<u>Example A.</u>   Find the Maclaurin series for $\dfrac{1}{(1-x)^3}$.

<u>Solution.</u>   Beginning with the known Maclaurin series

$$\frac{1}{1-x} = \sum_{k=0}^{\infty} x^k = 1 + x + x^2 + x^3 + x^4 + x^5 + \cdots \quad \text{when} \quad -1 < x < 1$$

(see Table 11. 10. 1), we differentiate both sides, using the theorem above to differentiate the

power series term-by-term:

$$\frac{1}{(1-x)^2} = \sum_{k=1}^{\infty} k x^{k-1} = 1 + 2x + 3x^2 + 4x^3 + 5x^4 + \cdots \quad \text{when} \quad -1 < x < 1$$

(note that the summation now begins with $k = 1$ since the $k = 0$ term has been differentiated

to zero). Then, differentiating once again, we obtain

$$\frac{2}{(1-x)^3} = \sum_{k=2}^{\infty} k(k-1) x^{k-2} = 2 + 6x + 12x^2 + 20x^3 + \cdots \quad \text{when} \quad -1 < x < 1$$

(now the summation begins with $k = 2$). Dividing by 2 yields the Maclaurin series for $\dfrac{1}{(1-x)^3}$:

$$\frac{1}{(1-x)^3} = \sum_{k=2}^{\infty} \frac{k(k-1)}{2} x^{k-2} = 1 + 3x + 6x^2 + 10x^3 + \cdots$$

which from our theorem must converge to $\dfrac{1}{(1-x)^3}$ for $-1 < x < 1$.

<u>Example B.</u>   Find the Maclaurin series for $\ln(1+x)$.

<u>Solution.</u>   Beginning with the known Maclaurin series

$$\frac{1}{1-x} = \sum_{k=0}^{\infty} x^k = 1 + x + x^2 + x^3 + \cdots \quad , \qquad -1 < x < 1$$

(see Table 11. 10. 1), we replace x by $-x$ and obtain

11. 12. 4

$$\frac{1}{1+x} = \frac{1}{1-(-x)} = \sum_{k=0}^{\infty} (-1)^k x^k = 1 - x + x^2 - x^3 + \cdots \quad \text{when} \quad -1 < -x < 1 \; ,$$

$$\text{i. e.,} \quad -1 < x < 1 \quad .$$

Then integrating both sides (using our theorem to integrate the power series term-by-term)

yields the Maclaurin series for $\ln(1+x)$:

$$\ln(1+x) = \int \frac{1}{1+x} \, dx = \sum_{k=0}^{\infty} (-1)^k \frac{x^{k+1}}{k+1} = x - \frac{x^2}{2} + \frac{x^3}{3} - \frac{x^4}{4} + \cdots$$

By our theorem, the interval of convergence of this series is still $-1 < x < 1$. $\qquad \square$

Example C. Use a Maclaurin series to approximate $\displaystyle\int_{.2}^{.3} \frac{\tan^{-1} x}{x} \, dx$ to four decimal place

accuracy.

Solution. Beginning with the known Maclaurin series

$$\tan^{-1} x = \sum_{k=0}^{\infty} (-1)^k \frac{x^{2k+1}}{2k+1} = x - \frac{x^3}{3} + \frac{x^5}{5} - \frac{x^7}{7} + \cdots \quad \text{when} \quad -1 \leq x \leq 1$$

(See Table 11. 10. 1), we multiply by $1/x$ to obtain

$$\frac{\tan^{-1} x}{x} = \sum_{k=0}^{\infty} (-1)^k \frac{x^{2k}}{2k+1} = 1 - \frac{x^2}{3} + \frac{x^4}{5} - \frac{x^6}{7} + \cdots \quad \text{when} \quad -1 \leq x < 0$$

$$\text{or} \quad 0 < x \leq 1$$

(Note that division by x removed $x = 0$ from the interval of convergence.) Integrating both

sides (using our theorem to integrate the power series term-by-term) yields

$$\int \frac{\tan^{-1} x}{x} \, dx = \sum_{k=0}^{\infty} (-1)^k \frac{x^{2k+1}}{(2k+1)^2} = x - \frac{x^3}{3^2} + \frac{x^5}{5^2} - \frac{x^7}{7^2} + \cdots \quad \text{when} \quad -1 \leq x < 0$$

$$\text{or} \quad 0 < x \leq 1$$

so that

$$\int_{.2}^{.3} \frac{\tan^{-1} x}{x} \, dx = \left[x - \frac{x^3}{3^2} + \frac{x^5}{5^2} - \frac{x^7}{7^2} + \cdots \right]_{.2}^{.3}$$

$$= \left((.3) - \frac{(.3)^3}{3^2} + \frac{(.3)^5}{5^2} - \frac{(.3)^7}{7^2} + \cdots \right) - \left((.2) - \frac{(.2)^3}{3^2} + \frac{(.2)^5}{5^2} - \frac{(.3)^7}{7^2} + \cdots \right)$$

We are asked to approximate this integral to 4 decimal place accuracy. But that will be relatively easy because the two series are both alternating series with terms whose absolute values approach zero. Thus the magnitude of the error in approximating the sum of each series by a partial sum will be at most the absolute value of the first term left out of that partial sum. Moreover, if ϵ_1 is the error made when the true value of the first series is approximated by its n-th partial sum and ϵ_2 is the corresponding error for the second series, the error made when the definite integral is approximated by the difference of those two partial sums will be <u>at most</u> $\epsilon_1 + \epsilon_2$, the <u>sum</u> of the two errors. So, the error in approximating $\int_{.2}^{.3} \frac{\tan^{-1} x}{x} \, dx$ by

$$\left((.3) - \frac{(.3)^3}{3^2} + \cdots + (-1)^{n-1} \frac{(.3)^{2n-1}}{(2n-1)^2} \right) - \left((.2) - \frac{(.2)^3}{3^2} + \cdots + (-1)^{n-1} \frac{(.2)^{2n-1}}{(2n-1)^2} \right)$$

will be at most

$$\left| \frac{(.3)^{2n+1}}{(2n+1)^2} \right| + \left| \frac{(.2)^{2n+1}}{(2n+1)^2} \right| = \frac{1}{(2n+1)^2} \left((.3)^{2n+1} + (.2)^{2n+1} \right)$$

Thus we want to find the first n such that

$$\frac{1}{(2n+1)^2} \left((.3)^{2n+1} + (.2)^{2n+1} \right) < .5 \times 10^{-4} = .00005$$

A few computations reveal that the first such n is n = 3 so that

$$\int_{.2}^{.3} \frac{\tan^{-1} x}{x} \, dx \approx \sum_{k=0}^{2} (-1)^k \frac{(.3)^{2k+1}}{(2k+1)^2} - \sum_{k=0}^{2} (-1)^k \frac{(.2)^{2k+1}}{(2k+1)^2}$$

$$= \left((.3) - \frac{(.3)^3}{3^2} + \frac{(.3)^5}{5^2} \right) - \left((.2) - \frac{(.2)^3}{3^2} + \frac{(.2)^5}{5^2} \right)$$

$$\approx (.3 - .003 + .00010) - (.2 - .00089 + .00001)$$

$$\approx .2971 - .1991 = .0980$$

to 4 decimal place accuracy. □

Note that the method of Example C gives us yet another method for approximating definite integrals which cannot be evaluated directly (§9.9 contains others). Of course, it is of use only when you have a definite integral $\int_a^b f(x) \, dx$ where $f(x)$ is the sum of its Taylor (Maclaurin) series on an interval of convergence which contains the closed interval $[a, b]$.

2. Intervals of convergence of differentiated and integrated power series. Anton's Theorem 11.12.1 and his Remark preceding Example 4 make some interesting points about the interval of convergence of a power series:

(i)   The radius of convergence R of a differentiated or integrated power series is the same as for the original series.

(ii) Convergence at either endpoint of the interval of convergence may be lost by differentiation.

(iii) Convergence at either endpoint of the interval of convergence may be gained by integration.

To illustrate this point, Anton invites his reader to compare the intervals of convergence of the power series for $\tan^{-1} x$ and $\dfrac{1}{1+x^2}$. We'll accept that invitation:

To handle $1/(1+x^2)$, consider the Maclaurin series

$$\frac{1}{1-x} = 1 + x + x^2 + x^3 + \cdots + x^n + \cdots$$

which is <u>valid for</u> $-1 < x < 1$. (See Table 11. 10. 1) Substituting $-x^2$ for x gives

$$\frac{1}{1+x^2} = 1 - x^2 + x^4 - x^6 + \cdots + (-1)^n x^{2n} + \cdots$$

which is valid whenever $-1 < -x^2 < 1$, i.e., <u>valid for</u> $-1 < x < 1$. The Divergence Test shows that we do not have convergence at the endpoints: if $x = \pm 1$, then $\lim\limits_{n \to \infty} (-1)^n x^{2n} =$

$\lim\limits_{n \to \infty} (-1)^n$ which is undefined. Thus

---

the Maclaurin series for $1/(1+x^2)$ has

$(-1, 1)$ as its interval of convergence.

In Example 3, the Maclaurin series for $\tan^{-1} x$,

$$\tan^{-1} x = x - \frac{x^3}{3} + \frac{x^5}{5} - \frac{x^7}{7} + \cdots + (-1)^n \frac{x^{2n+1}}{2n+1} + \cdots$$

is obtained by integrating the Maclaurin series for $1/(1+x^2)$ obtained above. However, Example 3 does not discuss the radius of convergence for the series. From Theorem 11. 12. 1 this radius is $R = 1$ since integrating a Maclaurin series leaves the radius of convergence unchanged (we noted that $R = 1$ for the series for $1/(1+x^2)$). Hence the Maclaurin series for $\tan^{-1} x$ converges for $-1 < x < 1$.

But what about the endpoints $x = \pm 1$? Unlike the series for $1/(1+x^2)$, the series for

$\tan^{-1} x$ <u>does converge to something</u> for $x = \pm 1$ (that's a simple application of The Alternating Series Test; try it!). The only question is: does the Maclaurin series for $\tan^{-1} x$ converge to $\tan^{-1} x$ at $x = \pm 1$? The answer to this question is "yes," although we will not prove the result since it requires some techniques not developed in Anton. Thus

<blockquote>

the Maclaurin series for $\tan^{-1} x$ has

$[-1, 1]$ as its interval of convergence.

</blockquote>

Comparing the intervals of convergence for $1/(1 + x^2)$ and $\tan^{-1} x$, we have an example showing that

-- convergence at at endpoint may be gained by integration

(going from $1/(1 + x^2)$ to $\tan^{-1} x$)

-- convergence at an endpoint may be lost by differentiation

(going from $\tan^{-1} x$ to $1/(1 + x^2)$)

3. <u>The uniqueness of the Taylor series.</u> Anton's Theorem 11. 12. 2 fills in a gap in the theory (a gap which Anton pointed out in §11. 10 following Example 6). For instance, in the previous subsection we said that integrating the Maclaurin series for $\dfrac{1}{1 + x^2}$ produces <u>the Maclaurin series</u> for $\tan^{-1} x$. It is certainly true that integrating the $\dfrac{1}{1 + x^2}$ series produces <u>a power series</u> for $\tan^{-1} x$, but how can we be sure it is <u>the Maclaurin series</u> for $\tan^{-1} x$? We could try to check the result by calculating the Maclaurin series for $\tan^{-1} x$ with the formula

$\displaystyle\sum_{k = 0}^{\infty} \dfrac{f^{(k)}(0)}{k!} x^k$, and comparing it with the integrated $1/(1 + x^2)^2$ power series. However there are two problems with this:

(1) calculating the Maclaurin series for $\tan^{-1} x$ directly from the formula

   is unpleasant, and proving that it converges to $\tan^{-1} x$ directly from

   the Remainder formula is very difficult;

(2) the whole point of this section is to develop techniques for finding Taylor and

   Maclaurin series which do NOT use the defining formula.

Fortunately, Anton's Theorem 11.12.2 comes to the rescue. Informally stated it says:

The Uniqueness
Theorem for
Taylor Series

> If a power series expansion for $f(x)$ looks like the Taylor
>
> series about $x = a$, then it is the Taylor series about
>
> $x = a$!!

That is, if you find that

$$f(x) = c_0 + c_1 (x - a) + c_2 (x - a)^2 + \cdots \qquad \text{when } x \text{ is in an open} \atop \text{interval containing } a$$

then you can conclude that the series on the right is the Taylor series for f about $x = a$.

Applying this theorem to the example we discussed above, when we integrate the $1/(1 + x^2)$

series term-by-term and obtain the $\tan^{-1} x$ series

$$\tan^{-1} x = \sum_{k=0}^{\infty} (-1)^k \frac{x^{2k+1}}{2k+1} = x - \frac{x^3}{3} + \frac{x^5}{5} - \frac{x^7}{7} + \cdots$$

we see that this series has the form of a Maclaurin series: it's a series $\sum_{k=0}^{\infty} c_k (x - a)^k$

with $a = 0$, $c_0 = 0$, $c_1 = 1$, $c_2 = 0$, $c_3 = -\frac{1}{3}$, etc. Thus by the Uniqueness Theorem it is

the Maclaurin series for $\tan^{-1} x$!! The theorem makes any further check unnecessary. It

says that the form of a power series expansion for a function determines whether it is a Taylor or Maclaurin series, regardless of how the series was obtained. That's a very important, and nice, thing to know.

4. <u>Tricks and Techniques.</u> The Uniqueness Theorem just discussed in the previous subsection tells us in effect that, in obtaining Taylor series, the end justifies the means. That is, if we can expand $f(x)$ as a power series that has the form of a Taylor series, then the power series <u>is</u> a Taylor series for $f(x)$. It does not matter how the power series was originally obtained. That naturally leads us to consider the methods we might use to obtain power series expansions for functions. So far, we have discussed the following <u>methods for finding Taylor (Maclaurin)</u> <u>series</u>:

1. <u>Direct calculation</u> using the formula. (See <u>Companion</u> Examples B and C in §11.9.2.)

2. <u>Substitution.</u> (See <u>Companion</u> Example C in §11.10.3 where $-2x$ was substituted for x in the series for $\dfrac{1}{1-x}$ to obtain the series for $\dfrac{1}{1+2x}$.)

3. <u>Multiplication by a power of</u> x. (See <u>Companion</u> Example C in §11.10.3 where the series for $\dfrac{1}{1+2x}$ was multiplied by x^3 to obtain the series for $\dfrac{x^3}{1+2x}$.)

4. <u>Differentiation</u> of another series. (See <u>Companion</u> Example A in §11.12.1.)

5. <u>Integration</u> of another series. (See <u>Companion</u> Example B in §11.12.1 where the series for $1/(1+x)$ was integrated to obtain a series for $\ln(1+x)$.)

At the end of §11.12, Anton illustrates one additional family of techniques:

> 6. <u>Algebraic Manipulation.</u> Add, subtract, multiply or divide power series
>
> by treating them as polynomials.

Anton's Example 4 illustrates how two power series may be multiplied like polynomials to

obtain a new power series and his Example 5 illustrates how polynomial long division may be

used to obtain a new power series from two known ones. Here are two additional examples

illustrating method 6:

<u>Example D.</u>   Find the Maclaurin series for $f(x) = \sin x + \cos x$.

<u>Solution.</u>   Since

$$\sin x = \sum_{k=0}^{\infty} \frac{(-1)^k}{(2k+1)!} x^{2k+1} = x - \frac{x^3}{3!} + \frac{x^5}{5!} - \frac{x^7}{7!} + \cdots$$

and

$$\cos x = \sum_{k=0}^{\infty} \frac{(-1)^k}{(2k)!} x^{2k} = 1 - \frac{x^2}{2!} + \frac{x^4}{4!} - \frac{x^6}{6!} + \cdots$$

we add the series term-by-term as we would polynomials and obtain

$$\sin x + \cos x = \sum_{k=0}^{\infty} (-1)^k \left(\frac{x^{2k}}{(2k)!} + \frac{x^{2k+1}}{(2k+1)!} \right) = 1 + x - \frac{x^2}{2!} - \frac{x^3}{3!} + \frac{x^4}{4!} + \frac{x^5}{5!} - \cdots$$

(Note that the signs would be wrong if we wrote this series in summation notation as

$\sum_{k=0}^{\infty} (-1)^k \frac{x^k}{k!}$, as one is tempted to do.) Since the $\sin x$ and $\cos x$ series both have the

same interval of convergence $(-\infty, +\infty)$, the new series also has interval of convergence

$(-\infty, +\infty)$. By Theorem 11. 12. 2, this must be the Maclaurin series for $\sin x + \cos x$.   □

<u>Example E.</u> Find the first four nonzero terms of the Maclaurin series for $\cos^2 x$.

<u>Solution.</u> We know that $\cos x = \displaystyle\sum_{k=0}^{\infty} (-1)^k \frac{x^{2k}}{(2k)!}$ (see Table 11.10.1) so, multiplying

the series as we would polynomials, we obtain

$$\cos^2 x = (\cos x)(\cos x)$$

$$= \left(1 - \frac{x^2}{2!} + \frac{x^4}{4!} - \frac{x^6}{6!} + \cdots\right)\left(1 - \frac{x^2}{2!} + \frac{x^4}{4!} - \frac{x^6}{6!} + \cdots\right)$$

$$= 1\left(1 - \frac{x^2}{2!} + \frac{x^4}{4!} - \frac{x^6}{6!} + \cdots\right) - \frac{x^2}{2!}\left(1 - \frac{x^2}{2!} + \frac{x^4}{4!} - \frac{x^6}{6!} + \cdots\right) + \frac{x^4}{4!}\left(1 - \frac{x^2}{2!} + \frac{x^4}{4!} - \frac{x^6}{6!} + \cdots\right)$$

$$= \left(1 - \frac{x^2}{2!} + \frac{x^4}{4!} - \frac{x^6}{6!} + \cdots\right) - \left(\frac{x^2}{2!} - \frac{x^4}{2!\,2!} + \frac{x^6}{2!\,4!} - \frac{x^8}{2!\,6!} + \cdots\right) + \left(\frac{x^4}{4!} - \frac{x^6}{4!\,2!} + \cdots\right)$$

$$= 1 + \left(\frac{-x^2}{2!} - \frac{x^2}{2!}\right) + \left(\frac{x^4}{4!} + \frac{x^4}{2!\,2!} + \frac{x^4}{4!}\right) + \left(\frac{-x^6}{6!} - \frac{x^6}{2!\,4!} - \frac{x^6}{4!\,2!} - \frac{x^6}{6!}\right) + \cdots$$

$$= 1 - x^2 + \frac{1}{3} x^4 - \frac{2}{45} x^6 + \cdots$$

By Theorem 11.12.2, this is the Maclaurin series for $\cos^2 x$. Since the interval of convergence

for the Maclaurin series for $\cos x$ is $(-\infty, +\infty)$, the interval of convergence for the

Maclaurin series for $\cos^2 x$ is $(-\infty, +\infty)$ as well. $\square$

5. <u>The Binomial Series.</u> Anton ends this section with the introduction of one more family of

Maclaurin series: those for $(1+x)^m$ where m is any real number. These are called

<u>binomial series</u>:

Maclaurin series for $(1+x)^m$	$$(1+x)^m = 1 + mx + \frac{m(m-1)}{2!}x^2 + \frac{m(m-1)(m-2)}{3!}x^3 + \cdots$$ $$= 1 + \sum_{k=1}^{\infty} \frac{m(m-1)\cdots(m-k+1)}{k!}x^k \quad \text{when} \quad -1 < x < 1$$

Notice that if m is a positive integer, and $k \le m$, then the coefficient of the x^k term can be written as

$$\frac{m(m-1)\cdots(m-k+1)}{k!} = \frac{m(m-1)\cdots(m-k+1)}{k!} \cdot \frac{(m-k)(m-k-1)\cdots \cdot 2 \cdot 1}{(m-k)(m-k-1)\cdots \cdot 2 \cdot 1}$$

$$= \frac{m!}{k!\,(m-k)!}$$

Moreover, when m is a positive integer, all the coefficients are zero for $k > m$. Then we get the familiar series

The usual Binomial Series	$$(1+x)^m = 1 + \sum_{k=1}^{m} \frac{m!}{k!\,(m-k)!}x^k = \sum_{k=0}^{m} \frac{m!}{k!\,(m-k)!}x^k$$ for m a positive integer.

Even when m is not a positive integer, using $\frac{m!}{k!\,(m-k)!}$ as a device for remembering the coefficient formula is useful. Notice, however, that it actually is an abuse of the notation since factorials of non-integers are not defined! However, by taking a little license, we can handle them this way:

$$\frac{\left(-\frac{1}{2}\right)!}{3!\left(-\frac{1}{2}-3\right)!} = \frac{\left(-\frac{1}{2}\right)\left(-\frac{1}{2}-1\right)\left(-\frac{1}{2}-2\right)\left(-\frac{1}{2}-3\right)\left(-\frac{1}{2}-4\right)\cdots}{3\cdot2\cdot1\cdot\left(-\frac{1}{2}-3\right)\left(-\frac{1}{2}-4\right)\cdots}$$

$$= \frac{\left(-\frac{1}{2}\right)\left(-\frac{3}{2}\right)\left(-\frac{5}{2}\right)}{3\cdot2\cdot1} = \frac{-\frac{15}{8}}{6} = -\frac{5}{16}$$

The Binomial series are very useful, especially when combined with the other techniques discussed in this chapter. There are two reasons for this:

1. expressions such as $\sqrt[k]{1+x}$ or $1/\sqrt{1+x}$ occur often, and

2. the binomial series are <u>complicated</u> Maclaurin series, i.e., it is not an easy matter to prove their convergence to $(1+x)^m$.

Anton's Example 6 illustrates the use of the binomial series. Here are two more illustrations:

<u>Example F.</u> Find the Maclaurin Series for $\sqrt[4]{1-2x}$.

<u>Solution.</u> Letting $m = \frac{1}{4}$ and replacing x by $-2x$ in the binomial series above yields

$$\sqrt[4]{1-2x} = (1+(-2x))^{1/4} = 1 + \sum_{k=1}^{\infty} \frac{(1/4)!}{k!\,(1/4-k)!}\,(-2x)^k$$

$$= 1 + \sum_{k=1}^{\infty} \frac{(-1)^k 2^k}{k!}\left(\frac{1}{4}\right)\left(\frac{1}{4}-1\right)\left(\frac{1}{4}-2\right)\cdots\left(\frac{1}{4}-k+1\right)x^k$$

$$= 1 - 2\left(\frac{1}{4}\right)x + \frac{2^2}{2}\left(\frac{1}{4}\right)\left(-\frac{3}{4}\right)x^2 - \frac{2^3}{6}\left(\frac{1}{4}\right)\left(\frac{-3}{4}\right)\left(\frac{-7}{4}\right)x^3 + \cdots$$

$$\cdots + \frac{(-1)^k 2^k}{k!}\left(\frac{1}{4}\right)\left(\frac{-3}{4}\right)\left(\frac{-7}{4}\right)\left(\frac{-11}{4}\right)\cdots\left(\frac{5-4k}{4}\right)x^k + \cdots$$

$$= 1 - \frac{1}{2}x - \frac{3}{8}x^2 - \frac{7}{16}x^3 - \cdots$$

$$\cdots + \frac{(-1)^k 2^k}{k!} \frac{(-1)^{k-1} 1 \cdot 3 \cdot 7 \cdot 11 \cdot 15 \cdots (5-4k)}{4^k} x^k + \cdots$$

$$= 1 - \frac{1}{2}x - \frac{3}{8}x^2 - \frac{7}{16}x^3 - \cdots - \frac{1 \cdot 3 \cdot 7 \cdot 11 \cdot 15 \cdots (5-4k)}{k! \, 2^k} x^k - \cdots$$

$$\text{if} \quad |-2x| < 1$$

$$\left(\text{i. e., if} \quad |x| < \frac{1}{2}\right)$$

□

Example G. Suppose $f(x)$ is a function such that

$$\text{i.} \quad f(0) = 1$$

$$\text{ii.} \quad \int \frac{dx}{(1+x^2)^{2/3}} = f(x) + C$$

(a) Find the first four nonzero terms in the Maclaurin series for $f(x)$.

(b) Express the series in sigma notation.

(c) What is the radius of convergence?

Solution. Letting $m = -2/3$ and replacing x by x^2 in the binomial series yields

$$(1+x^2)^{-2/3} = 1 + \sum_{k=1}^{\infty} \frac{(-2/3)!}{k!\,(-2/3-k)!} (x^2)^k$$

$$= 1 + \sum_{k=1}^{\infty} \frac{(-2/3)(-2/3-1)(-2/3-2)\cdots(-2/3-k+1)}{k!} x^{2k}$$

$$= 1 + \sum_{k=1}^{\infty} \frac{(-2/3)(-5/3)(-8/3)\cdots((1-3k)/3)}{k!} x^{2k}$$

Thus, integrating the power series term-by-term (Theorem 11. 12. 1)

$$f(x) + C = \int (1+x^2)^{-2/3}\, dx = x + \sum_{k=1}^{\infty} \frac{(-2/3)\,(-5/3)\,(-8/3)\ldots((1-3k)/3)}{(2k+1)\,k!}\, x^{2k+1}$$

We use the given condition $f(0) = 1$ to evaluate C:

$$1 + C = f(0) + C = (0) + \sum_{k=1}^{\infty} \frac{(-2/3)\,(-5/3)\,(-8/3)\ldots((1-3k)/3)}{(2k+1)k!}\,(0)^{2k+1} = 0$$

so $C = -1$. Thus

$$f(x) = 1 + x + \sum_{k=1}^{\infty} \frac{(-2/3)\,(-5/3)\,(-8/3)\ldots((1-3k)/3)}{(2k+1)k!}\, x^{2k+1}$$

By the Uniqueness Theorem (Theorem 11. 12. 2), this is the Maclaurin series for $f(x)$, answering (b). Moreover, its interval of convergence is $-1 < x^2 < 1$ or $-1 < x < 1$ (since the binomial series has interval of convergence $(-1, 1)$); thus the radius of convergence is $R = 1$, answering (c). Finally, answering (a), the first four terms of the Maclaurin series for f are

$$f(x) = 1 + x + \sum_{k=1}^{2} \frac{(-2/3)\,(-5/3)\,(-8/3)\ldots((1-3k)/3)}{(2k+1)k!}\, x^{2k+1} = 1 + x - \frac{2}{9}\,x^3 + \frac{1}{9}\,x^5 + \cdots$$

$\square$

6. Euler's formula; e, π and i (optional).   You have probably noticed some similarities between the Maclaurin series for e^x and the Maclaurin series for $\sin x$ and $\cos x$:

$$e^x = 1 + x + \frac{x^2}{2!} + \frac{x^3}{3!} + \cdots + \frac{x^n}{n!} + \cdots$$

$$\sin x = x - \frac{x^3}{3!} + \frac{x^5}{5!} - \cdots + (-1)^n \frac{x^{2n+1}}{(2n+1)!} + \cdots$$

$$\cos x = 1 - \frac{x^2}{2!} + \frac{x^4}{4!} - \cdots + (-1)^n \frac{x^{2n}}{(2n)!} + \cdots$$

If we could somehow change all the minus signs to plus signs, then adding the series for $\sin x$ and $\cos x$ together would yield the series for e^x. Do you suppose that there is some significance to this?

Yes!... but we have to use the complex numbers to see it. Let i denote the "imaginary" number $\sqrt{-1}$, i.e., i is a number whose square equals -1,

$$i^2 = -1$$

We can even consider expressions like e^{ix}. How? Simple: just substitute ix for x in the Maclaurin series for e^x. Playing fast and loose with series manipulations, here's what we get

$$e^{ix} = 1 + ix + \frac{i^2 x^2}{2!} + \frac{i^3 x^3}{3!} + \frac{i^4 x^4}{4!} + \frac{i^5 x^5}{5!} + \frac{i^6 x^6}{6!} + \cdots$$

$$= 1 + ix - \frac{x^2}{2!} - i\frac{x^3}{3!} + \frac{x^4}{4!} + i\frac{x^5}{5!} - \frac{x^6}{6!} + \cdots$$

since $i^2 = -1$, $i^3 = -i$, $i^4 = 1$, $i^5 = i$, $i^6 = -1$, etc.

Now combining the terms with the i's gives

$$e^{ix} = \left(\underbrace{1 - \frac{x^2}{2!} + \frac{x^4}{x!} - \frac{x^6}{6!} + \cdots}_{\cos x} \right) + i \left(\underbrace{x - \frac{x^3}{3!} + \frac{x^5}{5!} - \cdots}_{\sin x} \right)$$

Thus

$$e^{ix} = \cos x + i \sin x$$

This is known as <u>Euler's Formula</u>, and it is valid for all x. Euler's formula gives a particularly startling result when $x = \pi$:

$$e^{i\pi} = \cos \pi + i \sin \pi = -1 \ , \qquad \text{so}$$

$$e^{i\pi} = -1$$

And thus we have, in one small formula, a relationship between three of the most important constants in mathematics: e, π and i.

Chapter 12.   Topics In Analytic Geometry

Section 12. 1. Introduction to the Conic Sections

Don't rush through this little section. If you take the few minutes it requires to read it and to study the figures, you will have an overview of the conic sections which will serve you well throughout this chapter.

An elementary treatment of conic sections can be found in Appendix E of The Companion. You might read through that appendix if you have not studied conics in the past.

The conic sections provide a classic example of how mathematics which is developed for one purpose, or which is developed only because someone is interested in a mathematical question and wants to know the answer, can have surprising and important applications in other fields. The Greeks studied conic sections in order to solve construction problems like "doubling the cube. " Then about 2000 years later Kepler discovered that they were useful in studying planetary motion. No mathematics can ever be dismissed as being "useless" or "unimportant;" in 2000 or so years someone might have to eat those words!

Section 12. 2. The Parabola; Translation of Coordinate Axes

1.   The standard and general equations.   An interplay between geometry and algebra lies at the heart of this section. Understanding this interplay helps organize what might otherwise seem to be just a random collection of facts about parabolas. The essential idea is:

> To every PARABOLA we can associate its STANDARD EQUATION and its GENERAL EQUATION.
>
> Conversely, every equation in one of these two forms has a parabola for its graph.

Hence there is a correspondence between parabolas (geometric objects) and certain types of equations (algebraic objects). In this way geometric questions concerning parabolas can be turned into algebraic questions.

Here are the specific definitions:

A <u>PARABOLA</u> is the set of all points in the plane which are equidistant from a given line (the <u>DIRECTRIX</u>) * and a given point (the <u>FOCUS</u>) not on the line.

Moreover, the <u>AXIS</u> is the line of symmetry for the parabola, and the <u>VERTEX</u> is the intersection point of the parabola and the axis.

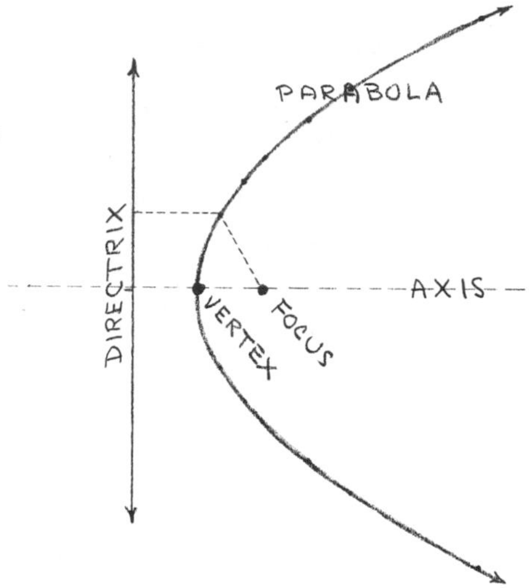

---

* For simplicity we will only consider parabolas whose directrices are lines parallel to the x-axis or the y-axis. The general situation will be considered in §12.5.

A <u>STANDARD EQUATION</u> for a parabola is an equation of one of the following two forms:

$$(x - h)^2 = 4p(y - k)$$ The graph of this equation is a parabola for which:

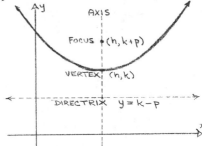

   i.   The <u>axis</u> is parallel to the y-axis

   ii.   The <u>vertex</u> has coordinates (h, k)

   iii. The <u>focus</u> has coordinates $(h, k + p)$

   iv.   The <u>directrix</u> is given by $y = k - p$

$$(y - k)^2 = 4p(x - h)$$ The graph of this equation is a parabola for which:

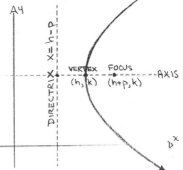

   i.   The <u>axis</u> is parallel to the x-axis

   ii.   The <u>vertex</u> has coordinates (h, k)

   iii. The <u>focus</u> has coordinates $(h + p, k)$

   iv.   The <u>directrix</u> is given by $x = h - p$

A <u>GENERAL EQUATION</u> for a parabola is an equation of one of the following two forms:

$$y = Ax^2 + Dx + F, \quad A \neq 0$$ The graph of this equation is a parabola with axis parallel

   to the y-axis .

$$x = Cy^2 + Ey + F, \quad C \neq 0$$ The graph of this equation is a parabola with axis parallel

   to the x-axis.

Unfortunately there are a lot of coordinates and formulas to remember in the two versions of the standard equation. However, noting the role of the <u>non-squared variable</u> can help: In the first standard equation

$$(x - h)^2 = 4p(y - k)$$

the y-variable is not squared. Then

- the axis of symmetry is parallel to the <u>y-axis</u>
- add p to the <u>y-coordinate</u> of the vertex to get the focus $(h, k + p)$
- subtract p from the <u>y-coordinate</u> of the vertex to get the directrix $y = k - p$

In the second standard equation

$$(y - k)^2 = 4p(x - h)$$

the <u>x-variable</u> is not squared, and all the above statements hold with x replacing y :

- the axis of symmetry is parallel to the <u>x-axis</u>
- add p to the <u>x-coordinate</u> of the vertex to get the focus $(h + p, k)$
- subtract p from the <u>x-coordinate</u> of the vertex to get the directrix $x = h - p$

2. <u>Applications: the standard equation is crucial.</u> Many of the questions about parabolas involve the interplay between geometry and algebra discussed in the previous subsection. There are essentially two types of these questions:

       i.   Given a parabola described geometrically, determine either

             its standard or general equation.

      ii.   Given an equation for a parabola, describe the parabola

             geometrically (e. g. , find its focus and directrix).

Anton's Examples 2 and 5 are of the first type, while Examples 1, 3 and 4 are of the second type. Here are some additional examples:

<u>Example A.</u>   Sketch the parabola given by the equation

$$y^2 + 4y - 6x + 22 = 0$$

Show the focus, vertex and directrix.

<u>Solution.</u>   We can rewrite the given equation as

$$x = \frac{1}{6} y^2 + \frac{2}{3} y + \frac{11}{3}$$

Thus the question is one of the form: "Given the general equation of a parabola, describe it geometrically. "

The trick is to find the STANDARD EQUATION first. To go from the general equation to the standard equation, we merely <u>complete the square</u> (see Appendix D. 7 if you are rusty on this technique):

$$x = \frac{1}{6}(y^2 + 4y) + \frac{11}{3}$$

$$= \frac{1}{6}\left(y^2 + 4y + \left(\frac{4}{2}\right)^2\right) - \frac{1}{6}\left(\frac{4}{2}\right)^2 + \frac{11}{3}$$

$$= \frac{1}{6}(y^2 + 4y + 4) - \frac{2}{3} + \frac{11}{3}$$

$$= \frac{1}{6}(y + 2)^2 + 3$$

Thus $\boxed{(y + 2)^2 = 6(x - 3)}$ is the standard equation.

Now the desired geometric description of the parabola can be obtained easily. Recalling our "memory trick" of Subsection 1, we observe that

$$\boxed{\begin{array}{l} x \quad \text{is the \underline{non}-squared variable, and} \\ 4p = 6, \quad \text{so that} \quad p = 3/2 \end{array}}$$

Thus

i. the axis of symmetry is parallel to the x-axis

ii. the <u>vertex</u> has coordinates $(3, -2)$, these being the numbers subtracted from x and y, respectively, in the standard equation

iii. by adding $p = 3/2$ to the x-coordinate of the vertex, we get the <u>focus</u> $\left(3 + \frac{3}{2}, -2\right) = \left(\frac{9}{2}, -2\right)$

iv. by subtracting $p = 3/2$ from the x-coordinate of the vertex, we get the <u>directrix</u> $x = 3 - \frac{3}{2} = \frac{3}{2}$

Placing all of this information into a graph yields the following sketch:

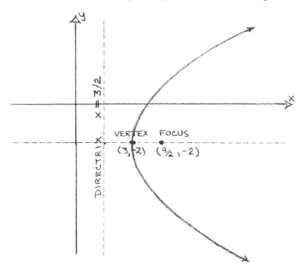

Example B.   Find an equation for the parabola with focus (1,5) and vertex (1,3).

Solution.   The question is one of the form: "Given a geometric description of a parabola,

find an equation for it." The type of equation is not specified, but the easiest type to find is

usually the standard equation. To do this <u>we must determine if the axis of symmetry is</u>

<u>parallel to the   x-axis   or to the   y-axis,</u> and then <u>determine the constants   h, k   and   p.</u>

Since the focus and the vertex both lie on

the line   x = 1 , this is the axis of symmetry

and <u>it is parallel to the   y-axis</u>. Hence y

must be the <u>non</u>-squared variable, and our

standard equation must be of the form

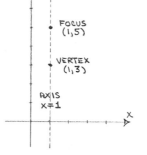

$$(x - h)^2 = 4p(y - k)$$

The constants   h   and   k   are merely the coordinates of the vertex, which are given as (1,3).

Thus

$$h = 1 \qquad \text{and} \qquad k = 3$$

Finally,

$$\begin{bmatrix} \text{y-coordinate} \\ \text{of vertex} \end{bmatrix} + p = \begin{bmatrix} \text{y-coordinate} \\ \text{of focus} \end{bmatrix}$$

$$3 + p = 5$$

$$p = 2$$

Thus the standard equation of the parabola is

$$(x - 1)^2 = 8 (y - 3)$$

If you prefer the general equation, you have only to multiply out the terms in the standard equation:

$$x^2 - 2x + 1 = 8y - 24$$

$$x^2 - 2x + 25 = 8y$$

Thus

$$y = \frac{1}{8} x^2 - \frac{1}{4} x + \frac{25}{8}$$ $\square$

Example C.   Find an equation for the parabola with axis $y = 2$ which passes through the points

$(0, 1)$ and $(-3, 4)$.

Solution.   As in Example B, our question is of the form: "Given a geometric description of a parabola, find an equation for it." However, in this example the geometric description of the parabola is a bit more unusual.

Since we are given the axis of symmetry

to be $y = 2$, which is <u>parallel to the</u> x-axis,

then x must be the <u>non</u>-squared variable. Our

standard equation is therefore of the form

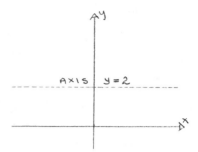

$$(y - k)^2 = 4p(x - h)$$

Moreover, since the vertex (h, k) lies on the axis of symmetry $y = 2$, then $k = 2$.

Hence our equation becomes

$$\boxed{(y - 2)^2 = 4p(x - h)}$$

We must now determine h and p. Since $(0, 1)$ and $(-3, 4)$ are both points on

the parabola, we have

$$\begin{cases} (1 - 2)^2 = 4p(0 - h) \\ (4 - 2)^2 = 4p(-3 \quad h) \end{cases}$$

so that

$$\begin{cases} 1 = -4ph \\ 4 = -4p(3 + h) \end{cases}$$

We now have two simultaneous equations in two unknowns (as studied in Appendix F.1).

Solving the first equation for p yields

$$p = -1/4h$$

Then plugging into the second equation gives

$$4 = -4\left(-\frac{1}{4h}\right)(3 + h)$$

$$4 = \frac{3}{h} + 1$$

$$3 = 3/h$$

Thus   h = 1   and   p = - 1/4 ,   and our equation becomes

$$(y - 2)^2 = - (x - 1)$$

☐

3.   <u>Translation of axes (Optional).</u>   The formulas associated with the standard equation for a parabola discussed in Subsection 1 can be difficult to memorize.   Anton provides an easy way to obtain these formulas via "translation of axes. "

The procedure starts by memorizing the STANDARD EQUATIONS of a parabola whose <u>vertex is at the origin</u>:

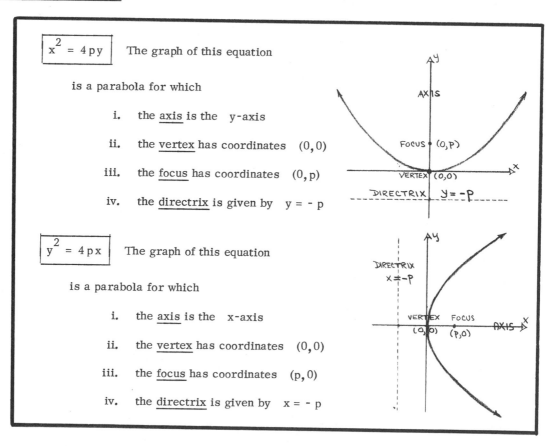

$\boxed{x^2 = 4py}$ The graph of this equation

    is a parabola for which

      i.   the <u>axis</u> is the   y-axis

      ii.   the <u>vertex</u> has coordinates   (0,0)

      iii. the <u>focus</u> has coordinates   (0,p)

      iv. the <u>directrix</u> is given by   y = - p

$\boxed{y^2 = 4px}$ The graph of this equation

    is a parabola for which

      i.   the <u>axis</u> is the   x-axis

      ii.   the <u>vertex</u> has coordinates   (0,0)

      iii. the <u>focus</u> has coordinates   (p, 0)

      iv. the <u>directrix</u> is given by   x = - p

With just one constant p to contend with, these equations are not hard to memorize.   In fact,

the two simple graphs (drawn for $p > 0$)   summarize all the geometric information quite

nicely.

 To apply these equations to a parabola whose vertex is at a point (h, k) , we use

translation of axes:

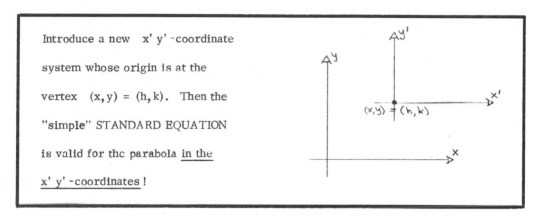

Introduce a new $x'\,y'$-coordinate system whose origin is at the vertex $(x, y) = (h, k)$.   Then the "simple" STANDARD EQUATION is valid for the parabola in the $x'\,y'$-coordinates !

To change the formulas so obtained from the $x'\,y'$-coordinates into the xy-coordinates , we

merely use the translation equations:

$$x' = x - h$$
$$y' = y - k$$

or

$$x = x' + h$$
$$y = y' + k$$

 To illustrate how this procedure can be used in computations, we will redo Examples A

and B by applying translation of axes:

Example A (revisited).   Sketch the parabola given by the equation

$$y^2 + 4y - 6x + 22 = 0$$

Show the focus, vertex and directrix.

12. 2. 11

Solution. As in the original Example A, we determine that the standard equation for this parabola is

$$(y + 2)^2 = 6(x - 3)$$

However, now we introduce x' y' -coordinates by using the translation equations

$$\boxed{\begin{array}{l} x' = x - 3 \\[2mm] y' = y + 2 \end{array}}$$ or $$\boxed{\begin{array}{l} x = x' + 3 \\[2mm] y = y' - 2 \end{array}}$$

so that our standard equation becomes

$$\boxed{(y')^2 = 6x'}$$

This puts us into the simple situation of the <u>vertex at the origin</u> with the standard equation $(y')^2 = 4px'$. Thus $4p = 6$, which yields $p = 3/2$, and

    i.   the <u>axis</u> is the x'-axis (the <u>non</u>-squared variable)

    ii.   the <u>vertex</u> has coordinates $(x', y') = (0, 0)$

    iii.   the <u>focus</u> has coordinates $(x', y') = (p, 0) = (3/2, 0)$

    iv.   the <u>directrix</u> is given by $x' = -p = -3/2$

To change back into xy-coordinates, we use the translation equations again:

    i.   the <u>axis</u> is parallel to the x-axis

    ii.   the <u>vertex</u> has coordinates $(x, y) = (0 + 3, 0 - 2) = (3, -2)$

    iii.   the <u>focus</u> has coordinates $(x, y) = \left(\frac{3}{2} + 3, 0 - 2\right) = \left(\frac{9}{2}, -2\right)$

    iv.   the <u>directrix</u> is given by $x = -\frac{3}{2} + 3 = \frac{3}{2}$

These answers agree with those obtained in the original Example A , and hence the graph will be the same as given there.   ☐

Example B (revisited).   Find an equation for the parabola with focus (1, 5) and vertex (1, 3).

Solution.   Since we wish the vertex to have $x'y'$-coordinates $(0, 0)$, we introduce the $x'y'$-coordinates by using the translation equations:

$$
\begin{array}{c}
x' = x - 1 \\[6pt]
y' = y - 3
\end{array}
\qquad \text{or} \qquad
\begin{array}{c}
x = x' + 1 \\[6pt]
y = y' + 3
\end{array}
$$

In terms of these new coordinates, the focus is given by

$$(x', y') = (1 - 1 , 5 - 3) = (0, 2)$$

Thus $p = 2$ (since this is the difference in the y'-coordinates of the vertex and focus) and the axis of symmetry is the y'-axis (since the vertex and focus both lie on this axis). This means that y' is the non-squared variable, so our standard equation is just

$$(x')^2 = 4py' = 8y'$$

To convert this back into xy-coordinates, we merely plug in the translation equations:

$$(x - 1)^2 = 8(y - 3)$$

As expected, this agrees with the answer we obtained in the original Example B.   ☐

4.   The Reflection Property (Optional).   As Anton explains at the end of the section, some of the important uses of parabolas stem from the reflection property given in Theorem 12. 2. 3.

12. 2. 13

We'll prove that theorem:

THEOREM 12. 2. 3.   The tangent line at a point   P   on a parabola makes equal angles with the line ℓ through   P   parallel to the axis of symmetry and the line through   P   and the focus.   (Figure 12. 2. 8.)

Proof.   We choose coordinate axes so the parabola has equation $x^2 = 4py$ with   p   positive. Then if   P ,   a point on the parabola, has coordinates (x_0, y_0) ,   the equation of the tangent line at   P   is the equation of the line through (x_0, y_0) with slope

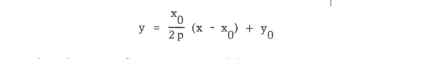

$$\frac{dy}{dx}\bigg|_{x = x_0} = \frac{x}{2p}\bigg|_{x = x_0} = \frac{x_0}{2p}$$

i. e. ,   the line $y - y_0 = \frac{x_0}{2p}(x - x_0)$

$$y = \frac{x_0}{2p}(x - x_0) + y_0$$

We wish to determine the   y-intercept of this tangent line.   To do so, set $x = 0$:

$$y = \frac{x_0}{2p}(0 - x_0) + y_0$$

$$= \frac{-x_0^2}{2p} + y_0$$

$$= \frac{-4py_0}{2p} + y_0 \qquad (\text{since} \quad x_0^2 = 4py_0)$$

$$= -2y_0 + y_0$$

$$= -y_0$$

Thus the tangent line has y-intercept $Q(0, -y_0)$; the distance from Q to the focus

$F(0, p)$ is thus $y_0 + p$.

    The directrix of the parabola is the

line $y = -p$. The distance from $P(x_0, y_0)$

to this directrix is $y_0 + p$. Hence the

distance from P to the focus F is also

$y_0 + p$ (by the definition of a parabola)!!

Hence $\triangle QFP$ is an isosceles triangle

and therefore the angles α and β

shown in the diagram to the right must be

equal.

    If we draw in the line ℓ "through

P parallel to the axis of symmetry" (which

is the y-axis), the angle γ that ℓ

makes with the tangent PQ equals the

angle α (because ℓ and the y-axis

are parallel). Thus, since $\alpha = \beta$, we

have that $\gamma = \beta$, the equality of angles

that we wished to show.

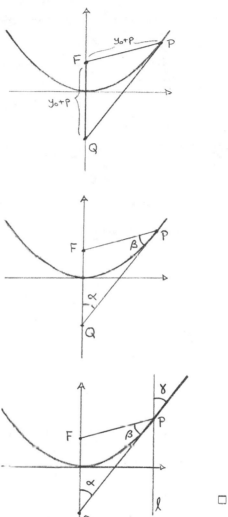

Section 12.3. The Ellipse

1. The standard and general equations. As in the previous section on the parabola, an interplay

between geometry and algebra lies at the heart of this material. The essential idea is:

To every ELLIPSE we can associate its STANDARD EQUATION and its GENERAL EQUATION. Conversely, every equation in standard equation form, and many in general form, have ellipses for graphs.

Here are the specific definitions:

An <u>ELLIPSE</u> is the set of all points in the plane, the sum of whose distances from two fixed points (the <u>FOCI</u>) is a given constant. *

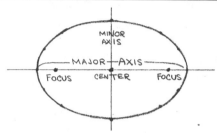

As shown in the diagram, the <u>CENTER</u> is the point half-way between the two foci, the <u>MAJOR AXIS</u> is the diameter line segment containing the two foci, and the <u>MINOR AXIS</u> is the diameter line segment through the center and perpendicular to the major axis.

To any ellipse we associate three numbers:

a = half the length of the major axis

b = half the length of the minor axis

c = half the distance between the foci

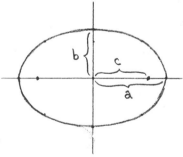

The numbers a and b are called the <u>SEMIAXES</u> of the ellipse.

---

* For simplicity we will only consider ellipses whose major axis is parallel to either the x-axis or the y-axis. The general situation will be considered in §12.5.

Anton proves that   a , b   and   c   form the

sides of a right triangle (as shown to the right),

and hence we have the basic relationship

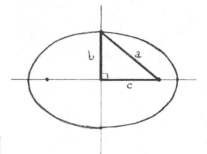

$$a^2 = b^2 + c^2$$

A <u>STANDARD EQUATION</u> for an ellipse is an equation

of one of the following two forms:

$$\frac{(x - h)^2}{a^2} + \frac{(y - k)^2}{b^2} = 1 \quad \text{where} \quad a > b > 0$$

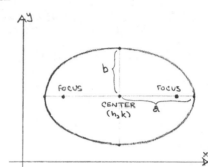

The graph of this equation is an ellipse

for which

     i.   the <u>major axis</u> is parallel to the <u>x-axis</u>

     ii.   the <u>center</u> has coordinates (h, k)

     iii.   the <u>semiaxes</u> are a and b

     iv.   the <u>foci</u> have coordinates (h ± c , k) , where $a^2 = b^2 + c^2$

$$\frac{(x - h)^2}{b^2} + \frac{(y - k)^2}{a^2} = 1 \quad \text{where} \quad a > b > 0$$

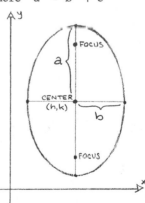

The graph of this equation is an ellipse

for which

     i.   the <u>major axis</u> is parallel to the <u>y-axis</u>

     ii.   the <u>center</u> has coordinates (h, k)

     iii.   the <u>semiaxes</u> are a and b

     iv.   the <u>foci</u> have coordinates (h , k ± c) , where $a^2 = b^2 + c^2$

A GENERAL EQUATION for an ellipse is an equation of the form

$$A x^2 + C y^2 + Dx + Ey + F = 0$$

where A and C are non-zero numbers of the same sign

The graph of this equation will either be an ellipse, a circle, a single point or the empty set (i.e., no solutions).

There are two important observations to be made about the STANDARD and GENERAL EQUATIONS:

1. In the standard equation the larger denominator is always denoted by a^2, and there is a special role played by the corresponding variable (i.e., the variable whose square is divided by a^2). In the first standard equation

$$\frac{(x - h)^2}{a^2} + \frac{(y - k)^2}{b^2} = 1$$

the x-variable corresponds to the larger denominator. Then

- the major axis is parallel to the x-axis
- add and subtract c to the x-coordinate of the center to get the two foci $(h \pm c, k)$

In the second standard equation,

$$\frac{(x - h)^2}{b^2} + \frac{(y - k)^2}{a^2} = 1$$

the y-variable corresponds to the larger denominator, and all the above statements hold with y replacing x.

2.   As noted in the description of the general equation for an ellipse, not every equation of this form has an ellipse for its graph. Given such an equation, you must attempt to put it into one of the two standard equation forms. If you succeed, then the graph is an ellipse; if you fail, then the graph is not an ellipse. See Anton's Examples 2 and 4 , and Example A below.

2.   Applications: the standard equation is crucial.   As with parabolas, many of the questions in this section involve the interplay between geometry and algebra discussed in Subsection 1. There are essentially two types of these questions:

    i.   Given an ellipse described geometrically, determine

       either its standard or general equation.

    ii.   Given an equation for an ellipse, describe the ellipse

       geometrically (e. g. , find its foci, semiaxes, etc.).

Anton's Example 3 is of the first type, while Examples 1, 2 and 4 are of the second type. Here are some additional examples:

Example A.   Sketch the graph of the equation $2x^2 + y^2 - 4x + 4y - 0$.

Solution.   This is a problem of the second type. The given equation is in the form of a general equation of an ellipse, so its graph might be an ellipse. To determine whether it is, we must attempt to put the equation into standard form. First we complete the square in both x and y (see Appendix D. 7 if completing the square gives you trouble):

$$2\left(x^2 - 2x + (-1)^2\right) + \left(y^2 + 4y + (2)^2\right) - 2(-1)^2 - (2)^2 = 0$$

$$2(x - 1)^2 + (y + 2)^2 = 6$$

$$\frac{(x - 1)^2}{3} + \frac{(y + 2)^2}{6} = 1$$ ← Making the constant 1 is absolutely essential

$$\frac{(x - 1)^2}{(\sqrt{3})^2} + \frac{(y + 2)^2}{(\sqrt{6})^2} = 1$$ ← The denominators must be squares

This is the STANDARD EQUATION of an ellipse with center $(1, -2)$. Since the y-variable corresponds to the larger denominator, the major axis is parallel to the y-axis and the semiaxes are $a = \sqrt{6}$ and $b = \sqrt{3}$. Hence adding $\pm a = \pm \sqrt{6}$ to the y-coordinate of the center will yield the endpoints of the major axis, $(1, -2 \pm \sqrt{6})$, and adding $\pm b = \pm \sqrt{3}$ to the x-coordinate of the center will yield the endpoints of the minor axis, $(1 \pm \sqrt{3}, -2)$.

The endpoints of the major and minor axes are generally enough to sketch the graph of an ellipse. In this case, however, it is also easy to find the points where the ellipse intersects the coordinate axes: $(0, -4)$, $(0,0)$ and $(2,0)$. Our graph then looks like this:

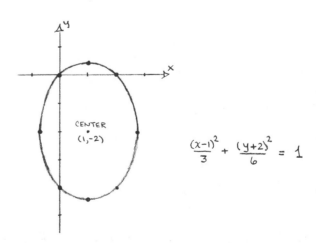

CENTER
(1,-2)

$$\frac{(x-1)^2}{3} + \frac{(y+2)^2}{6} = 1$$

Although we were not specifically asked to find the foci, doing so is easy: We first calculate

c , the distance of a focus from the center, by

$$c = \sqrt{a^2 - b^2} = \sqrt{6 - 3} = \sqrt{3}$$

Then (since y corresponds to the larger denominator) we merely add $\pm c = \pm \sqrt{3}$ to the

y-coordinates of the center: $(1 , -2 \pm \sqrt{3})$. These are the coordinates of the two foci.   □

Example B.   Find the equation of the ellipse which passes through the point $\left(\dfrac{2}{3} , 2\sqrt{2}\right)$ and has

$(0 , \pm 3)$ as the endpoints of its major axis.

Solution.   This is a problem of the first type described above. Since $(0 , \pm 3)$ are the end-

points of the major axis, the major axis lines on the y-axis and hence y must be the

variable which corresponds to the larger denominator a^2. Moreover, $a = 3$ and $(0,0)$

is the center of the ellipse (since it lies

half-way between $(0,3)$ and $(0, -3)$).

Thus the STANDARD EQUATION of the

ellipse is of the form

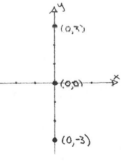

$$\frac{(x - 0)^2}{b^2} + \frac{(y - 0)^2}{(3)^2} = 1$$

where $b < 3$ is a number yet to be determined. Since $\left(\dfrac{2}{3} , 2\sqrt{2}\right)$ is on the ellipse, we

have

$$\frac{\left(\frac{2}{3}\right)^2}{b^2} + \frac{\left(2\sqrt{2}\right)^2}{9} = 1$$

which simplifies to

$$b^2 = 4$$

Hence the minor semiaxis is b = 2 and the equation of the ellipse is

$$\boxed{\dfrac{x^2}{4} + \dfrac{y^2}{9} = 1}$$ □

Here is an ellipse problem of a somewhat more unusual nature:

<u>Example C.</u>   Determine all values of k for which the line y = kx is tangent to the ellipse

$$\frac{(x-5)^2}{16} + \frac{y^2}{9} = 1 \ .$$

<u>Solution.</u>   Although the word "tangent" might suggest using a derivative, that is not necessary
in this problem: a line is tangent to an ellipse
if and only if it intersects the ellipse in exactly
one point (as shown in the diagram to the right).
Hence we have only to determine those values of
k for which y = kx intersects the given ellipse
in exactly one point.

To find the points of intersection of y = kx and the ellipse, we substitute y = kx
into the ellipse equation:

$$9(x-5)^2 + 16(kx)^2 = 144$$

$$9x^2 - 90x + 225 + 16k^2x^2 = 144$$

$$(9 + 16k^2)x^2 - 90x + 81 = 0$$

Thus

$$x = \frac{90 \pm \sqrt{(-90)^2 - 4(9 + 16 k^2)(81)}}{2(9 + 16 k^2)} \qquad \text{(by the quadratic formula)}$$

There will be exactly one intersection point when the term under the radical sign equals zero, i. e. , when

$$90^2 - 4(9 + 16 k^2)(81) = 0$$

Solving this equation for k yields $k = \pm 1$. Hence the lines $y = x$ and $y = -x$ are tangent to the given ellipse.   □

3.   Translation of axes (Optional).   As in the case of the parabola (see §12.2.3 of The Companion), the use of "translation of axes" can help in remembering the formulas associated with the standard equation of an ellipse, and can also be an aid in many computational problems.

The procedure starts by memorizing the STANDARD EQUATIONS of an ellipse whose center is at the origin:

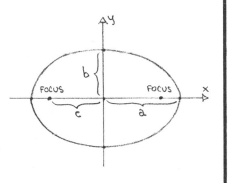

$$\frac{x^2}{a^2} + \frac{y^2}{b^2} = 1 \quad \text{where} \quad a > b > 0$$

The graph of this equation is an ellipse

for which

   i. the major axis lies on the   x-axis

   ii. the center has coordinates $(0, 0)$

   iii. the semiaxes are   a   and   b

   iv. the foci have coordinates $(\pm c, 0)$, where $a^2 = b^2 + c^2$

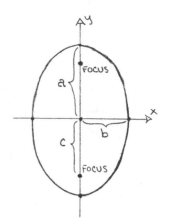

$$\frac{x^2}{b^2} + \frac{y^2}{a^2} = 1 \quad \text{where} \quad a > b > 0$$

The graph of this equation is an ellipse

for which

    i. the <u>major axis</u> lies on the   y-axis

    ii. the <u>center</u> has coordinates $(0,0)$

    iii. the <u>semiaxes</u> are   a   and   b

    iv. the <u>foci</u> have coordinates $(0, \pm c)$,   where $a^2 = b^2 + c^2$

To apply these equations to an ellipse whose center is at a point (h,k) , we can use "translation of axes" in exactly the way described for the parabola in §12.2.3. Here is an example:

<u>Example D.</u>   Find an equation for the ellipse whose foci are $(1,2)$ and $(5,2)$ and whose major axis has length 6.

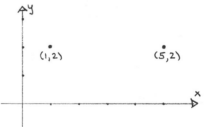

<u>Solution.</u>   The center of the ellipse lies half-way between the foci, so in this case the center has coordinates

$$\left(\frac{1+5}{2}, \, 2\right) = (3, 2)$$

We want this center to have $x' y'$-coordinates $(0,0)$ so we introduce the $x' y'$-coordinates by using the <u>translation equations:</u>

$$\begin{array}{l} x' = x - 3 \\ y' = y - 2 \end{array} \qquad \text{or} \qquad \begin{array}{l} x = x' + 3 \\ y = y' + 2 \end{array}$$

In terms of these new coordinates, the foci are given by

$$(x', y') = (1 - 3, 2 - 2) = (-2, 0)$$

and $$(x', y') = (5 - 3, 2 - 2) = (2, 0)$$

We are now in the simple situation of center at the origin, so the major axis lies on the

x'-axis, the distance of the focus from the center is c = 2, and the STANDARD EQUATION

is of the form

$$\frac{(x')^2}{a^2} + \frac{(y')^2}{b^2} = 1$$

We must determine a and b.

We are given that the length of the major axis is 6 ; since this distance is also 2a,

we have a = 3. Then b can be determined from

$$a^2 = b^2 + c^2$$

$$9 = b^2 + 4$$

$$b = \sqrt{5}$$

Thus our standard equation in x' y'-coordinates is

$$\frac{(x')^2}{9} + \frac{(y')^2}{5} = 1$$

To change back into xy-coordinates, we use the translation equations again and obtain

$$\boxed{\frac{(x - 3)^2}{9} + \frac{(y - 2)^2}{5} = 1}$$

$\square$

4.   The Reflection Property (Optional).   An interesting and useful property of ellipses is the

reflection property given in Anton's Theorem 12.3.2. Here is a sketch of the proof of that

result:

THEOREM 12.3.2. A line tangent to an ellipse at a point P makes equal angles with the line though P and the foci.

Proof. (The details necessary to fill in the following outline are left to you.) Choose coordinate axes so that the ellipse has the origin as its center and its major axis lies along the x-axis. Therefore its standard equation is $\frac{x^2}{a^2} + \frac{y^2}{b^2} = 1$ where $a > b > 0$ and its foci are $(c, 0)$ and $(-c, 0)$ where $c^2 = a^2 - b^2$. Then, with ℓ_T , $P(x_0, y_0)$, ℓ_1 , ℓ_2 , α and β as shown in the figure,

1.   The slope of the tangent line ℓ_T is

$$m_T = -\left(\frac{b^2}{a^2}\right)\left(\frac{x_0}{y_0}\right)$$

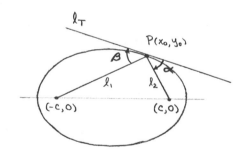

$$\left[\text{Differentiate the equation} \quad \frac{x^2}{a^2} + \frac{y^2}{b^2} = 1 \right.$$

implicitly to find $\frac{dy}{dx}$ and evaluate it

$$\left. \text{at} \quad (x_0, y_0) . \right]$$

2.   The slopes of ℓ_1 and ℓ_2 are

$$m_1 = \frac{y_0}{x_0 - c} \quad \text{and} \quad m_2 = \frac{y_0}{x_0 + c}$$

[Compute the slopes directly using the coordinates

of P and the foci .]

3.   From Exercise 21 of Exercise Set 1.4 ,

$$\tan \alpha = \frac{m_2 - m_T}{1 + m_T m_2} = \frac{b^2}{c y_0}$$

$$\tan \beta = \frac{m_T - m_1}{1 + m_1 m_T} = \frac{b^2}{c y_0}$$

These involve unpleasant computations

using the equations

$$b^2 x_0^2 + a^2 y_0^2 = a^2 b^2 , \quad \text{and}$$

$$a^2 - b^2 = c^2$$

so that $\alpha = \beta$, as desired.   □

5.   The focus-directrix definition of an ellipse (Optional).   An ellipse can also be defined in a way

which is very similar to the definition of a parabola:

The constant   e   is called the ECCENTRICITY of the ellipse.   Note that this definition with

e = 1   is the definition of the parabola.

Example E.   Find the standard equation of the ellipse with focus   (5/2, 3) ,   directrix   x = 7

and eccentricity $e = \frac{1}{2}$.

Solution. If $P(x,y)$ is a point on the ellipse, then

$$\begin{bmatrix} \text{distance from} & P \\ \text{to focus} \end{bmatrix} = \sqrt{(x - 5/2)^2 + (y - 3)^2} \ , \qquad \text{and}$$

$$\begin{bmatrix} \text{distance from} & P \\ \text{to directrix} \end{bmatrix} = \begin{bmatrix} \text{distance from} & (x,y) \\ \text{to} & (7,y) \end{bmatrix} = |x - 7|$$

Hence by the focus-directrix definition of the ellipse

$$\sqrt{(x - 5/2)^2 + (y - 3)^2} = \frac{1}{2} |x - 7|$$

$$4(x - 5/2)^2 + 4(y - 3)^2 = (x - 7)^2$$

$$3x^2 - 6x + 4(y - 3)^2 = 24$$

Completing the square in x (see Appendix D.7 of The Companion) yields

$$3(x^2 - 2x) + 4(y - 3)^2 = 24$$

$$3\left(x^2 - 2x + \left(\frac{-2}{2}\right)^2\right) + 4(y - 3)^2 = 24 + 3\left(\frac{-2}{2}\right)^2$$

$$3(x - 1)^2 + 4(y - 3)^2 = 27$$

$$\boxed{\frac{(x - 1)^2}{(3)^2} + \frac{(y - 3)^2}{(3\sqrt{3}/2)^2} = 1}$$

This is the desired standard equation. It tells us that the ellipse has major axis parallel to the x-axis, center $(1,3)$, semiaxes $a = 3$ and $b = 3\sqrt{3}/2$, and (since $c = \sqrt{a^2 - b^2} = 3/2$) foci $(1 + 3/2, 3) = (5/2, 3)$ and $(1 - 3/2, 3) = (-1/2, 3)$. $\qquad \square$

Exercises.   (These exercises use the focus-directrix definition of the ellipse covered in

§12.3.5 above.)

1.   Find the standard equation of the ellipse centered at the origin with focus $(0, -2)$

and directrix $y = -6$.

2.   Find the focus, directrix and eccentricity of the ellipse

$$\frac{x^2}{25} + \frac{y^2}{9} = 1$$

3.   Find the focus, directrix and eccentricity of the ellipse

$$3x^2 + 4y^2 - 16y = 92$$

(Hint: Begin by completing the squre in y .)

Answers.

1. $\dfrac{x^2}{8} + \dfrac{y^2}{12} = 1$

2. $(4, 0)$ [and $(-4, 0)$]; $x = \dfrac{25}{4}$ [and $x = \dfrac{-25}{4}$]; $e = \dfrac{4}{5}$

3. $(3, 2)$ [and $(-3, 2)$]; $x = 12$ [and $x = -12$]; $e = \dfrac{1}{2}$

Section 12.4. The Hyperbola

1. The standard and general equations.   The essential idea of this section is similar to those of

the two previous sections, an interplay between geometry and algebra:

> To every hyperbola we can associate its STANDARD EQUATION and its GENERAL EQUATION.
>
> Conversely, every equation of the standard form, and many of the general form, have
>
> hyperbolas for graphs.

Here are the specific definitions:

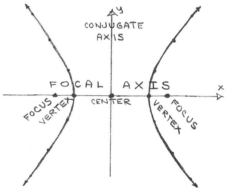

A HYPERBOLA is the set of all points

in the plane, the difference of whose

distances from two fixed points (the

FOCI) is a given constant. *

As shown in the diagram, the CENTER is the

point half-way between the two foci, the

FOCAL AXIS is the line through the two foci, the CONJUGATE AXIS is the line

through the center and perpendicular to the focal axis, and the VERTICES are

the two points where the hyperbola intersects the focal axis.

---

* For simplicity we will only consider hyperbolas whose focal axis is parallel to either the
x-axis or the y-axis. The general situation will be considered in §12.5.

To any hyperbola we associate three numbers:

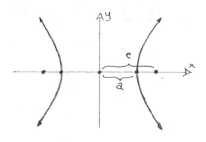

a = half the distance between the vertices

c = half the distance between the foci

$b = \sqrt{c^2 - a^2}$

From these definitions we see that   a , b   and   c   form

the sides of a right triangle (as shown to the right)

and we have the basic relationship

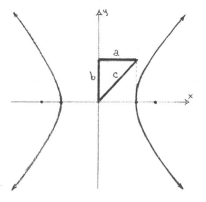

$$\boxed{a^2 + b^2 = c^2}$$ *

For any hyperbola we can also find a pair of lines

intersecting at the center which are extremely good

approximations to the hyperbola as it moves away from

the center.   These lines are called the ASYMPTOTES

of the hyperbola. **

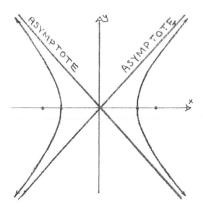

---

*
    Be careful! This is a different relationship between   a , b   and   c   than we had with the ellipse.
**
    Asymptotes were discussed earlier in §4.4 .

A <u>STANDARD EQUATION</u> for a hyperbola is an equation

of one of the following two forms:

$$\frac{(x - h)^2}{a^2} - \frac{(y - k)^2}{b^2} = 1$$

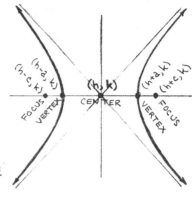

The graph of this equation is a

hyperbola for which

    i.   the <u>focal axis</u> is parallel to the   <u>x-axis</u>

    ii.   the <u>center</u> has coordinates (h, k)

    iii. the <u>vertices</u> have coordinates $(h \pm a, k)$

    iv. the <u>foci</u> have coordinates $(h \pm c, k)$

          where $c^2 = a^2 + b^2$

$$\frac{(y - k)^2}{a^2} - \frac{(x - h)^2}{b^2} = 1$$

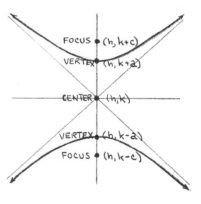

The graph of this equation is a

hyperbola for which

    i.   the <u>focal axis</u> is parallel to the   <u>y-axis</u>

    ii.   the <u>center</u> has coordinates (h, k)

    iii. the <u>vertices</u> have coordinates $(h, k \pm a)$

    iv. the <u>foci</u> have coordinates $(h, k \pm c)$

          where $c^2 = a^2 + b^2$

The equations for the asymptotes of a hyperbola are very easy to compute using the standard equation: <u>simply replace the 1 with a 0</u>. For example, the asymptotes of

$$\frac{(x-3)^2}{4} - y^2 = 1$$

are given by

$$\frac{(x-3)^2}{4} - y^2 = 0$$

$$y = \pm \frac{(x-3)}{2}$$

$$y = \frac{1}{2}x - \frac{3}{2} \quad \text{and} \quad y = -\frac{1}{2}x + \frac{3}{2}$$

A <u>GENERAL EQUATION</u> for a hyperbola is an equation of the form

$$A x^2 + C y^2 + Dx + Ey + F = 0$$

where A and C are non-zero numbers of opposite sign.

The graph of this equation will either be a hyperbola or a pair of intersecting lines.

There is an <u>important observation</u> to be made about the standard equation of a hyperbola, one which is similar to a remark made in the previous section about the ellipse. In the standard equation the <u>denominator of the positive term is always denoted by</u> a^2, and there is a special role played by the corresponding variable (i.e., the variable of the positive term). In the first standard equation

$$\frac{(x - h)^2}{a^2} - \frac{(y - k)^2}{b^2} = 1$$

the x-variable is the variable of the positive term. Then

- the focal axis is parallel to the x-axis

- add and subtract a to the x-coordinate of the center

  to get the two vertices $(h \pm a , k)$

- add and subtract c to the x-coordinate of the center

  to get the two foci $(h \pm c , k)$.

In the second standard equation

$$\frac{(y - k)^2}{a^2} - \frac{(x - h)^2}{b^2} = 1$$

the y-variable is the variable of the positive term, and all the above statements hold with y

replacing x .

2. Applications: the standard equation is crucial. As with parabolas and ellipses, many questions
about hyperbolas involve the interplay between geometry and algebra discussed in Subsection 1.
There are essentially two types of these questions:

    i)   Given a hyperbola described geometrically, determine either its standard

        or general equation.

    ii)   Given an equation for a hyperbola, describe the hyperbola geometrically

        (e. g. , find its focal axis, conjugate axis, asymptotes, vertices, etc.).

Anton's Example 3 is of the first type, while Examples 1, 2 and 4 are of the second type.
Here are some additional examples:

Example A.   Sketch the graph of the equation

$$9x^2 - 25y^2 + 36x - 50y + 236 = 0$$

Solution.   This is a problem of the second type.   The given equation is in the form of a GENERAL EQUATION of a hyperbola, so its graph <u>might</u> be a hyperbola.   To determine whether it is, we must attempt to put the equation into standard form.   First we complete the square in both   x   and   y :

$$9\left(x^2 + 4x + (2)^2\right) - 25\left(y^2 + 2y + (1)^2\right) - 9(2)^2 + 25(1)^2 = -236$$

$$9(x + 2)^2 - 25(y + 1)^2 = -225$$

$$\frac{(y + 1)^2}{9} - \frac{(x + 2)^2}{25} = 1 \quad \longleftarrow \boxed{\text{Making the constant } 1 \text{ is absolutely essential}}$$

$$\boxed{\frac{(y + 1)^2}{3^2} - \frac{(x + 2)^2}{5^2} = 1} \quad \longleftarrow \boxed{\text{The denominators must be squares}}$$

This is the standard equation of a hyperbola with center $(-2, -1)$, $a = 3$ and $b = 5$.   Since   y   is the variable of the positive term, we know that the focal axis is parallel to the y-axis,   and we add and subtract $a = 3$ to the   <u>y-coordinate</u>   of the center to get the two vertices

$$(-2, -1 \pm 3) = (-2, 2) \quad \text{and} \quad (-2, -4)$$

In order to draw a good sketch of the hyperbola, we need to determine the asymptotes.   We find these by replacing 1   with   0   in the standard equation:

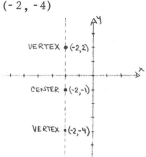

$$\frac{(y + 1)^2}{3^2} - \frac{(x + 2)^2}{5^2} = 0$$

$$\frac{y + 1}{3} = \pm \frac{x + 2}{5}$$

$$y = \frac{3}{5} x + \frac{1}{5} \quad \text{and} \quad y = -\frac{3}{5} x - \frac{11}{5}$$

To draw these lines most easily, start at the center $(-2, -1)$ (which must lie on each asymptote), go 5 units in the x and $-x$ direction, and then 3 units in the y and $-y$ direction. The four points found in this way, plus the center point, determine an "✗" formed by our desired asymptotes!

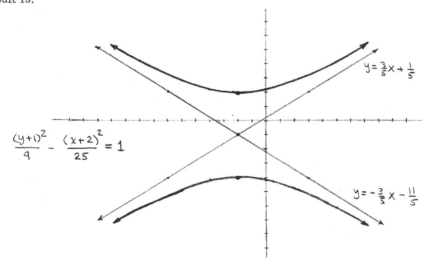

Using the asymptotes and the two vertices makes it easy to sketch the graph. The result is:

$$\frac{(y+1)^2}{9} - \frac{(x+2)^2}{25} = 1$$

Although we were not required to find the foci, it is easy to do so. We first calculate c, the distance of a focus from the center, by

$$c = \sqrt{a^2 + b^2} = \sqrt{9 + 25} = \sqrt{34} \approx 5.83$$

Then (since y is the variable of the positive term) we merely add $\pm c$ to the y-coordinate of the center: $(-2, -1 \pm 5.83) = (-2, 4.83)$ and $(-2, -6.83)$. These are the coordinates of the two foci.   □

Example B. Find the equation of the hyperbola which passes through the point $(6\sqrt{2} - 1, 7)$ and has vertices $(-1, 3)$ and $(-1, -1)$.

<u>Solution.</u> This is a problem of the first type described above. The vertices lie on the vertical line $x = -1$ so that line must be the focal axis. Since $x = -1$ is parallel to the <u>y-axis</u>, then y is the variable in the positive term of the standard equation. The center is also easy to determine: it lies halfway between the two vertices so its coordinates must be

$\left(-1, \dfrac{3 + (-1)}{2} \right) = (-1, 1)$. Hence the hyperbola has a standard equation of the form

$$\boxed{\dfrac{(y - 1)^2}{a^2} - \dfrac{(x + 1)^2}{b^2} = 1}$$

where a and b are numbers yet to be determined.

By definition, a is half the distance between the vertices. Since this distance is $3 - (-1) = 4$, then $a = 2$ and our standard equation becomes

$$\dfrac{(y - 1)^2}{4} - \dfrac{(x + 1)^2}{b^2} = 1$$

To determine b, we must use the fact that $(6\sqrt{2} - 1, 7)$ is on the hyperbola. Thus

$$1 = \frac{(7-1)^2}{4} - \frac{(6\sqrt{2})^2}{b^2} = \frac{36}{4} - \frac{72}{b^2}$$

$$b^2 = 9$$

and the standard equation of the hyperbola is

$$\frac{(y-1)^2}{4} - \frac{(x+1)^2}{9} = 1$$

□

Anton's Exercises 31-42 illustrate more challenging types of hyperbola problems. Here is one such example; its full significance is explained in Subsection 5:

Example C.   A point P moves so that its distance from the point $F(4,0)$ is twice its distance to the line $x = 1$. Show that the point moves along a hyperbola.

Solution.   We must compute the distance of $P(x,y)$ from both $F(4,0)$ and the line $x = 1$. As we can see from the diagram, the distances are

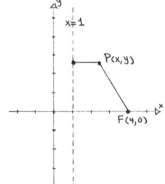

$$\left[\begin{array}{c} \text{distance} \\ \text{to } x = 1 \end{array} \right] = |x - 1|$$

$$\left[\begin{array}{c} \text{distance} \\ \text{to } F(4,0) \end{array} \right] = \sqrt{(x-4)^2 + (y-0)^2}$$

We thus obtain the following equation for x and y :

$$\sqrt{(x-4)^2 + y^2} = 2\,|\,x-1\,|$$

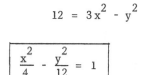

"... is twice its distance to ... "

$$(x-4)^2 + y^2 = 4(x-1)^2$$

$$x^2 - 8x + 16 + y^2 = 4x^2 - 8x + 4$$

$$12 = 3x^2 - y^2$$

$$\boxed{\dfrac{x^2}{4} - \dfrac{y^2}{12} = 1}$$

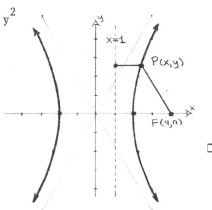

Sure enough, we have the equation of a

hyperbola, as shown to the right!

3. Translation of axes (Optional).   As done previously for both the parabola and the ellipse, the

use of "translation of axes" can help in remembering the formulas associated with the standard

equation of a hyperbola, and can also be useful in many computational problems.

The procedure starts by memorizing the STANDARD EQUATIONS of a hyperbola whose

center is at the origin:

$$\boxed{\dfrac{x^2}{a^2} - \dfrac{y^2}{b^2} = 1}$$ The graph of this equation is a hyperbola for which

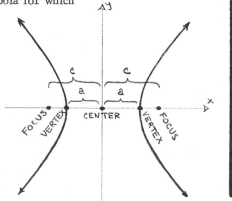

i.   The focal axis lies on the   x-axis

ii.   The center has coordinates $(0,0)$

iii. The vertices have coordinates $(\pm a, 0)$

iv.   The foci have coordinates $(\pm c, 0)$

where $c^2 = a^2 + b^2$

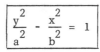
$$\frac{y^2}{a^2} - \frac{x^2}{b^2} = 1$$

The graph of this equation is a hyperbola for which

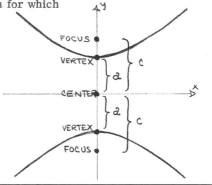

   i.   The <u>focal axis</u> lies on the y-axis

   ii.   The <u>center</u> has coordinates $(0,0)$

   iii. The <u>vertices</u> have coordinates $(0, \pm a)$

   iv. The <u>foci</u> have coordinates $(0, \pm c)$

        where $c^2 = a^2 + b^2$

To apply these equations to a hyperbola whose center is at a point (h, k) we can use "translation of axes" in exactly the way described for the parabola in §12.2.3. Here is an example:

<u>Example D.</u>   Sketch the hyperbola given by the equation

$$\frac{(x + 1)^2}{4} - (y - 2)^2 = 1$$

and determine the two foci.

<u>Solution.</u>   To use translation of axes we introduce x' y'-coordinates by using the <u>translation equations</u>:

$$\begin{array}{|l|}\hline x' = x + 1 \\[1ex] y' = y - 2 \\\hline\end{array} \qquad \text{or} \qquad \begin{array}{|l|}\hline x = x' - 1 \\[1ex] y = y' + 2 \\\hline\end{array}$$

so that our equation becomes

$$\boxed{\frac{(x')^2}{4} - (y')^2 = 1}$$

Hence we are now in the simple situation of center at the origin, so the center is at

$(x', y') = (0,0)$, the focal axis lies on the x'-axis, and

$$a = \sqrt{4} = 2$$

$$b = \sqrt{1} = 1$$

$$c = \sqrt{a^2 + b^2} = \sqrt{5}$$

The vertices are therefore

$$(x', y') = (\pm a, 0) = (\pm 2, 0)$$

the foci are

$$(x', y') = (\pm c, 0) = (\pm\sqrt{5}, 0)$$

and the asymptotes are

$$\frac{(x')^2}{4} - (y')^2 = 0$$

i.e., $y' = \pm x'/2$

To change back into xy-coordinates, we use the translation equations:

center: $(x, y) = (0 - 1, 0 + 2) = (-1, 2)$

focal axis: parallel to x-axis

vertices: $(x, y) = (\pm 2 - 1, 0 + 2) = (1, 2)$ and $(-3, 2)$

foci: $(x, y) = (\pm\sqrt{5} - 1, 0 + 2) = (\sqrt{5} - 1, 2)$ and $(-\sqrt{5} - 1, 2)$

asymptotes: $(y - 2) = \pm(x + 1)/2$

This information produces the following graph:

12. 4. 13

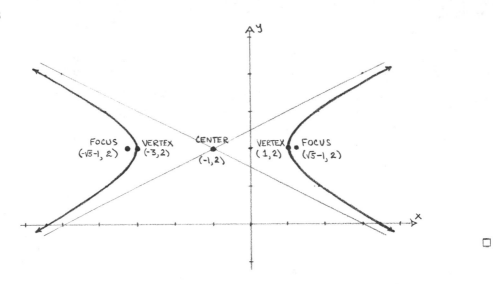

4. <u>The Reflection Property (Optional).</u>   As he did for both the parabola and the ellipse, Anton ends

this section with an interesting and useful reflection property of the hyperbola.   We'll sketch

the proof of the result, leaving the details to you:

<u>THEOREM 12.4.3.</u>   A line tangent to a hyperbola at a

point P makes equal angles with the lines through

P and the foci   (i. e. , $\alpha = \beta$ in the diagram

to the right).

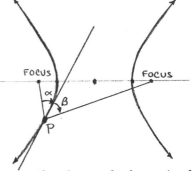

<u>Proof Outline.</u>   Choose the   xy-coordinate axes   in such a way that the standard equation for

the hyperbola is

$$\frac{x^2}{a^2} - \frac{y^2}{b^2} = 1$$

i. e. , in this coordinate system the center is the origin, the vertices are $(\pm a, 0)$ and the

foci are $(\pm c, 0)$ where $c^2 = a^2 + b^2$.   Then let $P(x_0, y_0)$ be any fixed point on the

hyperbola.

Step 1.   The slope of the tangent line ℓ_T at $P(x_0, y_0)$ is given by *

$$m_T = \left(\frac{b^2}{a^2}\right)\left(\frac{x_0}{y_0}\right)$$

$\Big[$ Use implicit differentiation on the equation for the hyperbola, and then solve for

$$m_T = \frac{dy}{dx}\bigg|_{(x_0, y_0)}\Big]$$

Step 2.   Let ℓ_1 be the line joining the

focus $(-c, 0)$ to $P(x_0, y_0)$, and ℓ_2

the line joining the focus $(c, 0)$ to

$P(x_0, y_0)$. Then

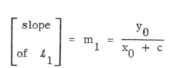

$$\begin{bmatrix} \text{slope} \\ \text{of } \ell_1 \end{bmatrix} = m_1 = \frac{y_0}{x_0 + c}$$

$$\begin{bmatrix} \text{slope} \\ \text{of } \ell_2 \end{bmatrix} = m_2 = \frac{y_0}{x_0 - c}$$

[Compute the slopes directly using the coordinates of these points.]

Step 3.   By Exercise 21 of §1.4,

$$\tan \alpha = \frac{m_T - m_1}{1 + m_1 m_T} = \frac{b^2}{c y_0}$$

$$\tan \beta = \frac{m_2 - m_T}{1 + m_T m_2} = \frac{h^2}{c y_0}$$

* Assuming $y_0 \neq 0$ of course! If y_0 does equal zero, then it is easy to see that $\alpha = \pi/2$ and $\beta = \pi/2$, so that the Theorem is true in that situation.

so that $\alpha = \beta$ since α and β are both acute angles. [The computations are messy and involve the equations $a^2 + b^2 = c^2$ and $b^2 x_0^2 - a^2 y_0^2 = a^2 b^2$.]   □

5. The focus-directrix definition of a hyperbola (Optional). Just as for the ellipse, there is a focus-directrix definition of a hyperbola which is very similar to the definition of a parabola:

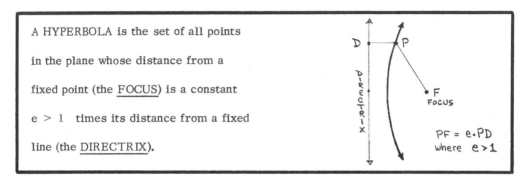

A HYPERBOLA is the set of all points in the plane whose distance from a fixed point (the FOCUS) is a constant $e > 1$ times its distance from a fixed line (the DIRECTRIX).

$PF = e \cdot PD$ where $e > 1$

Once again, the constant e is called the ECCENTRICITY of the hyperbola. This definition with $0 < e < 1$ is the focus-directrix definition of the ellipse and with $e = 1$ is the definition of the parabola.

Example C above illustrates this definition:

taking the focus $(4, 0)$,

the directrix $x = 1$, and

the eccentricity $e = 2$,

then we obtained the hyperbola

$$\frac{x^2}{4} - \frac{y^2}{12} = 1$$

Using the same procedure used in that example, you can prove the following:

taking the focus $(c, 0)$, $(a > c > 0)$

the directrix $x = a^2/c$, and

the eccentricity $e = c/a$,

then we obtain the hyperbola

$$\frac{x^2}{a^2} - \frac{y^2}{b^2} = 1 \qquad \text{where} \quad a^2 + b^2 = c^2$$

Try this proof - it's not difficult!

Exercises.   (These exercises use the focus-directrix definition of the hyperbola covered in §12.4.5 above.)

1.   Find the standard equation of the hyperbola centered at the origin with focus $(3, 0)$ and directrix $x = 4/3$.

2.   Find the focus, directrix and eccentricity of the hyperbola

$$\frac{x^2}{16} - \frac{y^2}{9} = 1$$

3.   Find the focus, directrix and eccentricity of the hyperbola

$$3x^2 - y^2 - 18x + 2y = 22$$

(Hint: Begin by completing the square in both x and y.)

Answers.

1. $\dfrac{x^2}{4} - \dfrac{y^2}{5} = 1$

2. $(5, 0)$ [and $(-5, 0)$]; $x = 16/5$ [and $x = -16/5$]; $e = 5/4$

3. $(11, 1)$ [and $(-5, 1)$]; $x = 5$ [and $x = 1$]; $e = 2$

Section 12.5.   Rotation of Axes; Second Degree Equations

1.   Quadratic equations in two variables and conic sections.   The following terminology is used

for subsets of the plane:

a conic section is any circle, ellipse, parabola or hyperbola;

a degenerate conic section is any single point, line, pair of lines

or the empty set (i. e. , the set with no elements).

Using this terminology, we can combine the statements of the last three sections concerning

general equations for parabolas, ellipses and hyperbolas, and obtain the major part of the

following result:

Theorem A.   Every conic section whose axis is parallel to either the   x-axis   or the

y-axis   can be represented by an equation of the form

$$A x^2 + C y^2 + Dx + Ey + F = 0$$

Conversely, every such equation represents either a conic section whose

axis is parallel to either the   x-axis   or the   y-axis , or a degenerate

conic section.

It is natural to ask "What can be said about conic sections whose axes are not necessarily

parallel to either the   x-axis   or the   y-axis ?" The answer is amazingly simple: the above

theorem remains valid in the general non-parallel case if we simply add a "mixed" term   Bxy

into the equation. The statement then reads as follows:

Theorem B.   Every conic section can be represented by an equation of the form

$$A x^2 + Bxy + Cy^2 + Dx + Ey + F = 0$$

Conversely, every such equation represents either a conic section or a

degenerate conic section.

Later in this section we will show that Theorem B is really an easy corollary of

Theorem A. For the moment, however, we will assume that Theorem B is valid, and ask

the next-most-obvious question:

Given a specific equation of the form

$$A x^2 + Bxy + Cy^2 + Dx + Ey + F = 0$$

how do we recognize what conic section

(or degenerate conic section) this equation

represents? Furthermore, how do we determine

the vertices, foci, etc. , and sketch the graph?

This is the problem which Anton tackles (and completely solves) in this section. The essence

of his solution is to <u>rotate</u> the xy-coordinates into new x' y'-coordinates in such a way that

the "mixed" term Bxy "disappears." Therefore we must discuss ...

2.   <u>Rotation of axes.</u>   Suppose the xy-coordinate system
is rotated counterclockwise about the origin through an
angle θ to produce a new x' y'-coordinate system.
Then Anton proves that the xy-coordinates are

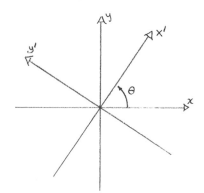

related to the $x'y'$-coordinates via the <u>rotation equations</u>:

$$x = x' \cos\theta - y' \sin\theta$$
$$y = x' \sin\theta + y' \cos\theta$$

or

$$x' = x \cos\theta + y \sin\theta$$
$$y' = -x \sin\theta + y \cos\theta$$

The essence of these equations is captured in the array

$$\begin{bmatrix} \cos\theta & -\sin\theta \\ \sin\theta & \cos\theta \end{bmatrix}$$

For the second set of equations $(x'y'$ in terms of $xy)$ the minus sign simply shifts over to the other $\sin\theta$ term. Anton's Examples 1 and 2 show how these equations can be used in simple computations.

Our primary use of these equations in this section will be in the ...

3. <u>Classification of quadratic equations: a 5-step method.</u> Given a quadratic equation of the form

$$Ax^2 + Bxy + Cy^2 + Dx + Ey + F = 0$$

with $B \neq 0$, we will determine an angle θ which is such that rotating the xy-coordinate system by θ to a new $x'y'$-coordinate system will "eliminate" the mixed term Bxy. Here's the procedure:

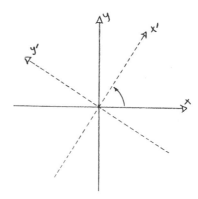

To classify a quadratic equation

$$Ax^2 + Bxy + Cy^2 + Dx + Ey + F = 0, \qquad B \neq 0,$$

<u>Step 1.</u> Determine $\sin \theta$ and $\cos \theta$ for the angle θ defined by

$$\cot 2\theta = \frac{A - C}{B}$$

This is generally done by:

       i. finding $\cos 2\theta$ by The Triangle Method (see

         <u>The Companion</u> §8.2.3) and then

      ii. calculating $\sin \theta$ and $\cos \theta$ by

$$\sin \theta = \sqrt{\frac{1 - \cos 2\theta}{2}} \quad \text{and} \quad \cos \theta = \sqrt{\frac{1 + \cos 2\theta}{2}}$$

<u>Step 2.</u> Write the rotation equations using the values of $\sin \theta$ and $\cos \theta$ found in Step 1.

<u>Step 3.</u> Use the rotation equations to transform the given xy-equation into an $x'y'$-equation. <u>This $x'y'$-equation will have no mixed term $B'x'y'$!</u>

<u>Step 4.</u> Identify the graph of the $x'y'$-equation using the results from §§12.2 - 12.4. Find the vertices, foci, etc., in terms of the $x'y'$-coordinates.

<u>Step 5.</u> Use the rotation equations to express the vertices, foci, etc., in terms of the original xy-coordinates.

Anton's Examples 3 and 4 illustrate this procedure, although the steps are not labeled in his solutions. Here is another example:

Example A. Identify and sketch the curve

$$6x^2 + 4\sqrt{2}\,xy - y^2 + 84x + 98 = 0$$

Step 1: Determine $\sin\theta$ and $\cos\theta$. Since $A = 6$, $B = 4\sqrt{2}$ and $C = -1$, we have

$$\cot 2\theta = \frac{A - C}{B} = \frac{6 + 1}{4\sqrt{2}} = \frac{7}{4\sqrt{2}}$$

In order to find $\sin\theta$ and $\cos\theta$ we first determine $\cos 2\theta$ by The Triangle Method of §8.2.3 of The Companion. Since

$$\frac{7}{4\sqrt{2}} = \cot 2\theta = \frac{\text{adjacent}}{\text{opposite}}$$

we have the right triangle to the right. Thus

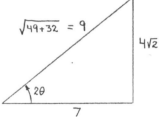

$$\cos 2\theta = \frac{\text{adjacent}}{\text{hypotenuse}} = \frac{7}{9}$$

and

$$\sin\theta = \sqrt{\frac{1 - \cos 2\theta}{2}} = \sqrt{\frac{2/9}{2}} = \frac{1}{3}$$

$$\cos\theta = \sqrt{\frac{1 + \cos 2\theta}{2}} = \sqrt{\frac{16/9}{2}} = \frac{2}{3}\sqrt{2}$$

Step 2: Rotation equations.

$$\begin{aligned} x &= \frac{2}{3}\sqrt{2}\,x' - \frac{1}{3}y' \\ y &= \frac{1}{3}x' + \frac{2}{3}\sqrt{2}\,y' \end{aligned}$$

and

$$\begin{aligned} x' &= \frac{2}{3}\sqrt{2}\,x + \frac{1}{3}y \\ y' &= -\frac{1}{3}x + \frac{2}{3}\sqrt{2}\,y \end{aligned}$$

Step 3: Transform into x'y'-equation. Substituting the first pair of rotation equations into the original quadratic equation yields

$$6\left(\frac{2}{3}\sqrt{2}\ x' - \frac{1}{3}\ y'\right)^2 + 4\sqrt{2}\left(\frac{2}{3}\sqrt{2}\ x' - \frac{1}{3}\ y'\right)\left(\frac{1}{3}\ x' + \frac{2}{3}\sqrt{2}\ y'\right)$$

$$-\left(\frac{1}{3}\ x' + \frac{2}{3}\sqrt{2}\ y'\right)^2 + 84\left(\frac{2}{3}\sqrt{2}\ x' - \frac{1}{3}\ y'\right) + 98 = 0$$

The process of multiplying out all the terms and gathering the like terms together is long and boring, but it's elementary. When it's done we are left with

$$7(x')^2 - 2(y')^2 + 56\sqrt{2}\ x' - 28 y' + 98 = 0$$

Note that (as predicted) this equation has no "mixed" (i.e., x'y') term!

Step 4: Identify the graph. To put this equation into one of the standard forms we must complete the squares in the x' and y' terms (see Appendix D.7):

$$7[(x')^2 + 8\sqrt{2}\ x' + 32] - 2[(y')^2 + 14 y' + 49] - 224 + 98 + 98 = 0$$

$$7(x' + 4\sqrt{2})^2 - 2(y' + 7)^2 = 28$$

$$\boxed{\frac{(x' + 4\sqrt{2})^2}{4} - \frac{(y' + 7)^2}{14} = 1}$$

This is the standard equation for a hyperbola. Since x' is the variable of the positive term, the focal axis is parallel to the x'-axis, and

$$a = 2$$
$$b = \sqrt{14}$$
$$c = \sqrt{4 + 14} = \sqrt{18} = 3\sqrt{2}$$

Thus:

- the <u>center</u> is $(x', y') = (-4\sqrt{2}, -7)$

- the <u>vertices</u> are $(x', y') = (-4\sqrt{2} \pm 2, -7)$

- the <u>foci</u> are $(x', y') = (-4\sqrt{2} \pm 3\sqrt{2}, -7)$

Finally, the <u>asymptotes</u> are given by

$$\frac{(x' + 4\sqrt{2})^2}{4} - \frac{(y' + 7)^2}{14} = 0$$

or
$$y' + 7 = \pm \frac{\sqrt{14}}{2}(x' + 4\sqrt{2})$$

<u>Step 5</u>: <u>Rotate back to xy-coordinates</u>.   Here we make use of the rotation equations found in Step 2:

- the <u>axis of symmetry</u> is parallel to $y' = 0$ (the x'-axis), a line whose

  xy-equation is

$$-\frac{1}{3}x + \frac{2}{3}\sqrt{2}\, y = 0$$

or
$$y = \frac{\sqrt{2}}{4}y \approx .35\,x$$

- the <u>center</u> is $(x, y) = \left(\frac{2}{3}\sqrt{2}(-4\sqrt{2}) - \frac{1}{3}(-7), \frac{1}{3}(-4\sqrt{2}) + \frac{2}{3}\sqrt{2}(-7)\right)$

$$= (-3, -6\sqrt{2})$$

$$\approx (-3, -8.49)$$

In a similar fashion,

- the <u>vertices</u> are $(x,y) = ((-9 + 4\sqrt{2})/3 , -6\sqrt{2} + 2/3)$ and

$$((-9 - 4\sqrt{2})/3 , -6\sqrt{2} - 2/3)$$

$$\approx (-1.11 , -7.82)   \text{ and }   (-4.89 , -9.15)$$

- the <u>foci</u> are $(x,y) = (-3 + 4 , -6\sqrt{2} + \sqrt{2})$ and $(-3 - 4 , -6\sqrt{2} - \sqrt{2})$

$$\approx (1 , -7.07)   \text{ and }   (-7 , -9.90)$$

Finally, the <u>asymptotes</u> are given by

$$\left(-\frac{1}{3}x + \frac{2}{3}\sqrt{2}\,y\right) + 7 = \pm \frac{\sqrt{14}}{2}\left(\left(\frac{2}{3}\sqrt{2}\,x + \frac{1}{3}y\right) + 4\sqrt{2}\right)$$

which simplify to

$$y \approx -.91x - 11.23 , \quad \text{ and }$$

$$y \approx 6.57x + 11.23$$

<u>Summary</u>:   axis of symmetry parallel to $y \approx .35x$

center $(x,y) \approx (-3 , -8.49)$

vertices $(x,y) \approx (-1.11 , 7.02)$ and $(-4.89 , -9.15)$

foci $(x,y) \approx (1 , -7.07)$ and $(-7 , -9.90)$

asymptotes $y \approx -.91x - 11.23$

$$y \approx 6.57x + 11.23$$

With all this information we can sketch the graph:

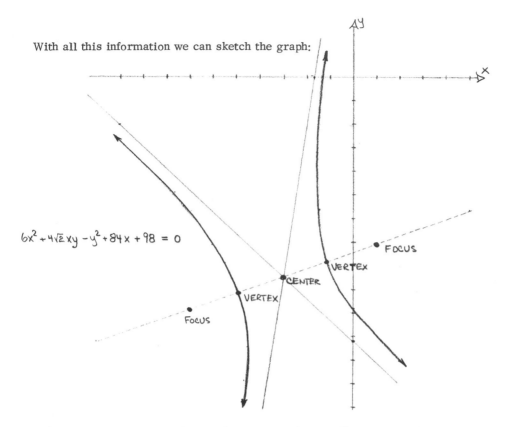

$$6x^2 + 4\sqrt{2}\,xy - y^2 + 84x + 98 = 0$$

4.   <u>The discriminant: a quick identification tool (Optional).</u>   You may be wondering

"isn't there an easier way to identify a conic

section when given its equation?"

The answer is a qualified "yes." There is an easy procedure which can be used to <u>narrow the</u>

<u>possibilities</u> significantly, although it does not separate out all the degenerate cases.

This easy identification method, described in Anton's optional section at the end of

§12.5, goes as follows:

To identify a quadratic equation $A x^2 + B x y + C y^2 + D x + E y + F = 0$:

Compute the <u>discriminant</u> $B^2 - 4 A C$. Then if the discriminant is

    a) <u>negative,</u> the equation is an ellipse, a circle,

        a point or the empty set

    b) <u>positive,</u> the equation is a hyperbola or a pair

        of intersecting lines

    c) <u>zero,</u> the equation is a parabola, a line, a pair

        of parallel lines or the empty set.

For example, suppose we are given the equation

$$6 x y + 8 y^2 - 12 x - 26 y + 11 = 0$$

Here $A = 0$, $B = 6$ and $C = 8$ so the discriminant is

$$B^2 - 4 A C = 6^2 - 4 (0) (8) = 36$$

Since the discriminant is positive, we know the conic section is either a hyperbola or a pair of

intersecting lines. This does not completely answer the question of course, but it narrows the

possibilities and it tells us what to look for as we go through the 5-step procedure of the

previous subsection.

    The only problem with the discriminant technique is that it is hard to remember all the

various possibilities. However, if we restrict ourselves to the "non-degenerate" conic sections,

then we can summarize the discriminant procedure in the easier-to-remember table:

$B^2 - 4AC$	If not degenerate, $Ax^2 + Bxy + Cy^2 + Dx + Ey + F = 0$ is
negative	ellipse or circle
positive	hyperbola
zero	parabola

5. <u>Proof of Theorem B (Optional).</u>   Recall the statement of Theorem B from Subsection 1 above:

Every conic section can be represented by an equation of the form

$$Ax^2 + Bxy + Cy^2 + Dx + Ey + F = 0$$

Conversely, every such equation represents either a conic section

or a degenerate conic section.

<u>Proof.</u>   First suppose that $\mathcal{C}$ is any conic section in the plane.

We must prove that $\mathcal{C}$ can be represented by an

equation of the form

(*) $Ax^2 + Bxy + Cy^2 + Dx + Ey + F = 0$

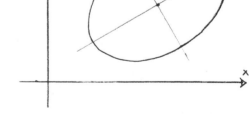

Let ℓ be an axis for $\mathcal{C}$.

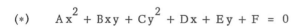

Case 1: If ℓ is parallel to the x-axis or the y-axis, then we know by

Theorem A (i. e., by the results from §§12. 2 - 12. 4) that $\mathcal{C}$ can be represented by an equation

of the form

$$Ax^2 + Cy^2 + Dx + Ey + F = 0$$

which is the desired result with B = 0 .

Case 2: Suppose ℓ is not parallel to the x-axis or the y-axis.

Then we rotate the xy-coordinate system

through an angle θ to a new x' y'-coordinate

system, so that ℓ is parallel to the x'-axis

or the y'-axis. Then, by Theorem A, we know

that C can be represented by an equation of the form

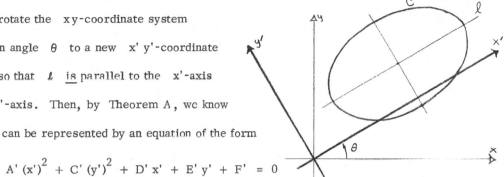

(**) $A' (x')^2 + C' (y')^2 + D' x' + E' y' + F' = 0$

To switch this into an xy-equation, we use the rotation equations

$$x' =\ \ x \cos \theta + y \sin \theta$$
$$y' =\ - x \sin \theta + y \cos \theta$$

Plugging these equations into (**) and simplifying will yield an equation of the form (*), as

desired. This proves that C is represented by an equation of the form (*).

Conversely, suppose we start with an equation of form (*). We must prove that the graph

of this equation is a conic section or degenerate conic section. If B = 0 the result is

immediate from Theorem A. If B ≠ 0, then take the angle θ defined by $\cot 2\theta = (A - C)/B$

and rotate the xy-coordinate system by θ to a new x' y'-coordinate system. By Theorem

12. 5. 1 this will take our equation (*) and transform it into the form of (**), i.e., the mixed

term is eliminated. Ah ha! Now Theorem A tells us that the graph of (**) in the

x' y'-coordinate system (which equals the graph of (*) in the xy-coordinate system!) is a

conic section or a degenerate conic section. And that's what we wanted to prove!   □

Chapter 13: Polar Coordinates and Parametric Equations

<u>Section 13. 1.</u> <u>Polar Coordinates.</u>

1. <u>Plotting points.</u> Suppose you are giving directions on how to get to a certain point. The direc-

tions you give may very well depend on where you are. If you are on a streetcorner in a city,

for instance, you might say "go 3 blocks east and 4 blocks north. " But if you are in the middle

of a flat field, you might say "go 500 feet in a direction of 53° 7' 5" from this row of corn. "

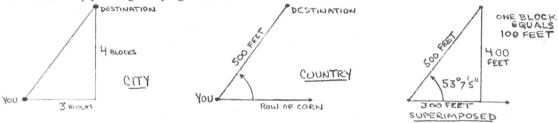

As the superimposed figure shows, both sets of directions will get you from where you are to

the same destination. Which set of directions is better depends on the circumstances!

The "city" directions correspond to locating a point via <u>rectangular</u> coordinates; the

"country" directions correspond to locating a point via <u>polar</u> coordinates. If you are standing

at the origin, the destination point has rectangular coordinates (3 , 4) and polar coordinates

$(5 , 53^{\circ})$. The two sets of coordinates are equally good for locating the point. And, as with

directions, <u>you have a choice; which is better depends on the circumstances.</u>   In particular,

one of the major concerns of calculus is graphing functions. Sometimes it is easier to graph

a function using rectangular coordinates and sometimes it is easier to graph it using polar

coordinates. Knowing how to use both coordinate systems, and how to get from one to the other,

can be a major advantage. (We'll return to this below.)

The key <u>conversion formulas</u> for going back and forth between the rectangular coordinates

(x,y) and the polar coordinates (r , θ) of a point are

From Polar to Rectangular	given (r, θ), the rectangular coordinates are (x, y) where $$x = r \cos \theta \quad \text{and} \quad y = r \sin \theta$$

and

From Rectangular to Polar	given (x, y), the polar coordinates are (r, θ) where $$r^2 = x^2 + y^2 \quad \text{and} \quad \tan \theta = \frac{y}{x} \quad \text{when} \quad x \neq 0 \quad *$$

These are Anton's formulas (1) and (2ab) and it is <u>vital</u> that you remember them.

Fortunately, that is very easy if you draw the following figure (Anton's Figure 13.1.4):

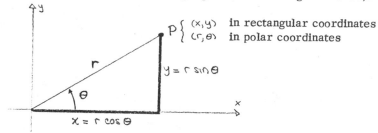

Key Points to Remember.

1. A minus sign on r indicates that the point is on the <u>backwards extension</u> of the terminal side of θ rather than on the terminal side itself.

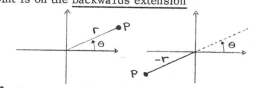

2. A negative value for θ indicates that θ is measured in a <u>clockwise direction</u> rather than in a counterclockwise direction.

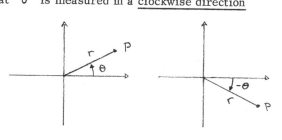

---

*   If $x = 0$, then
$$\theta = \begin{cases} \pi/2 & \text{when } y > 0 \\ 3\pi/2 & \text{when } y < 0 \\ \text{arbitrary} & \text{when } y = 0 \end{cases}$$

3. The polar coordinates of a point are <u>not</u> unique. A point with polar coordinates

    (r, θ) will also have polar coordinates

    $(-r, \theta + \pi)$. Moreover, adding <u>any</u>

    integer multiple of 2π to an angle

    leaves the angle unchanged. Thus (r, θ)

    also has polar coordinates $(r, \theta + 2n\pi)$ and $(-r, (\theta + \pi) + 2n\pi)$ for any

    positive or negative integer n .

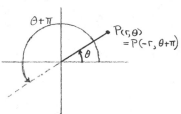

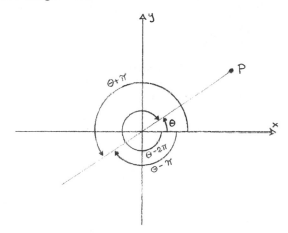

4. When converting from given rectangular coordinates (x, y) to polar coordinates

    (r, θ) , there will be <u>more than one r</u> which satisfies $r^2 = x^2 + y^2$ and

    <u>more than one θ</u> which satisfies $\tan \theta = \frac{y}{x}$. The quadrant in which (x, y)

    lies should be used to obtain more information about r and θ . (See Anton's

    Example 2).

<u>Example A.</u>   Find <u>all</u> polar coordinates of the point given in polar form by $(2, \frac{\pi}{6})$.

<u>Solution.</u>   By Key Point 3 , $(2, \frac{\pi}{6})$ can be written as $(2, \frac{\pi}{6} + 2n\pi)$ or $(-2, \frac{7\pi}{6} + 2n\pi)$

for $n = 0, \pm 1, \pm 2, \pm 3, \ldots$. $\square$

<u>Example B.</u> Find the rectangular coordinates of the point P given in polar form by $(-2, \frac{\pi}{3})$.

<u>Solution.</u> The conversion formulas $x = r \cos \theta$, $y = \sin \theta$ give $x = -2 \cos \frac{\pi}{3} = -1$

and $y = -2 \sin \frac{\pi}{3} = -\sqrt{3}$ so the rectangular coordinates of P are $(-1, -\sqrt{3})$. □

<u>Example C.</u> Find the polar coordinates of the point P with rectangular coordinates $(-2, 2\sqrt{3})$.

<u>Solution.</u> The conversion formula $r^2 = x^2 + y^2$ gives $r^2 = (-2)^2 + (2\sqrt{3})^2 = 16$, so

$r = \pm 4$. The conversion formula $\tan \theta = \frac{y}{x}$ gives $\tan \theta = \frac{2\sqrt{3}}{-2} = -\sqrt{3}$, so $\theta = \frac{2}{3}\pi$

or $\frac{5}{3}\pi$. But $(-2, 2\sqrt{3})$ is in the <u>second quadrant</u>. Hence $(4, \frac{2}{3}\pi)$ and $(-4, \frac{5}{3}\pi)$ are

the correct combinations of r and θ . In general, $(4, \frac{2}{3}\pi + 2n\pi)$ and $(-4, \frac{5}{3}\pi + 2n\pi)$

for any integer n will be polar coordinates of P . □

2. <u>Polar curve sketching: two methods.</u> We have said that sometimes it is easier to graph a function

using polar coordinates than it is using rectangular coordinates and sometimes the converse is

true.

<u>Example D.</u> Sketch the graph of the polar curve $r^2 \left[\cos^2 \theta + \frac{4}{\csc^2 \theta} \right] = 4$.

<u>Solution.</u> Using some simple trigonometry, we obtain

$$4 = r^2 \left[\cos^2 \theta + \frac{4}{\csc^2 \theta} \right] = r^2 \left[\cos^2 \theta + 4 \sin^2 \theta \right]$$

$$= (r \cos \theta)^2 + 4 (r \sin \theta)^2$$

But, by the conversion formulas $x = r \cos \theta$ and $y = r \sin \theta$, this says that the rectangular

coordinate form of our curve is $x^2 + 4y^2 = 4$ or $x^2/4 + y^2 = 1$. You should recognize

this as an ellipse (see Appendix E.3)

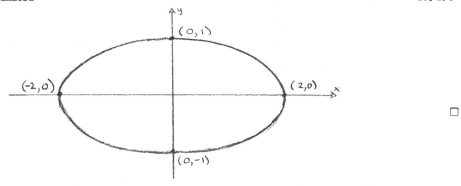

Example E.   Sketch the graph of the curve $r = \theta$, $\theta \geq 0$.

Solution.   We begin by making a table of values of   r   and θ:

θ	0	$\pi/4$	$\pi/2$	π	$3\pi/2$	2π	$5\pi/2$
r	0	$\pi/4$	$\pi/2$	π	$3\pi/2$	2π	$5\pi/2$

We can graph this by first graphing $r = \theta$ on a $\theta - r$ <u>rectangular axis</u>

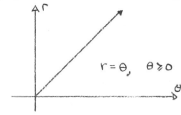

and then translating it to   x - y -axes :

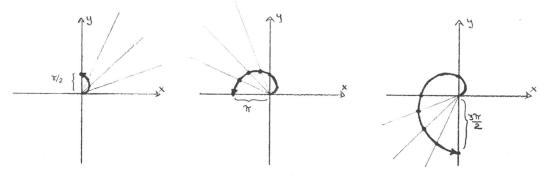

The final graph is a spiral (sometimes called "the spiral of Archimedes".)

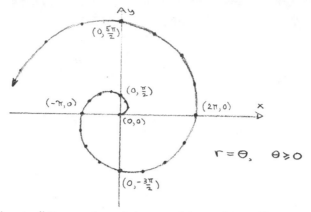

$r = \theta, \quad \theta \geqslant 0$

A good way of "seeing" the graph as it is traced is to picture Howard Ant crawling along a long (very long!) flat stick which is nailed loosely at one end so that it can be revolved.

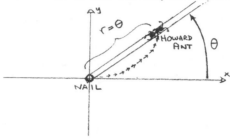

To start, place Howard Ant on the nail (which has polar coordinates $(0,0)$). As the stick is revolved around and around the nail, he crawls out along the stick at just the right place to assure that his distance r on the stick from the nail is equal to the angle θ (measured in radians) through which the stick has been revolved. That is, he moves so that $r = \theta$ (our equation!). It is not hard to see that his path is a spiral. □

Both these examples were done by <u>converting from one coordinate system to another.</u> In particular, Example D illustrates that one way of graphing a polar curve is to

The Conversion Method	transform the polar curve into xy coordinates

However, our Howard Ant approach to graphing $r = \theta$ in Example E also illustrates another method. That is, to

The Interval Method

graph the polar curve on rectangular $r\theta$ axes

and then transform to the usual xy axes considering

successive θ-intervals .

It's not as complicated as it sounds. Anton's figure 13. 1. 9 is a good illustration of it and

here is another.

Example F. Sketch the graph of the polar curve $r = -1 + 2\cos\theta$.

Solution. We "build up" the graph, first treating r and θ as rectangular coordinates and

then converting to the polar graph:

STAGE 1

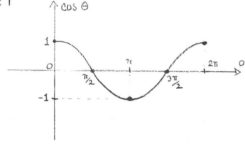

◁—————— First graph $\cos\theta$

STAGE 2

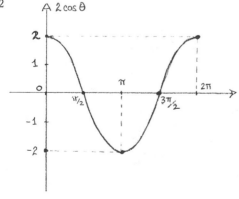

◁—————— Then graph $2\cos\theta$. When <u>you</u> do this,
a separate graph is <u>not</u> necessary ... simply
change the scale on the vertical axis of your
<u>first</u> graph.

13. 1. 8

STAGE 3

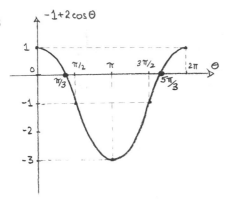

-1 + 2 cos θ

◁——— Then graph - 1 + 2 cos θ . This corresponds to shifting the horizontal θ-axis up one unit on the vertical axis. Again, a "new" graph is not really needed.

We also should compute those θ's which give - 1 + 2 cos θ = 0 , i.e., cos θ = 1/2 , or θ = π/3 , $\frac{5\pi}{3}$.

We now use our "rectangular" sketch for - 1 + 2 cos θ to sketch a polar graph for r = - 1 + 2 cos θ . We do this by considering the θ intervals defined by those values of θ which make r a maximum value, a minimum value, or zero. In the case at hand we have

$$[\, 0 \, , \, \pi/3 \,] \quad , \quad [\, \pi/3 \, , \, \pi \,] \quad , \quad \left[\, \pi \, , \, \frac{5\pi}{3} \,\right] \quad \text{and} \quad \left[\, \frac{5\pi}{3} \, , \, 2\pi \,\right]$$

Notice that we simply read these intervals off of the last sketch given above.

So we now sketch our graph in four steps using our Howard Ant-on-the-revolving-stick approach:

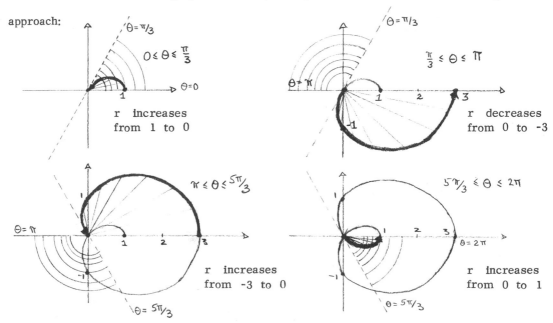

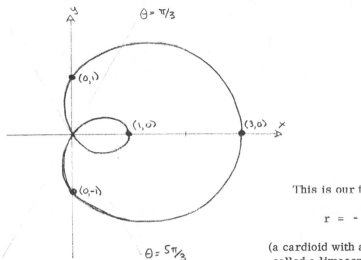

This is our finished graph for

$$r = -1 + 2\cos\theta$$

(a cardioid with an inner loop, sometimes called a limaçon).   □

Section 13. 2.   Graphs In Polar Coordinates.

1.   Recognizing polar curves: it depends on memorization.   In the last section, we described two

methods for graphing polar curves:

    1.   conversion to xy coordinates, and

    2.   the interval method.

There is also a third method:

The Recognition Method	
	3. recognize the graph

(if this can be called a "method"). That is, depending on how much you are willing to memorize

you can recognize equations as being of a certain type and you can use this information to graph

them. This is precisely what you did for circles, ellipses, parabolas and hyperbolas in rectangular coordinates (e. g. , when you see the equation $\dfrac{x^2}{6} + \dfrac{y^2}{4} = 3$, you think "ellipse" and, even if you can't remember the details about the semi-major and semi-minor axes, you know the curve has an oval shape and is centered around the origin). The purpose of Section 13.2 is to present similar information about polar curves.

The key phrase in the paragraph above is "depending on how much you are willing to memorize. " Obviously, the more you memorize, the easier it will be to recognize graphs. If you do not memorize very much, the recognition method will be of limited use to you and you will depend on the conversion and interval methods. We recommend that you study the curves in Section 13.2 carefully and that you know their names and their shapes, but we do not believe it is essential to memorize the specifics of all of the graphs associated with the formulas in blue boxes. After all, you should always be able to figure them out! In Subsections 3 - 5 below, we describe the minimum amount you should know.

2. <u>Symmetry: find it and use it.</u> The <u>symmetry tests</u> (Anton's Theorem 13.2.1) are particularly useful when graphing polar curves. Here is a summary:

The Symmetry Tests	Symmetric about	if equivalent equation is obtained when you replace
	x-axis	θ by $-\theta$
	y-axis	θ by $\pi - \theta$
	origin	r by $-r$

Anton's Figure 13.2.9 shows why these symmetry conditions are true. You should study it carefully.

<u>Example A</u>. Sketch the graph of $r^2 = 4 \cos 2\theta$.

<u>Solution.</u>   Replacing θ by $-\theta$ yields

$$r^2 = 4\cos(-2\theta) = 4\cos 2\theta \qquad\qquad (\text{because } \cos(-\theta) = \cos\theta)$$

Replacing θ by $\pi - \theta$ yields

$$r^2 = 4\cos 2(\pi - \theta) = 4\cos(2\pi - 2\theta)$$

$$= 4\cos(-2\theta) \qquad\qquad (\text{because } \cos(\theta + 2\pi) = \cos\theta)$$

$$= 4\cos 2\theta$$

Replacing r by $-r$ yields

$$4\cos 2\theta = (-r)^2 = r^2$$

Hence $r^2 = 4\cos 2\theta$ is symmetric about the x-axis, the y-axis and the origin. Knowing this, it is sufficient to graph only part of the curve, in this case the part where θ satisfies $0 \leq \theta \leq \frac{\pi}{4}$. [There are no points on the curve with θ in the range $\frac{\pi}{4} < \theta < \frac{\pi}{2}$ for the following reason: if $\frac{\pi}{4} < \theta < \frac{\pi}{2}$, then $\cos 2\theta < 0$ and this would yield the impossible condition $r^2 = 4\cos 2\theta < 0$.]

Then we use the symmetry to determine the rest of the graph (see Anton's Figure 13.2.13)

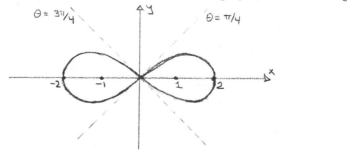

3.   <u>Lines in polar coordinates: the minimum amount you should know.</u>   It is easy to see that the graph of the equation (Anton's Equation (3))

$$\theta = \theta_0, \qquad \theta_0 \text{ a constant}$$

is a line which passes through the origin and makes an angle θ_0 with the positive x-axis.

The general <u>rectangular coordinate</u> equation of a line is $Ax + By + C = 0$. Using the conversion formulas $x = r \cos \theta$ and $y = r \sin \theta$, this becomes

General
polar form
of a line

$$r(A \cos \theta + B \sin \theta) + C = 0$$

Anton's Equations (1), (2) and (3) are all special cases of this general formula. For example, when the line is parallel to the x-axis, its equation (Anton's Equation (2)) has this form with $A = 0$ and $B = 1$.

4. <u>Circles in polar coordinates: the minimum amount you should know.</u> It is easy to see that the graph of the equation (Anton's Equation (5))

$$r = a, \qquad a \geqq 0$$

is a circle of radius a centered at the origin. Anton's other equations for circles (Equations (6ab) and (7ab)) are best remembered by <u>remembering the xy coordinate form of the equation for a circle and using the conversion formulas.</u> For example, the equation of a circle of radius a and center with <u>polar</u> coordinates $(a, -\frac{\pi}{2})$ can be obtained as follows:

The rectangular coordinates of the point

$(a, -\frac{\pi}{2})$ are $(0, -a)$ [using the conversion

formulas $x = r \cos \theta$, $y = r \sin \theta$].

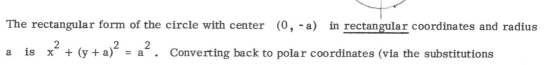

The rectangular form of the circle with center $(0, -a)$ in <u>rectangular</u> coordinates and radius a is $x^2 + (y+a)^2 = a^2$. Converting back to polar coordinates (via the substitutions x = r cos θ , y = r sin θ) yields $r(r + 2a \sin \theta) = 0$ or, since $r \neq 0$ (why?),

$r = -2a \sin \theta$. This is Anton's Equation (7b).

Reasoning in this manner will show that

> $r = \pm 2a \sin \theta$ is the circle with center
> $(0, \pm a)$ which passes through the origin;
>
> $r = \pm 2a \cos \theta$ is the circle with center
> $(\pm a, 0)$ which passes through the origin.

5. Cardiods, limaçons, lemniscates, spirals, rose curves: the minimum amount you should know.

The spiral of Archimedes (Anton's Equations (10 ab))

> $r = a\theta$, $a > 0$, $\theta \geqq 0$ (or $\theta \leq 0$)

is almost unforgettable. (Review Example E of Section 13.1 of the Companion.) Otherwise,

it is probably enough to remember that

> limaçons   look like dented circles (sometimes with an inner loop)
> cardioids   look like valentines
> lemniscates look like infinity signs ("lazy 8's")
> rose curves look like flowers with petals

and to remember how they are traced (by Howard Ant on his revolving stick). You will be

surprised by how much knowing just this much will facilitate the use of the interval method

used in Example F of the previous section.

Section 13.3.   Area in Polar Coordinates.

1.   The basic polar coordinate formula.   The basic formula is Anton's Definition 13.3.2:

The area   A   enclosed by the polar curve $r = f(\theta)$ and the lines $\theta = \alpha$ and $\theta = \beta$ is given by

<table>
<tr><td>Definition
of Radial
Area</td><td>$$A = \frac{1}{2} \int_\alpha^\beta [f(\theta)]^2 \, d\theta = \frac{1}{2} \int_\alpha^\beta r^2 \, d\theta$$ 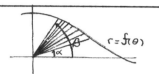</td></tr>
</table>

where   f   is continuous and nonnegative for $\alpha \leq \theta \leq \beta$.

Remembering Definition 13.3.2 is easy if you remember (or can derive quickly) the formula for the area of a sector with angle θ and radius   r   (see figure)

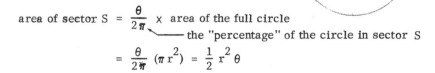

$$\text{area of sector S} = \frac{1}{2} r^2 \theta$$

This is easily derived as follows:

$$\text{area of sector S} = \frac{\theta}{2\pi} \times \text{area of the full circle}$$
the "percentage" of the circle in sector S

$$= \frac{\theta}{2\pi} (\pi r^2) = \frac{1}{2} r^2 \theta$$

If you regard $d\theta$ as the width of an infinitesimal slice of the region   R

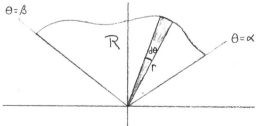

then the area of the infinitesimal slice (which is a circular sector) is $\frac{1}{2} r^2 \, d\theta$ and the area

A   of   R   is the "(infinite) sum of the areas of infinitesimal slices," i.e., $A = \int_\alpha^\beta \frac{1}{2} r^2 \, d\theta$.

2.   <u>Intersection points and collision points.</u>   Take careful note of Anton's four steps for applying

Definition 13.3.2.   In particular, note that

    <u>Step 1</u> requires that you use all the sketching techniques developed in

          Sections 13.1 and 13.2, and

    <u>Step 3</u> frequently requires finding intersection points of two polar curves.

Do not take the business of finding intersection points too lightly! There is a twist in finding

intersection points of polar curves that we did not encounter when we found the intersection

points of rectangular curves.   This "twist" is described briefly in Anton's "Warning"

following Example 3.

      We can illustrate the problem by looking at Anton's

Figure 13.3.6. Imagine that Howard Ant is traveling

around the blue curve ($r = 1 + \cos\theta$, where θ

represents <u>time</u>) and his wife Pat is traveling around

the white curve ($r = 1 - \cos\theta$). Both start when

$\theta = 0$ (time equals zero) so Howard's starting

point is $(r,\theta) = (2,0)$, while Pat's is

$(r,\theta) = (0,0)$. A point (r,θ) on the curves is called

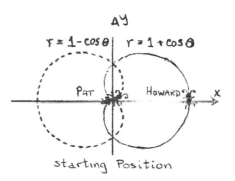

starting Position

---

    (1)   an <u>intersection point</u> if the Ants' paths cross there, and

    (2)   a <u>collision point</u> if the Ants' paths cross there <u>and</u>

           if the Ants are there <u>simultaneously</u>.

From our diagram (Anton's Figure 13.3.6) we
see that there are three points where the curves
cross:

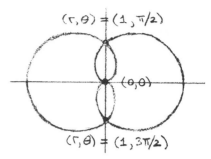

$$(r, \theta) = (0, 0), \quad (1, \pi/2), \quad (1, 3\pi/2)$$

These points are therefore at least <u>intersection points.</u>   The question is, are any of them also

<u>collision points?</u> To determine this, let's watch the Ants run around the curves:

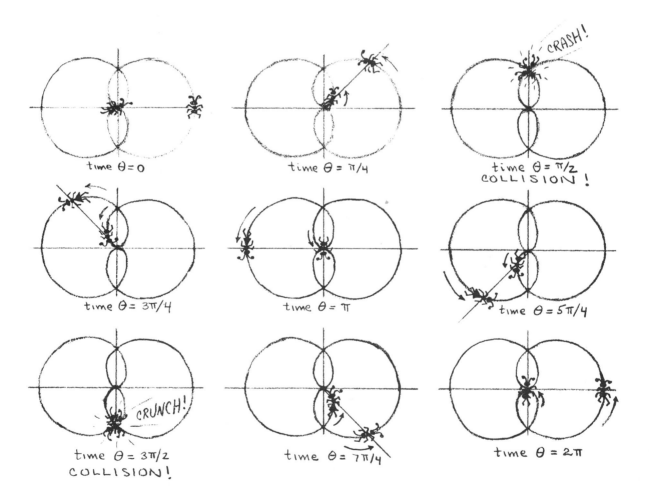

time $\theta = 0$

time $\theta = \pi/4$

time $\theta = \pi/2$
COLLISION !

time $\theta = 3\pi/4$

time $\theta = \pi$

time $\theta = 5\pi/4$

time $\theta = 3\pi/2$
COLLISION !

time $\theta = 7\pi/4$

time $\theta = 2\pi$

As you can see, there are collisions at $\theta = \pi/2$ and $3\pi/2$, but no collisions else-where. Hence

> $(r, \theta) = (1, \pi/2)$ and $(1, 3\pi/2)$ are collision points, while
>
> $(r, \theta) = (0, 0)$ is just an intersection point

Anton's "Warning" makes the point that

---

collision points of two polar curves can be found by solving their equations

simultaneously, but intersection points which are not also collision points

are best determined by graphing the equations.

Here is another example, one in which every intersection point is a collision point:

Example A.   Determine all the collision points and all the intersection points for the polar curves
$r \cos \theta = 1$ and $-2 + 3 \cos \theta = r$.

Solution.   First we determine the collision points. It is convenient to multiply the second
equation by r:

$$r^2 = -2r + 3r \cos \theta = -2r + 3 \qquad \text{since} \qquad r \cos \theta = 1$$

$$\text{Thus} \qquad r^2 + 2r - 3 = 0$$

$$\text{or} \qquad (r + 3)(r - 1) = 0$$

$$\text{Hence} \quad r = 1 \quad \text{or} \quad r = -3$$

We now need the θ-values

r = 1   gives $\cos \theta = 1$,   so $\theta = 0$

r = -3   gives $\cos \theta = -\dfrac{1}{3}$,   so $\theta = \pm \cos^{-1}\left(-\dfrac{1}{3}\right) \approx \pm 110^{\circ}$

Thus the collision points of our curves are

$$r = 1, \quad \theta = 0$$

$$r = -3, \quad \theta = \cos^{-1}(-1/3) \approx 110^{\circ}$$

$$r = -3, \quad \theta = -\cos^{-1}(-1/3) \approx -110^{\circ}$$

To determine whether there are any intersection points which are not also collision points, we graph the curves:

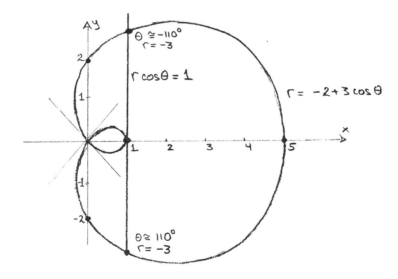

Thus every intersection point is a collision point since there are no intersection points other than the collision points already found.

3.   <u>Computing areas between two curves.</u>   Anton's Example 3 makes it clear that

> to compute the area <u>between</u> two polar curves
>
> you may have to compute areas enclosed by
>
> pieces of the curves and then add or subtract them.

Here is another example:

<u>Example B.</u>   The polar curves $r = \cos \theta$ and $r = 1 - \cos \theta$ have the following graphs:

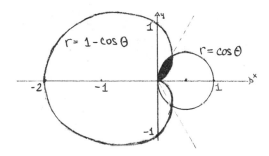

Determine the area of the shaded region.

<u>Solution.</u>   We first need to determine the θ-value   for the intersection point of interest to us.
Equating our two expressions for r yields:

$$\cos \theta = 1 - \cos \theta$$

$$\cos \theta = \frac{1}{2}$$

$$\text{or} \qquad \theta = \pm \ \pi/3$$

13.3.7

Hence the desired area is given by

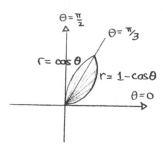

$$A = \int_0^{\pi/3} \frac{1}{2} (1 - \cos\theta)^2 \, d\theta + \int_{\pi/3}^{\pi/2} \frac{1}{2} (\cos\theta)^2 \, d\theta$$

or

$$A = \frac{1}{2} \left[\int_0^{\pi/3} (1 - 2\cos\theta + \cos^2\theta) \, d\theta + \int_{\pi/3}^{\pi/2} \cos^2\theta \, d\theta \right]$$

$$= \frac{1}{2} \left[\theta - 2\sin\theta + \left(\frac{\theta}{2} + \frac{\sin 2\theta}{4} \right) \Big]_0^{\pi/3} + \left(\frac{\theta}{2} + \frac{\sin 2\theta}{4} \right) \Big|_{\pi/3}^{\pi/2} \right]$$

$$= \frac{1}{2} \left[\frac{\pi}{3} - 2\frac{\sqrt{3}}{2} + \pi/4 \right] = \frac{1}{2} \left[\frac{7\pi}{12} - \sqrt{3} \right]$$

$$A = \frac{7\pi}{24} - \frac{\sqrt{3}}{2} \approx .05027$$

$\square$

But BEWARE of a common mistake! In finding the area between two curves in <u>rectangular</u> coordinates, we used the formula

$$A = \int_a^b | f(x) - g(x) | \, dx$$

However, it is WRONG to try to use this pattern for polar coordinates and calculate the area between $r_1 = f(\theta_1)$ and $r_2 = g(\theta_2)$ using

<u>Incorrect
Formula</u>

$$A \overset{\boxed{\text{FALSE}}}{=} \frac{1}{2} \int_\alpha^\beta [f(\theta_1) - g(\theta_2)]^2 \, d\theta = \frac{1}{2} \int_\alpha^\beta [r_1 - r_2]^2 \, d\theta$$

Instead, you must calculate the area <u>inside</u> of $r_1 = f(\theta_1)$ and <u>subtract</u> from it the area <u>outside</u> of $r_2 = g(\theta_2)$, i.e.,

$$A \underset{\boxed{\text{CORRECT}}}{=} \frac{1}{2} \int_{\alpha}^{\beta} r_1^2 \, d\theta \; - \; \frac{1}{2} \int_{\alpha}^{\beta} r_2^2 \, d\theta$$

Example C. (Anton's Example 2). Find the area of the region that is inside the cardiod

$r = 4 + 4 \cos \theta$ and outside the circle $r = 6$.

Incorrect But Common Solution.

$$A \underset{\boxed{\text{FALSE}}}{=} \frac{1}{2} \int_{-\pi/3}^{\pi/3} [(4 + 4 \cos \theta) - 6]^2 \, d\theta = \ldots \quad \text{a wrong answer!}$$

See Anton's Example 2 for a correct solution:

$$A = \frac{1}{2} \int_{-\pi/3}^{\pi/3} [(4 + 4 \cos \theta)^2 - (6)^2] \, d\theta$$

$$= \frac{1}{2} \int_{-\pi/3}^{\pi/3} (4 + 4 \cos \theta)^2 \, d\theta - \frac{1}{2} \int_{-\pi/3}^{\pi/3} (6)^2 \, d\theta = \ldots = 18\sqrt{3} - 4\pi \qquad \square$$

Section 13.4. Parametric Equations

1. <u>Why parameters? They give more information!</u> A common reaction to this material is,

   "Why use parameters? We are getting along just fine without them." The answer

   is simple: <u>parameters are used because they give more information.</u> For instance, consider

   the line given by the equations

$$y = 3x + 2 \qquad \text{or} \qquad \begin{cases} x = 2t - 3 \\ y = 6t - 7 \end{cases}$$

   (xy equation of the line)   (parametric equations of the line)

As Anton's Example 3 makes clear, these are equations of the same line. But the xy equation of the line gives you only the <u>path</u> followed by a point moving along it. If t is time (which is, in general, a good way to interpret the parameter), the parametric equations not only give you the path, they also tell you exactly where on the path the point is at time t! At t = 2, for example, the point is at (1, 5). Frequently the extra information contained in parametric equations is extremely important.

2. <u>Graphing parametric equations.</u>   There are essentially two methods which can be used to graph parametric equations in the plane:

    1.   Make a 4-row table with the parameter values as the first row (see Anton's Example 1), or

    2.   Eliminate the parameter and graph the resulting equation (see Anton's Examples 3 and 4).

Here is an illustration of both methods:

<u>Example A.</u>   Graph the parametric equations

$$x = \sin^2 t$$

$$y = \cos t$$

<u>First Solution (4-row table approach).</u>   We make the following 4-row table

t	0	$\pi/6$	$\pi/4$	$\pi/3$	$\pi/2$	$2\pi/3$	$3\pi/4$	$5\pi/6$	π
$x = \sin^2 t$	0	1/4	1/2	3/4	1	3/4	1/2	1/4	0
$y = \cos t$	1	$\sqrt{3}/2$	$\sqrt{2}/2$	1/2	0	$-1/2$	$-\sqrt{2}/2$	$-\sqrt{3}/2$	-1
(x, y)	$(0, 1)$	$\left(\frac{1}{4}, \frac{\sqrt{3}}{2}\right)$	$\left(\frac{1}{2}, \frac{\sqrt{2}}{2}\right)$	$\left(\frac{3}{4}, \frac{1}{2}\right)$	$(1, 0)$	$\left(\frac{3}{4}, -\frac{1}{2}\right)$	$\left(\frac{1}{2}, -\frac{\sqrt{2}}{2}\right)$	$\left(\frac{1}{4}, -\frac{\sqrt{3}}{2}\right)$	$(0, -1)$

Plotting the nine points (x, y) yields the following graph

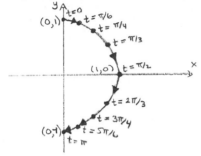

The arrow indicates the <u>orientation</u> of the curve (see Anton's discussion preceding Example 3).

<u>Second Solution.</u> (Eliminate the parameter).   In this example, it is easy to eliminate t by

taking advantage of the identity $\sin^2 t + \cos^2 t = 1$ as follows:

$$x = \sin^2 t = 1 - \cos^2 t = 1 - y^2$$

Plotting this curve we get

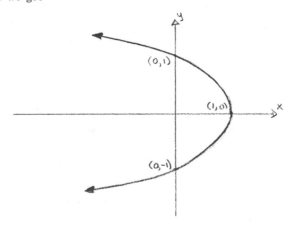

When we eliminate the parameter in this manner we frequently introduce some extra "unwanted" points into our graph (compare our second unbounded graph with the first bounded graph). In order to eliminate these "unwanted" points, we first compute our starting point (at $t = 0$) and our terminal point (at $t = \pi$):

$$t = 0 \quad \text{gives} \quad (x, y) = (0, 1)$$

$$t = \pi \quad \text{gives} \quad (x, y) = (0, -1)$$

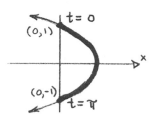

Since the curve is continuous, all the points on the curve between $(0, 1)$ and $(0, -1)$ must be on the parameterized curve. Moreover, since $x = \sin^2 t \geq 0$, then no other point on the unbounded curve can be on the parameterized curve. Why? Because all the remaining points have $x < 0 \ldots$ which would contradict $x = \sin^2 t \geq 0$. Thus we must cut back the unbounded curve to the one piece shown to the right. This graph now agrees with the one obtained in our First Solution.

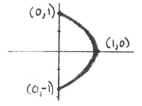

The previous example underscores the importance of the remark Anton makes following Example 7:

> whenever the parameter is eliminated from parametric equations, check that the graph of the resulting equation does not include too much.

A foolproof way of making this "check" is to make a 4-row table and see whether there are any points on the unparameterized graph that are not on the parameterized graph. However, you can usually avoid making the table and check the unparameterized graph simply by

i) looking for maximum and minimum values of x and y

(In Example A , $0 \leq \sin^2 t \leq 1$ implies $0 \leq x \leq 1$

$-1 \leq \cos t \leq 1$ implies $-1 \leq y \leq 1$)

ii) checking the signs of x and y

(in Example A , $0 \leq \sin^2 t$ implies $0 \leq x$)

iii) checking where x and y are increasing or decreasing as functions of t

3. Switching between parameterized and unparameterized equations.   Sometimes it is desirable (even necessary!) to switch from parameterized equations to unparameterized ones or vice versa.

To change from parameterized to unparameterized equations,

$$\boxed{\text{eliminate the parameter}}$$

using techniques such as are found in the second solution to our Example A , in Anton's Examples 3 and 4 or in the paragraph following Anton's Example 6 . However, keep two points in mind:

1. It may be difficult (or even impossible) to eliminate the parameter completely, and the unparameterized equation may not be easier to graph; and

2. The graph of the unparameterized curve may include too much.

Example A illustrates the second point. Here is an example to illustrate the first:

Example B.   Eliminate the parameter from the parametric equations

$$x = \cos^3 t$$

$$y = \sin^3 t$$

Solution.   Solving each equation for t and equating them we get

$$\sin^{-1} y^{1/3} = \cos^{-1} x^{1/3}$$

$$y^{1/3} = \sin(\cos^{-1} x^{1/3})$$

$$y^{1/3} = \sqrt{1 - x^{2/3}} \qquad\qquad \text{using the right triangle}$$

$$y = (1 - x^{2/3})^{3/2}$$

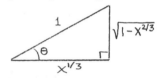

Out of the frying pan and into the fire! This unparameterized equation is certainly not easy to

graph, so we are probably better off to use the 4-row table approach and graph the

parameterized equations. We'll leave that to you. (Hint: This curve is sometimes called an

astroid. It is shown in Example F.)   □

To change from unparameterized to parameterized equations, the key is to

> choose the right parameter.

The point to keep in mind is that there are <u>lots of choices</u>. Unfortunately, no one can give you

a method for deciding in advance which one is the <u>best</u> choice; that will depend on the equation

and on the use(s) you want to make of it. Anton's Examples 5 , 6 and 7 illustrate three

different choices of parameters and you should take careful note of them. In particular, Anton's

Examples 6 and 7 give <u>two sure-fire methods of choosing a parameter</u>:

1. If $r = f(\theta)$ is a polar curve, the standard equations

$$x = r \cos \theta = f(\theta) \cos \theta$$

$$y = r \sin \theta = f(\theta) \sin \theta$$

will parameterize it in terms of θ .

2. If $y = f(x)$ is a rectangular curve, the equations

$$x = t$$

$$y = f(t)$$

will parameterize it in terms of t .

4. <u>Derivatives of parametric equations.</u>   The chain rule

$$\boxed{\dfrac{dy}{dx} = \dfrac{dy \,/\, dt}{dx \,/\, dt}}$$

(Anton's Equation (3)) is certainly easy to remember! And, as Anton's Example 8 illustrates,

it can be used to <u>find derivatives of y with respect to   x   without eliminating the parameters.</u>

However, when $dx/dt = 0$, the division in the chain rule is not valid. Anton's remark

following his Example 8 tells what happens in this case:

1. If $dx/dt = 0$ and $dy/dt \neq 0$ at a point there is generally (but not always) a <u>vertical tangent</u> line at that point.

2. If $dx/dt = 0$ and $dy/dt = 0$ at a point, the point is called a <u>singular point</u>. No general statement can be made about the behavior of a curve at a singular point.

The following examples illustrate these statements:

<u>Example C.</u> Find dy/dx at the origin for the curve

$$x = t^3$$
$$y = t^2$$

<u>Solution.</u> Clearly the origin is the point where $t = 0$, and

$$\left.\frac{dx}{dt}\right|_{t=0} = \left.3t^2\right|_{t=0} = 3(0)^2 = 0$$

$$\left.\frac{dy}{dt}\right|_{t=0} = \left.2t\right|_{t=0} = 2(0) = 0$$

Thus $(0, 0)$ is a <u>singular point.</u> Eliminating the parameter, we find $t = \sqrt[3]{x}$ and $t = \sqrt{y}$ so $\sqrt[3]{x} = \sqrt{y}$ or

$$y = x^{2/3}$$

The graph of this equation is

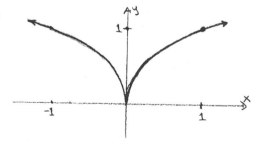

Thus the curve has a cusp at the singular point $(0, 0)$ and hence neither dy/dx nor the tangent line exists at that point.   □

Example D.   Find $\dfrac{dy}{dx}$ at the origin for the curve

$$x = t^3$$
$$y = t^6$$

Solution.   The origin is the point given by $t = 0$ and

$$\left.\frac{dx}{dt}\right|_{t = 0} = 3t^2\Big|_{t = 0} = 0$$

$$\left.\frac{dy}{dt}\right|_{t = 0} = 6t^5\Big|_{t = 0} = 0$$

Thus $(0, 0)$ is a singular point. Eliminating the parameter, we find that the curve is the parabola $y = x^2$. Thus $\left.\dfrac{dy}{dx}\right|_{x = 0} = 2x\Big|_{x = 0} = 0$ (i.e., the tangent line at the singular point $(0, 0)$ is horizontal).

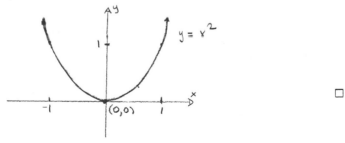

□

Example E.   Find dy/dx at $t = 1$ for the curve

$$x = (t - 1)^2$$
$$y = t$$

Solution. We find

$$\left.\frac{dx}{dt}\right|_{t=1} = 2(t-1)\Big|_{t=1} = 0$$

$$\left.\frac{dy}{dt}\right|_{t=1} = 1$$

so (by 1) we expect a vertical tangent line at $(0,1)$. Eliminating t (by substituting), we obtain

$$x = (y-1)^2$$

the graph of which is

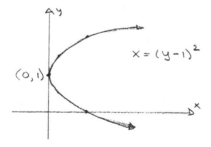

Thus there is a vertical tangent line at $(0,1)$ (i.e., $dy/dx\big|_{t=1} = \infty$). ☐

5. Arc length of parametric curves. The formula for the arc length L of a parametric curve $x = x(t)$, $y = y(t)$ (where $x'(t)$ and $y'(t)$ are continuous for $a \le t \le b$)

$$\boxed{L = \int_a^b \sqrt{\left(\frac{dx}{dt}\right)^2 + \left(\frac{dy}{dt}\right)^2}\ dt}$$

(Anton's equation (6) or (7)) is both easy to remember and very useful.

Remembering it: Think of dL, dx and dy as infinitesimal pieces of the curve and the x- and y-axes , respectively. That is, remember the following picture:

an infinitesimal piece
of the curve, so small
that it can be considered
to be a straight line.

From this picture, $dL = \sqrt{(dx)^2 + (dy)^2} = \sqrt{\left(\dfrac{dx}{dt}\right)^2 + \left(\dfrac{dy}{dt}\right)^2}\ dt$.   "Summing up" all these

infinitesimal pieces, i.e., integrating our last expression, yields the formula for arc length.

Using it: Remember that you do not have to convert the parameterized equations to

unparameterized equations to find arc length. Anton's Example 9 illustrates this. Here is

another example:

Example F.   Find the length of the part of the astroid

$$x = \cos^3 t$$

$$y = \sin^3 t$$

corresponding to the   t-interval $[0, 2\pi]$.

Solution.   As we saw in Example B above, the rectangular coordinate equation of the astroid

is unpleasant, to say the least. Fortunately, we can avoid it altogether. If you haven't already

graphed it for yourself, its graph looks like this:

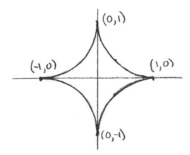

We find

$$\frac{dx}{dt} = -3\cos^2 t \sin t , \quad \text{and} \quad \frac{dy}{dt} = 3\sin^2 t \cos t$$

so that the desired arc length L is 4 times the arc length in the first quadrant $(0 \le t \le \frac{\pi}{2})$ or

$$L = \int_a^b \sqrt{\left(\frac{dx}{dt}\right)^2 + \left(\frac{dy}{dt}\right)^2} \, dt$$

$$= 4 \int_0^{\pi/2} \sqrt{(-3\cos^2 t \sin t)^2 + (3\sin^2 t \cos t)^2} \, dt$$

$$= 12 \int_0^{\pi/2} \sin t \cos t \sqrt{\sin^2 + \cos^2 t} \, dt$$

$$= 6 \sin^2 t \Big|_0^{\pi/2} = 6$$

☐

We conclude this section with two brief comments:

1. Most of the arc length examples given in this section, both in Anton and in the Companion, are artificial in that the integrals work out nicely. In reality, most arc length integrals cannot be evaluated in "closed form," i.e., they require numerical methods.

2. However, even if most arc length integrals aren't so nice, we can still use arc length as a parameter for curves to our great advantage! This may seem weird ... but we do it in Section 14.3.!

Section 13.5. Tangent Lines and Arc Length in Polar Coordinates (Optional). Anton includes this section simply to present and justify three polar coordinate formulas:

1. A formula for the slope m of $r = f(\theta)$ at $P(r, \theta)$ [Anton's Equation (5)].

2. A formula for the angle ψ between the radial line   OP   and the tangent line

   to $r = f(\theta)$ at $P(r, \theta)$ [Anton's Equation (14)].

3. A formula for the arc length of $r = f(\theta)$ [Anton's Equations (15) and (16)].

1. <u>Computing slopes of polar curves: an alternate method.</u>   Anton's Formula (5) for the slope

m   of $r = f(\theta)$ at $P(r, \theta)$ can be hard to remember. Here is another way of finding   m :

$$m \;=\; \frac{dy}{dx} \;=\; \frac{dy/d\theta}{dx/d\theta} \qquad \text{(chain rule)}$$

where

$$x \;=\; r \cos \theta \;=\; f(\theta) \cos \theta$$

$$y \;=\; r \sin \theta \;=\; f(\theta) \sin \theta$$

You already know these formulas (or you should!) and hence you do not need to memorize a

complicated formula such as Anton's Formula (5).

Example A. [Anton's Example 1 done without using Formula (5)].   Find the slope of the

tangent line to the circle

$$r \;=\; 4 \cos \theta$$

at the point where $\theta = \pi/4$.

<u>(Alternate) Solution.</u> $r = f(\theta) = 4 \cos \theta$,   so

$$x \;=\; r \cos \theta \;=\; f(\theta) \cos \theta \;=\; 4 \cos^2 \theta \, , \quad \text{and}$$

$$y \;=\; r \sin \theta \;=\; f(\theta) \sin \theta \;=\; 4 \cos \theta \sin \theta$$

13.5.3

Thus

$$\frac{dx}{d\theta} = -8 \cos \theta \sin \theta$$

$$\frac{dy}{d\theta} = 4 \cos^2 \theta - 4 \sin^2 \theta$$

so that

$$m = \frac{dy}{dx} = \frac{dy/d\theta}{dx/d\theta} = \frac{\cos^2 \theta - \sin^2 \theta}{-2 \cos \theta \sin \theta}$$

Hence

$$m\bigg|_{\theta = \pi/4} = \frac{(1/2) - (1/2)}{-1} = 0 \qquad \square$$

2. <u>Computing the angle ψ between the radial and tangent lines.</u>   While Anton's formula (5) is probably easier to derive than to remember (hence our alternative method above), his formula (14)

$$\boxed{\tan \psi = \frac{r}{dr/d\theta}}$$

for the angle ψ between the radial line OP and the tangent line to $r = f(\theta)$ at $P(r, \theta)$ is probably <u>easier to remember than to derive.</u>   However, as Anton's Example 3 illustrates, using it effectively involves remembering the assumptions under which it is derived:

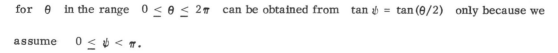

i. ψ is measured counterclockwise from OP

ii. $0 \le \psi < \pi$

For instance, in Example 3, the equation $\psi = \theta/2$

for θ in the range $0 \le \theta \le 2\pi$ can be obtained from $\tan \psi = \tan(\theta/2)$ only because we assume $0 \le \psi < \pi$.

3. <u>Arc length in polar coordinates.</u> Anton's Formula (16)

$$L = \int_a^b \sqrt{r^2 + \left(\frac{dr}{d\theta}\right)^2} \, d\theta$$

for the arc length L of a polar equation $r = f(\theta)$ from $\theta = a$ to $\theta = b$ (where $f'(\theta)$

is continuous for $a \le \theta \le b$) can be very useful. Moreover, it is not hard to remember if

we think "infinitesimally":

Think of dL, $d\theta$ and dr as infinitesimal pieces of the curve, the polar angle and

the radial distance, respectively. That is, remember the following picture:*

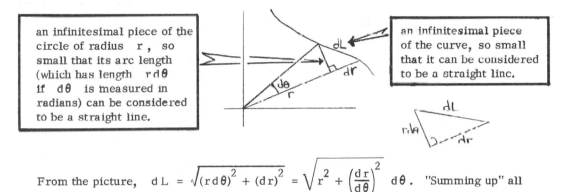

an infinitesimal piece of the circle of radius r, so small that its arc length (which has length $r\,d\theta$ if $d\theta$ is measured in radians) can be considered to be a straight line.

an infinitesimal piece of the curve, so small that it can be considered to be a straight line.

From the picture, $dL = \sqrt{(r\,d\theta)^2 + (dr)^2} = \sqrt{r^2 + \left(\frac{dr}{d\theta}\right)^2} \, d\theta$. "Summing up" all

these infinitesimal pieces, i.e., integrating our last expression, yields the formula for

arc length.

Anton's Examples 4 and 5 illustrate the use of the polar coordinate arc length formula.

Here is one more example:

<u>Example B.</u> Find the arc length of the curve

$$r = \cos^2(\theta/2) \quad \text{from} \quad \theta = 0 \quad \text{to} \quad \theta = \pi/3$$

------

*
  Engineers and physicists call this the "differential triangle" and they use it often.

13.5.5

Solution. We will apply Anton's Formula (16). To do this we need to compute $dr/d\theta$:

$$\frac{dr}{d\theta} = 2\cos(\theta/2)(-\sin(\theta/2))(1/2) = -\cos(\theta/2)\sin(\theta/2)$$

Hence

$$L = \int_0^{\pi/3} \sqrt{r^2 + \left(\frac{dr}{d\theta}\right)^2}\ d\theta$$

$$= \int_0^{\pi/3} \sqrt{\cos^4(\theta/2) + \cos^2(\theta/2)\sin^2(\theta/2)}\ d\theta$$

$$= \int_0^{\pi/3} \cos(\theta/2)\sqrt{\cos^2(\theta/2) + \sin^2(\theta/2)}\ d\theta$$
$$\text{since} \quad \cos(\theta/2) \text{ is positive when } 0 \leq \theta \leq \pi/3$$

$$= \int_0^{\pi/3} \cos(\theta/2)\ d\theta$$
$$\text{since} \quad \cos^2 x + \sin^2 x = 1 \text{ for all } x$$

$$= 2\sin(\theta/2)\ \Big|_0^{\pi/3}$$

$$= 2\sin(\pi/6) - 2\sin 0$$

$$= 2(1/2) - 0 = \boxed{1}$$

□

INDEX TO VOLUME I